DISCRETE/TRANSISTOR CIRCUIT SOURCEMASTER

DISCRETE/TRANSISTOR CIRCUIT SOURCEMASTER
Kendall Webster Sessions

John Wiley & Sons

New York　　Santa Barbara　　Chichester　　Brisbane　　Toronto

Copyright © 1978, by John Wiley & Sons, Inc.

All rights reserved. Published simultaneously in Canada.

Reproduction or translation of any part of this work beyond that permitted by Sections 107 and 108 of the 1976 United States Copyright Act without the permission of the copyright owner is unlawful. Requests for permission or further information should be addressed to the Permissions Department, John Wiley & Sons.

Library of Congress Cataloging in Publication Data:

Sessions, Ken W
  Discrete/transistor circuit sourcemaster.

  Includes index.
  1. Transistor circuits. 2. Electronic circuits.
3. Semiconductors. I. Title.
TK7871.9.S459 621.3815'3'0422 78-6774
ISBN 0-471-02626-3

Printed in the United States
10 9 8 7 6 5 4 3 2 1

PREFACE

A well-planned digest of schematics is to the engineer and technician what a cookbook is to the chef. It is a timesaver in that it pays for itself many times over if it is used at all. The enterprising circuit designer need not start each project from scratch—he can begin his effort by employing the devices and circuit arrangements that have been pioneered and exploited by those who have "been there before." He can concentrate the bulk of his preliminary work in selection of time-tested circuit blocks or stages and then devote some of that precious time saved to the rational piecing together of those blocks.

That in itself is sufficient justification for keeping a circuit book on the workbench, as near as the tools.

But I wanted this book to be something more—to offer something not offered by other books that might seem comparable from the standpoint of the circuit content alone. It seemed to me during the planning of this volume that, although circuit "recipes" alone could offer the kind of building-block information the experimenter needs to transform his ideas into hardware quickly, to be equally valuable to student and engineer there should be a kind of theoretical cohesiveness to the whole; there should be sufficient information to explain at least in a general way how things work. Such treatment, I reasoned, would result in a more innovative utilization of the circuits and the devices they employ.

Obviously, however, a detailed theoretical discussion couldn't be presented for every included circuit, for the book would wind up as a Brobdignagian book of words, most of which would be tiresomely repetitious. The alternative was to present *representative* theoretical discussions for circuits of a given genre; but even this approach posed problems, not the least of which would be the "production" restrictions on layout and order of circuit presentation.

The approach finally selected was to offer descriptions of what the devices are doing in their respective circuits but to avoid repetitive explanations when a device is used the same way in a variety of comparable circuits. What you *won't* see in this book are those kinds of general theory to be found in any available textbook—details, for example, on why an audio amplifier has to be biased to operate in the center of its linear region or why an electronic switching transistor is made to spend most of its time in cutoff or saturation. Instead, I've tried to include explanations of the reasons for including certain devices in circuits when their functions are not particularly obvious.

The circuits themselves have been supplied by the engineering staffs of the firms who make the devices featured. In the main, the text that accompanies the circuits has been distilled from application notes which contain considerably more information than is required for understanding and duplicating the schematic diagrams. My involvement with this book has been more of overseer, compiler, and editor than of author.

It is my conviction that this book can impart more important and meaningful technical information to the interested reader than many times the volume in textbook theory. The reason? Simply this: In a book of theory the reader must be exposed to *general* and usually quite basic circuits that very likely have no relation to any specific problem at hand. Studies show that facts fed to students shotgun fashion aren't well remembered. But give the student a quest, a *need* to learn, and the job's half done; give him sorted-out facts, facts that relate to his own particular problem, and he'll respond by retaining a *substantially* larger percentage of their total.

So it is that this book is not *just* a book of circuits; it is rather a collection of circuits in 9 broad categories, with salient facts presented for the reader who wants or needs more information than a given diagram can impart by itself. Even the practicing engineer, who very likely will already know the how and why of these circuits and the devices they use, will appreciate having this kind of information available for reference; it should help him know when and where to employ certain devices as he adapts these circuits to his own needs. And, of course, there exists a strong likelihood that he will be able to become better acquainted with some of those active devices that have come along since his scholastic days.

In most cases the circuits in this book contain references to manufacturers' part numbers rather than to generic number assignments. Despite the fact that most "house" catalog numbers could have been changed to the "2N" equivalents, I refrained from making this alteration—for three reasons:

First, the maker of the device in most cases paid an engineer to design, prototype, and perfect the circuits featuring that device, so it seems only reasonable to expect those who want to duplicate such circuits to do it with the semiconductors of that manufacturer. After all, isn't this why the engineer devoted his time to the design effort in the first place?

The second reason is really more important than the first, and this is one that seasoned experimenters will be able to identify with readily: the *unrestricted* use of equivalents in circuit construction can lead to plenty of trouble. I'll always remember the pride of my early days in ham radio—a Multi-Elmac tube-type mobile transceiver with an instruction booklet that stipulated replacement of the driver stage with a Tung-Sol 6AQ5. A tube is a tube, I reasoned, and when replacement time came, I purchased whatever brand the nearest electronics distributor carried. When the set exhibited insufficient drive, I attributed the problem to just about everything but the obvious one. I dug into the chassis, knifed apart the appropriate tank coil, played with the coupling, and diddled endlessly, trying vainly to squeeze out of that rig the kind of performance I got "when it was new." But it wasn't until I replaced the stage with a Tung-Sol 6AQ5 that the final amplifier received enough drive to pack out a respectable signal.

Accuracy is the final reason for leaving the original diagrams alone—retaining the original part numbers and values. Few of us can take a schematic and build the circuit it represents so that it works correctly the first time. And we're experienced constructionists who have a personal interest in getting it right. We'll omit a connection. We'll solder a wire to the wrong terminal. We'll forget a bypass capacitor or use the wrong value of resistor. We make mistakes despite our huge incentive for accuracy. Well, consider this: The draftsman who copies a circuit or substitutes generic numbers for the identifiers employed by the original designers has to go through essentially the same process as the builder. The process is faster for him, of course, because he doesn't have to actually dress the leads or apply solder or worry about what part goes where. But he does have to route the "wires" correctly, to apply "solder dots" at the proper circuit junctions, and to correctly identify the devices and their positions. Is not the draftsman, conscientious though he is, many times more likely to err than the individual who builds a circuit to satisfy his own needs?

Unfortunately, there are nearly as many device symbols and standards as there are manufacturers. So you're apt to see a *diac* pictured in one diagram as a baseless, collectorless transistor with two emitters, and in another as a gateless *triac*. Forgive the inconsistencies and bear in mind that these circuits are, for the most part, photographic reproductions of the diagrams contributed by the semiconductor makers. A compromise? To be sure—but I think it's one you'd want us to make, for it virtually eliminates what would otherwise certainly be the largest single source of error.

I am extremely grateful to the manufacturers who cooperated with me in preparing this circuit collection; in alphabetical order they are:

Amperex, Solid State and Active Devices Division
Delco Electronics, Division of General Motors
Fairchild, Semiconductor Division
General Electric, Semiconductor Products Department
General Instrument, Semiconductor Components Division
GTE/Sylvania
Motorola Semiconductor Products Inc.
National Semiconductors, Inc.
RCA, Solid State Division
Siliconix Incorporated
Sprague Electronics
Texas Instruments Incorporated
Workman Electronics (source of devices preceded by IR designator)

Kendall Webster Sessions

CONTENTS

SECTION 1 *POWER CONVERSION AND REGULATION* — **1**
1.1 Regulators and Current Sources — **3**
1.2 Power Supplies and Rectifiers — **17**
1.3 Direct Current Converters — **33**
1.4 Inverters — **41**

SECTION 2 *CONTROL AND SENSING* — **59**
2.1 Solid-State Relays and Switches — **60**
2.2 Photoelectric Switches — **76**
2.3 Dimmers and Motor Speed Controllers — **84**
2.4 Temperature Regulation Devices — **132**
2.5 Value, Rate, and Process Monitors — **147**
2.6 Intrusion and Hazard Detectors — **160**
2.7 Phone-Line Sensors — **167**
2.8 Remote Control and Servos — **169**

SECTION 3 *TIMING CIRCUITS* — **177**
3.1 Flashers — **178**
3.2 Ring Counters — **185**
3.3 Timers — **201**

SECTION 4 *HOBBY* — **213**
4.1 Photographic Circuits — **213**
4.2 Amateur Radio and CB — **218**
4.3 Communicators — **223**
4.4 Electronic Novelties and Toys — **227**

SECTION 5 *RADIO/TV* — **239**
5.1 AM and FM Radio — **239**
5.2 Video and Deflection — **247**
5.3 Black and White TV — **254**
5.4 Color TV — **263**

SECTION 6 *COMMUNICATIONS* — **269**
6.1 Oscillators — **269**
6.2 Multipliers — **274**
6.3 Transmitters — **278**
6.4 RF Power Amplifiers and Modulators — **288**
6.5 RF Preamplifiers and Converters — **297**
6.6 Front Ends, RF/IF Amplifiers, and Mixers — **302**

SECTION 7 *AUDIO* **313**
7.1 Preamplifiers, Followers, and Mixers **313**
7.2 Amplifiers **324**
7.3 Special-Purpose Circuits **361**

SECTION 8 *TEST AND MEASUREMENT* **365**
8.1 Go/No-Go Testers **365**
8.2 Measurement Devices **368**
8.3 Miscellaneous Circuits **374**

SECTION 9 *AUTOMOTIVE CIRCUITS* **379**
9.1 In-Car Circuits **379**
9.2 In-Garage Circuits **390**

Index **393**

SECTION 1
POWER CONVERSION AND REGULATION

Every electronic circuit requires power—even the simple "crystal set," which draws its energy from the radio signal it detects. It is fitting, then, that the first section contains those circuit arrangements that have been designed to provide a source of electrical power for operating other circuits. To make it easier for you to browse through appropriate circuit candidates for your own applications, the circuits in this section are subclassified into four divisions: regulators and current sources, power supplies and rectifiers, dc-to-dc converters, and dc-to-ac inverters.

The characteristics of a rectifier system contained in a power source depend to a large extent on the electrical characteristics of the load or filter it drives. The following table, adapted from a large and comprehensive wall chart prepared and distributed by the General Electric Semiconductor Products Department, gives all the specification values you need to ascertain diode and transformer requirements on a per-volt-of-output basis. Although the original of this chart lists values for multiphase and interphase circuitry, the information here applies only to the "basic" rectifiers: half-wave, full-wave, and bridge. Bear in mind as you use this table for power supply design that the information is necessarily based on a zero forward voltage drop and zero reverse current in rectifiers, and no ac line or source reactance. Also, remember that all figures are based on a dc output voltage of 1.0, which means that values must be scaled upward to reflect the circuit conditions imposed by your own requirements.

The chart for rectifier design does not include information for selecting diodes for circuit applications where voltage transients might be encountered. Transient voltage surges as high as 440V are not uncommon in 117V ac lines, the result of switching highly inductive circuits or circuits that use a phase-controlling element such as silicon controlled switches and rectifiers.

Probably the most common approach to compensating for these occasional transients is to employ diodes with a peak reverse voltage rating high enough to permit passage of such transients without diode destruction. In practice, however, diodes rated for higher voltages cost more than low-voltage types; so some engineers choose an approach that allows automatic shunting of voltage transients around the rectifying devices. The following alternatives have been suggested by the engineering staff at National Semiconductors; although application of the techniques presupposes that the ICs will be powered, the information is generally applicable to any power conversion circuit.

Type of Circuit			Single Phase Half Wave (1-1-1-H)	Single Phase Center-Tap (2-1-1-C)	Single Phase Bridge (4-1-1-B)
Primary					
Secondary					
One Cycle Wave of Rectifier Output Voltage (No Overlap)					
Number of Rectifier Elements in Circuit			1	2	4
Average D.C. Volts Output		=	1.00	1.00	1.00
RMS D.C. Volts Output		=	1.57	1.11	1.11
Peak D.C. Volts Output		=	3.14	1.57	1.57
Peak Reverse Volts per Rectifier Element		=	3.14	3.14	1.57
		=	1.41	2.82	1.41
		=	1.41	1.41	1.41
Average D.C. Output Current		=	1.00	1.00	1.00
Average D.C. Output Current per Rectifier Element		=	1.00	0.500	0.500
RMS Current per Rectifier Element	Resistive Load	=	1.57	0.785	0.785
	Inductive Load	=	—	0.707	0.707
Peak Current per Rectifier Element	Resistive Load	=	3.14	1.57	1.57
	Inductive Load	=	—	1.00	1.00
Ratio: Peak to Average Current Per Element	Resistive Load		3.14	3.14	3.14
	Inductive Load		—	2.00	2.00
% Ripple (RMS of Ripple / Average Output Voltage)			121%	48%	48%
			Resistive Load		Inductive
Transformer Secondary RMS Volts per Leg		=	2.22	1.11 (To Center-Tap)	1.11 (Total)
Transformer Secondary RMS Volts Line-to-Line		=	2.22	2.22	1.11
Secondary Line Current		=	1.57	0.707	1.00
Transformer Secondary Volt-Amperes per Leg		=	3.49	1.57	1.11
Transformer Primary RMS Amperes per Leg		=	1.57	1.00	1.00
Transformer Primary Volt-Amperes per Leg		=	3.49	1.11	1.11
Average of Primary and Secondary Volt-Amperes		=	3.49	1.34	1.11
Primary Line Current		=	1.57	1.00	1.00
Line Power Factor			—	0.900	0.900

DISCRETE / TRANSISTOR CIRCUIT SOURCEMASTER

1. Insert a low-value resistor in series with one of the transformer primary windings and a shunt capacitor across the primary.

2. Insert a series inductance in the primary winding of the power transformer along with a shunt capacitor across the primary.

3. Install a shunt or buffer capacitor across the secondary winding of the power transformer.

4. Install a capacitor across the half-wave rectifying diode so that the power represented by the transient is dissipated in the total circuit series resistance.

5. Install a shunt varistor such as a GE MOV device across the secondary, in the same manner as a buffer capacitor.

6. Other schemes, such as the installation of a dual-diode shunt clipper, zener, or diode-and-capacitor combination across the rectifying elements.

Unless the transient problem is quite severe, any of the first four alternative methods should prove satisfactory, even though these approaches do not offer full protection against high-voltage surges.

1.1 Regulators and Current Sources

The 25 or 30 circuits featured in this section are representative of the regulation methods typically employed in solid-state power systems; additional regulators may be found as part of the power supply circuits in the subsequent subsections.

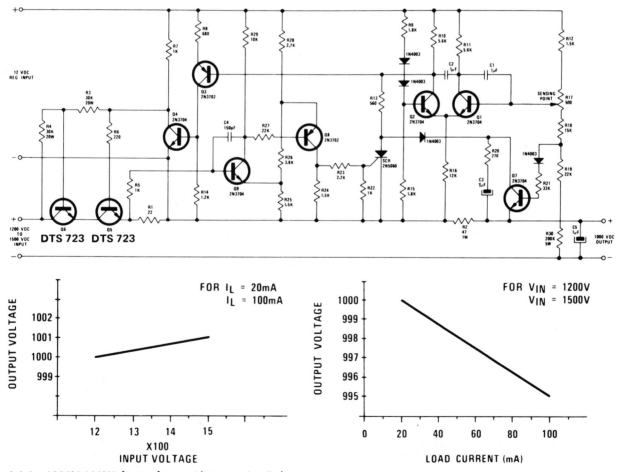

1.1.1 1000V 100W dc regulator with two series Delco DTS-723 transistors functioning as the pass element. The input voltage of the regulator may vary from 1200 to 1500V with 0.1% regulation at full load.

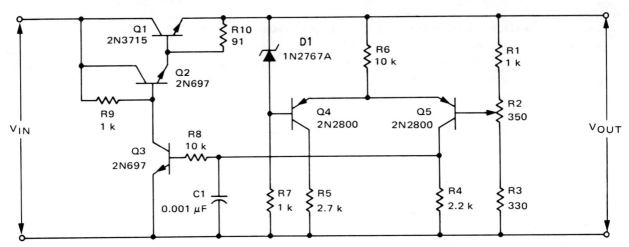

1.1.2 Motorola basic series-pass regulator. The drive for the series-pass transistor is derived by sampling the output voltage of the regulator with the voltage divider R1, R2, and R3. This sampled voltage is compared to the reference voltage provided by zener D1. The difference between the reference voltage and the sampled voltage is amplified and provides drive to the series-pass transistor, which compensates for any change in the output voltage.

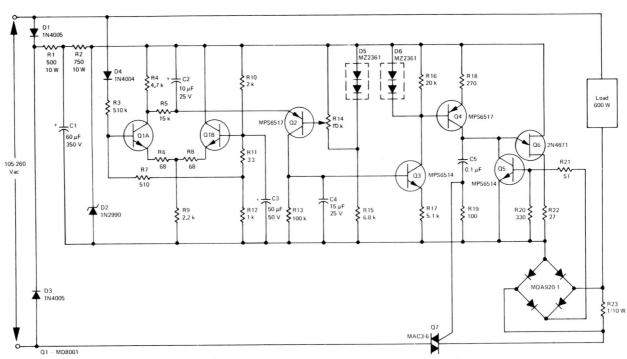

1.1.3 Motorola closed-loop rms regulator will hold the output voltage at 90V (±2V) rms with any input voltage between 105 and 260V. The sensing circuit uses a differential amplifier employing a dual transistor shown as Q1A and Q1B. Q1A is biased to operate in the nonlinear cutoff region of the differential amplifier transfer function, where the output current varies approximately as the square of the input voltage. The signal input magnitude is necessary for the sensing circuit to operate properly above and below cutoff. R11 is provided to hold the base of Q1A 300 mV below the base of Q1B, which is sufficient to insure that Q1A does operate about cutoff.

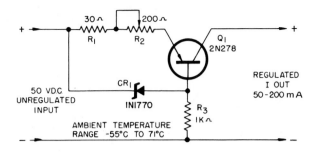

COMPONENT PARTS LIST

R_1 — 30 Ω, 5 W, wire wound
R_2 — 200 Ω, 4 W, wire wound
R_3 — 1000 Ω, 5 W, wire wound
Q_1 — 2N278 Delco power transistor

CR_1 — high power 8 volt Zener diode, G.E. type 1N1770 or equivalent
Heat sink — Delco Radio 7281366 or equivalent (This size permits opration in ambient temperatures up to 71°C.)

1.1.4 Delco current regulator uses a grounded-base circuit. The collector current is essentially equal to the emitter current because of the high current-gain characteristic of the transistor. When the emitter current is held constant, the collector current will remain approximately constant. The emitter current is determined by the voltage V_1 applied across the resistors R1 and R2. This voltage is equal to $V_1 = V_{VR} - V_{EB}$, where V_{VR} is the voltage across the voltage reference diode CR1, and V_{EB} is the voltage appearing between the emitter and base on the transistor. For the current and voltage range considered, V_{EB} is small. A typical value for V_{EB} is 0.2V. In this circuit, V_{VR} is approximately 8V. Voltage V_1 appearing across resistances R1 and R2 is approximately equal to V_{VR} and we have for the emitter current:

$$I_E \frac{V_{VR}}{R1 + R2}$$

As noted on the schematic diagram, R2 is variable. By changing the value of R2, a change in the emitter current is possible. R3 provides a keep-alive current for the reference diode. With loading, the transistor supplies a small portion of the current drawn by R3; therefore, less current will flow through the reference diode. R3 must draw enough current through the reference diode so that the voltage drop across the diode will remain at 8V as the current regulator is loaded.

The current supplied to the load will remain essentially constant until R_L is increased in size to the point where the voltage drop across R_L is as large as the voltage drop across R3.

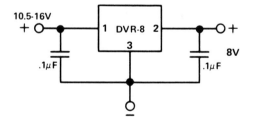

8V Regulator

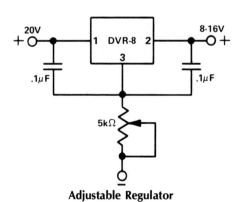

Adjustable Regulator

1.1.5 Delco dc voltage regulators employ 100 mA monolithic devices. The circuit at the right provides control of the output from 8 to 16V.

SECTION **1** *Power Conversion and Regulation*

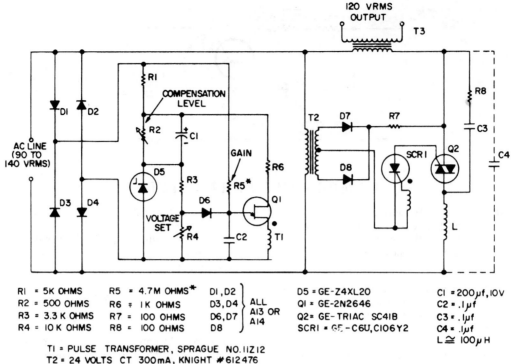

1.1.6 GE 500W switching type ac line-voltage regulator

provides ±2% regulation for a fixed load as line voltage varies over ±20%. The regulator employs a phase-controlled triac to control voltage applied to transformer T3. The addition of 24V transformer T2, which drives the gate of the triac when pilot SCR1 is fired, provides a continuous gate signal for the triac, avoiding possible half-wave operation whenever the output of the circuit feeds an inductive load.

Adjusting the regulation: After applying 90V at the input, adjust R4 to obtain 120V on the output for the desired load conditions. Now raise the input voltage to 140V. If the output voltage increases, adjust R2 to bring the voltage down to its correct value. Now adjust the input to 90V and repeat the adjustment there.

Either root-mean-square or *average* voltage can be regulated, depending on the type of load. The adjustments should be made with a suitable voltmeter across the load. Each time the load impedance is changed, the circuit must be readjusted.

Since this type of circuit regulates by phase control, some loads (e.g., certain types of ac motors) may not function properly on the output waveform because of its harmonic content.

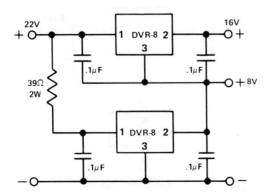

1.1.7 Delco dual-level regulator supplies 8 and 16V dc at 100 mA. Short-circuit output current: 200 mA minimum.

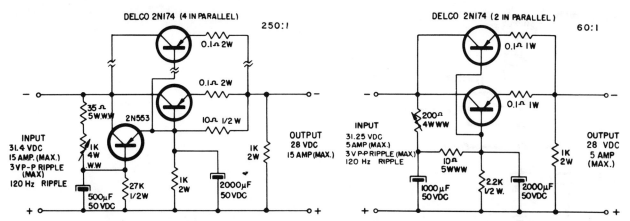

1.1.8 Delco electronic ripple filters provide 28V output. Electronic filters do not take the place of conventional LC filter networks; rather, they supplement the LC filter to achieve very low values of power supply ripple. To apply these circuits efficiently, reduce the ripple to a value of 3V or less before using the electronic filter.

The circuits shown will reduce ripple by a factor of about 250:1. If the ripple is reduced to 250 mV by LC filters, the output from the electronic filter will contain only about 1 mV ripple. The simpler circuit shown at right will reduce ripple by a factor of about 60:1.

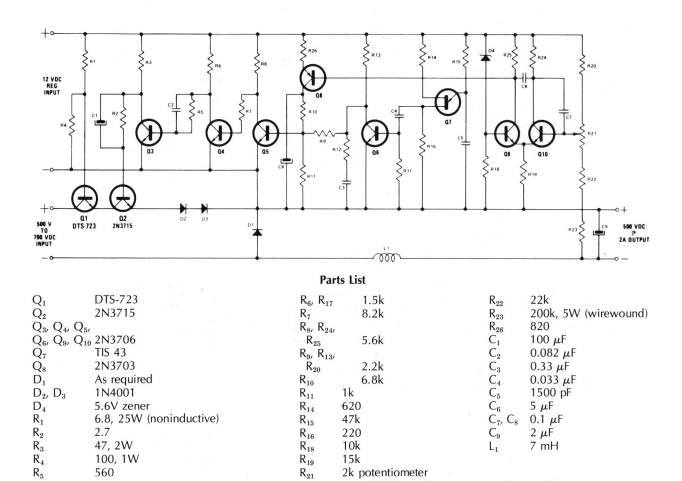

Parts List

Q_1	DTS-723	R_6, R_{17}	1.5k	R_{22}	22k
Q_2	2N3715	R_7	8.2k	R_{23}	200k, 5W (wirewound)
$Q_3, Q_4, Q_5,$		$R_8, R_{24},$		R_{26}	820
Q_6, Q_9, Q_{10}	2N3706	R_{25}	5.6k	C_1	100 μF
Q_7	TIS 43	$R_9, R_{13},$		C_2	0.082 μF
Q_8	2N3703	R_{20}	2.2k	C_3	0.33 μF
D_1	As required	R_{10}	6.8k	C_4	0.033 μF
D_2, D_3	1N4001	R_{11}	1k	C_5	1500 pF
D_4	5.6V zener	R_{14}	620	C_6	5 μF
R_1	6.8, 25W (noninductive)	R_{15}	47k	C_7, C_8	0.1 μF
R_2	2.7	R_{16}	220	C_9	2 μF
R_3	47, 2W	R_{18}	10k	L_1	7 mH
R_4	100, 1W	R_{19}	15k		
R_5	560	R_{21}	2k potentiometer		

1.1.9 Delco low-cost switching regulator circuit useful in the 1 kW range employs a single DTS-723 silicon transistor as a series switching element. At 500V and 1A output, regulation of 0.4% can be achieved; at 2A output, the regulation is better than 1%. Regulation is accomplished by pulse-width modulation at a constant switching frequency. Efficiency is more than 90% at a switching frequency of 13 kHz. At output levels up to 500W, switching frequencies up to 20 kHz would be practical with small size components at slightly reduced efficiency.

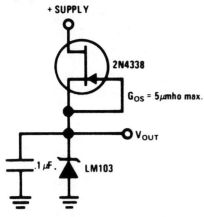

1.1.10 National low-power regulator reference circuit provides a stable voltage reference almost totally free of supply voltage hash. Typical power supply rejection exceeds 100 dB.

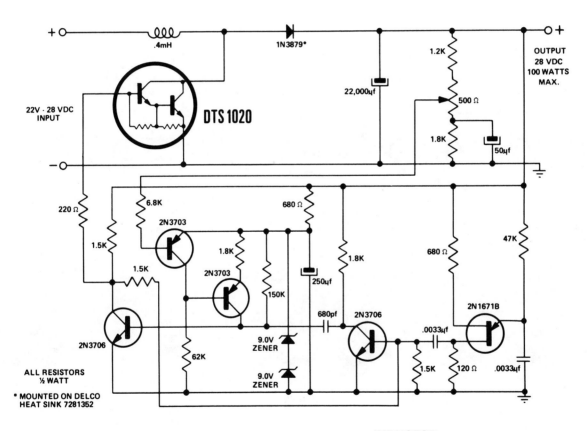

INDUCTOR
Core: Powdered Iron
   Arnold B079024-3
Wire: 124 Turns, No. 17

1.1.11 Delco miniature 100W flyback switching regulator uses a DTS-1020 Darlington silicon power transistor. Efficient regulation by pulse-width modulation is achieved at an output voltage as high as 6V above the input voltage. The switching rate is 9 kHz; it operates with an input of 22 to 28V. Regulation and ripple are less than 1% at full output.

8 DISCRETE / TRANSISTOR CIRCUIT SOURCEMASTER

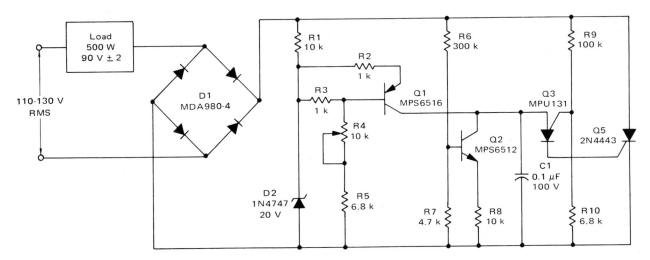

1.1.12 Motorola open-loop rms voltage regulator provides 500W 90V rms with good regulation for an input voltage range of 110 to 130V rms.

With the input voltage applied, capacitor C1 charges until the firing point of Q3 is reached, causing it to fire. This turns Q5 on, which allows current to flow through the load. As the input voltage increases, the voltage across R10 increases, which increases the firing point of Q3. This delays the firing of Q3 because C1 now has to charge to a higher voltage before the peak-point voltage is reached. Thus the output voltage is held fairly constant by delaying the firing of Q5 as the input voltage increases. For a decrease in the input voltage, the reverse occurs.

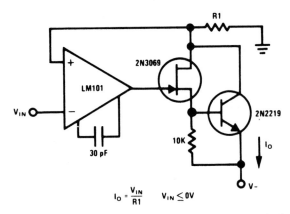

$I_O = \dfrac{V_{IN}}{R1}$ $V_{IN} \leq 0V$

1.1.13 National precision current source in which the 2N3069 JFET and 2N2219 bipolar serve as voltage isolation devices between the output and the current sensing resistor, R1. The LM101 provides a large amount of loop gain to assure that the circuit acts as a current source. For small values of current, the 2N2219 and 10K resistor may be eliminated with the output appearing at the source of the 2N3069.

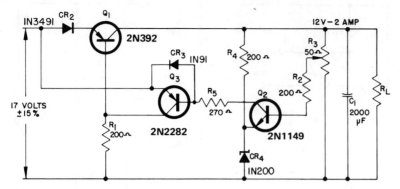

1.1.14 Delco proportional voltage regulator. A voltage proportional to the output voltage is compared to the voltage of reference diode CR4. The differential signal is then amplified by transistor Q3. The output signal of this transistor is used to control the current in the series transistor Q1 in such a way that the voltage across the load remains constant.

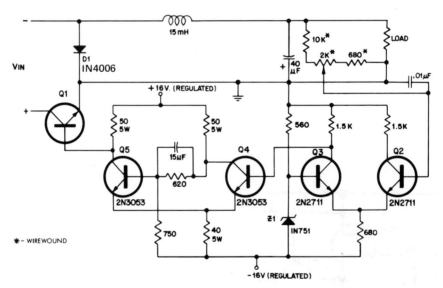

1.1.15 Delco pulse-width-modulated switching regulator uses a single high-voltage silicon transistor as a series element. The output voltage is manually adjustable within the range of the potentiometer in the error voltage divider. This method of voltage regulation has several definite advantages over conventional regulation:

1. Higher efficiency.
2. Fewer series elements.
3. Less heatsink area.

By using a transistor (for Q1) from the Delco DTS-410, DTS-411, DTS-423, DTS-430, or DTS-431 series, various combinations of output voltages and currents may be obtained. The following table shows the maximum regulator conditions for these devices.

Q1	Max Input	Max Output	
DTS-410	200V	150V,	2A
DTS-411	300V	225V,	2A
DTS-423	325V	250V,	2A
DTS-430	300V	225V,	3A
DTS-431	325V	250V,	3A

Efficiencies as high as 92% can be obtained under full load. Total regulation for minimum-to-maximum inputs and minimum-to-maximum loads is less than 0.6%, and ripple is 0.75V peak at full load.

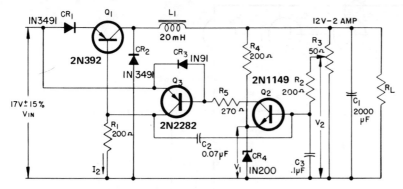

1.1.16 Delco pulse-width-modulated voltage regulator is a series regulator operating in the switching mode. High efficiency is the outstanding advantage to be gained from this mode of operation. Efficiencies as high as 80% can be realized even with the input voltage several times larger than the output voltage. The main disadvantage, poor frequency response, may be lessened by using a higher switching frequency.

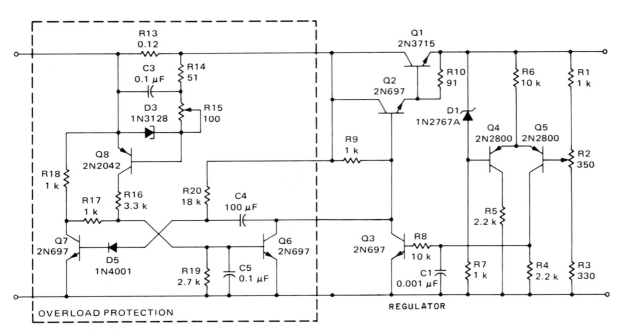

1.1.17 Motorola series regulator with overload protection. Overload sensing circuit triggers the monostable multivibrator, which in turn removes the drive from the series-pass driver transistor. This turns the regulator circuit off until the monostable multivibrator resets. If the overload still exists after reset, the regulator is again turned off. This type of protection would be adequate for resistive loads; however, for large capacitive loads, the surge current charging the capacitor would cause the overload circuit to turn the regulator off.

SECTION 1 *Power Conversion and Regulation* 11

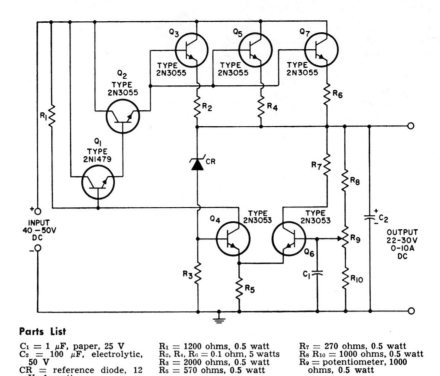

Parts List

- C_1 = 1 µF, paper, 25 V
- C_2 = 100 µF, electrolytic, 50 V
- CR = reference diode, 12 V, 1 watt
- R_1 = 1200 ohms, 0.5 watt
- R_2, R_4, R_6 = 0.1 ohm, 5 watts
- R_3 = 2000 ohms, 0.5 watt
- R_5 = 570 ohms, 0.5 watt
- R_7 = 270 ohms, 0.5 watt
- R_8, R_{10} = 1000 ohms, 0.5 watt
- R_9 = potentiometer, 1000 ohms, 0.5 watt

1.1.18 RCA series voltage regulator has adjustable output, line regulation within 1.0%, and load regulation within 0.5%. Regulation is accomplished by varying the current through three parallel 2N3055 transistors connected in series with the load. A zener provides the reference voltage for the circuit.

If the output voltage tends to rise for any reason, the total increase in voltage is distributed across bleeder resistors R8, R9, and R10. If R9 is set to midpoint, half the increase in output voltage is applied to the base of the 2N3053 transistor Q6. This increased voltage is coupled to the base of Q4 by R5, the emitter resistor for the two transistors. Because the voltage drop across CR remains constant, the full increase in voltage is developed across R3 and thus is applied directly to the base of Q4. Because the increase in voltage at the base is higher than that at the emitter, the collector current of transistor Q4 increases.

As the collector current of Q4 increases, the base voltage of Q1 decreases by the amount of the increased drop across R1. The resultant decrease in current through Q1 causes a decrease in the emitter voltage of this transistor. The resultant decrease in current through Q1 causes a decrease in the emitter voltage and thus in the base voltage of Q2.

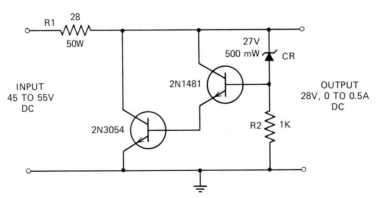

1.1.19 Simple two-transistor shunt voltage regulator can provide a constant (within 0.5%) dc output of 28V for load currents up to 0.5A and dc inputs from 45 to 55V. The two transistors operate as variable resistors to provide the output regulation. A 27V zener is used as the control element.

The output voltage tends to rise with an increase in either the applied voltage or the load-circuit impedance. The current through R2 and CR then increases. However, the voltage drop across CR remains constant at 27V and the full increase in the output voltage is developed across R2. This increased voltage across R2 is direct-coupled to the base of the 2N1481 transistor and increases the forward bias so that the 2N1481 conducts more heavily. The rise in the emitter current of the 2N1481 increases the forward bias on the 2N3054, and the current through this transistor also increases.

As the increased currents of the transistors flow through resistor R1, which is in series with the load impedance, the voltage drop across R1 becomes a larger proportion of the total applied voltage. In this way, any tendency for an increase in the output voltage is immediately reflected as an increased voltage drop across R1 so that the output voltage delivered to the load circuit remains constant.

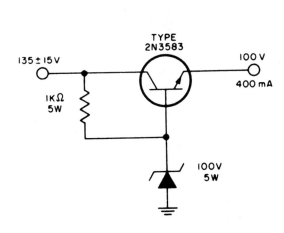

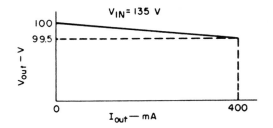

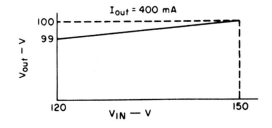

1.1.20 RCA simple transistor voltage regulator.

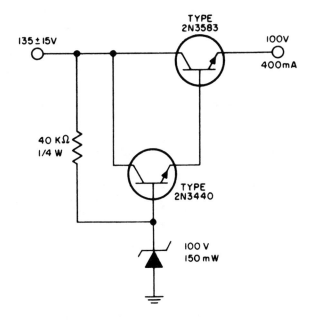

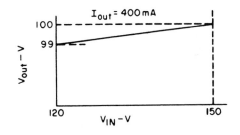

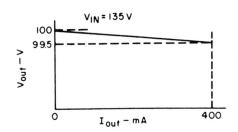

1.1.21 RCA series voltage regulator using the Darlington driver.

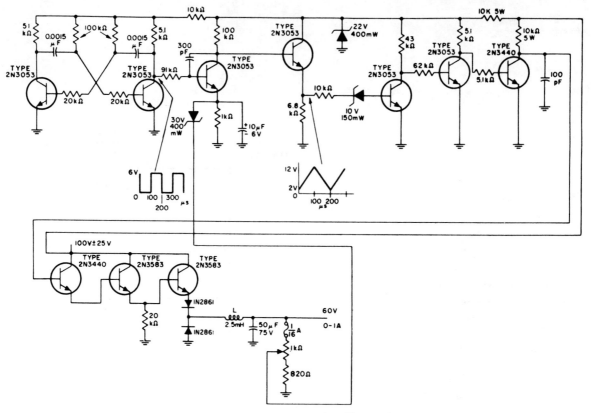

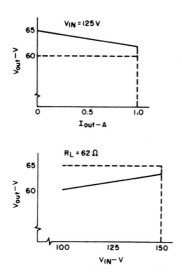

1.1.22 RCA stepdown switching regulator.

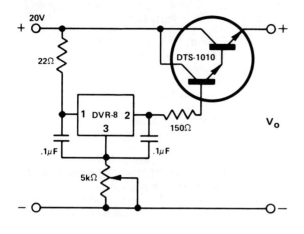
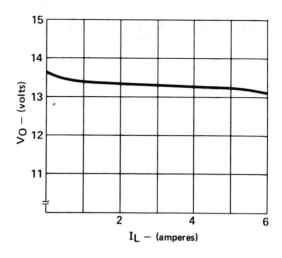

V_O vs I_L

1.1.23 Use of 100 mA Delco regulator as a controlling element increases regulation capability. Circuit provides 13.2 to 13.6 volts dc over a current range of a few milliamperes to a full 6.2 amperes.

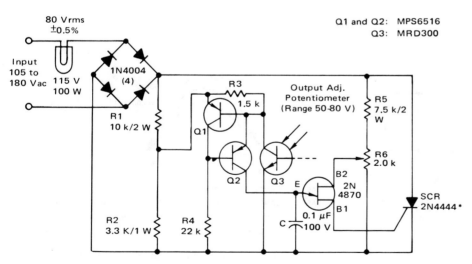

1.1.24 Motorola voltage regulator for projection lamp. The asterisk on the 2N4444 is Motorola's reminder to use a heatsink with this transistor.

SECTION 1 *Power Conversion and Regulation* 15

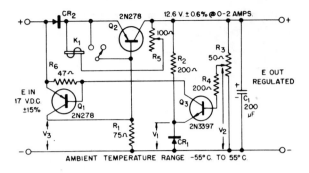

Parts List

R1	75-ohm 10W wirewound
R2	100-ohm 2W carbon
R3	50-ohm 4W pot
R4	200-ohm ½W carbon
R5	100-ohm ½W pot
R6	47-ohm ½W carbon
K1	90-ohm 5V voltage relay
C1	200 μF 50V dc tantalum
Q1, Q2	Delco 2N278
CR1	GE 1N1770 or equivalent
CR2	Delco 1N3491, 1N3208, or equivalent
Q3	GE 2N3397 or equivalent
Heatsink	Delco 7281366 or equivalent (permits operation in ambient temperatures up to 55°C)

1.1.25 Delco voltage regulator provides 12.6V at up to 2A for driving "automotive" loads such as tape players, CB radios, etc. Should the voltage across transistor Q2 rise above a preset value, voltage-sensitive relay K1 operates. The excess voltage could be caused by a large increase in input voltage, an overload, or a short-circuited output. When relay K1 operates, the emitter and base of transistor Q2 are shorted together. This causes the collector current of transistor Q2 to be turned off. Resistor R2 provides a bias current for voltage reference diode CR1, and resistor R4 protects transistor Q3 and diode CR1 in the event transistor Q2 should short from emitter to collector. Resistor R5 provides an adjustment to control operation of voltage relay K1. Capacitor C1 prevents high-frequency oscillations by reducing the loop gain to a safe level for frequencies at which the internal phase shift of the transistors would be significant. The forward voltage drop across diode CR2 provides additional collector bias for transistor Q1. Thus transistor Q1 is able to cut off transistor Q2 collector current. This diode is particularly useful at high temperatures. Resistor R6 reduces the collector leakage current of transistor Q1.

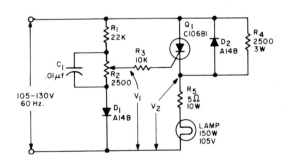

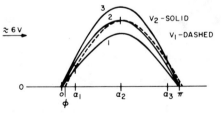

CASE 1: LAMP R LOW, TRIGGERING ADVANCES TO α_1; LOW LINE VOLTAGE.
CASE 2: LAMP R NORMAL, TRIGGERING AT α_2; MEDIAN LINE VOLTAGE.
CASE 3: LAMP R HIGH, TRIGGERING AT α_3; HIGH LINE VOLTAGE.
NOTE: GAIN & STABILITY CONTROLLED BY C_1 (PHASE SHIFT ϕ)

1.1.26 GE lamp voltage regulator operates a specific lamp and measures lamp power by measuring the filament resistance. During the negative half-cycle, the lamp receives full current through diode D2. At the beginning of the positive half-cycle, R4 provides a bias current to develop a voltage, V2, that is dependent on lamp resistance. This voltage is compared with reference voltage V1 by the gate–cathode junction of the SCR, and phase-controlled triggering results when V1 becomes more positive than V2.

The waveforms of V1 and V2 are shown for low, medium, and high line voltages, with the effect of line voltage alone subtracted to show only the effect of lamp resistance change. The slight phase shift that capacitor C1 produces on V1 controls the gain of this feedback control system. R4 and R5 dissipate 13W, which requires some care in handling.

1.2 Power Supplies and Rectifiers

It is probably safe to say that most of us, when we want to construct a power supply for a given application, just "throw it together" from whatever parts we have on hand. We approach power supply design more empirically than we would the design of an audio stage or a receiver. We start with the most appropriate transformer we happen to have. We tend to use unmarked diodes for the rectifier elements—hoping they'll still function when the line voltage flares momentarily or when we need to draw more current from the line than our previous circuit using the same devices. Unless we're building a regulated supply, we'll use the largest value of filter capacitor we can find in the junkbox, then go with a capacitor-input design rather than the more expensive choke-input arrangement.

There aren't many permutations of a basic power supply circuit as long as we're dealing with the 60 Hz ac line and the requirements of our load are not unusual. Except for the few special-purpose variations, the circuits presented here are simply elaborations on the basic half- and full-wave themes we're accustomed to.

If the power supply you're looking for is to be used as a source for another of the circuits included in this book, it might prove worth your while to examine those respective sections first, since many have either a recommended power supply circuit accompanying them or an adequate source of power as an integral part of the schematic.

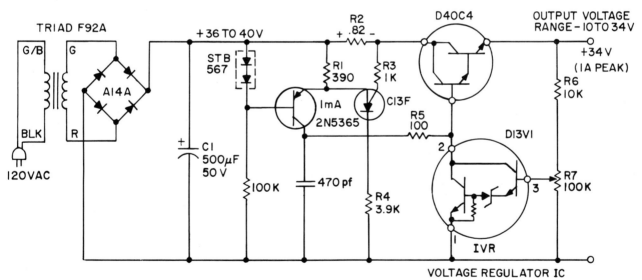

NOTE: .82Ω LIMITS CURRENT AT 1.0 TO 2.0A
DYNAMIC OUTPUT RESISTANCE = .06V / .95A = 0.063 Ω

1.2.1 GE 10 to 34V regulated supply with electronic circuit protection. The current-limiting circuit breaker protects both the power supply and the load.

The GE C13F is the breaker device. This complementary SCR operates similarly to the conventional SCR; the difference is that this device is turned on by forward-biasing the junction between the anode and the anode gate by making the anode gate negative with respect to the voltage on the anode. Conventional SCRs are turned on by a positive pulse on the cathode gate. The C13F is triggered when a 1 to 2.0A current flows through R2. This turns off the 2N5365 by reverse-biasing its emitter–base junction and, therefore, interrupts the bias current to the D40C4. Consequently, it turns off the power Darlington, preventing it or an amplifier load from being burned out by excessive current.

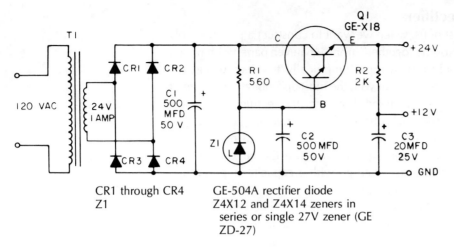

1.2.2 GE 24V dc regulated supply provides as much as a half-ampere of current.

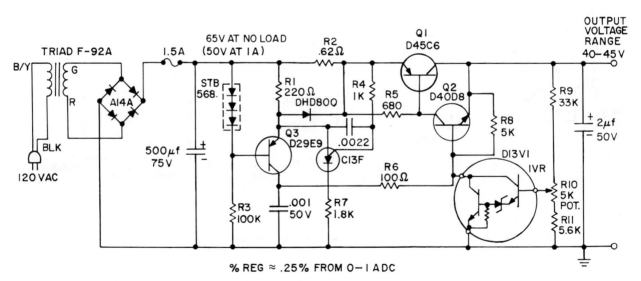

1.2.3 GE 34 to 45V regulated power supply delivers up to 1A. This supply has a current-limiting circuit and an electronic circuit breaker to prevent current overload pulses from destroying series-pass transistor Q1.

The GE D29E9 is used as a constant-current source for the D13V, Q2, and Q1. Current limiting is accomplished with a diode (DHD800) and current-sensing resistor R2. When a 3A current flows through R2, the voltage across the diode is enough to turn it on; this turns off the D29E9.

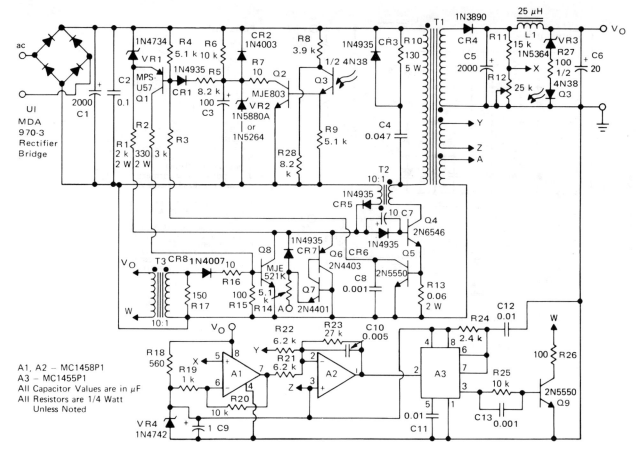

1.2.4 Motorola 80W CATV switching supply. Control circuitry (A1, A2, and A3) is powered from the output circuit, eliminating the need for a separate regulated supply.

At startup, the base drive supplied through R1 saturates the 2N6546 and the full input voltage is applied across the primary winding of the power transformer. With this voltage constant, the current ramps up linearly until the 2N6546 is switched off. While the power transistor is on, the current transformer provides base drive to keep the 2N6546 in saturation ($I_C/I_B = 10$). The secondary is phased so that diode CR4 is reversed-biased and no current flows in the secondary. This causes all of the energy absorbed during the on time of the power transistor to be stored in the primary magnetic field.

When the power transistor is switched off, the transformer polarities reverse, diode CR4 is forward-biased, and the energy stored in the primary is transferred to the secondary. Output capacitor C_0 charges to the required output voltage and must be large enough to supply worst-case load current during the time that diode CR4 is in the blocking state (on time of the power transistor).

The output regulation is accomplished by sensing both the output and input voltages and varying the on time of the power transistor to supply more or less energy depending on the output and input conditions. The output voltage is sensed via a resistor divider network and applied to the noninverting input of A1 (½MC1458). This feedback voltage is compared to the voltage reference (V_{ref}); the difference is amplified by A1 and A2 (½MC1458); and the result appears at the output of A2 as a positive dc level.

The positive portion of the signal from the Y–Z winding is proportional to V_I ($K2V_I$) and is integrated by A2, producing a negative ramp at the output of A2. The slope of this ramp is proportional to the input voltage, and the starting point is proportional to the input and output voltage. As the slope and the starting point vary, the time required to reach the threshold voltage of the MC1455 varies, thereby varying the on time of the 2N6546.

When the output voltage of A2 goes below the threshold voltage of the MC1455, its output goes high and turns on a control transistor that pulls the base of the 2N6546 to ground and turns it off. A pulse transformer is used between the control circuitry and the power transistor to maintain input–output isolation. The off time, fixed by the MC1455, is set to allow the complete transfer of energy under worst-case conditions (heavy load and low input voltage). When the MC1455 times out, the 2N6546 is allowed to turn on again and the cycle repeats.

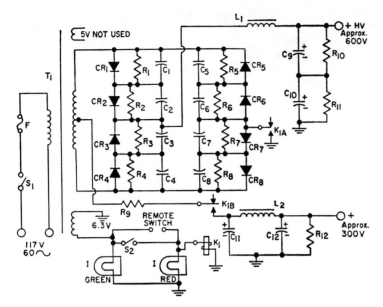

1.2.5 RCA 600V/300V power supply uses eight 1N2864 silicon diodes in series-connected pairs in a bridge-rectifier circuit to supply 300 mA of power.

Parts List

C_1 C_2 C_3 C_4 C_5 C_6 C_7 C_8 = 0.001 μF, ceramic disc, 1000 V
C_9, C_{10}, C_{11}, C_{12} = 40 μF, electrolytic, 450 V
CR_1 CR_2 CR_3 CR_4 CR_5 CR_6 CR_7 CR_8 = RCA-1N2864
F = fuse, 5 amperes
I = indicator lamp

R_9 = 47 ohms, 1 watt
R_{10} R_{11} = 15000 ohms, 10 watts
R_{12} = 47000 ohms, 2 watts
S_1 S_2 = toggle switch, single-pole single-throw
T = power transformer; Stancor P-8166 or equiv.
K_1 = relay; Potter and Brumfield KA11AY or equiv.
L_1 = 2.8 henries, 300 mA; Stancor C-2334 or equiv.
L_2 = 4 henries, 175 mA; Stancor C-1410 or equiv.
R_1 R_2 R_3 R_4 R_5 R_6 R_7 R_8 = 0.47 megohm, 0.5 watt

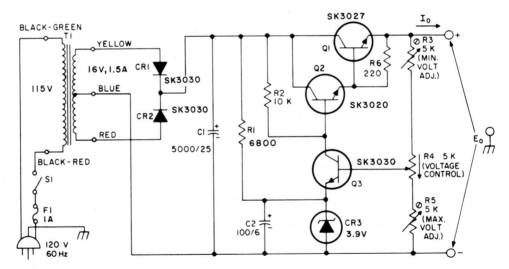

1.2.6 Adjustable supply is capable of delivering 4.5 to 12V at a maximum current of 1A. The voltage-regulating circuit in this supply uses a transistor (Q3) in conjunction with a zener. The base of Q3 is connected to voltage-control resistor R4, which, along with trimmer controls R3 and R5, is in parallel with the load. Therefore, any change in load or output voltage affects the voltage at the base of transistor Q3. If the output voltage tends to increase, the base of Q3 becomes more positive and more collector current flows. The increased collector current makes the base of Q2 less positive and reduces the collector–emitter current supplied to the base of transistor Q1. Reduced base current in Q1 results in an increased collector–emitter voltage drop in Q1; this voltage drop maintains the load voltage at the desired level. The opposite effect occurs when the load voltage tends to decrease.

Resistors R3 and R5 are used to set the upper and lower voltage limits of the supply; these values normally need be set only once. Screwdriver-adjust trimmer-type potentiometers are the best type for R3 and R5.

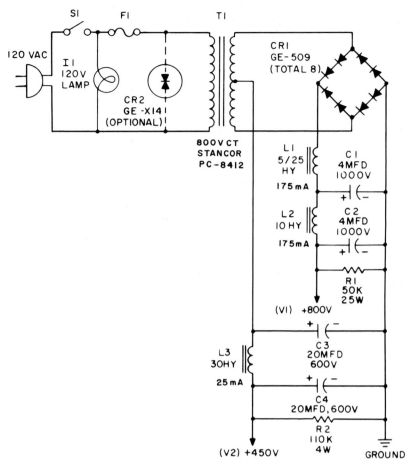

1.2.7 GE dual-voltage transmitter power supply provides +800V at 175 mA, intermittent duty, 1% ripple, 16% load regulation, for a final amplifier, and +450V at 25 mA, 0.02% ripple, for preamplifier and oscillator circuits.

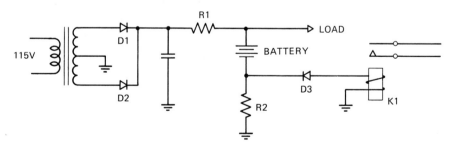

1.2.8 General Instrument failsafe power supply for security systems and other circuitry that includes a standby battery for emergency situations. During line-powered operation, dc power is supplied from a conventional full-wave centertap power supply constructed with 1N4001GP diodes. Resistors R1 and R2 permit a continuous trickle current to flow into the battery; the value of these resistors should be selected after determining the ampere-hour rating of the battery to be used: the charge rate should be 1% of the ampere-hour rating. Rectifier D3 keeps the trickle current from flowing through relay K1, which triggers the alarm. When the power fails, no current can flow through R1, so the battery assumes the full load as the battery current reverses. Diode D3 now in series with the battery and relay, conducts and causes current to flow through the relay coil, pulling in the contacts to signal the battery-operation condition.

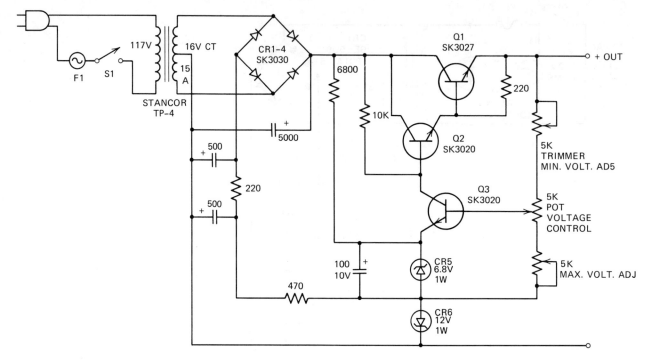

1.2.9 RCA full-range variable-voltage power supply is capable of delivering up to 12V at a maximum current of 1A. The regulator circuit in this supply receives full-wave rectified ac from a bridge, an arrangement that provides the regulator with a higher than typical input voltage. The load voltage is equal to the regulator voltage minus the voltage across zener CR6. When the two voltages are equal, the load voltage is zero. If the circuit voltage falls below 12V, the base–emitter junction of transistor Q3 becomes reverse-biased and the transistor turns off. As a result, Q2 and Q1 also turn off and prevent the load voltage from reversing polarity (becoming negative).

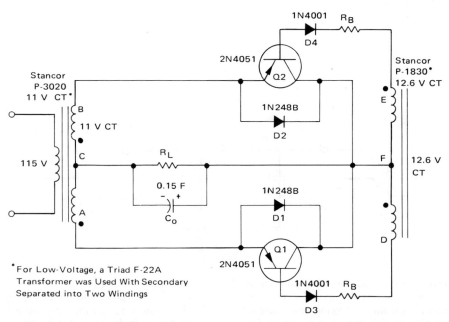

1.2.10 Motorola full-wave synchronous rectification circuit takes advantage of the low voltage drop of germanium transistors. A silicon diode across the transistor from collector to emitter permits use with capacitive loads. Because of the low collector–emitter saturation voltage of the transistor, the diodes do not conduct until the transistors start to pull out of saturation. Then the diodes conduct, supplying part of the increased load current and clamping the collector–emitter voltage of the transistors at a diode drop. As the capacitor charges, the collector current decreases, allowing the transistors to saturate, which turns the diodes off.

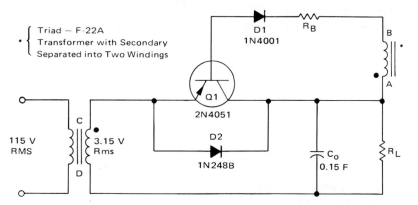

1.2.11 Motorola half-wave synchronous rectification system, in which a transistor is synchronously biased on by the ac input voltage (replaces a standard rectifier). Using a germanium transistor with a very low collector–emitter saturation voltage results in a more efficient low-voltage rectification than is possible with conventional rectifiers. The 2N4052 has only a 0.3V drop at 60A collector current.

When points **A** and **C** are positive with respect to points **B** and **D**, the base–emitter junction of Q1 is forward-biased and collector current flows through load resistor R_L. On the negative alternation, when points **B** and **D** are positive, the base–emitter junction of Q1 is reverse-biased, causing the transistor to block the voltage.

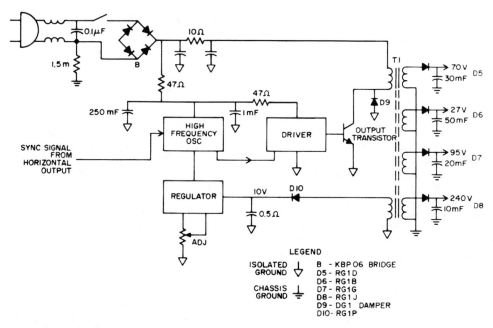

1.2.12 General Instrument high-frequency regulated supply for developing TV's low voltage. The initial voltage is developed by bridge B. This powers a high-frequency oscillator that is synchronized to the horizontal output circuit. The oscillator signal is built up by the driver to the output transistor. Rectifier D9 is a General Instrument DG1 damper. Rectifier D10 develops a control voltage for the regulator that is proportional to the output voltage. Diodes D5 to D8 are fast-recovery rectifiers to furnish low-voltage operating potentials for the entire set. High-voltage transformer T1 isolates the power line from the rest of the set. This circuit, although containing more components than a 60 Hz transformer-isolated TV power supply, offers a 3:1 reduction in size and almost 10:1 reduction in weight.

SECTION 1 Power Conversion and Regulation

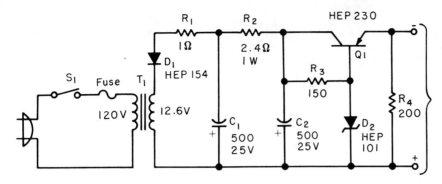

1.2.13 Motorola general-purpose regulated power supply furnishes a well-filtered voltage in the 9 to 10V range at a continuous current drain of up to 250 mA.

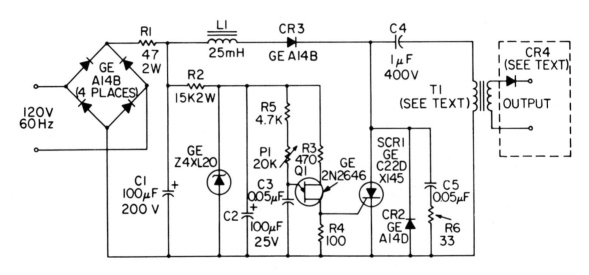

HIGH VOLTAGE POWER SUPPLY

1.2.14 GE high-voltage power supply is ideal for powering neon advertising signs, electrostatic precipitators, copying machines, etc. Conventionally, a high-voltage output is obtained by using a 60 Hz stepup transformer. But at 60 Hz the size and weight restrictions can be prohibitive. Frequency is one of the main factors that determine the size and weight of a transformer; increasing the operating frequency of a transformer can result in a reduction of its physical size. Here increased frequency is obtained by rectifying the 60 Hz ac line and "inverting" the resultant dc at higher frequency with an SCR inverter. The inverter utilizes the capacitor discharge principle, in which the energy stored in a capacitor is "dumped" by an SCR through the transformer primary winding at a high repetition rate.

In this circuit, the main filter capacitor C1 is charged approximately to the peak of the line voltage. The energy stored in C4 is dumped by SCR1 through the primary of the transformer at the rate of 1000 Hz. Trigger signals are derived from the free-running unijunction oscillator Q1. The SCR is turned off by the resonant action between C4 and the primary inductance of the transformer. Diode CR2 provides the return path in this resonant loop; while CR2 conducts, it reverse-biases SCR1 and turns it off. Capacitor C4 starts charging again, and the cycle repeats as explained above.

The transformer used in this circuit requires some attention. The stepup ratio should be selected to give the output voltage desired, and the core should be of a type suitable for the frequency chosen. At 1000 Hz operation, ferrite is suitable. The secondary winding should be well insulated from the primary and from the core. There should also be sufficient insulation between the layers of the secondary winding.

When a dc output voltage is required, the high voltage of T1 is rectified by CR4, a GE device (6RS9PH250PHB1) with nominal output of 7500V dc. The output voltage can be regulated by taking a feedback signal from the dc output to the triggering circuit.

For sequential neon signs, the same dc power supplies can be used for each step. However, it would be necessary to have separate stepup transformers, SCRs, and associated triggering networks for each step.

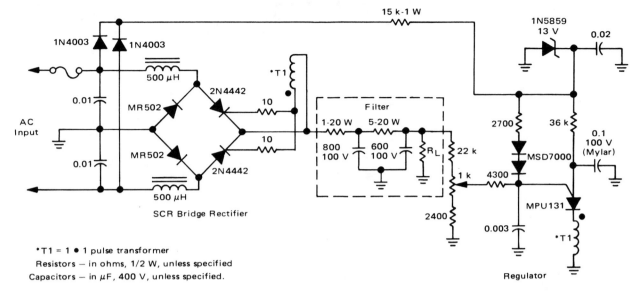

1.2.15 Motorola low-cost 80V, 1.5A power supply will maintain output voltage within 2% over a line-voltage range of 105 to 140V. This is important because of the lower line voltages that may be encountered during brownouts. SCRs are used because they are more rugged than transistors and offer cost savings as well. The SCR bridge actually does two functions at once: it rectifies the ac line and controls the amount of output voltage. The control function is accomplished by a variable duty cycle.

The line-filter chokes and 0.01 µF capacitors filter out the fast voltage spikes caused by the quick turnon of the SCRs; this filter network is also used to keep the horizontal scan frequencies from radiating back into the ac power line. These chokes should be constructed with a minimum wire size of No. 20 (AWG).

A programmable unijunction transistor (PUT) is used for the regulator and SCR gating function. The PUT is a negative-resistance device that is easily used in low-frequency sawtooth oscillators. The time constant of the 36K resistor and 0.1 µF capacitor determines the angle or pitch of the sawtooth waveform. The firing point of the PUT is set by the gate voltage. By increasing the gate voltage, the PUT will fire or trigger later. This action causes the SCRs to fire later in the ac cycle, which decreases the output voltage. The opposite effect occurs when the PUT's gate voltage decreases; the PUT triggers quickly and turns on the SCRs for a longer period, allowing more voltage to the output.

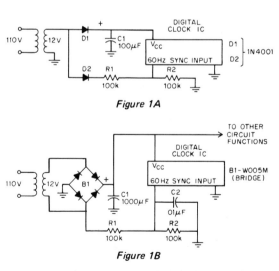

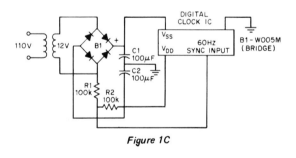

1.2.16 General Instrument digital-clock sync and rectifier circuits for both bipolar and MOS IC units. In Fig. 1A, rectifier D1 and capacitor C1 are used to supply dc voltage for V_{CC}. Rectifier D2 and resistors R1 and R2 develop a positive 60 Hz half-wave synchronizing signal. In Fig. 1B, a bridge circuit using a General Instrument W005M supplies both the dc voltage for V_{CC} and a positive synchronizing signal. The small size of the W005M (approximately TO-5) makes it ideal to use where full-wave rectification is needed. A W005M bridge is being used in Fig. 1C to supply both positive and negative voltages for a MOS IC. A synchronizing signal is also supplied by the bridge.

SECTION 1 *Power Conversion and Regulation*

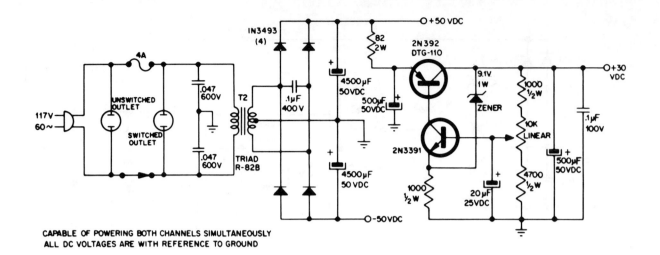

CAPABLE OF POWERING BOTH CHANNELS SIMULTANEOUSLY
ALL DC VOLTAGES ARE WITH REFERENCE TO GROUND

T-2 POWER TRANSFORMER*

RATINGS
117 v to 70 v c.t.; f = 60 Hz
Secondary load current: 5A rms max.

SPECIFICATIONS
Core: EI-15 (1½"), fabricated with laminations of M-19 grade silicon steel interleaved 2 × 2.
Winding: Standard design on paper layers.
Primary consists of 294 turns of #19 enameled wire.
Secondary consists of 186 turns of #17 enameled wire, tapped halfway.
Test Measurements:
$R_{pri} \simeq 1.4$ ohms
$R_{sec} \simeq 0.7$ ohms
*Triad Transformer Corp. P/N assigned: No. R-82B

1.2.17 Delco power supply can drive stereo amplifiers of up to 80W per channel.

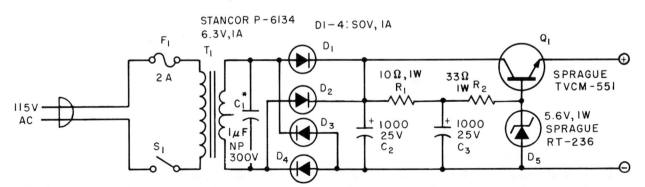

1.2.18 Sprague power supply for light-emitting diodes delivers regulated 5V. Since LEDs perform much like zeners, in that a constant voltage is developed across them regardless of the amount of current flowing in the circuit, it is important to employ a current-limiting resistor in series with one of the supply leads. The value of the resistor can be determined using Ohm's law after selecting the LEDs to be used. The supply shown will deliver a full ampere of current at 5V, making it suitable for IC applications as well.

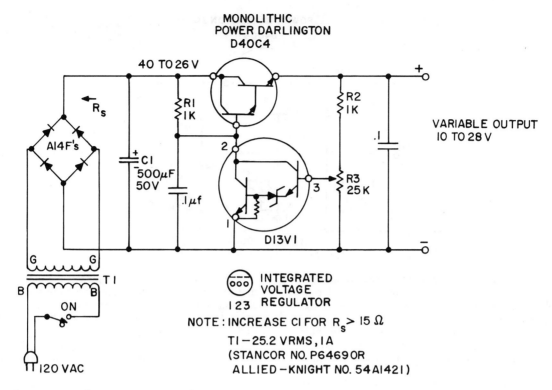

1.2.19 GE power supply incorporates an integrated voltage regulator (IVR), a three-terminal integrated circuit that contains two transistors, a voltage reference diode, and a resistor. The circuit is a 10 to 28V supply with a maximum dc output current of 300 mA. The output can be increased to one-half ampere with an adequate heatsink.

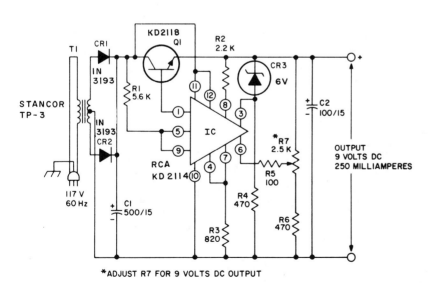

*ADJUST R7 FOR 9 VOLTS DC OUTPUT

1.2.20 RCA regulated 9V power supply is useful as a transistor-radio battery eliminator and as a power source for experimenter projects and for test-bench applications, such as the servicing of portable transistorized equipment. It can deliver a dc output of 9V with a voltage regulation of 3% at 250 mA.

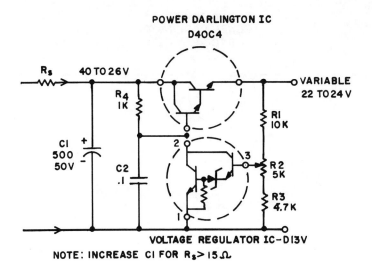

1.2.21 GE regulated 24V supply has a low component cost for its performance and is inherently reliable due to its simplicity. The output voltage is easily adjusted to the exact value desired.

The rectified power is supplied to the filter capacitor through R_s (the Thevenin equivalent impedance associated with the transformer and rectifiers). The GE D40C4 is a series-pass element that regulates the output voltage, and the D13V is a voltage-controlled current shunt. The amount of current shunted from the base of the D40C4 varies to maintain a constant output voltage. Resistor R2 in the feedback loop is adjusted to provide the exact output voltage desired. The purpose of the 0.1 μF capacitor is to provide high-frequency stability to the power supply.

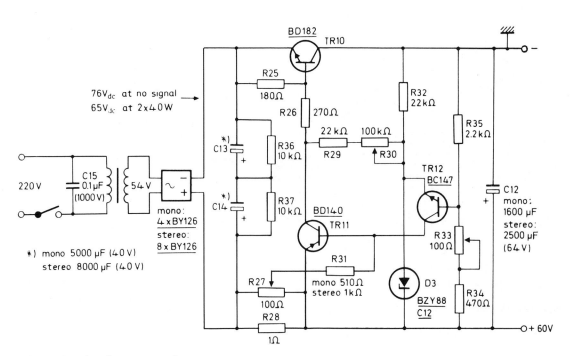

1.2.22 Regulated power supply using an Amperex one-piece bridge and discrete transistors for regulation.

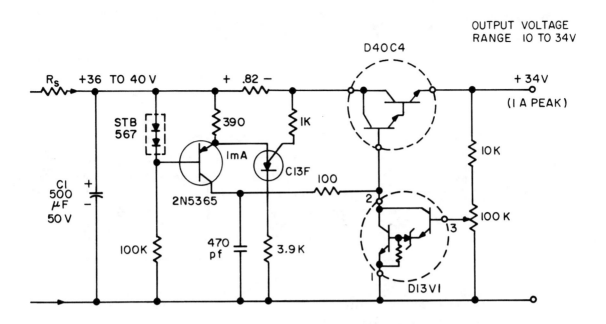

NOTE: .82Ω LIMITS CURRENT AT 1.5 TO 2.0 A
INCREASE C_1 IF $R_s > 15\Omega$
DYNAMIC OUTPUT RESISTANCE = .06V/.95A = 0.063 Ω

1.2.23 GE regulated power supply with foldback current limiting. In this circuit the GE CSCR acts as a switch, turning the 2N5365 off when the maximum current of 2A is reached. The voltage drop across the 0.82-ohm resistor forward-baises the anode–gate junction of the C13F, turning the device on. This turns the 2N5365 off and removes base bias from the GE D40C4. This causes the output voltage to go to zero. This circuit is versatile enough to adapt to most series power supply designs. By changing the value of the current-sensing resistor, this current limit is adjusted to any desired value.

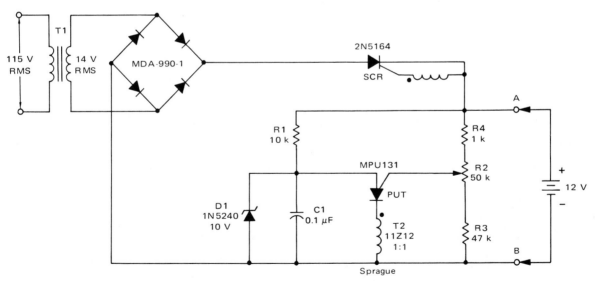

1.2.24 Motorola short-circuit-proof battery charger provides an average charging current of about 8A to a 12V lead–acid storage battery. It will not function nor will it be damaged by improperly connecting the battery to the circuit.

With 115V at the input, the circuit commences to function when the battery is properly attached. The battery provides the current to charge the timing capacitor C1 used in the PUT relaxation oscillator. When C1 charges to the peak-point voltage of the PUT, the PUT fires, turning the SCR on. This applies charging current to the battery. As the battery charges, the voltage increases slightly, which increases the peak-point voltage of the PUT. This means that C1 has to charge to a slightly higher voltage to fire the PUT. The voltage on C1 increases until the zener voltage of D1 is reached, which clamps the voltage on C1 and prevents the PUT oscillator from oscillating; charging ceases. The maximum battery voltage is set by potentiometer R2, which sets the peak-point firing voltage of the PUT.

SECTION 1 *Power Conversion and Regulation*

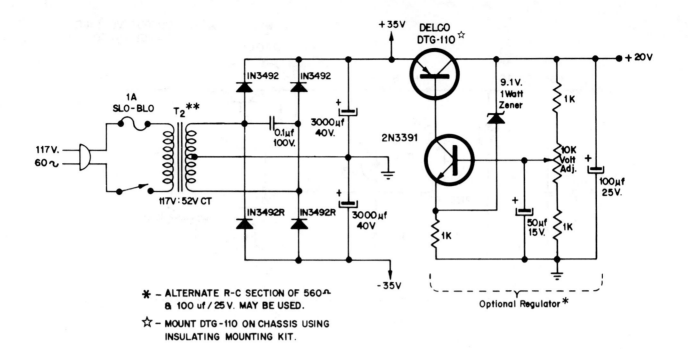

* — ALTERNATE R-C SECTION OF 560Ω & 100 uf / 25 V. MAY BE USED.

☆ — MOUNT DTG-110 ON CHASSIS USING INSULATING MOUNTING KIT.

**T-2 POWER TRANSFORMER
   Triad Transformer P/N assigned: No. R-206B
   RATINGS
      117 v to 52 v c.t.; f = 60 Hz
      Secondary load current: 2A rms max.
   SPECIFICATIONS
      Core: EI 125 laminations of M-19 grade silicon steel interleaved 2x2, stack 1¾".
      Winding: Standard design on paper layers. Wind secondaries first.
      Secondary consists of dual windings of #17 formavar wire, 64 turns on each winding. Connect windings in series.

1.2.25 Delco regulated supply can power stereo amplifiers of 50W per channel.

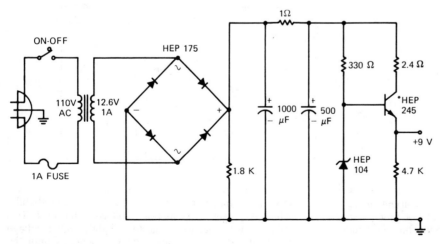

1.2.26 Motorola simple regulated power supply for transistor radios and experimenter circuits that require a 9V power source. The HEP 245 regulator transistor should be mounted on a Motorola HEP 500 heatsink to allow adequate heat dissipation.

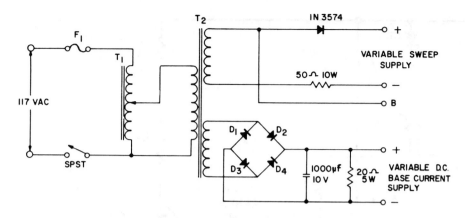

T₁ — Powerstat variable transformer Type 10B
T₂ — Power transformer secondaries 135 volts at 200 mA 6.3 volts at 5.5 amperes — Triad R-73B
D_1, D_2, D_3, D_4 — Delco 1N3491 or 1N3208
F₁ — Fuse 3AG, ¾ ampere

1.2.27 Delco universal power supply is designed to supply variable-amplitude pulses for diode voltage sweep tests. The low-voltage portion of the supply provides dc base current for gain measurements. The two sections of the supply should not be used simultaneously.

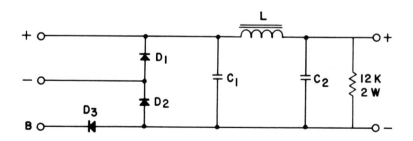

D_1, D_2, D_3 — 1N3574
L — 2.8 H at 300 mA   Stancor C2334
C_1, C_2 — Dual 80 μF, 300 W VDC Sprague TVL 2585

1.2.28 Sprague adapter unit. Added to the variable sweep supply, the unit converts the variable-sweep supply into a full-wave bridge supply complete with filtered dc output.

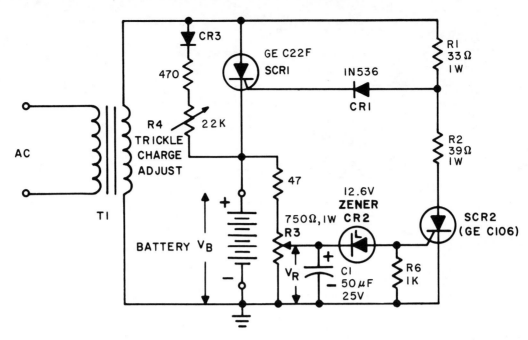

1.2.29 GE half-wave 12V SCR battery charger. The secondary winding of the charging transformer, SCR1, and the battery to be charged are connected in a series string. If gate current is applied to SCR1 during positive half-cycles of the input ac, SCR1 will trigger and supply half-wave charging current to the battery. If gate current to SCR1 is stopped once the battery has become charged, the charging process will cease. With the battery voltage low (discharged), SCR1 is gated on each cycle via resistor R1 and diode CR1. Under these conditions pickoff voltage V_R at the wiper of potentiometer R3 is adjusted to be less than the breakdown voltage of zener CR2, and SCR2 cannot fire. As the battery approaches full charge its terminal voltage rises; SCR2, which must have a sensitive-gate triggering characteristic like the GE C5, starts to fire each cycle. At first SCR2 fires $\pi/2$ radians after the start of each positive half-cycle, coincident with the peak supply voltage, peak charging current, and maximum battery voltage. As the battery terminal voltage climbs, the firing angle of SCR2 advances until eventually SCR2 is firing before the input sine wave has risen far enough to trigger SCR1. With SCR2 on first in a cycle, the voltage-divider action of R1 and R2 keeps CR1 back-biased, SCR1 stays off, and heavy charging ceases.

1.3 Direct-Current Converters

Since dc cannot be transformed from one voltage value to another for power applications, when a load requires a higher or a lower value of direct current than our dc source can provide, we must employ some form of dc conversion. Most generally, this involves inversion or commutation of the dc into a string of pulses, which we can then feed into an appropriate transformer whose output can be rectified and filtered like that of any power supply.

In the process of conversion, however, we try to get the fastest possible commutation possible, since we know that the size of a power transformer (which governs its cost) is inversely proportional to the frequency of the commutated input signal.

The owner of an expensive piece of entertainment equipment designed for operation from the U.S.-standard 60 Hz power line is usually dismayed to find that the unit overheats considerably when he tries to use it in a country that has a 50 Hz standard. He learns the hard way that lower line frequencies mean heavier, bulkier transformer cores.

A dc-to-dc converter requires ac for only a small portion of the total circuit, and it doesn't really matter to anyone whether the switching frequency is fast or slow, since the stage following ac transformation involved rectification of the ac anyway. Designers who care about package size and weight, component cost, and ease of filtering will select a high-frequency converter circuit every time. But that doesn't mean that there are no disadvantages with this approach.

There are two disadvantages worth considering: the sound wave created by the system switching at an audio rate and the interference that is likely to result from the nonsinusoidal waveform of the switching transistors. Both these anomalies can be corrected with simple engineering efforts, however. Various sound-deadening techniques can be employed to curtail the radiated wave from the converter unit, from use of special chassis mounts for mobile applications to packing the chassis (or wrapping it) with sound-absorbing material. Conventional shielding techniques can overcome the signal-radiating problem when it causes objectionable interference to other equipment.

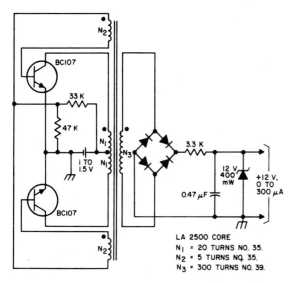

1.3.1 RCA dc-to-dc converter for powering CMOS from a single cell.

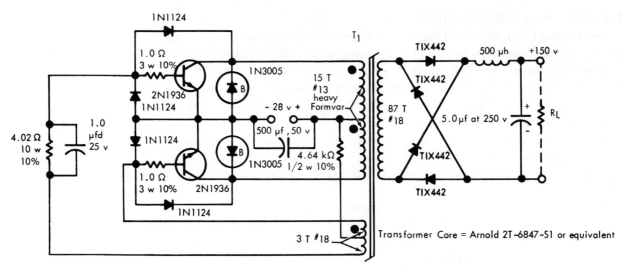

NOTE: Minimum usable $h_{FE} = 10$ at $V_{CE} = 3v$, $I_C = 10a$, $T_A = 25°C$
Both transistors use heat sinks, $\theta_{C-HS} + \theta_{HS-A} \leq 2°C/w$ each. All diodes must have adequate heat sinks.

TYPICAL PERFORMANCE AT $T_A = 25°C$
Efficiency = 80%
Output ripple 1.0 volt

1.3.2 225W, 10 kHz dc-to-dc converter operates from −55 to +125°C. Semiconductors are Texas Instrument devices.

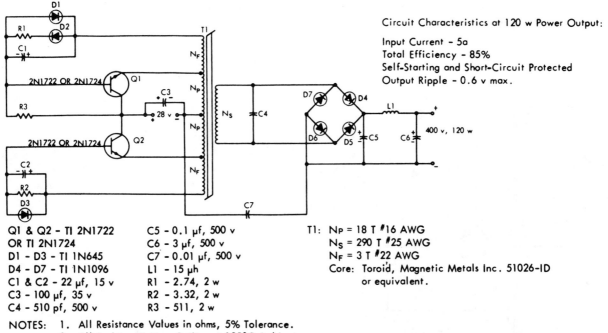

Q1 & Q2 - TI 2N1722
OR TI 2N1724
D1 - D3 - TI 1N645
D4 - D7 - TI 1N1096
C1 & C2 - 22 µf, 15 v
C3 - 100 µf, 35 v
C4 - 510 pf, 500 v

C5 - 0.1 µf, 500 v
C6 - 3 µf, 500 v
C7 - 0.01 µf, 500 v
L1 - 15 µh
R1 - 2.74, 2 w
R2 - 3.32, 2 w
R3 - 511, 2 w

T1: N_P = 18 T #16 AWG
N_S = 290 T #25 AWG
N_F = 3 T #22 AWG
Core: Toroid, Magnetic Metals Inc. 51026-1D or equivalent.

NOTES: 1. All Resistance Values in ohms, 5% Tolerance.
2. All Resistor Wattage Ratings at 125°C Ambient.
3. Capacitor Voltage Ratings at 125°C Ambient.
4. Q1 and Q2 on Same Heat Sink, $\theta_{C-HS} + \theta_{HS-A} \leq 4 C°/w$ each.

1.3.3 Texas Instruments 10 kHz dc-to-dc converter delivers 400V at up to 120W.

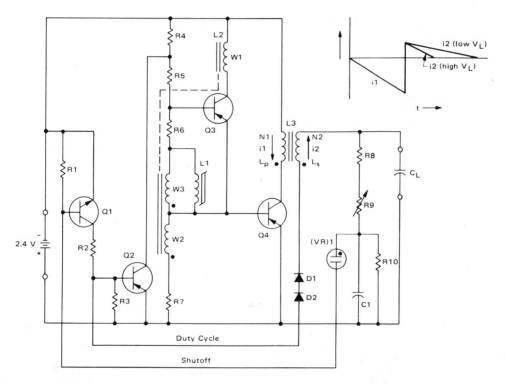

1.3.4 Motorola dc-to-dc converter circuit for capacitor charging of a photoflash includes energy-storage capacitor C_L for the converter load and a 2.4V battery power supply. The converter is a variable-duty-cycle circuit; it stops and starts itself to maintain the voltage on the load capacitor within narrow limits. It is also self-compensating for battery voltage variation and tends to charge the load capacitor in the same length of time, regardless of decreasing battery voltage.

During the on portion of the cycle, inductor L3 stores energy. When Q4 is switched off, this stored energy is transferred to the load capacitor, and the cycle is repeated.

The driver-stage oscillator has approximately a 75% duty cycle when free-running. However, the time required for L3 to transfer its stored energy to C_L is dependent on the existing voltage across C_L, the turns ratio of L3, and the battery voltage. It is undesirable to start a new cycle before L3 is discharged; consequently, Q2 is used to prevent a new cycle from starting until the output current decreases to zero at the end of each off period.

The free-running frequency of the driver-stage blocking oscillator is determined by the square-loop timing core of L1. Switch Q3 is on until this core saturates from the constant-voltage winding of W3. When L1 saturates, it shunts the base drive and Q3 turns off. During the off period W3 voltage is initially several times higher than its on value, and the timing core can be reset much more rapidly than it can be saturated. As soon as L1 is reset and the stored energy in L2 is dissipated, the driver stage is ready to begin a new cycle. Resistor R4 supplies the starting current to begin each cycle.

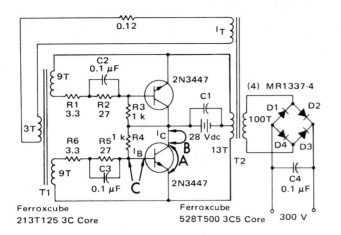
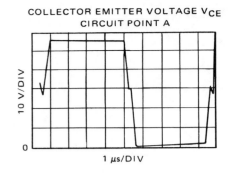
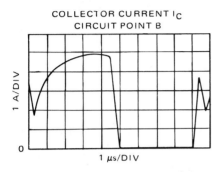
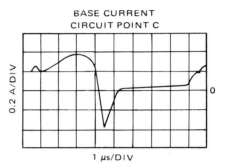

1.3.5 Motorola dc-to-dc converter operates at 100 kHz to provide 100W of output power at approximately 200V, from a 28V input source. Operation at such high frequencies affords considerable component size and weight reduction.

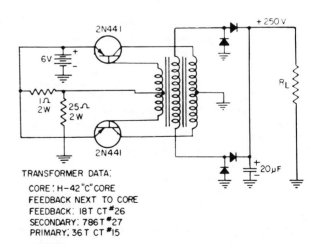

1.3.6 Delco power converters. Transistorized square-wave oscillator power converters are being used in many applications as replacements for vibrator and rotary-type power converters. Delco has compiled a group of transistor power converter circuits to aid the design engineer in the solution of conversion problems. These two representative circuits utilize lower priced transistors to convert power at the 6, 12, and 24V level to power at higher potentials.

36 DISCRETE / TRANSISTOR CIRCUIT SOURCEMASTER

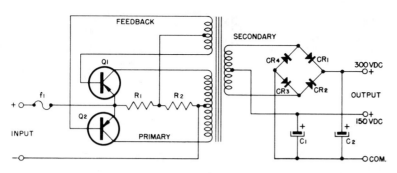

Parts List

Part	14V Converter	28V Converter
Q1, Q2	Delco 2N174	Delco 2N1412
R1	1.5 ohms, 10W wirewound	3 ohms, 5W wirewound
R2	40 ohms, 10W wirewound	150 ohms, 10W
CR1, 2, 3, 4	RCA 1N3195	RCA 1N3195
C1	10 μF 200V dc	10 μF 200V dc
C2	30 μF 350V dc	30 μF 350V dc
f1	20A fuse	15A fuse
Toroid core	Magnetics Inc. 50001-2A or 51001-2A	Magnetics Inc. 50001-2A or 51001-2A
Transformer Winding		
Primary*	40t tap at 20t No. 10 AWG	80 t tap at 40t No. 12 AWG
Feedback*	12t tap at 6t No. 18 AWG	12t tap at 6t No. 18 AWG
Secondary	448t tap at 224 No. 23 AWG	440t tap at 220t No. 23 AWG

*Bifilar wound.

1.3.7 Delco dc-to-dc converter supplies 145W at 14V input or 225W with 28V input. The square-wave output is rectified by CR1-4 and filtered by C1, C2. The parts list is different for 28V and 12V input. Choose the parts needed from the appropriate list.

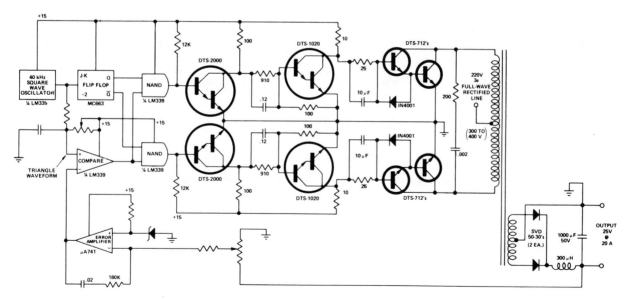

1.3.8 GE low-cost industrial regulating converter uses DTS-712 transistors in a Darlington configuration and takes advantage of their high voltage capability. A method of pulse-width modulation on the push-pull inverter allows regulation, while the small 20 kHz transformer steps the voltage down to a useful level. Operation from the 220V, 3-phase, full-wave rectifier line is permitted, eliminating the need for bulky 60 Hz filters. Small size and light weight are advantages of switching at 20 kHz, while overall efficiency is 70 to 80%.

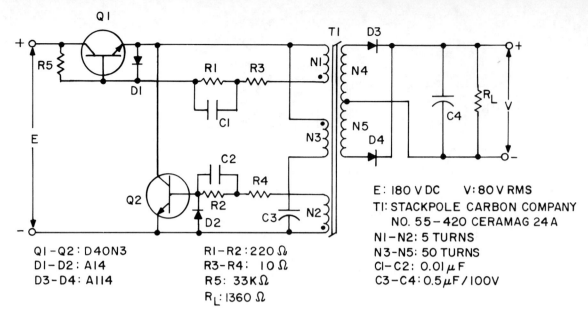

Q1–Q2: D40N3
D1–D2: A14
D3–D4: A114
R1–R2: 220 Ω
R3–R4: 10 Ω
R5: 33 KΩ
R_L: 1360 Ω
E: 180 V DC   V: 80 V RMS
T1: STACKPOLE CARBON COMPANY NO. 55-420 CERAMAG 24 A
N1–N2: 5 TURNS
N3–N5: 50 TURNS
C1–C2: 0.01 μF
C3–C4: 0.5 μF / 100V

1.3.9 GE high-speed converter allows the use of a small ferrite transformer and requires a transistor whose rating is only slightly larger than the input voltage. Because diodes D1 and D2 reverse-bias the transistor during turnoff, GE high-voltage transistors (D40N, D44R) can be used with voltage supplies up to 500V.

When Q1 turns on, half the input voltage appears across winding N3 of the transformer. The induced voltage on the base winding, N1, keeps Q1 on and in saturation until the transformer core saturates. At this point the reduced base drive turns Q1 off, reducing the current in N3 and reversing the voltage on all the windings. Q2 turns on, applying the capacitor voltage to winding N3 until the core saturates in the opposite direction, and the cycle repeats. If C3 were not charged to half the supply voltage, more current would flow into the output filter and load when Q1 is on than when Q2 is on because of the larger voltage across winding N3. The load therefore tends to keep C3 charged to half the supply voltage. To insure reliable starting, resistor R5 must be small enough to raise the base of Q1 above the trigger voltage of the transistor when the converter is not operating. In addition, a large resistor may be placed across C3 to allow C3 to discharge should it inadvertently get charged too high to allow the converter to start.

Using D40N transistors with the components shown, an output power of 4.6W was attained at 14 kHz, with a 60% efficiency for a load of 1360 ohms. When R1 and R2 were changed to 1000 ohms and capacitor C3 was increased to 20 μF, two D44R transistors provided 40W of output power into a 200-ohm load at 12.5 kHz with 82% efficiency. More output could be attained at a higher efficiency by going to a higher input voltage.

By varying the number of output turns, any voltage output can be attained from a full-wave rectified line voltage of 110 or 220V. Because of the high frequency of operation, the circuit without the dc filter on the output may also see applications in fluorescent lighting with proper ballasting of the fluorescent lamp.

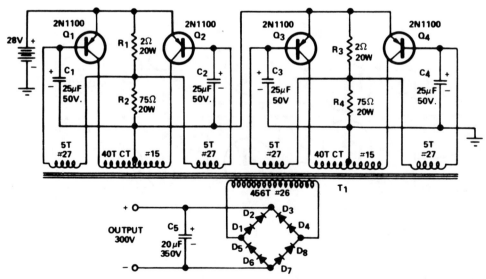

1.3.10 Delco series-connected dc-to-dc converter allows operation at high supply voltages. To maintain proper phasing and equal power distribution, it is necessary that all of the transformer windings be on the same core: the voltages applied to each transistor will be equal and inversely proportional to the number of pairs of transistors in the circuit. Maximum efficiency of operation will be achieved by using a core material whose hysteresis loop is as nearly rectangular as possible.

The four transistors operate as switches controlled by the bias voltages induced in the feedback windings of the transformer and applied to the transistor bases.

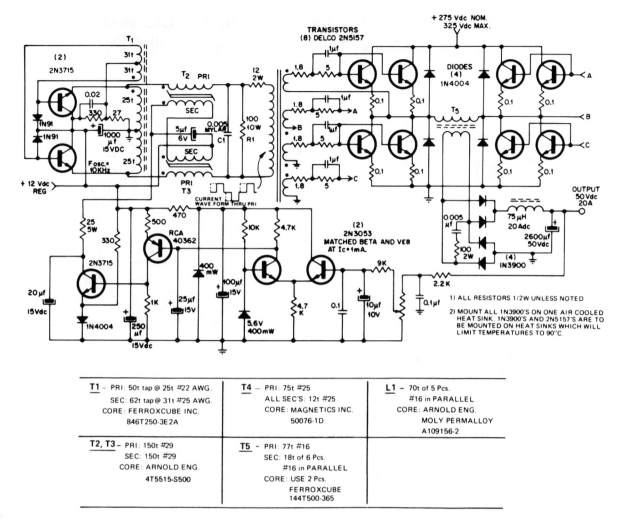

1.3.11 Delco pulse-width-modulated power converter capable of 1 kW output at 87.5% efficiency. Input and output dc voltages are 275 and 50V, respectively. A voltage regulator is used to compensate for changes in load or input voltage. The switching frequency is 10 kHz. Tape-wound toroid cores are used with magnetic amplifier circuitry to create a controllable delay in the base-drive waveforms of the bridge output circuit. The converter occupies a volume of ¼ cubic foot (0.325m) and weighs less than 6 lb (2.72 kg).

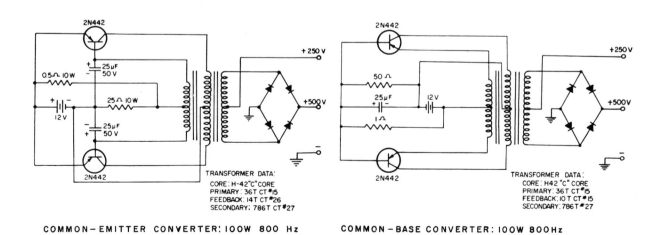

1.3.12 Delco dc-to-dc converters.

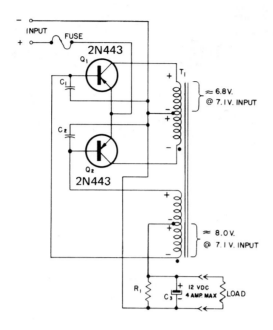

Q_1 Q_2 — Delco 2N443
C_1 C_2 — .5µF @ 100V Paper
C_3 — 1500 µF 15Vdc
R_1 — 470Ω 1watt Carbon
Fuse — 10 A Slo-Blo
T_1 — Primary
    48 Turns #17
    tap @ 24 t
    Feedback
    56 Turns #17
    tap @ 28 t
    Core: H-3 Hypersil
        Type "C" Core

NOTES:

Nominal frequency of oscillation = 300 Hz
Mount both transistors on one Delco 7281352 Heat Sink
Ambient temperature = −60°C to + 71°C

1.3.13 Delco rectifierless 6-to-12V dc-to-dc converter. Voltage multiplication is achieved by adding the feedback voltage to the supply voltage. Assume that Q1 is turning on. Voltages will be generated by the windings of the transformer. A negative voltage is being applied to the base of Q1. Since this is positive feedback, Q1 will continue to conduct until the transformer core saturates. With Q1 fully turned on, approximately 6.8V is applied across the upper half of the primary winding. This develops 8V in each half of the feedback winding. The voltage developed in the lower half of the feedback winding is of a polarity that adds algebraically to the supply voltage to develop 14V across the load resistor. The current path is from the lower end of the feedback winding through the emitter diode to the positive terminal of the input voltage source; thus the voltage in the feedback winding is added to the 6V source. Of the 16V developed across the feedback winding, 15V appears as reverse bias across the emitter diode of Q2. The 2V drop is the forward conducting voltage of the Q1 emitter. As the transformer core saturates, the magnetic field collapses and voltages of the opposite polarity will be generated in both windings. Thus Q1 will be cut off and Q2 will begin conducting. The emitter diodes may be considered as diodes rectifying in a full-wave circuit with the dc output being taken from the centertap of the feedback winding. The polarity across the load resistor does not change.

1.4 Inverters

An inverter is a circuit that changes a direct-current source into an alternating voltage. Where inversion was once the province of such mechanical contraptions as the vibrating reed, today it is accomplished almost universally with switching transistors. So economical has the process of voltage inversion become that it is now widely employed in industry, even to the extent that alternating line voltages are used to obtain the direct current needed to power the inverter, as some of the circuits included here show.

The inverter becomes a dc converter when a rectifier is driven with the inverter output; similarly, converters may be used to make inverters by removing the rectifying element from the final stage. Keep this in mind as you search for the ideal circuit for your application, and examine the circuits of Section 1.3 along with those shown in this subsection.

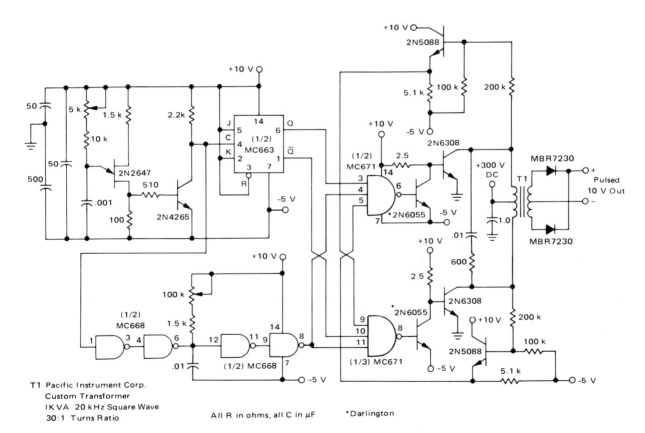

1.4.1 Motorola 20 kHz, 1 kW line-operated inverter could become the heart of a small computer power supply. The list of products includes high threshold logic (HTL), Darlington power transistors, high-voltage power transistor, and Schottky barrier diodes. Using an ultrasonic inverter instead of a transformer for power conversion yields the design features of small size and high efficiency. Other features inherent in this design are low component count, crossover inhibit of the push-pull output, and internal means of regulation. Even though the output is intended for computer mainframes, it can readily be adapted to any voltage–current requirements within the power constraint.

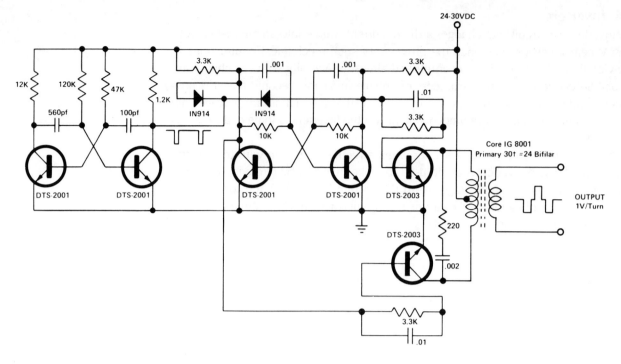

1.4.2 Delco 15-watt inverter requires 24 to 30V dc input and switches at a rate of 25 kHz.

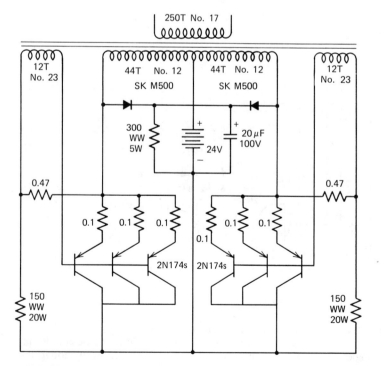

1.4.3 RCA power inverter provides 400W at 117V, 60 Hz when driven from 12V car battery. Transformer data: lamination type 36 EI, 1⁵/₁₆-in. (3.34 cm) stack, 14 mil silicon steel.

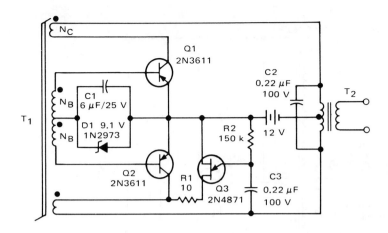

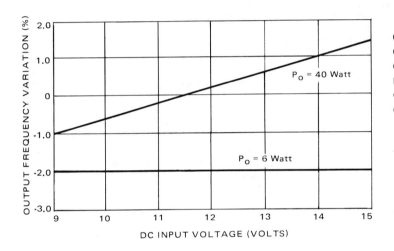

PARTS LIST

C1	Capacitor	6 μF, 25 Vdc	
C2	Capacitor	0.22 μF, 100 V	
C3	Capacitor	0.22 μF, 100 V	
D1	Zener	9.1 V, 10 W	1N2973
Q1, Q2	Transistor	7 A, 40 V	2N3611
Q3	UJT	300 mW, 35 V	2N4871
R1	Resistor	10 Ω, 1/2 W	
R2	Resistor	150 k, 1/2 W	
T1	Transformer		
	Arnold Core 4T − 4179 D1		
	N_B 200 turns #20 Wire		
	N_C 20 turns three #16 wires in parallel		
T2	Transformer	110-24 VCT Felco VJ-1	

1.4.4 Motorola 40-watt 400 Hz inverter for 12V input. The output is a 110V square wave whose frequency is extremely stable and almost unaffected by changes in input voltage and output power. Nominal input current is 4A but can vary from 240 mA to 7A, depending on the input voltage and the output load. If the input voltage is 12V, the permissible range of output current will be from 25 to 760 mA. Capacitive loads are permissible if the load surge current is less than 1.5A.

Circuit efficiency is approximately 80%. The zener voltage is an order of magnitude greater than the base–emitter voltage of the power transistor and provides excellent frequency stability. The zener current is about 0.4A and dissipates approximately 4W. As the zener voltage is reduced, this power will decrease and efficiency can be improved by sacrificing frequency stability.

Unijunction transistor Q3 is used to start the inverter. Its output voltage pulse causes current to flow into the lower collector winding of transformer T1.

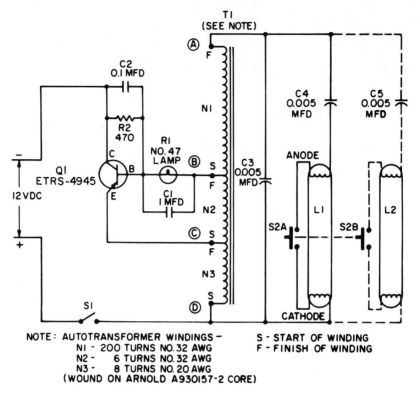

NOTE: AUTOTRANSFORMER WINDINGS –
N1 – 200 TURNS NO. 32 AWG
N2 – 6 TURNS NO. 32 AWG
N3 – 8 TURNS NO. 20 AWG
(WOUND ON ARNOLD A930157-2 CORE)

S – START OF WINDING
F – FINISH OF WINDING

1.4.5 GE battery-operated fluorescent light. This patented circuit uses a single transistor inverter to operate from a 12V battery. This compact circuit uses only three primary components: a power transistor, Q1: an autotransformer, T1; and a capacitor ballast, C4 or C5. Capacitor C3 regulates the output voltage and frequency, while the other components provide base drive for the switching power transistor. This switching action induces a high ac voltage across the N3 winding of T1 by the autotransformer action of the transformer's total turns.

For T1, start with 35 ft (10.66m) of 32 AWG magnet wire and place 200 turns on the core with two layers of 100 turns each. After the first layer is wound, cover it with a single layer of electrical tape to prevent shorting between turns. The second layer of 100 turns is wound over the tape, and this layer is again covered with a single layer of electrical tape. Mark the two ends of the winding as to start (S) and finish (F) and identify it as the N1 winding. Apply the N2 winding made up to 6 turns of 32 AWG magnet wire wound in the same direction as the N1 winding. Identify the winding and mark the start and finish wires. The N3 winding, consisting of 8 turns of No. 20 AWG magnet wire, is applied last and wound in the same direction as the other two windings. After the N3 winding is marked, connect all three windings in series and mark the four leads extending from the transformer. Apply a final layer of electrical tape over the outer windings.

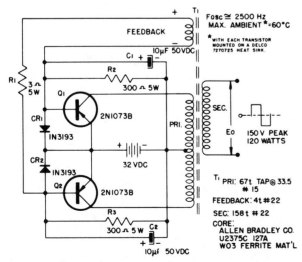

1.4.6 Delco ac inverter switches at 2500 Hz and operates with a dc input of 32V to provide 150V peak output with a power of 120W.

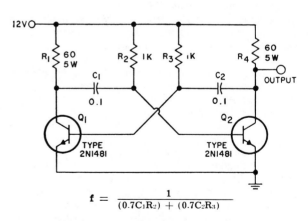

$$f = \frac{1}{(0.7 C_1 R_2) + (0.7 C_2 R_3)}$$

1.4.7 RCA astable multivibrator develops a square wave that has a peak value equal to the +12V supply and a minimum value equal to the collector saturation voltage of the transistors. The circuit employs two transistors operated in identical common-emitter amplifier stages with regenerative feedback from the collector of one transistor to the base of the other.

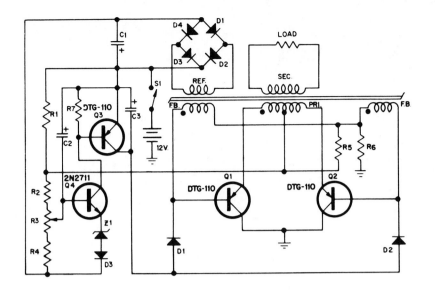

Q₁, Q₂, Q₃, — Delco DTG-110's
Q₄ — 2N2711
$D_1, D_2, D_3, D_4, D_5, D_6, D_7$ — Silicon, DRS-102
Z_1 — 1N3018 or Equiv., 8.2V
C_1, C_2 — 50μF, 25V
C_3 — 1000μF, 25V
R_1 — .25 Ω 2W
R_2, R_4 — 100 Ω 1W
R_3 — 400 Ω 2W
R_5 — 68 Ω
R_6 — 1.5K Ω
R_7 — 15 Ω
Toroid — Magnetics Inc.
  51001-2A
  Windings — Pri
    60T #18 C.T.
  Feedback
    92T #30 C.T.
  Reference
    49T #29
  Sec.
    310 #24
Heat Sinks — For Q_1, & Q_2, Delco
  7270725
  For Q_3, Delco
    7271357

1.4.8 In this 400 Hz dc-to-ac inverter using Delco DTG-110 transistors, both output voltage and frequency are regulated. Base drive to the output transistors is controlled by a series regulator that senses the output of an additional reference winding. This inverter is capable of providing 10W output and is intended to satisfy the requirements of a low-cost regulated ac power source for automotive applications. Output is 115V, 400 Hz; efficiency at full load (10W) is 72%.

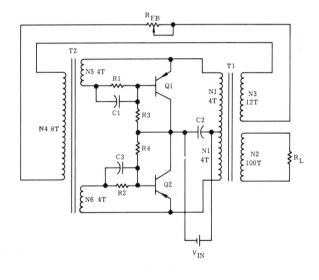

R_1	0.75 Ω, 5 w
R_2	0.75 Ω, 5 w
R_3	7.5 Ω, 5 w
R_4	7.5 Ω, 5 w
R_{FB}	1 Ω, 5 w
C_1	20 μf, 6 v
C_2	10,000 μf, 6 v
C_3	20 μf, 6 v
T_1	Phoenix Transformer PX2127
T_2	Phoenix Transformer PX2126
Q_1, Q_2	2N2728
VIN	2V 50A

1.4.9 Motorola inverter circuit converts low voltages to high. The extremely low saturation voltage of the 2N2728 power transistor enables switching currents up to 50A efficiently. The saturation voltage of the transistor will be less than 0.1V at 50 amperes if the transistor is driven with a forced gain of 10. This is a resistance of 0.002 ohm, the equivalent of a 12 in. (30.48 cm) length of 12-gage wire. The frequency is set by adjusting the feedback voltage. For this application, the core was set to oscillate at 1 kHz.

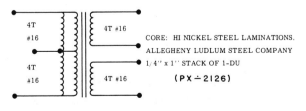

SECTION 1 *Power Conversion and Regulation*

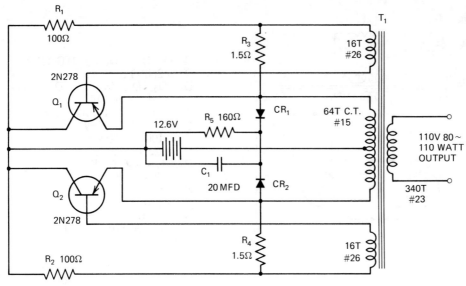

AMBIENT TEMPERATURE RANGE 71°C TO −55°C

Parts List

R1, R2	100 ohms, 5W wirewound
R3, R4	1.5 ohms, 5W wirewound
R5	150 ohms, 5W wirewound
CR1, 2	1N4004
C1	20 μF 50V dc electrolytic
Q1, Q2	2N278 Delco power transistor
T1 core	1 5/16 in. (3.34 cm) stack of 125 E.I. 0.014 in. (1.5 mm) silicon iron
Heatsink	Delco 7270725

1.4.10 Delco dc-to-ac inverter operates from car battery and delivers 110W at 117V, 60 Hz.

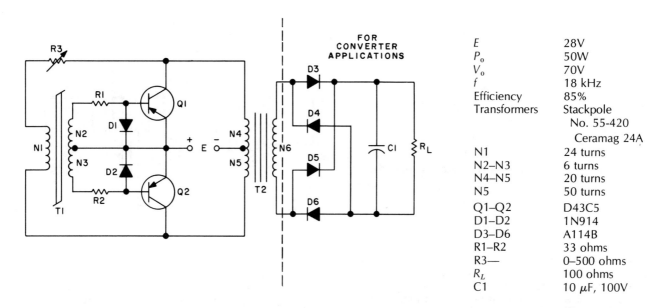

1.4.11 GE two-transformer inverter for use at higher power levels. A more efficient, nonsaturating transformer (T2) can be used on the output, and the saturating base-drive transformer (T1) can be made much smaller, thus reducing the losses in the saturating core. The saturation current is now limited by resistor R3 and, because the voltage on input winding N1 decreases as the saturation current increases, the base drive falls off more rapidly so that switching is enhanced and power dissipation excursions are not as severe as in one-transformer inverters.

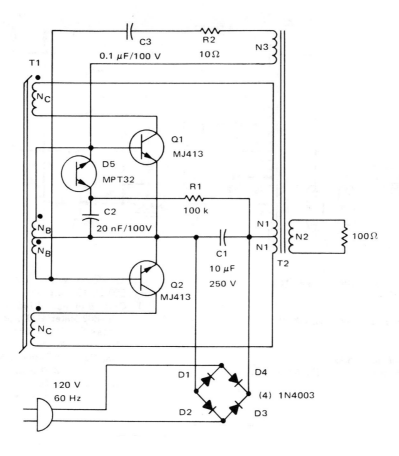

TRANSFORMERS
T1 — CORE — MAGNETICS INC. #80623-1/2D-080
    N_B 15 TURNS OF AWG #26 WIRE
    N_C 3 TURNS OF AWG #22 WIRE
T2 — CORE — ARNOLD GT-5800-D1
    N1 100 TURNS OF 3 AWG #22 WIRES
    N2 104 TURNS OF 3 AWG #19 WIRES
    N3 7 TURNS OF AWG #26 WIRE

1.4.12 Motorola fluorescent-lamp driver offers 80% light-conversion efficiency. It is essentially a line-operated hybrid-feedback inverter that operates at 15 kHz. The input voltage is 120V at 60 Hz; the output is 120V at 200W.

The dc input is obtained by placing a bridge rectifier across the ac line. To keep the size to a minimum, there is very little filtering of the rectified input. This means that the voltage across C1 varies from 10 to 150V and back to 10V during each alternation of the input voltage, making the output voltage an amplitude-modulated 15 kHz square wave.

The power common-emitter transistors are driven from a saturating timing transformer. This transformer produces a base drive proportional to the collector current and provides good efficiency even though the input voltage swings from 10 to 150V.

The output transformer transfers power to the load and does not saturate. The output voltage is almost unaffected by changes in the load but it does vary with changing input voltage.

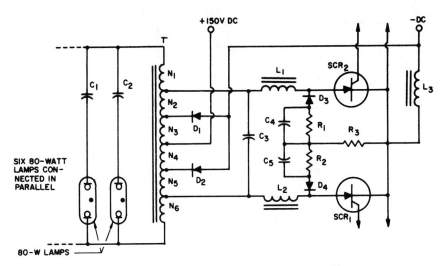

C_1, C_2: 0.01 μF, 1200 V (Ballast Capacitors)
C_3: 0.01 μF, 600 V
C_4, C_5: 0.02 μF, 600 V
D_1, D_2: Fast-Recovery Diodes, 6 A, 600 V
D_3, D_4: 1N574
L_1, L_2: 32 μH
L_3: 131 Turns of No.15 Magnet Wire on Arnold Engineering Core No.A4-04117, or equivalent

R_1, R_2: 1.2 kΩ, 5 watt
R_3: 200 Ω, 10 watt
T: Core, 8 pieces of Indiana General No. CF-602 Material 05, or equivalent. Cross Section, 8 cm^2
N_1, N_6 – 30 Turns of No.18 Magnet Wire
N_2, N_5 – 13 Turns of No.18 Magnet Wire, 2 Strands
N_3, N_4 – 52 Turns of No.18 Magnet Wire, 2 Strands

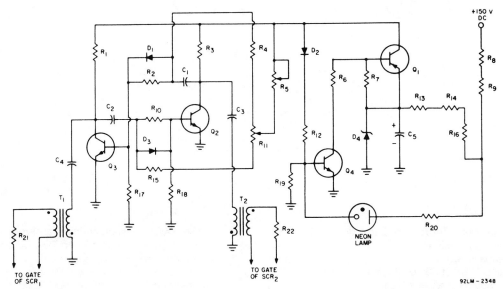

Q_1: RCA-40438
Q_2, Q_3, Q_4: RCA-2N3053
C_1, C_2: 0.003 μF, 100 V
C_3, C_4: 0.02 μF, 100 V
C_5: 25 μF, 25 V, electrolytic
D_1, D_2, D_3: Transitron type T1G, or equivalent
D_4: Motorola type 1M20Z10, or equivalent
Neon Lamp: GE type NE-83, or equivalent
R_1, R_3: 1 kΩ, 1/4 watt
R_2, R_{10}: 180 kΩ, 1/4 watt

R_4, R_{12}, R_{15}, R_{17}, R_{18}: 22 kΩ, 1/4 watt
R_5, R_{11}: 10 kΩ potentiometer
R_6: 10 kΩ, 1/4 watt
R_7: 1.5 kΩ, 1/4 watt
R_8, R_9, R_{13}, R_{14}, R_{16}: 680 Ω, 2 watts
R_{19}: 5.6 kΩ, 1/4 watt
R_{20}: 33 kΩ, 1/4 watt
R_{21}, R_{22}: 10 Ω, 1/4 watt
T_1, T_2: Sprague Pulse Transformer 42Z109, or equivalent

1.4.14 RCA fluorescent-light inverter circuit, permits operation of up to 500W of lamps at a switching frequency of 8 kHz with an input of 150V dc. The lower circuit shows a typical trigger-pulse generator for gating inverter.

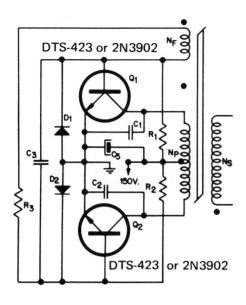

PARTS LIST

Q_1, Q_2	Delco DTS-423
D_1, D_2	Delco DRS-100
C_1, C_2, C_3	.1µF at 400 V
C_4	100µF at 150 VDC
C_5	200µF at 150 VDC
R_1, R_2	6000 Ω, 5 W
R_3	39 Ω, 1 W
D_3, D_4, D_5, D_6	= 1N3491
R_L	= 4Ω, 250 W

TRANSFORMER

Core: Magnetics, Inc. 51001-2A
Pri: 157 t ea. half of pri., #20 AWG
Bifilar wind the halves.
Feedback: 7t #23 AWG
Secondary: 1.05 turns per volt. Note: Secondary output power = 180 W max.

1.4.14 Delco superefficient 180W inverter delivers more than 94% of its input power to the load. The output voltage may be increased by winding more turns on the toroidal core, but the output power will remain at 180W maximum; for every volt desired in the secondary, wind 1.05 turns.

Diodes D1 and D2 provide a return path for the feedback winding and insure the proper reverse bias (600 or 700 mV) on the transistors at the right time; also, they protect the emitter diode of the switched-off transistor against excessive reverse voltage. Resistors R1 and R2 provide about 20 mA of quiescent collector current to insure easy startup at low temperatures. The base resistor, R3, serves to limit the base drive current to about 150 mA.

Despiking is accomplished by capacitors C1, C2, and C3. Capacitor C5 is used to bypass the power lead, a technique that helps considerably to eliminate waveform spikes. (Don't look for C4—it was a dc filter on the prototype, which was used for dc conversion rather than ac inversion.)

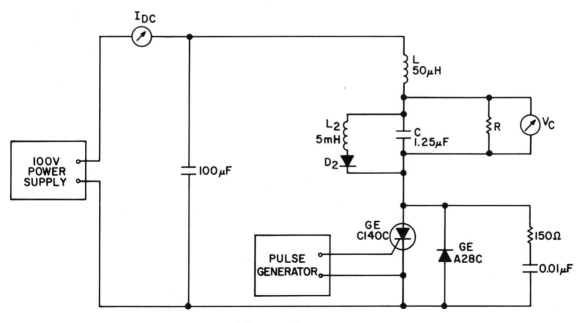

1.4.15 GE ultrasonic inverter generates kilowatts of power over the frequency range of 400 Hz to 30 kHz. The circuit features good waveform and good regulation over a wide range of load magnitude and phase angle.

SECTION 1 *Power Conversion and Regulation*

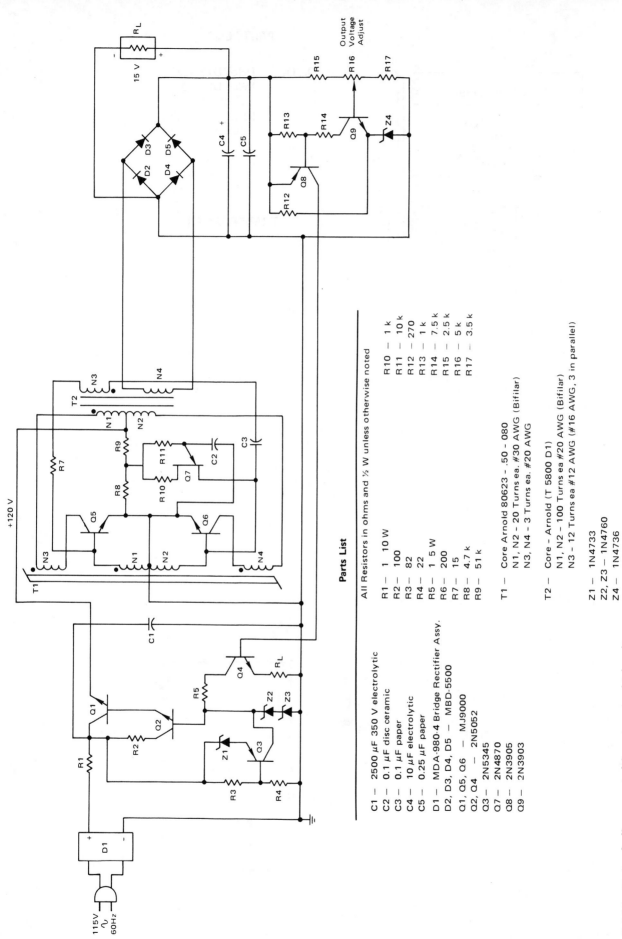

Parts List

All Resistors in ohms and ½ W unless otherwise noted

R1 — 1 10 W	R10 — 1 k
R2 — 100	R11 — 10 k
R3 — 82	R12 — 270
R4 — 22	R13 — 1 k
R5 — 1 5 W	R14 — 7.5 k
R6 — 200	R15 — 2.5 k
R7 — 15	R16 — 5 k
R8 — 4.7 k	R17 — 3.5 k
R9 — 51 k	

C1 — 2500 µF 350 V electrolytic
C2 — 0.1 µF disc ceramic
C3 — 0.1 µF paper
C4 — 10 µF electrolytic
C5 — 0.25 µF paper

D1 — MDA-980-4 Bridge Rectifier Assy.
D2, D3, D4, D5 — MBD-5500

Q1, Q5, Q6 — MJ9000
Q2, Q4 — 2N5052
Q3 — 2N5345
Q7 — 2N4870
Q8 — 2N3905
Q9 — 2N3903

T1 — Core Arnold 80623 - .50 - 080
N1, N2 - 20 Turns ea. #30 AWG (Bifilar)
N3, N4 - 3 Turns ea. #20 AWG

T2 — Core - Arnold (T 5800 D1)
N1, N2 - 100 Turns ea #20 AWG (Bifilar)
N3 - 12 Turns ea #12 AWG (#16 AWG, 3 in parallel)

Z1 — 1N4733
Z2, Z3 — 1N4760
Z4 — 1N4736

1.4.16 Motorola line-operated inverter uses high-voltage power transistors and hot-carrier rectifiers. It provides 15V, 15A to resistive loads. This circuit makes use of a frequency-changing technique in order to reduce the size of transformers and filtering components.

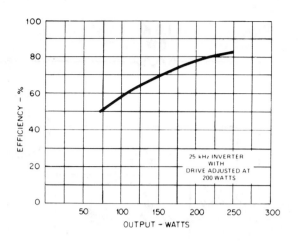

OUTPUT vs EFFICIENCY

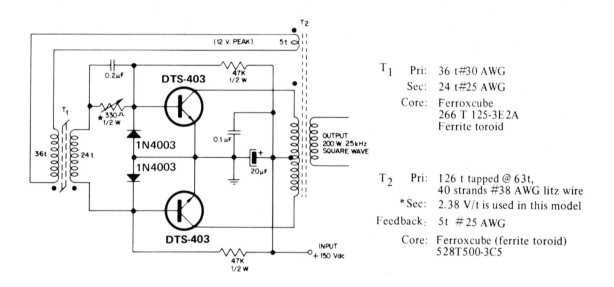

T_1 Pri: 36 t #30 AWG
   Sec: 24 t #25 AWG
   Core: Ferroxcube
         266 T 125-3E2A
         Ferrite toroid

T_2 Pri: 126 t tapped @ 63t,
            40 strands #38 AWG litz wire
  *Sec: 2.38 V/t is used in this model
  Feedback: 5t #25 AWG
  Core: Ferroxcube (ferrite toroid)
        528T500-3C5

★ Adjust value of resistor for maximum efficiency at full load.

* To be determined by individual requirements.

1.4.17 Delco miniature 25 kHz, 200W inverter. Two DTS-403 high-voltage silicon transistors are used in a push-pull oscillator that is biased with 150V. The circuit features a separate drive transformer and a linear output transformer. The transistor drive level should be adjusted for maximum efficiency at the desired load. An adjustable 330-ohm resistor is shown in the schematic for this purpose. With drive optimized at 200W, the efficiency of the inverter is 78%. The graph shows that the efficiency is less at lower output levels. A single winding serves as the drive to each transistor by the use of two 1N4003 diodes that serve alternately for steering and clamping.

SECTION 1 *Power Conversion and Regulation*

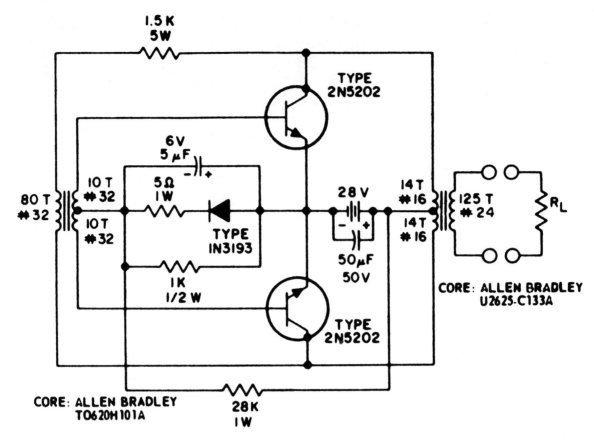

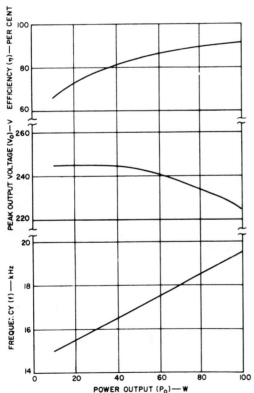

1.4.18 Practical 100W, 18 kHz inverter using RCA 2N5202 transistors. Performance characteristics for this inverter are shown.

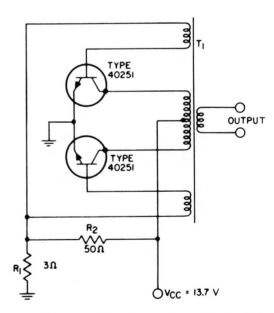

1.4.19 This GE 3.5 kHz inverter has a circuit efficiency of 83% when the input power is approximately 110W. The toroidal core of T1 is an Allen Bradley T3000H 106B or equivalent. The outside diameter of the core is 3 in. (7.62 cm) and the cross section of ½ x 1 in. (1.27 x 2.54 cm) (RO-3 material). For the primary winding use 16 turns of 20-gage wire (centertapped). Use 8 turns of 22-gage wire for each feedback winding. Both resistors used in this circuit should have a rating not less than 5W.

52 DISCRETE / TRANSISTOR CIRCUIT SOURCEMASTER

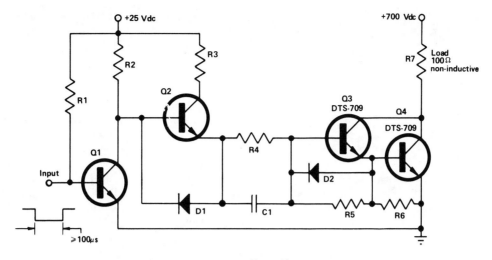

Parts List

Switching time
(as measured
from 10 to 90%
of final I_C value):

V_{CE} (sat) = 3V
t delay + rise = 1 μs
t storage = 3.5 μs
t fall = 1 μs

1.4.20 Delco high-speed, high-voltage switcher can switch loads of up to 700V at 7A at frequencies as high as 10 kHz. This circuit is not an inverter, but incorporation of a simple astable multivibrator as the input element converts it into one.

Capacitor C1 in this circuit is used to enhance the switching operation. Resistors R3 and R4 determine the voltage to which capacitor C1 charges (it also determines the base drive); these resistors should be selected for optimum switching at the current level being switched. At turnon, C1 appears as a short circuit across R4, which effects a rapid rise time of I_B that is limited by R3. During on times, capacitor C1 charges to 12V; and at turnoff this charge is applied through diode D1 and transistors Q1, Q3, and Q4 in a direction that reverse-biases the base–emitter junctions of the Darlington pair.

Resistors R5 and R6 determine the V_{CER} of the Darlington pair—lower values permit higher voltages but sacrifice gain.

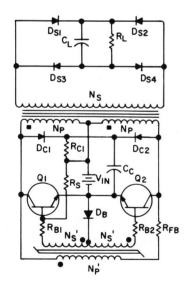

Q_1, Q_2 : D44H5
D_B : A14F
R_{B1}, R_{B2} : 4.7 Ω
R : 22 Ω
R_S : 1.5KΩ
D_{S1-4} : A114B
C_L : 2μf/200V
   GE BA15A 205B
D_{C1}, D_{C2} : DT230A
C_C : .01μf
R_C : 2.2K

V_{IN} = 12 VOLTS
V_{OUT} = 160 VOLTS
P = 60 WATTS NOMINAL
f ≈ 20 kHz
N_P: 7 TURNS #16 BIFILAR
N_S: 102 TURNS #24
NON-SATURATING CORE:
   INDIANA GENERAL IR8004-1
N_P': 26 TURNS #34
N_S': 5 TURNS 28 BIFILAR
SATURATING CORE:
   FAIRRITE 568314082

1.4.21 GE voltage-feedback inverter for driving a portable tachometer strobe lamp from a 12V battery. A minimum of 160V across 2 μF is needed to deliver the required light output at frequencies from 0 to 1 kHz. This circuit delivers 60+ watts with high efficiency where open-circuit capability is a must. The efficiency peaks at a maximum load, but is much less at lighter loads because the base current (and therefore the base-drive loss) is relatively constant relative to the load current.

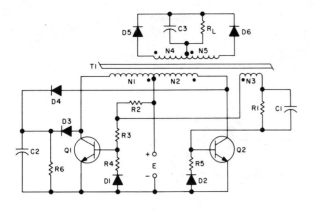

E:	180 V	28 V
P_{out}:	60 W	17 W
V_{out} Designed:	180 V	70 V
Actual:	172 V	69 V
f Designed:	28 KHz	10 KHz
Actual:	28 KHz	10 KHz
Efficiency:	88%	81%
Transformer:	Stackpole #55-420	Ceramag 24A
N_1-N_2:	50 Turns	20 Turns
N_3:	5 Turns	12 Turns
N_4-N_5:	50 Turns	50 Turns
Q_1-Q_2:	D44R1	D43C5
D_1-D_2:	A14F	1N914
D_3-D_4:	A114B	Not Used
D_5-D_6:	A114B	A114
R_1:	270 Ω	560 Ω
R_2:	650 KΩ	100 KΩ
R_3:	48 Ω	Not Used
R_4-R_5:	10 Ω	Not Used
R_6:	100 KΩ	Not Used
C_1:	.0068 μf	Not Used
C_2:	.01 μf/200 V	Not Used
C_3:	0.5 μf/200 V	0.5 μf/200 V
R_L:	500 Ω	280 Ω

1.4.22 GE basic inverter could be used with low- or medium-voltage inputs and in either low- or high-power applications up to the limit of the transistors. This circuit was operated with the values shown. The parts lists show how frequency and power can vary according to the components selected. In the left column are particulars and component listings for the 60W, high-frequency version. The right column shows values for a 17W unit that switches at 10 kHz.

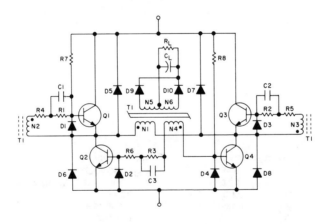

E:	180 V	N_1:	50 Turns	R_4-R_5:	33 Ω
V_o Designed:	180 V	N_2-N_3:	3 Turns	R_6:	82 Ω
Actual:	167 V	N_4:	5 Turns	R_7:	100 KΩ
f Designed:	28 KHz	N_5-N_6:	50 Turns	R_8:	220 KΩ
Actual:	23 KHz	Q_1-Q_4:	D44R1	R_L:	360 Ω
P_o:	77.4 W	D_1-D_4:	A14F	C_1-C_2:	.01 μf
Efficiency:	84%	D_5-D_{10}:	A114B	C_3:	.0068 μf
Transformer:	Stackpole #55-420 Ceramag 24A	R_1-R_2:	470 Ω	C_L:	0.5 μf/200 V
		R_3:	1 KΩ		

1.4.23 GE bridge inverter circuit can be used with a high-input voltage, since the transistors require a breakdown voltage rating only slightly larger than the input voltage. However, this circuit does not require a large capacitor as other circuits do at high power levels, although the additional transistor saturation losses will make the bridge circuit slightly less efficient.

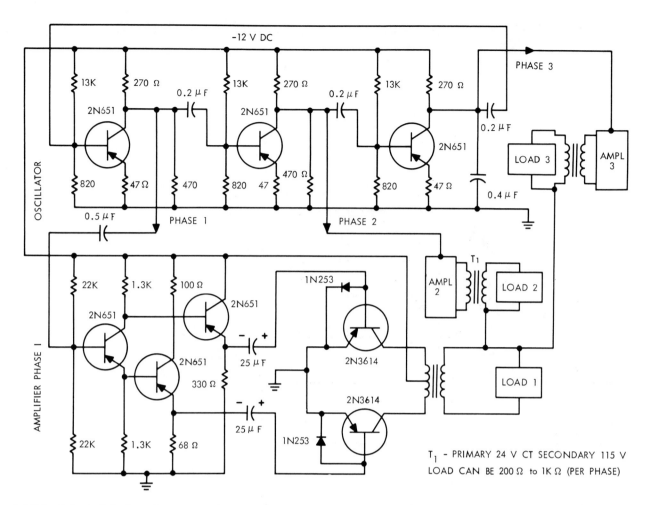

1.4.24 A typical 3-phase inverter is this combination oscillator and amplifier. The 3-phase oscillations are supplied by an oscillator RC coupled so that a 120° phase difference exists at the collectors of the Motorola 2N651 transistors. An emitter-follower amplifier drives the output power transistors. The power transistors are operated in their saturated switching mode.

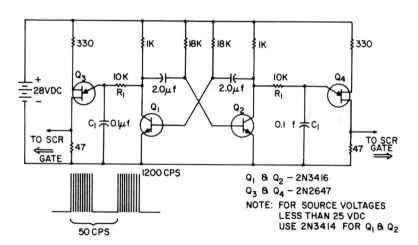

1.4.25 GE pulse shaper for SCR inverter circuits maintains trigger drive during the conduction period in the form of a pulse train, thus reducing the average gate dissipation. The transistor multivibrator provides alternate driving voltages to the two GE unijunction transistor oscillators. The outputs of these oscillators provide the alternating pulse train sequence as required for inverter circuits.

SECTION 1 *Power Conversion and Regulation* 55

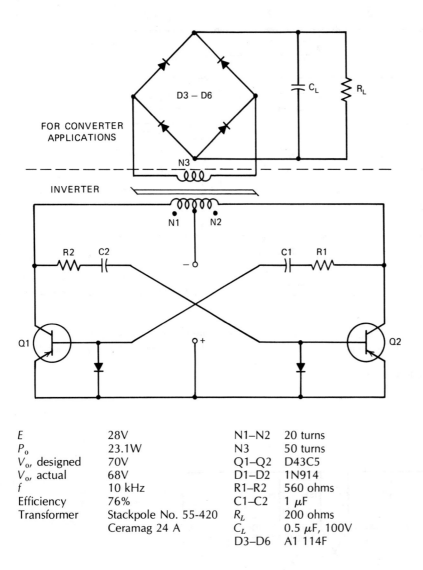

E	28V	N1–N2	20 turns
P_o	23.1W	N3	50 turns
V_o, designed	70V	Q1–Q2	D43C5
V_o, actual	68V	D1–D2	1N914
f	10 kHz	R1–R2	560 ohms
Efficiency	76%	C1–C2	1 μF
Transformer	Stackpole No. 55-420	R_L	200 ohms
	Ceramag 24 A	C_L	0.5 μF, 100V
		D3–D6	A1 114F

1.4.26 In this GE 10 kHz "flip-flop" inverter, the base windings have been eliminated and replaced by a series RC branch. However, for high supply voltages, there will be more power dissipated in the base resistors. Therefore, this circuit would be desirable only in applications utilizing a low input voltage.

Capacitors C1 and C2 are to prevent both transistors from getting turned on in a dc mode. During normal operation, C1 and C2 are normally charged to the supply voltage and base resistors R1 and R2 should be small enough to provide sufficient base current to keep the GE transistors saturated.

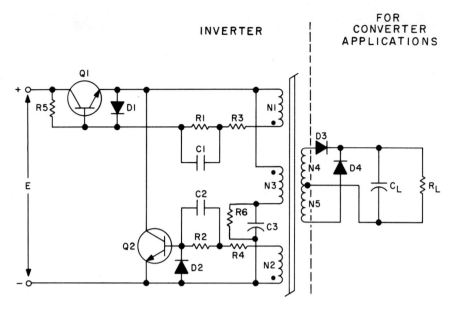

Transformer	Stackpole	
Core Material:	Ceramag 24A	
Size:	#55-420	
E:	180 V	180 V
P_o:	4.6 W	30 W
V_o Designed:	180 V	90 V
Actual:	150 V	77.5 V
f Designed:	14 KHz	14 KHz
Actual:	14 KHz	12.4 KHz
η:	60%	82%
Q_1-Q_2:	D40N1	D44R1
D_1-D_2:	A14F	A14F
D_3-D_4:	A114B	A114B

R_1-R_2:	2200 Ω	1000 Ω
R_3-R_4:	10 Ω	10 Ω
R_5:	50 KΩ	250 KΩ
C_1-C_2:	.01 μf	.01 μf
C_3:	.5 μf/100 V	20 μf
C_4:	.5 μf/100 V	.5 μf
N_1-N_2:	5 t	5 t
N_3:	50 t	50 t
N_4-N_5:	50 t	25 t
R_L:	1360 Ω	200 Ω
R_6:	39 KΩ	39 KΩ

1.4.27 GE high-input-voltage inverter (component values for two examples are given). This circuit requires a transistor breakdown voltage rating of only 1.2 times the input voltage. Because diodes D1 and D2 reverse-bias the transistors during turnoff, high-voltage transistors (D40N, D44R) can be used with voltage supplies up to 500V.

Under normal operation C3 is charged to half the input voltage. Then half the supply voltage is applied to winding N3 with alternating polarities as Q1 and Q2 alternately turn on. If during one cycle C3 were not charged to ½E, a larger voltage would be applied to N3 when Q1 is on than when Q2 is on. This would result in more current flowing into the output dc filter during that part of the cycle when Q1 is on. This imbalance in load current tends to charge C3 until equal currents flow into the filter during both half-cycles; i.e., the voltage on C3 is half the supply voltage. Since the load tends to keep C3 charged to ½E, under open output conditions C3 will not be charged to ½E and asymmetrical operation will result.

Since high-voltage transistors are to be used with their associated longer storage times, use slow diodes in the base circuits to delay transistor conduction slightly.

SECTION 2
CONTROL AND SENSING

Circuits that control—that do things like turn other circuits on and off at prespecified points or regulate some variable function or sense conditions and events—are becoming more plentiful as new devices emerge for such applications. There are many more solid-state devices that switch than devices that amplify; while even the amplifiers can switch, most switches cannot amplify. The circuits presented in the eight following subsections cover the entire gamut of devices new and old that have been designed for control applications.

Applications for the circuits in this section are not always obvious. This is the kind of material you approach with a "beforehand" problem. Many of the circuits in the first subsection—Solid-State Relays and Switches—will join nicely with those in other subsections. An amateur radio operator who is interested in building a remote control system for his station or in setting up an automatic repeating facility should find it easy to adapt circuits in several of these sections to his need. The engineer who wants to automate a production line will have no trouble locating the kinds of circuits he needs to get started on a prototype system.

Not all of these circuits are intended for automatic control functions. The advent of the four-layer diode has opened up an area of control that until recently was accomplished by huge and expensive variable transformers: light dimming and motor speed adjustment. With the variable transformer, the job was done by lowering the voltage applied to the load without modifying the waveform of the ac line. With triacs and silicon controlled switching devices, however, the average voltage to the load is controlled by switching the power on and off at a precise point in each ac cycle. You must bear this in mind as you adapt these phase-controlling circuits (Section 2.3), since the processed waveform is not sinusoidal and may contain voltage spikes that could be destructive in inductive circuits.

2.1 Solid-State Relays and Switches

Some of the solid-state switching devices developed over the past few years were conceived as ultimate replacements for the inherently unreliable electromechanical moving-contact switch we call the relay. But it hasn't happened yet, as many of the following circuits can attest. Circuits do still incorporate relays, but the prospect of replacing the coil and contact with the semiconductor is at least within reach of every designer. There aren't many applications in which such conversion couldn't be made if the rewards justify the effort.

In a sense, at least, a triac *is* a solid-state relay. But since the triac is a three-terminal device and the simplest relay is a four-terminal device, the circuit designer is faced with a compromise: one of the triac "contact" terminals must share a common connection with one of the "coil" terminals. This requirement can present a difficult situation to circuit designers who are interested in total isolation of the coil from the contacts.

When isolation is important between the switching signal and the power to be switched, you might consider one of the recent photodevices such as GE's optically coupled SCR, a fully isolated device that uses a light path rather than a hard-wired connection to transfer the control signal to the controlled power line.

One of the more interesting developments in electronic switching of recent years is the zero-voltage switch—a circuit built around a thyristor, which switches an ac power line during the instant the waveform on that line changes from positive to negative or from negative to positive. The zero-voltage switch cannot arc or create inductive voltage transients because the power is, for all practical purposes, off during the brief instant that semiconductor conduction is initiated. A glance at the final circuit in this subsection will show you how simply this zero-voltage switching technique can be accomplished.

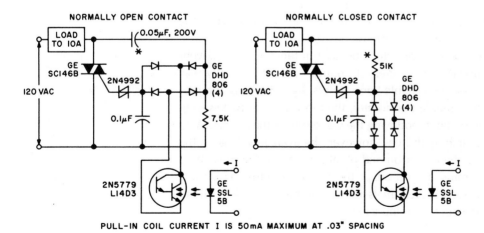

2.1.1 Solid-state "relays" use triacs to switch loads of 10A. These circuits use the GE SC146B triac as the load current "contact." These triacs are triggered by normal SBS (2N4992) trigger circuits controlled by the photo-Darlington acting through the DA806 bridge. To operate the relays at other line voltages, the asterisked components are scaled to supply identical current, and the ratings are changed as required. Incandescent lamps may be used in place of the light-emitting diodes if desired.

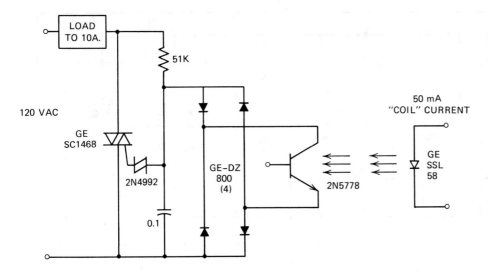

2.1.2 GE 10A solid-state normally closed relay uses a silicon bilateral switch (2N4992) to trigger the SC146B triac power switch operating as a normally closed contact. When the light-emitting diode is turned on by "coil" current, the photo-Darlington shunts the bias current from the 2N4992 and prevents triac triggering. The DZ800 diodes provide proper polarity biasing to the 2N5778. Circuit operation has been verified from −30 to +65°C.

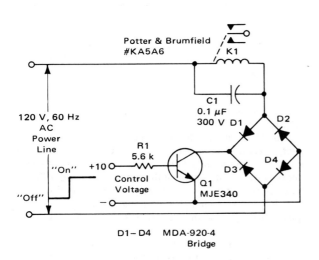

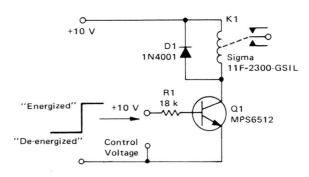

2.1.3 Motorola ac-operated relay control system. The bridge consisting of D1, D2, D3, and D4 provides dc to the transistor while the relay sees an ac voltage. When a dc control voltage is applied to R1, Q1 saturates and energizes the relay coil. An adequate base current must be provided to saturate Q1. A disadvantage of this circuit is that the control signal must be isolated from the power line. The prime advantage is, of course, that an ac relay can be controlled by a single transistor. A forced gain of 10 guarantees that Q1 will be saturated so that the base current of 1.6 mA will drive a relay coil requiring 16 mA. In this circuit, protection against voltage spikes must also be provided for the transistor when it is turned off. Capacitor C1 across the relay coil provides such protection.

2.1.4 Motorola dc relay can be controlled by an electronic signal that can supply a current of 0.5 mA. The relay coil is the load for transistor Q1. When a positive control voltage is applied to R1, Q1 receives base drive and saturates, thus connecting the relay coil to the supply voltage. R1 must allow enough base current to saturate Q1. For the components shown, the 0.5 mA of base current will assure this, since the relay requires 5 mA and the transistor has a minimum gain of 50. When the control voltage drops to zero the transistor turns off, deenergizing the relay. Since the relay coil is inductive, a voltage spike could occur at the collector of Q1; a protective circuit for the transistor should be included. The diode (D1) across the relay coil clamps the collector voltage to the supply voltage by providing a path for the current in the relay coil when the transistor turns off.

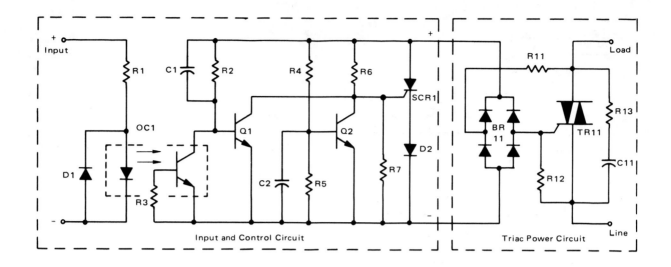

Voltage		120 Vrms				240 Vrms			
RMS Current Amperes		8.0	12	25	40	8.0	12	25	40
BR11		MDA102	MDA102	MDA102	MDA102	MDA104	MDA104	MDA104	MDA104
C11, µF (10%, line voltage ac rated)		0.047	0.047	0.1	0.1	0.047	0.047	0.1	0.1
R11 (10%, 1 W)		39	39	39	39	39	39	39	39
R12 (10%, 1/2 W)		18	18	18	18	18	18	18	18
R13 (10%, 1/2 W)		620	620	330	330	620	620	330	330
TR11	Plastic	2N6342	2N6342A	—	—	2N6343	2N6343A	—	—
	Metal	—	MAC40799	2N6163	MAC4688	—	MAC40800	2N6164	MAC4689

2.1.5 Motorola triac solid-state relay circuit. The input circuit is TTL-compatible. A sensitive-gate SCR (SCR1) is used to gate the power triac, and a transistor amplifier is used as an interface between the optocoupler and SCR1. (A sensitive-gate SCR and a diode bridge are used in preference to a sensitive-gate triac because of the higher sensitivity of the SCR.)

The AND function is performed by the wired-NOR collector configuration of Q1 and Q2. Q1 clamps the gate of SCR1 if optocoupler OC1 is off. Q2 clamps the gate if there is sufficient voltage at the junction of divider R4, R5 to overcome the V_{BE} of Q2. By judicious selection of R4 and R5, Q2 will clamp SCR1's gate if more than 5V appears at the anode of SCR1 (Q2 is the zero-crossing detector).

If OC1 is on, Q1 is clamped off and SCR1 can be turned on by current flowing down R6 only if Q2 is also off, which it is only at zero crossing.

The capacitors eliminate circuit race conditions and spurious firing in operation. A race condition exists on the up-slope of the second half-cycle in that SCR1 may be triggered via R6 before Q1 has enough base current via R2 to clamp SCR1's gate. C1 provides current by virtue of the rate of change of the supply voltage, and Q1 is turned on firmly as the supply voltage starts to rise, eliminating any possibility of unwanted firing.

This leaves the possibility of unwanted firing of the SSR on the down-slope of the first half-cycle shown. C2 provides a phase shift to the zero-voltage potential divider, and Q2 is held on through the real zero crossing. The resultant window is shown.

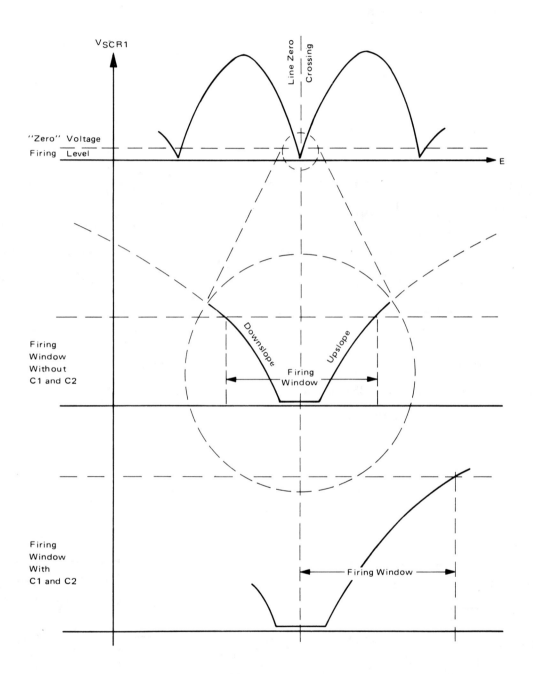

SECTION 2 *Control and Sensing*

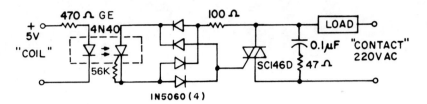

2.1.6 GE TTL-compatible solid-state relay can switch up to 10A at 220V ac.

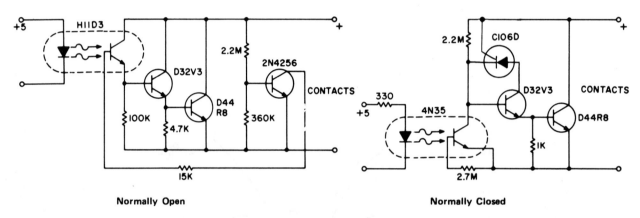

2.1.7 High-voltage (300V), low-current (250 mA) dc solid-state relays for both normally closed and normally open applications. Semiconductors are GE's.

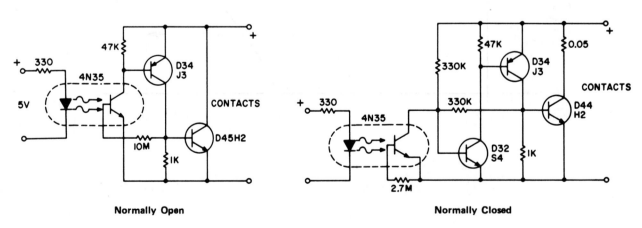

2.1.8 High-current (10A) 25V dc solid-state relays for both normally open and normally closed operations. Semiconductors are GE's.

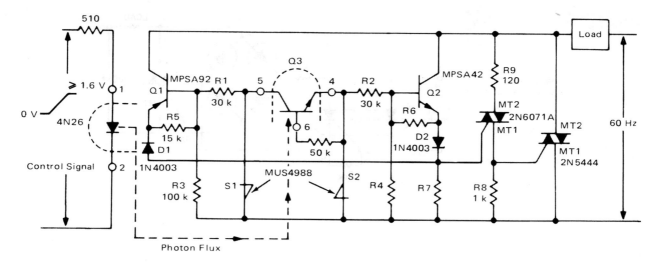

2.1.9 Motorola solid-state relay for ac control. The output is isolated from the input by means of photon coupling. To minimize electromagnetic interference (EMI), relay actuation is permitted only during the zero-voltage crossings of the ac line.

As the triac voltage passes through zero and becomes positive, the collector–base junction of Q1 will be forward-biased via the current path through R3. The divider, formed by R1 and S1, senses this voltage rise across R3. When the voltage exceeds a switching value, V_s, the MUS4988 will avalanche into conduction. Once triggered, this switch will stay on for the remainder of the alternation. This particular device has a nominal V_s of 8V and a conduction voltage, V_F, of about 1V. Notice that the series string consisting of Q3, R2, the base–emitter junction of Q2, D2, and the triac gate is in parallel with S1. The 2N6071A triac is used to increase the gate drive to the 2N5444. Once the voltage across S1 drops to V_F, the triacs will not turn on. In this manner, a relay-enabling voltage window is created. The lower limit of the window is that point where all the components involved in turning on the triacs can be forward-biased. The upper limit is the breakover voltage of the unilateral switch. In terms of voltage, the lower and upper limits are nominally 5 and 10V. The width of the window is about $80\mu s$ when the relay is operated from a 60 Hz power source.

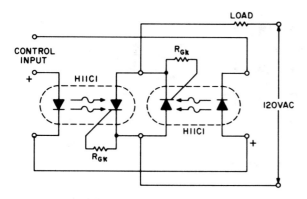

2.1.10 GE low-current ac relay. If load current requirements are less than one-half ampere, an ac solid-state relay can be constructed quite simply by connecting two optically coupled SCRs back-to-back.

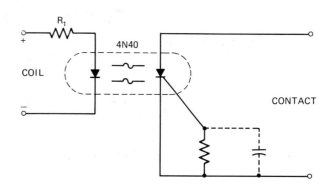

2.1.11 GE dc latching relay for currents up to 300 mA (depending on the ambient temperature). The gate cathode resistor may be supplemented by a capacitor to minimize transient and dv/dt sensitivity. For pulsating dc operation, the capacitor value must be designated to either retrigger the SCR at the application of the next pulse or prevent retriggering at the next power pulse. Otherwise, random operation may occur.

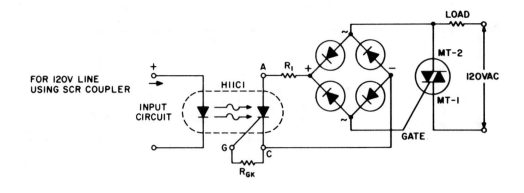

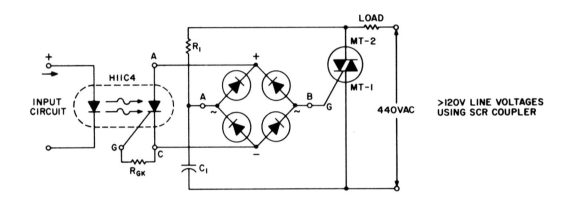

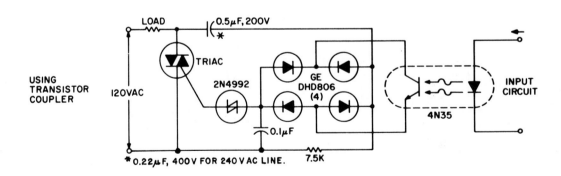

NORMALLY OPEN CONTACT RELAY CIRCUITS

2.1.12 Solid-state ac "relay" variations. When zero-voltage switching is not required, the "contact" circuitry can be simplified. Several methods of providing this function are illustrated. Note than an SCR coupler in a bridge, using a high gate resistor directly across the line voltage, can give commutating dv/dt and dv/dt triggering problems that are not evident in the ZVS circuits or at low voltages. Not all of these GE circuits are TTL-compatible at the input.

USING SCR COUPLER

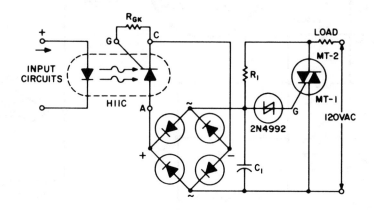

USING DARLINGTON COUPLER

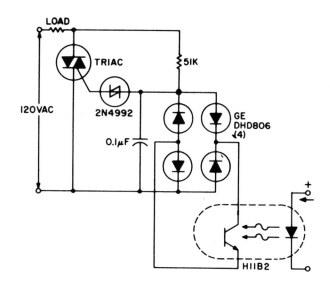

NORMALLY CLOSED CONTACT RELAY CIRCUITS

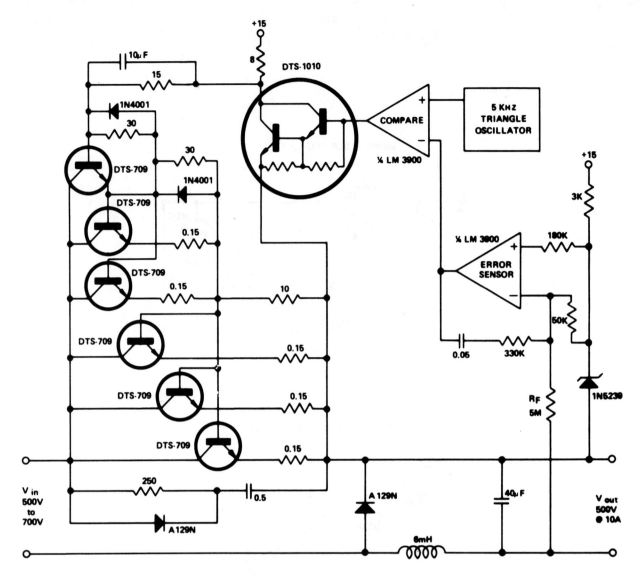

2.1.13 Delco DTS-709 transistors connected in a progressive Darlington configuration switch a level of power (5 kW) useful for many industrial and commercial applications. This 5 kW building block is stable, efficient, and dependable—features that lend themselves easily to higher power, high-voltage applications. Operation from a 480V, 3-phase, full-wave-rectified line reduces the filter cost.

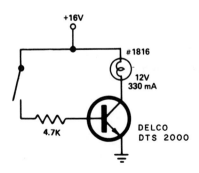

2.1.14 Lamp driver. The Delco switch shown can be relay contacts or the collector–emitter circuit of a small-signal transistor.

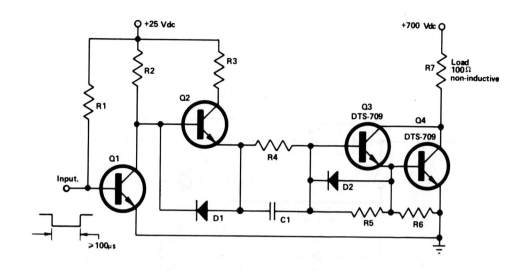

$V_{CE(sat)}$ = 3.0V
$t_{delay + rise}$ = 1.0μs
$t_{storage}$ = 3.5μs
t_{fall} = 1.0μs

R_1	510Ω, 1W	Q_1, Q_2	2N6100	D_1	1N4001
R_2	100Ω, 10W	Q_3, Q_4	DTS-709	D_2	1N4001
R_3	12Ω, 20W				
R_4	10Ω, 20W				
R_5	1kΩ, 0.5W	**CAPACITOR**			
R_6	47Ω, 0.5W				
R_7	100Ω	C_1	4μF, 15V		

2.1.15 High-speed Delco switching circuit uses Darlington configuration to switch 700V at up to 7A.

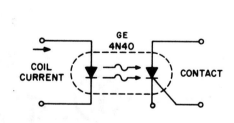

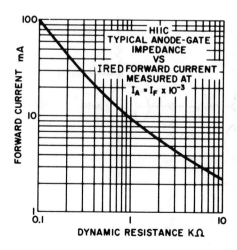

2.1.16 GE solid-state reed relay will switch low-level signals of unknown amplitude and polarity on command of an isolated logic signal. Resistive switching is desired to minimize changes in the signal caused by the switch. Although the current transfer ratio of about 1% limits the usefulness to fairly low-level signals, the 400V contact blocking capability, no-bounce and no-weld characteristics make it very attractive for acquiring audio signals, thermocouple and thermistor outputs, monitoring junction drops, strain gage outputs, and many other testing and control functions at electronic speeds.

SECTION **2** Control and Sensing

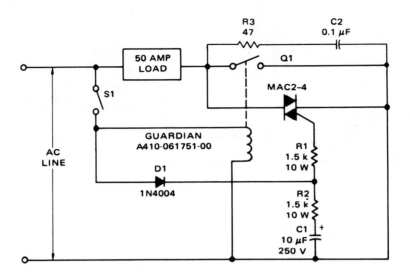

2.1.17 Motorola contact-arc preventer for loads up to 50A with a relay rated for as little as 5A.

There is some delay between the time in which a relay coil is energized and the time when the contacts close. There is also a delay between the time the coil is de-energized and the time the contacts open. For the relay used with this circuit, both of these times were about 15 ms. The triac across the relay contacts turns on as soon as sufficient gate current is present to fire it. This occurs after switch S1 is closed but before the relay contacts close. When the contacts close, the load current passes through them rather than through the triac, even though the triac has received gate current. If S1 should be closed during the negative half-cycle, the triac will not turn on immediately but will wait until the voltage begins to go positive, at which time diode D1 conducts, providing gate current through R1. The maximum time that could elapse before the triac turns on is $8\frac{1}{3}$ ms for a 60 Hz supply. This is adequate to assure that the triac will be on before the relay contact closes.

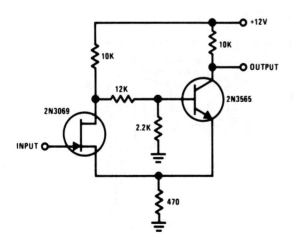

2.1.18 This Schmitt trigger circuit is "emitter-coupled" and provides a simple comparator action. The 2N3069 JFET places very little loading on the measured input. The 2N3565 bipolar is a high h_{FE} transistor so that the circuit has fast transition and a distinct hysteresis loop.

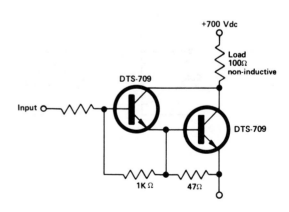

2.1.19 Low-speed power switch. There are numerous applications for high-power switching below 5 kHz. For lower frequencies the circuit shown here (from Delco) represents what might well be the most economical method to switch up to 700V at currents as high as 7A.

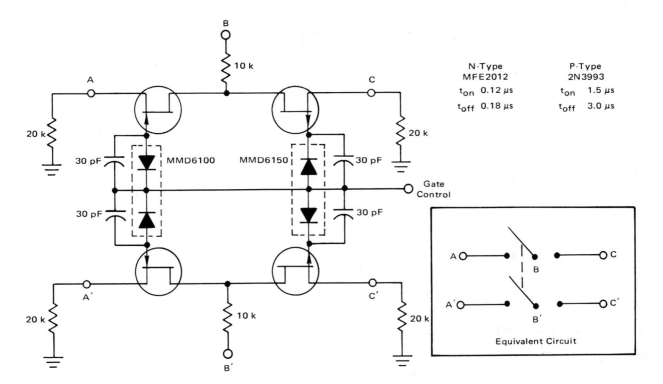

2.1.20 Motorola double-pole-double-throw switching circuit. There is nearly an order of magnitude of difference between the response times of the N- and P-channel devices due to the inherent mobility of the majority carriers in the N channel as compared to the P channel. (Carriers in the N channel, being more mobile, yield faster response times.)

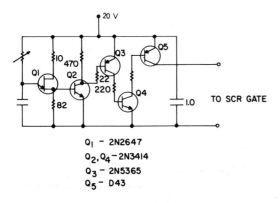

2.1.21 High-di/dt SCR trigger. For fast-rising current loads, an SCR may require a fast-rising high-level rectangular pulse to assure triggering. Rectangular pulses can be shaped by the use of reactive pulse-forming networks or by blocking oscillators. But these circuits are relatively costly and large. This circuit will generate rectangular pulses of 10 µs at repetition rates up to 20 kHz and does not require any inductive elements. With a 20V amplitude and 20-ohm source impedance, this circuit should adequately trigger most SCRs even under the most stringent di/dt conditions. The UJT output pulses drive a four-transistor amplifier circuit, which improves the rise time and extends the pulse width to 10 µs. (GE semiconductors are used throughout.)

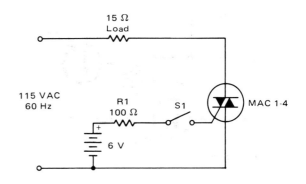

2.1.22 Motorola low-voltage-controlled triac switch. When switch S1 is closed, gate current is supplied to the triac from the 6V battery. The triac turns on and remains on until S1 is opened. This circuit switches at zero current.

SECTION **2** Control and Sensing **71**

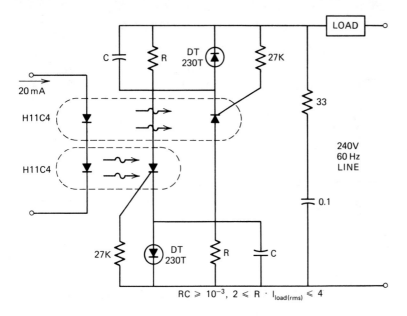

2.1.23 GE ac latching solid-state relay. When analog signals are being used as the logic control, hysteresis can be used to prevent "chatter" or half-wave power output. Latching is obtained by the storage of gate trigger energy from the preceding half-cycle in the capacitors. Power must be interrupted for more than one full cycle of the line to insure turnoff.

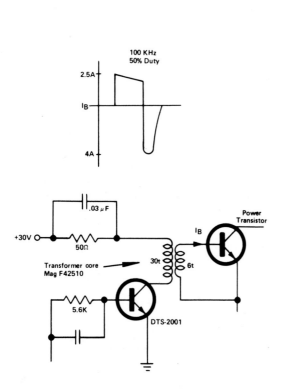

2.1.24 Delco power transistor driver requires very little operating current and may be driven from low-power logic.

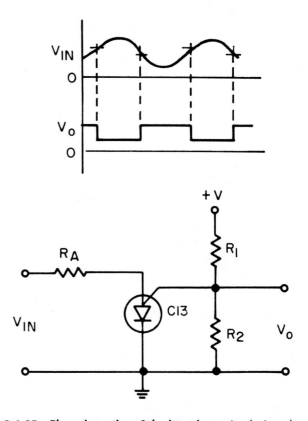

2.1.25 Phase-inverting Schmitt trigger is designed around the inherent hysteresis of a GE complementary SCR. Voltage-divider network R1–R2 determines the conduction voltage of the device while the turnoff level is a function of holding current. The turnoff level is set by varying the value of R_A.

72 DISCRETE / TRANSISTOR CIRCUIT SOURCEMASTER

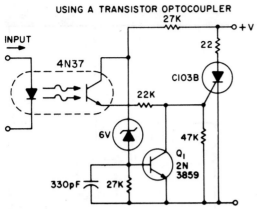
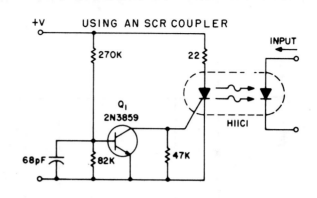

2.1.26 Simple circuits for zero-voltage switching can be used with full-wave bridges or in antiparallel to provide full-wave control and are normally used to trigger power thyristors. If an input signal is present during the time the ac voltage is between 0 and 7V, the SCR will turn on; but if the ac voltage has risen above 7.5V and the input signal is then applied, transistor Q1 will be biased on and will hold the SCR off. The circuit shown is from GE.

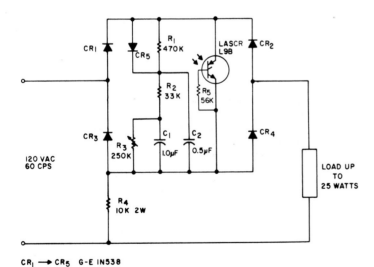

2.1.27 Impulse-actuated, variable-on-time switch. A random impulse of light fires the LASCR, applying current to the load. Capacitors C1 and C2 discharge through R2, R3, and through R1 and the LASCR. As long as this capacitor discharge current is higher than holding current, I_H, the LASCR cannot commutate, thus applying full-wave ac to the load. When the discharge current drops below I_H, the LASCR will turn off at the next succeeding current zero, assisted by R4 for inductive loads. Decreasing R3 reduces the time the switch remains on. The switch shown is from GE.

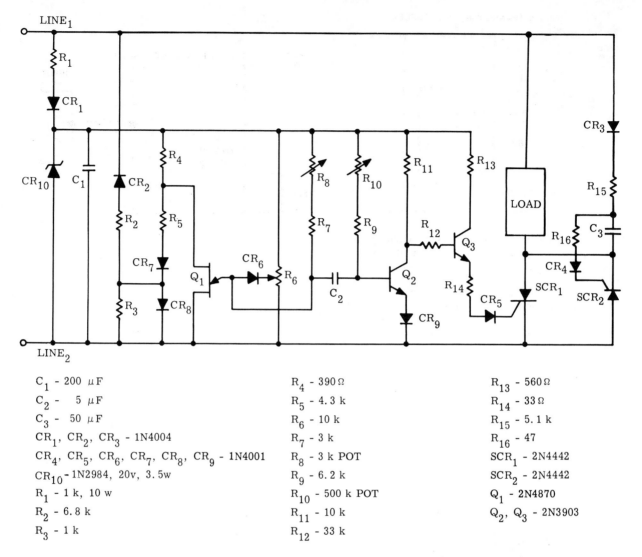

C_1 - 200 μF
C_2 - 5 μF
C_3 - 50 μF
CR_1, CR_2, CR_3 - 1N4004
CR_4, CR_5, CR_6, CR_7, CR_8, CR_9 - 1N4001
CR_{10} - 1N2984, 20v, 3.5w
R_1 - 1 k, 10 w
R_2 - 6.8 k
R_3 - 1 k

R_4 - 390 Ω
R_5 - 4.3 k
R_6 - 10 k
R_7 - 3 k
R_8 - 3 k POT
R_9 - 6.2 k
R_{10} - 500 k POT
R_{11} - 10 k
R_{12} - 33 k

R_{13} - 560 Ω
R_{14} - 33 Ω
R_{15} - 5.1 k
R_{16} - 47
SCR_1 - 2N4442
SCR_2 - 2N4442
Q_1 - 2N4870
Q_2, Q_3 - 2N3903

2.1.28 Motorola modulated SCR zero-point switcher. Every time an SCR is switched on at some phase angle above 0° or less than 180°, the switching transient will produce RFI. These transients can be detrimental either directly or indirectly to circuits operating in the same area. In order to hold these transients to a minimum, two conditions should be met; (1) the applied voltage should always be turned on as it passes through zero, and (2) the circuit should be switched off as the current passes through zero. In an SCR circuit controlling ac power, the latter of these two conditions is met automatically with the natural commutation of the SCR. Therefore, if the control circuitry is designed in a manner so as to always gate the SCR at the zero crossover point, the energy delivered to the load will be free of RFI due to SCR switching.

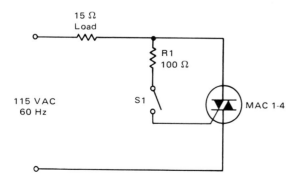

2.1.29 Motorola triac ac static contactor.

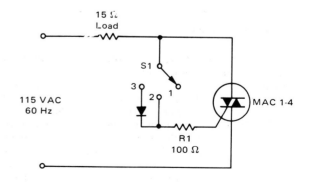

2.1.30 Motorola 3-position static switch.

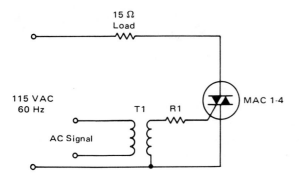

2.1.31 Motorola ac-controlled triac switch.

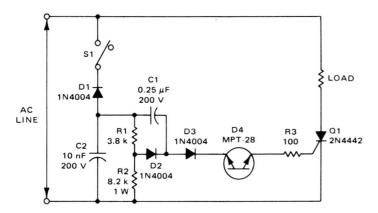

2.1.32 Motorola zero-point switch fires at zero voltage, thus eliminating interference from voltage spikes. Operation begins when switch S1 is closed. If the positive alternation is present, nothing will happen, since diode D1 is reverse-biased. When the negative alternation begins, capacitor C1 will charge through resistor R2 toward the limit of voltage set by the voltage divider consisting of resistors R1 and R2. As the negative alternation reaches its peak, C1 will have charged to about 40V. Line voltage will decrease, but C1 cannot discharge because diode D2 will be reverse-biased. It can be seen that C1 and three-layer diode D4 are effectively in series with the line. When the line drops to 10V, C1 will still be 40V positive with respect to the gate of Q1. At this time D4 will see about 30V and will trigger. This allows C1 to discharge through D3, D4, R3, the gate of Q1, and capacitor C2. C2 will quickly charge from this high pulse of current. This reduces the voltage across D4, causing it to turn off and again revert to its blocking state. Now C2 will discharge through R1 and R2 until the voltage on D4 again becomes sufficient to cause it to break back. This repetitive exchange of charge from C1 to C2 causes a series of gate-current pulses to flow as the line voltage crosses zero. This means that Q1 will again be turned on at the start of each positive alternation. Resistor R3 limits the peak gate current.

2.2 Photoelectric Switches

These circuits employ a variety of photodevices—resistive photocells, phototransistors, and integrated multidevice packages—to achieve their singular purpose: causing a circuit to switch either on or off in the presence of light (or the absence of it).

The photo switching circuits shown in this subsection are merely representative of the available methods of light control. Many of the functional circuits contained in other sections of this book incorporate similar circuitry. If your intended circuit application is no more complicated than using light to turn a power line on or off, then the approaches illustrated in this subsection should give you an adequate overview of how it is done; but if your application requires light-switching as but a single element of a more complicated circuit, examine the circuits offered in other, equally applicable sections of this book.

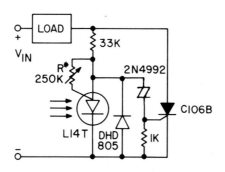

V_{IN}	V_{LOAD}
120V AC	HALF WAVE RECTIFIED
120V FWR	120V FWR

2.2.1 GE light-activated normally closed line switch. Adjust the asterisked resistor for the desired sensitivity. When light input is insufficient to trigger the L14T at V_{AK} = 8V, the GE 2N4992 (SBS) will trigger in the first 5° of the waveform, turning on the SCR and supplying load power. If the light input triggers the L14T, the anode of SBS is clamped to the ground and no load power is supplied. (This circuit will not phase-control.)

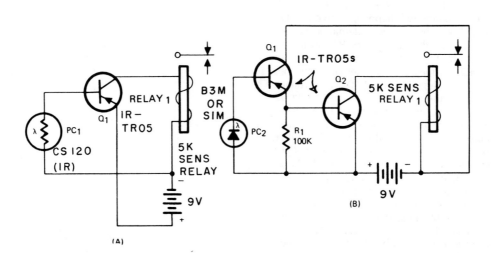

2.2.2 Workman electronic amplified solar relays. Circuit A uses a cadmium sulfide cell; circuit B uses selenium or silicon photocells. In both circuits, the relay can be connected to ring a bell, flash a light, or energize other types of alarm or demonstrator circuits. It can also be connected to turn on porch, store, or street lights whenever the sun drops below a specific point.

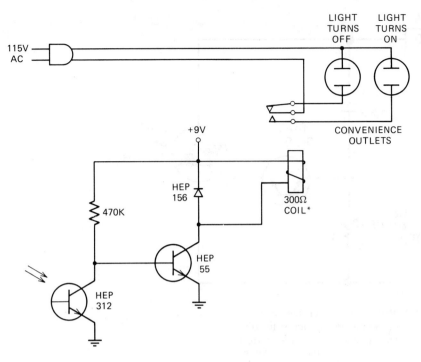

2.2.3 Motorola automatic night light provides multiple functions: on at dusk and off at sunrise, or vice versa. If contacts are used to switch a heavy-duty relay instead of 115V directly, the circuit can turn on such power-draining loads as coffeemakers and space heaters. Other uses are: to activate the electric business sign in the evening and shut it down in the morning automatically, to shut down office typewriters at night and turn them on in the morning. The relay for this application is a single-pole double-throw type with a 300- to 400-ohm coil.

*Do not exceed contact current rating.

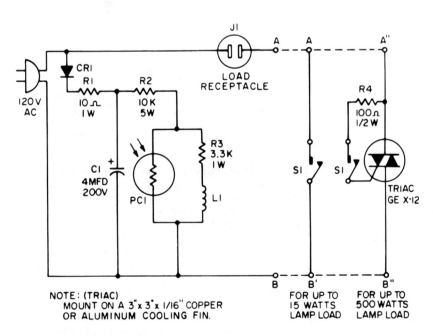

2.2.4 GE automatic night light turns off in daylight. If only a small lamp (up to 15W) is to be switched on and off, reed switch S1 can control the lamp directly. The right side of the circuit is used when controlling lamps up to 500W.

SECTION 2 *Control and Sensing* 77

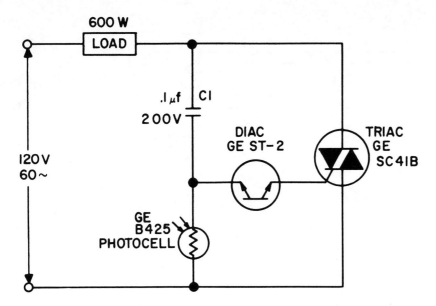

2.2.5 GE simple photoelectric ac power switch. For a dark photocell (high resistance) the voltage across the diac rises rapidly with the line voltage due to the current through C1, triggering the diac early in the cycle. When the photocell resistance is less than about 2000 ohms, the drop across it is limited to less than the diac triggering voltage and the load power is shut off.

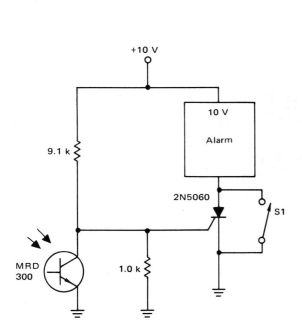

2.2.16 Motorola light-operated SCR alarm using a sensitive-gate SCR. The phototransistor holds the gate low as long as light is present but pulls the gate up to triggering level when the light is interrupted. The reset switch appears across the SCR.

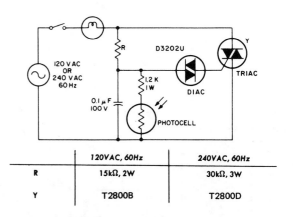

	120VAC, 60Hz	240VAC, 60Hz
R	15kΩ, 2W	30kΩ, 3W
Y	T2800B	T2800D

2.2.7 GE light-activated turnoff circuit functions in the same manner as light-dimming circuits, but the photocell controls its operation. When light impinges on the surface of the photocell, its resistance becomes low and prevents the voltage on the trigger capacitor from increasing to the breakover voltage of the trigger diode. The circuit is then inoperative. When the light source is removed, the photocell becomes a high resistance. The voltage on the trigger capacitor then increases to the breakover voltage of the trigger diode and causes the diode to fire. The trigger pulse formed by the capacitor discharge through the trigger diode makes the triac conduct and operates the circuit. The triac continues to be triggered on each half-cycle and supplies power to the load as long as the resistance of the photocell is high. When light again impinges on the surface of the photocell and reduces its resistance, the voltage on the capacitor can no longer reach the breakover voltage of the trigger diode, and the circuit turns off.

78 DISCRETE / TRANSISTOR CIRCUIT SOURCEMASTER

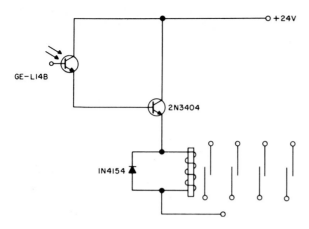

2.2.8 Light-activated relay uses GE's two-transistor photo-IC, the L14B. The relay used is the GE CR220G600D3 solenoid, which accommodates four reed switches. Any 24V relay with a 600-ohm coil will do, however.

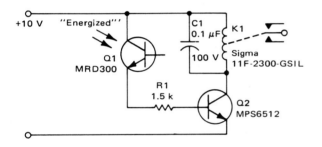

2.2.9 Motorola light-activated relay. When sufficient light is directed at Q1, it turns on. This drives Q2, which energizes the relay coil. A light magnitude of 220 footcandles is enough to push relay driver Q2 to saturation. When light is removed from Q1, base drive is removed to turn off Q2.

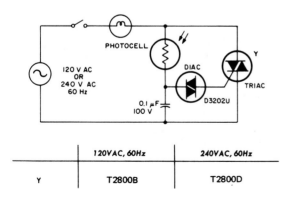

2.2.10 GE light-controlled turnon circuit, in which the low resistance of the photocell allows the triac to be triggered on. When light is removed from the photocell, the increased resistance of the photocell prevents the triac from being triggered and renders the circuit inoperative.

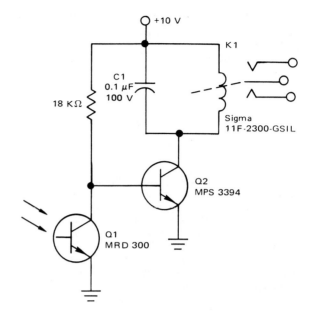

2.2.11 Motorola light-deenergized relay. The presence of light deenergizes this relay for automatic door activators, object or process counters, and intrusion alarms.

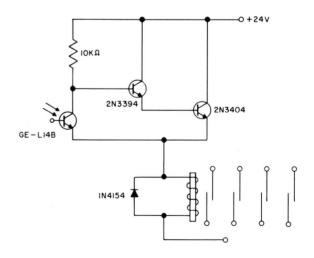

2.2.12 GE light-deenergized relay. Relay (24V, 600-ohm coil) will be energized so long as no light falls on the phototransistor. In the presence of light, the phototransistor robs power from the solenoid. The 10K series resistance drops the voltage too much to maintain activation.

SECTION 2 *Control and Sensing* 79

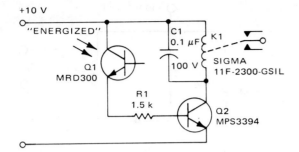

2.2.13 Motorola light-operated relay. The presence of light causes a relay to operate. The relay used in this circuit draws about 5 mA when Q2 is in saturation. Since h_{FE} (min) for the MPS3394 is 55 at a collector current of 2 mA, a base current of 0.5 mA is sufficient to insure saturation. Phototransistor Q1 provides the necessary base drive. If the MRD300 is used, the minimum illumination sensitivity is 4 µA/footcandle; therefore, sensitivity is sufficient for flashlight triggering.

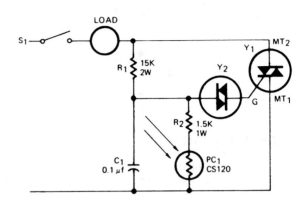

2.2.16 RCA light-controlled switch. Porch lights or night lights will turn on whether you are home or not. Burglar alarms may be installed using the on-at-night circuit which is also useful for alarms on ovens, heaters, etc.

When light hits photocell PC1, the resistance becomes low and the voltage does not reach the turnon voltage of trigger diac Y2. When the light is removed, the photocell resistance increases until diac Y2 triggers triac Y1. The triac is turned on each half-cycle while the resistance across PC1 remains high.

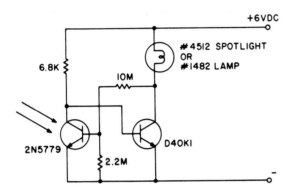

2.2.14 Turn on remote illumination such as warning or marker lights that operate from battery power supplies. The simplest circuit is one that provides illumination when darkness comes. By using the gain available in Darlington transistors, this GE circuit is simplified to use just a photo-Darlington sensor, a Darlington amplifier, and three resistors. The illumination level will be slightly lower than normal and longer bulb life can be expected because the GE D4OK saturation voltage lowers the lamp operating voltage slightly.

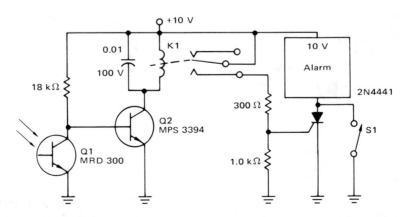

2.2.15 Motorola light-operated SCR alarm circuit. The presence of light keeps the relay deenergized, thus denying trigger current to the SCR gate. When the light is interrupted, the relay energizes, providing the SCR with trigger current. The SCR latches on so that only a momentary interruption of light is sufficient to cause the alarm to ring continuously. S1 is a momentary contact switch for resetting the system.

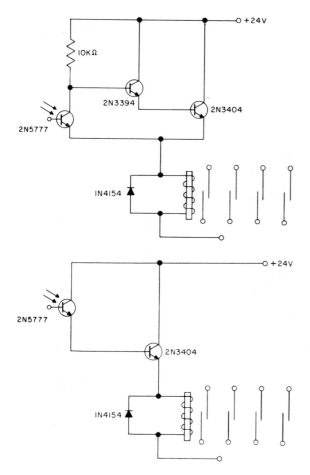

2.2.17 GE light interruption sensors detect major changes in light and can operate relays to control power. The first circuit needs light to energize the solenoid; the second needs light to deenergize the solenoid. This alternative is useful when making decisions on current consumption, failsafe operation, etc. The solenoid is GE CR220G600D3, rated at 24V, 600 ohms; it is designed to accommodate four reed switches.

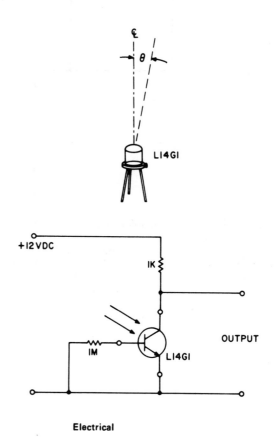

Electrical

2.2.19 GE sun tracker. In solar cell array applications and solar instrumentation, it is desired to know the position of the sun within 15° to allow efficient automatic alignment. The GE L14G1 lens can provide this type of accuracy in a simple level sensing circuit, and a full hemisphere can be monitored with about 150 phototransistors.

The sun provides about 80 mW/cm² to the L14G1 when on the centerline. The sky provides about 0.5 mW/cm² to the L14G1 and will keep the output above 10V when viewed. White clouds viewed from above could lower this voltage to 5V on some devices.

This circuit could directly drive TTL by clamping the output to the 5V logic supply with a signal diode. Different bright objects can also be located with the same type of circuitry by simply adjusting the resistor values to provide the desired sensitivity.

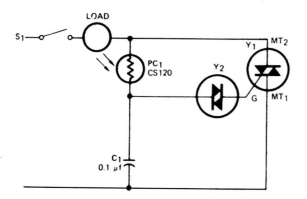

2.2.18 RCA off-at-dark circuit. The low resistance of photocell PC1 allows the voltage of diac Y2 to increase until it triggers triac Y1 on. When light is removed, the photocell resistance increases and prevents the diac from reaching its trigger voltage. The circuit will not operate unless the diac triggers.

SECTION **2** Control and Sensing **81**

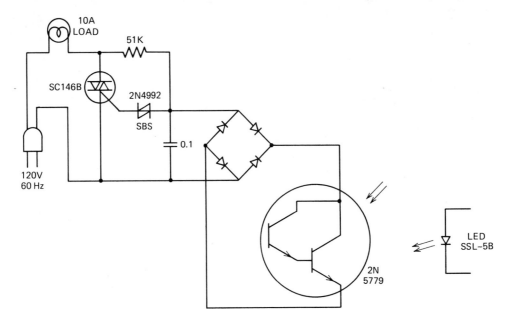

2.2.20 Light-deenergized relay requires an extremely small amount of impinging light to open the 120V circuit. The General Electric photo-Darlington can be held on with the radiated illumination of a GE SSL 5B light-emitting diode, which draws less than 50 mA at a very low dc voltage. The triac shown will switch 10A of current to a noninductive ac load.

When this circuit is suitably enclosed, it serves as a reliable object counter for assembly lines. Also, switching is fast enough to permit operation from the light passing through perforations in an opaque paper tape.

A silicon bilateral switch triggers the triac; the photo-Darlington shunts bias current from the SBS to prevent triggering when the LED is lit. (LED is actually not required, since triggering can be initiated from any light source).

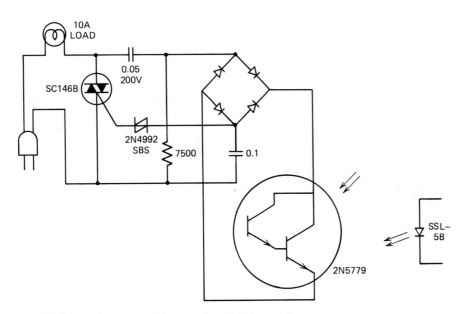

2.2.21 Normally open solid-state relay. A light-emitting diode, which draws no more than 50 mA at very low voltages, can be used to turn on this light-controlled solid-state switch from GE. This normally open circuit is the complement to the normally closed switch in the preceding schematic.

82 DISCRETE / TRANSISTOR CIRCUIT SOURCEMASTER

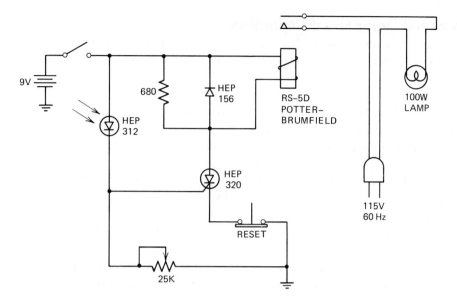

2.2.22 Garage-light control circuit. When the photo-diode is illuminated by the headlamps of a car entering the driveway, the Potter-Brumfield relay pulls in to apply 115V ac power to the 100W lamp. Position the light-sensitive circuitry in the garage so that daylight will not compete to cause false triggering. The sensitivity control will allow selective triggering so that the unit will not be activated in low-light daytime conditions or by other cars on the street. With proper shielding the unit can be positioned under the eave of the garage roof, but sensitivity adjustment under this condition becomes critical.

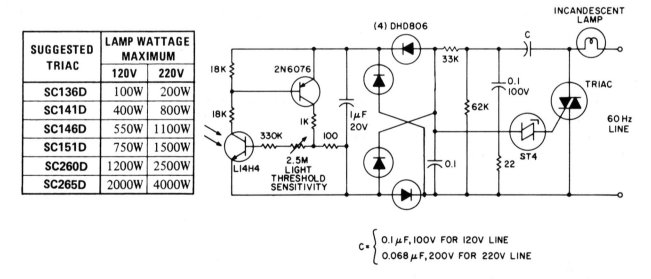

2.2.23 GE line-voltage-operated automatic night light has stable threshold characteristics due to its dependence on the photodiode current in the GE L14H4 generating a base–emitter drop across the sensitivity setting resistor. The double-phase-shift network supplying the voltage to the ST-4 trigger insures triac triggering at line-voltage phase angles small enough to minimize RFI problems with a lamp load. This eliminates the need for a large inductor, contains a snubber network, and utilizes lower voltage capacitors than the snubber or RFI suppression network normally would.

SECTION 2 *Control and Sensing*

2.3 Dimmers and Motor Speed Controllers

The hundred or so circuits in this subsection illustrate the ways in which we can employ modern semiconductor devices for power control. Although most of these circuits depend on phase control—periodic switching of the applied ac power at the frequency of the line—some rely on other methods. The fluorescent dimmers, for example, employ systems that operate on a completely different principle simply because a fluorescent lamp cannot be dimmed by the expedient of lowering the supply voltage. (Fluorescent tubes are filled with a gas that must be ionized before they ignite to emit light, and ignition takes a finite amount of time.)

Of the phase-control circuits, some use triacs and diacs, some use SCRs and SCSs, and still others employ programmable unijunctions and silicon unilateral switches as the principal element. The abundant variety of circuit types can be attributed to the various design engineers' approaches to getting around a multitude of inherent phase-control problems.

One example of a triac-dimmer anomaly is circuit hysteresis—the difference between the control setting that turns off the lamp and that which turns the lamp on again. The simpler circuits exhibit a considerable lag between these points, which can be annoying at the very least or, at the other extreme, totally unacceptable, depending on the specific application of the circuit.

Another legion problem with phase control is interference, the result of the harmonic-rich waveform caused by abrupt on-off switching during each incoming cycle of the alternating power-line voltage. This disconcerting characteristic of phase-controlling circuits has proved to be sufficient cause for incorporating a number of ingenious radio-frequency interference filtering methods as part of the circuits proper.

Triac-controlled lamp filaments have a tendency to "sing" or create an annoying whistlelike sound in the vicinity of the controlled light.

"Problems" abound. The cold resistance of an incandescent bulb is considerably lower than the resistance of the filament when it's hot. The inrush current caused by a voltage spike from the control circuitry can cause premature destruction of lamps in the very circuits that should prolong lamp life. So engineers attack the problem by designing in various "soft-start" elements (such as zero-voltage switching).

As you examine these circuits, you'll see one method after another for minimizing or eliminating the problems mentioned here; the circuit variations reflect an impressively diverse array of accomplishing the variable control function without yielding to the demands or restrictions imposed by the devices we use.

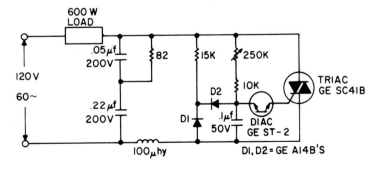

2.3.1 GE incandescent lamp dimmer with RF filter.

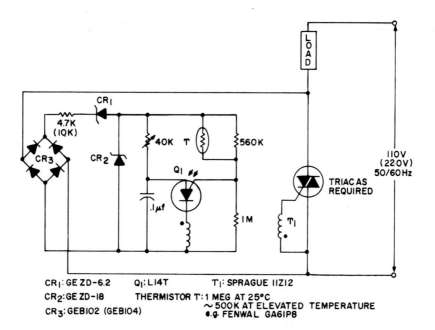

CR₁: GE ZD-6.2   Q₁: L14T   T₁: SPRAGUE 11Z12
CR₂: GE ZD-18   THERMISTOR T: 1 MEG AT 25°C
CR₃: GEB102 (GEB104)   ~500K AT ELEVATED TEMPERATURE
e.g. FENWAL GA61P8

2.3.2 GE 100% isolated full-wave phase control has built-in temperature compensation as well as 0 to 90° range. The negative-temperature-coefficient (NTC) thermistor in parallel with the 560K resistor provides temperature compensation. As the temperature rises, the decreasing resistance of the thermistor negates the increasing light sensitivity of the L14T. Power can be held constant within a few percent over the temperature range of 25 to 50°C.

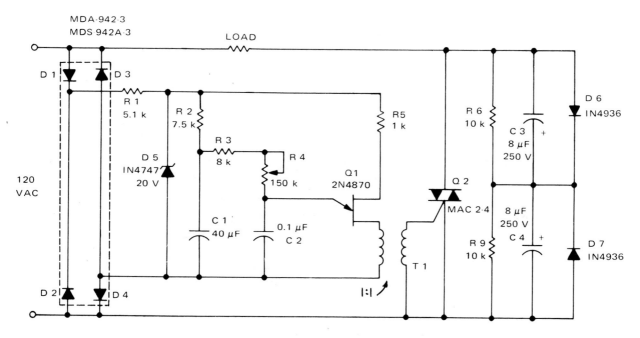

2.3.3 Motorola 1000W soft-start light dimmer. Soft starting is desirable because of the very low resistance of a cold filament compared to its hot resistance. This low resistance causes very high inrush currents when a lamp is first turned on, and this leads to short lamp life. Failures caused by high inrush currents are eliminated by the soft start feature, which applies current to the bulb slowly enough to eliminate high surges. Accidental turnon, which could nullify this advantage, is prevented by a special dv/dt network, which prevents the line voltage from triggering triac Q2 before the light has warmed up. This would occur at the instant power is first applied to the circuit if the instantaneous line voltage were sufficient to cause triggering. Capacitor C3 will delay a negative rise in voltage by charging through the load and diode D7; capacitor C4 does the same for a positive rise. Resistors R6 and R7 are used to discharge capacitors C3 and C4.

SECTION 2 *Control and Sensing* 85

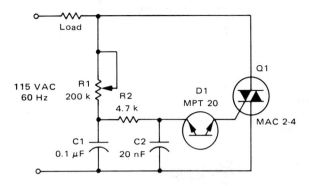
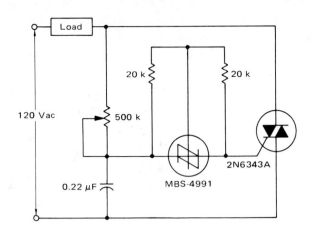

2.3.4 Motorola 800W triac light dimmer. On each half-cycle capacitor C2 charges through the two-section phase-shift circuit until it reaches the breakover potential of bilateral trigger D1. The potential of D1 then drops at least 5V, and the excess charge on C2 drains through the gate of the triac, turning it on. When the triac is turned on, the timing circuit is short-circuited by the triac, and no further pulsing can occur in that half-cycle. R1 varies the time constant of the trigger circuit, thus providing phase control for the triac. Since the same timing circuit is used for both half-cycles and the trigger device is symmetrical, the conduction angle of the triac is essentially identical for each succeeding half-cycle (assuming that R1 is fixed).

2.3.6 Motorola low-cost full-range lamp dimmer. Shunting the silicon bilateral switch (SBS) with two 20K resistors minimizes "flash-on" hysteresis. V_s of the SBS is reduced to about 4V, and since this is below the operating voltage of the internal zener diodes, the temperature sensitivity of the device is increased.

Hysteresis is eliminated in this circuit by two diodes and the 5.1K resistor connected to the MBS-4991 gate. At the end of each positive half-cycle (when the applied voltage drops below that of the capacitor) the gate current flows out of the SBS and it switches on, discharging the capacitor to near 0V.

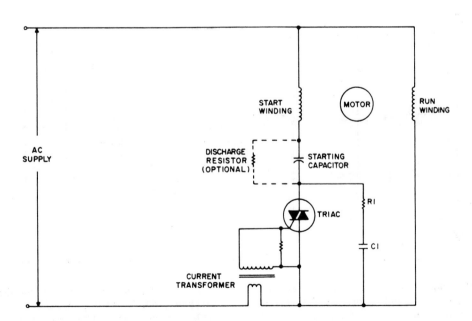

2.3.5 GE ac split-capacitor motor starter, in which the static triac switch replaces the conventional centrifugal switch or current relay. The turns ratio of the current transformer is designed to pick up the triac on the inrush current to the run winding. The triac drops out when the current settles to normal operating levels. A low-cost ferrite core is adequate for the current transformer. The entire circuit can be mounted inside the motor and is ideal when high reliability, frequent starting, or spark-free operation is necessary. R1 and C1 are for limiting the commutating change of voltage with time (dv/dt).

86 DISCRETE / TRANSISTOR CIRCUIT SOURCEMASTER

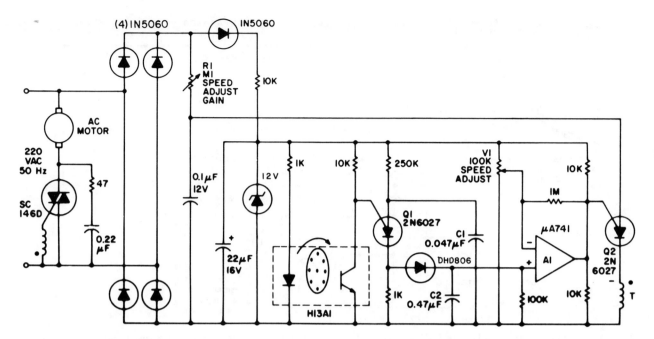

2.3.7 GE ac induction motor control illustrates feedback speed regulation, a function difficult to accomplish without a costly generator-type precision tachometer. When the apertured disc (attached to the motor shaft) interrupts the light beam across the coupler module, programmable unijunction transistor Q1 discharges capacitor C1 into C2. The voltage on C2 consequently is a direct function of the rotational speed of the motor. This speed-related potential is compared against an adjustable reference voltage, V_1, through monolithic operational amplifier A1, whose output establishes a dc control input to the second PUT, Q2. This latter device is synchronized to the ac mains and furnishes trigger pulses to the triac at a phase angle determined by speed control R1 and by the actual speed of the motor.

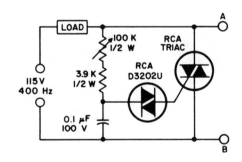
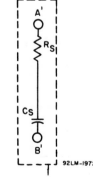

AC INPUT VOLTAGE		120 V 60 Hz	240 V 60 Hz
SNUBBER NETWORK FOR 2.5 A (RMS)• INDUCTIVE LOAD	C_S	0.068 µF 200 V	0.075 µF 400 V
	R_S	2.2 kΩ ½ W	2.5 kΩ ½ W
RCA TRIACS		2N5755 T2313B	2N5756 T2313D

SNUBBER NETWORK FOR INDUCTIVE LOADS.
CONNECT POINTS A¹ AND B¹ TO TERMINALS A AND B RESPECTIVELY.

2.3.8 RCA 400 Hz lamp dimmer can be used for loads of up to 6A. For incandescent lamps that produce burnout current surges with I^2t values greater than 2.5A² seconds, connect a 10-ohm resistor of appropriate power rating in series with the load. Determine the rating by $R = 10$ (rms load current)². For inductive loads up to 2.5A, build the snubber as shown and connect it to points **A** and **B** of dimmer.

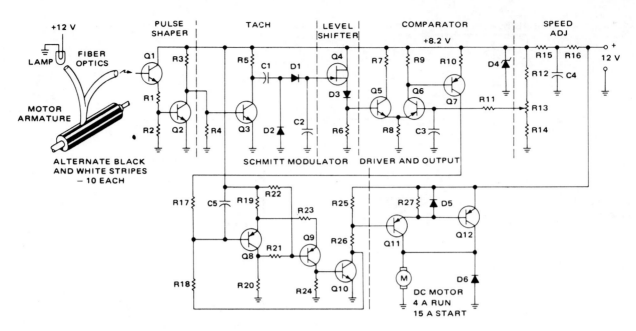

2.3.9 Motorola dc motor speed control using an optical pickoff to complete a closed-loop system.

The motor armature is painted with alternate black and white stripes. A total of 20 stripes was used for the circuit shown. A bundle of noncoherent fiber optics is used to transmit light from a 12V lamp to the armature and to transmit reflected light from the armature to the input of Q1, an MRD300 phototransistor. The light transmitted from bulb to armature to phototransistor is chopped at a frequency determined by the speed of the motor.

The output of the phototransistor is fed into a pulse-shaping circuit and then into a tachometer circuit whose dc output is proportional to the input frequency. The output of the tachometer drives JFET Q4. This transistor provides a high input impedance to minimize loading on the tachometer circuit and a reasonably low output impedance to drive the differential amplifier. It also acts as a level shifter to insure that there is sufficient output from the FET to bias the differential amplifier when the tachometer output is zero. The diode in series with the source of the FET is used for temperature compensation.

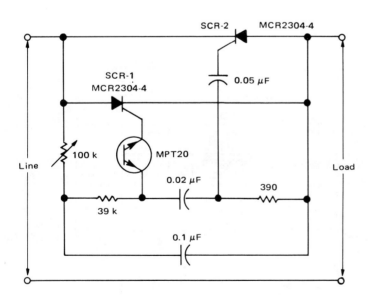

2.3.10 Motorola basic control circuit for SCRs with RC coupling, eliminating the need for a pulse transformer. In the negative half-cycle, the pulse developed across the 390-ohm resistor when the 0.02 µF trigger capacitor has discharged is capacitively coupled to SCR2 through a 0.05 µF capacitor. This circuit should be used with a filter in the power line, since the capacitive coupling makes it extremely sensitive to line transients.

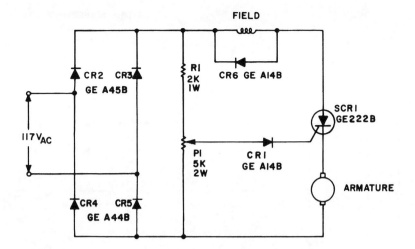

2.3.11 GE full-wave series motor speed control with feedback requires that a separate connection be available for the motor armature and field. The bridge supplies power to the series networks of motor field, SCR1 and armature, and R1 and P1. When the motor starts running, the SCR triggers as soon as the reference voltage across the arm of P1 exceeds the forward drop of CR1 and the gate-to-cathode drop of SCR1. The motor then builds up speed and, as the back emf increases, the speed of the motor adjusts to the setting of P1.

One of the drawbacks of this circuit is that at low-speed settings, because of the decreased back emf, the anode-to-cathode voltage of the SCR may not be negative for a sufficient time for the SCR to turn off. When this happens, the motor receives full power for the succeeding half-cycle and the motor starts hunting. An SCR selected for turnoff time should cure this. This circuit is limited by the fact that SCR1 cannot be triggered consistently later than 90°. A capacitor on the arm of P1 is not a cure because there will be no phase shift on the reference due to full-wave rectified charging.

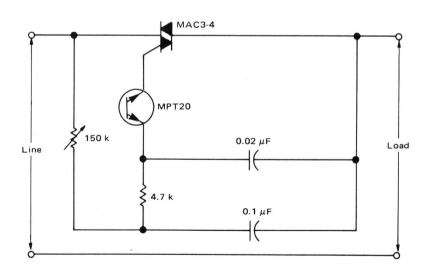

2.3.12 Motorola basic lamp brightness control circuit. In the positive half-cycle, the 0.02 μF capacitor charges through the dual-section phase-shift circuit until its voltage reaches the breakover potential of the MPT20 diac. The MPT20 potential then drops several volts, forcing the charge from the 0.02 μF capacitor through the gate of the MAC3-4 triac. This current turns the triac on. Once the triac has turned on, the timing circuit is shorted and no further pulsing can occur in this half-cycle. A similar action results in the negative half-cycle.

The time constant of the trigger circuit can be varied by the 150K variable, thus providing phase control for the triac.

SECTION 2 *Control and Sensing* **89**

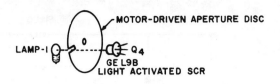

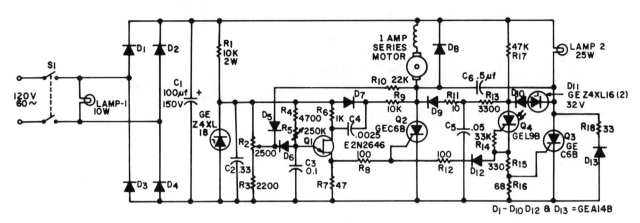

2.3.13 GE optical feedback control for high-precision motor speed control. Rather than use the mechanical commutating switch for a higher voltage series motor, this circuit uses a light beam and aperture disc to provide the off signal. The unijunction oscillator turns on the main SCR, Q2, which is then commutated by auxiliary SCR Q3 in the classical capacitor-commutated SCR flip-flop connection. General Electric's light-activated SCR Q4 is turned on by light when the motor shaft passes a certain position. If the main SCR, Q2, is conducting and Q3 is off, the current can flow through zener D11, diode D10, and SCR Q4 to the gate of Q3 to turn it on. When Q3 turns on, it commutates Q2 through capacitor C6 and also commutates Q4 through diode D9 and resistors R11 and R13.

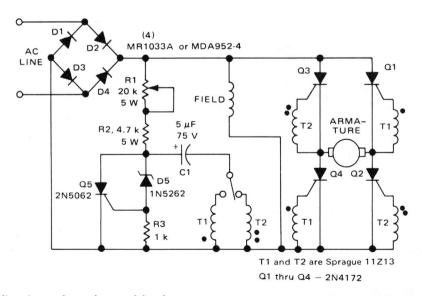

2.3.14 Motorola direction and speed control for shunt-wound motor. Here the field is placed across the rectified supply and the armature is placed in the SCR bridge. The field current is unidirectional but the armature current is reversible; consequently, the motor's direction of rotation is reversible. Potentiometer R1 controls the speed.

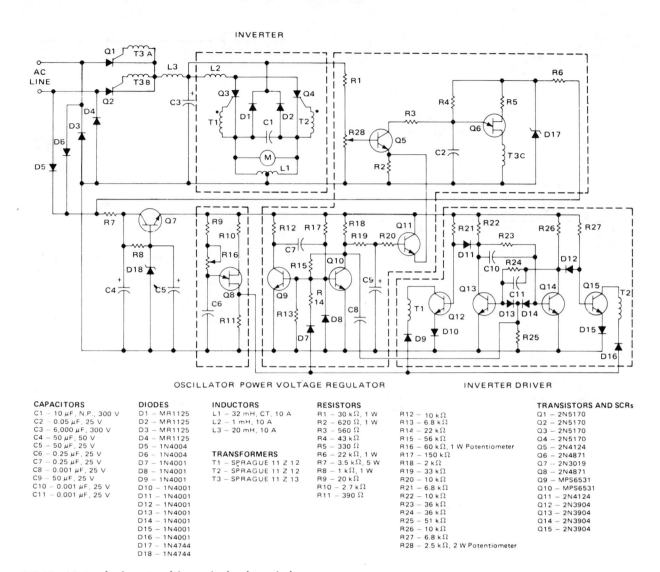

2.3.15 Motorola inverter-driven single-phase induction-motor speed control circuit.

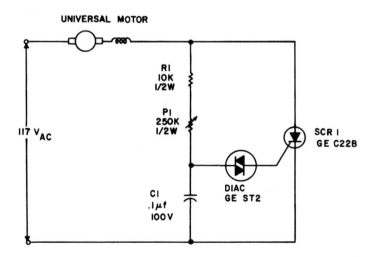

2.3.16 GE diac-triggered half-wave universal-motor speed control. Because of their lower trigger voltage, trigger devices give a wider control range than that obtainable using neons.

SECTION **2** *Control and Sensing* **91**

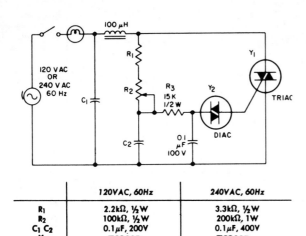

	120VAC, 60Hz	240VAC, 60Hz
R_1	2.2kΩ, ½W	3.3kΩ, ½W
R_2	100kΩ, ½W	200kΩ, 1W
$C_1 C_2$	0.1μF, 200V	0.1μF, 400V
Y_1	T2800B	T2800D
Y_2	D3202U	D3202U

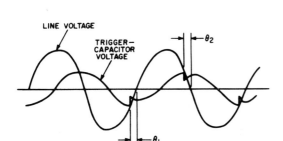

2.3.17 RCA double-time-constant lamp dimmer improves on the performance of the single-time-constant control. This circuit uses an additional RC network to extend the phase angle so that the triac can be triggered at small conduction angles. The additional RC network also minimizes the hysteresis effect. Because of the voltage drop across R3, input capacitor C2 charges to a higher voltage than trigger capacitor C3. When the voltage on C3 reaches the breakover voltage of the trigger diode (diac), the device conducts and causes the capacitor to discharge and produce the gate current pulse to trigger the triac. After the diac turns off, the charge on C3 is partially restored by the charge from input capacitor C2. The partial restoration of charge on C3 results in better circuit performance with a minimum of hysteresis.

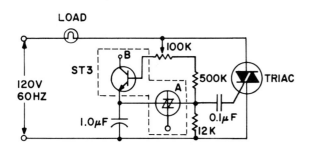

2.3.18 In this RCA hysteresis-free dimmer the circuit presents a gate pulse when V_A is greater than the breakover voltage of the ST3A. This point is adjusted mainly by the charging rate of the capacitor but is also affected to some extent by the change in voltage across the 12K resistor. This does not operate with symmetry.

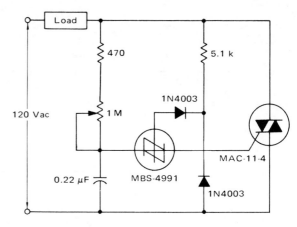

2.3.19 Motorola improved full-range power controller for lamp dimming and similar applications. For settings such that no power is delivered to the load, the timing capacitor would never discharge through the silicon bilateral switch (SBS). The result is an abnormal amount of apparent phase shift caused by the capacitor's starting to charge toward a source of voltage with a residual charge of the opposite sign. This is the cause of the hysteresis effect and is eliminated in this circuit by the addition of the two diodes and 5.1K resistor connected to the SBS gate. At the end of each positive half-cycle when the applied voltage drops below that of the capacitor, gate current flows out of the SBS and it switches on, discharging the capacitor to near 0V.

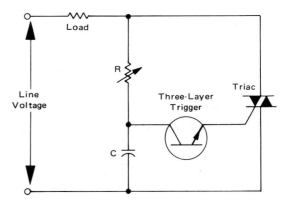

2.3.20 Motorola full-wave phase-control circuit. The thyristor is inserted in series with the load and power source. At the beginning of each half-cycle, the thyristor is in its blocking state, preventing power from reaching the load. At some point during the half-cycle, the thyristor changes to its conducting state, permitting power to reach the load.

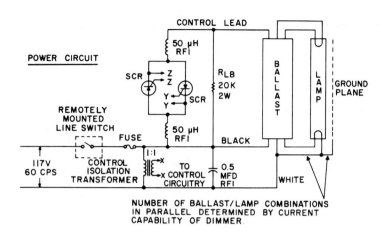

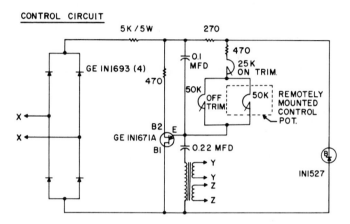

2.3.21 GE fluorescent lamp dimmer. The power circuit of the dimmer must have a symmetrical output to avoid magnetic core saturation of the ballast. The current rating of the SCRs and their associated heatsinking system must be selected on the basis of the current in the control lead required by the ballast(s) and the loading resistance. This current is not to be confused with the rated line current given on the nameplate of the ballast. The control lead current for one 89G718 ballast with one F40T12 lamp is about 285 mA. In view of the discontinuous current waveform the current was calculated on the basis of its oscilloscope waveform. Since the SCR in the circuit sees both half-cycles of the current waveform, the SCR and heatsink must be selected on the basis of the average value of this waveform, which calculated to about 230 mA average. On this basis, the GE C15 type of SCR on a 3 × 3 × 1/16 in. (7.62 × 7.62 × 0.47 cm) copper fin in a free-convection ambient temperature of 30°C could handle about 17 ballast/lamp combinations. A single-gang wall box or the inside of a lamp fixture is anything but that ideal an environment. The mechanical and thermal design of the dimmer, its operating environment, and the rating of the SCR used determine the dimmer rating. The C11 and C15 types of SCRs are adequate for all but the largest installations. General Electric type C35 (2N681 series) and type C50 SCRs are available for larger ratings.

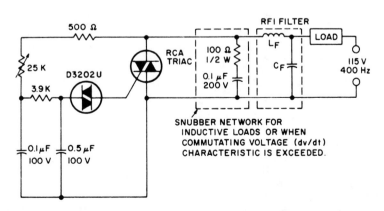

2.3.22 RCA typical phase-control circuit for operation at line frequency of 400 Hz, with a snubber which can be added for inductive loads or when the voltage characteristic of commutating input is exceeded. The voltage for the devices shown is 115V.

SECTION 2 *Control and Sensing* 93

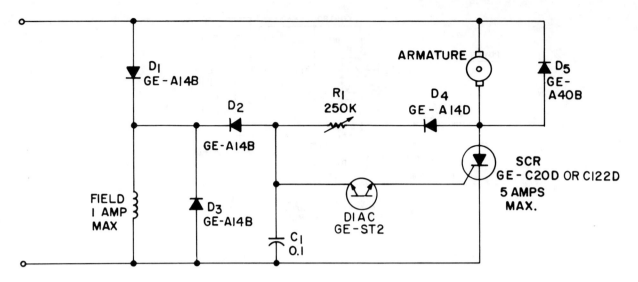

2.3.23 GE half-wave control for shunt-wound motors.
Field current is supplied by D1, and free-wheeling rectifier D3 provides a circulating current path for smoothing the field current waveform. The armature is supplied by current through the SCR. Voltage for the control circuit is derived from the voltage across the SCR. At the end of each positive half-cycle the voltage across the field drops to zero and control capacitor C1 is discharged through diode D2. This action insures that the voltage on C1 is always zero at the beginning of each positive half-cycle regardless of the setting of speed control resistor R1.

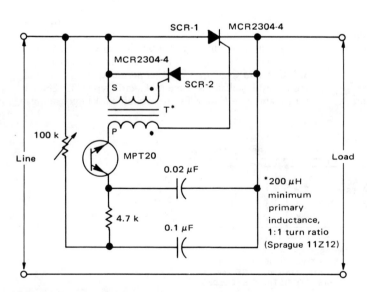

2.3.24 Motorola basic full-wave control circuit for SCRs. In the positive half-cycle, SCR1 is triggered through the primary of the pulse transformer. In the negative half-cycle, the 0.02 μF triggering capacitor discharges through the built-in shunt resistor of SCR1 and the primary of the pulse transformer, inducing a pulse in the secondary, which triggers SCR2.

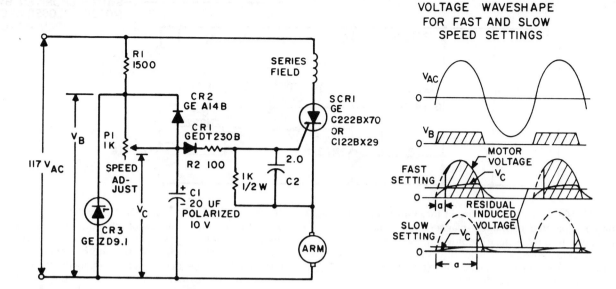

2.3.25 GE half-wave motor control with feedback relies on the residual induced field for a speed feedback signal but permits a very short conduction time for the SCR and hence a slow speed. During the negative half-cycle, C1 is discharged to zero. During the positive half-cycle C1 charges from a constant potential (zener voltage of CR3) at an exponential rate dependent on the time constant P1C1. If the motor armature is standing still, no voltage is induced in it by the residual field, and gate current to the SCR flows as soon as V_c exceeds the forward voltage drop of CR1 and the gate drop of SCR1. This will fire SCR1 early in the cycle, providing ample energy to accelerate the motor. As the motor approaches its preset speed, the residual induced voltage in the armature builds up. This voltage is positive on the top terminal of the armature and bucks the flow of gate current from capacitor C1 until V_c exceeds the armature voltage. This higher voltage requirement on C1 retards the firing angle and allows the motor to cease accelerating.

Once the motor has reached operating speed, the residual induced voltage provides automatic speed regulating action. If a heavy load starts to pull down the motor speed, the induced voltage decreases, and SCR1 therefore triggers earlier in the cycle. The additional energy thus furnished to the motor supplies the necessary torque to handle the increased load. Conversely, a light load with its tendency to increase speed raises the motor residual induced voltage, retarding the firing angle and reducing the voltage on the motor.

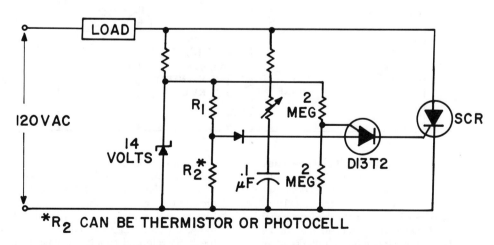

2.3.26 GE high-gain phase control circuit using cosine ramp-and-pedestal arrangement. The resistor divider formed by R1 and R2 quickly charges the 0.1 μF capacitor to a level just below the peak-point voltage of the D13T2. The capacitor is then charged via the potentiometer return to the 120V power line. Due to the high sensitivity of the D13T2, very high values of potentiometer resistance can be used, thereby achieving very high-gain phase control.

SECTION **2** Control and Sensing

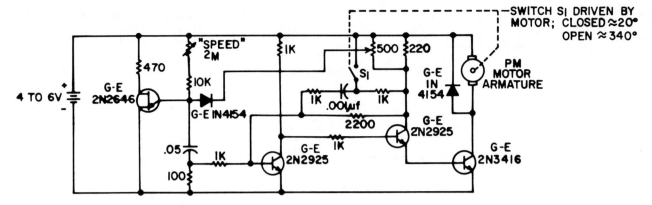

2.3.27 GE four-transistor circuit for controlling a small *battery-operated permanent-magnet motor* such as recording instruments, tape recorders, phonographs, etc. A unijunction transistor oscillator drives a transistor flip-flop to energize the motor. The motor is turned off when S1 closes and reverses the flip-flop.

The current through commutating switch S1 is very low, so this may be very small and with light actuating forces.

For best results at low speeds, the switch should close several times per revolution of the motor, thus providing multiple current pulses during each revolution. With a fixed oscillator frequency, the motor speed can be changed in discrete octave steps by changing the number of switch closures per revolution from 1 to 2 to 4 to 8, etc. This can be done either mechanically or electrically by transferring to various commutating switches.

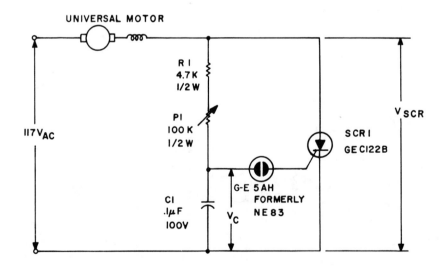

2.3.28 GE half-wave phase control without feedback uses one SCR with the minimum amount of components. The series network of R1, P1, and C1 supplies a phase-shift signal to the neon bulb, which triggers the SCR. Thus, by varying the setting of potentiometer P1, the gate signal of the SCR is phase-shifted with respect to the supply voltage to turn the SCR on at varying times in the positive ac half-cycles. The charge voltage V_c fires the neon bulb on both positive and negative half-cycles. The negative half-cycles can be disregarded, since both the trigger pulses and the anode voltage of the SCR are negative.

At a low resistance setting of P1, capacitor C1 tries to charge to the peak of the line voltage at a fast rate, but as soon as the voltage V_c across the capacitor reaches the breakdown voltage of the neon (around 60V), the bulb triggers and produces a positive pulse that is applied to the gate of the SCR. This pulse triggers the SCR early in the half-cycle and supplies voltage to the motor for the remainder of that half-cycle. The motor will then operate at a high speed.

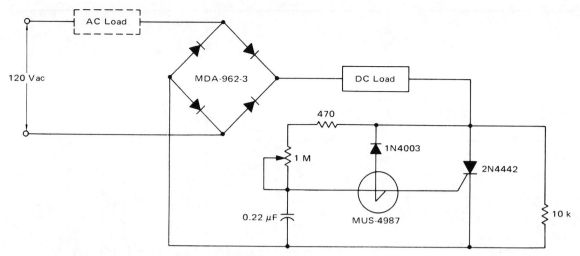

2.3.29 Motorola full-range power controller. A load requiring pulsating direct current may be connected between the bridge rectifier and the SCR. If the load requires an alternating voltage, it may be connected in series with either side of the ac power line. The addition of the diode and 10K resistor across the SCR will guarantee the discharge of the capacitor near the end of each half-cycle.

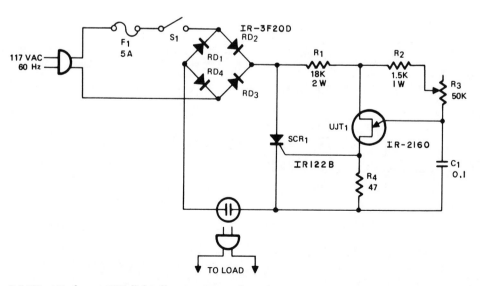

2.3.30 Workman SCR light dimmer uses unijunction to control the load voltage at powers up to 400W. The entire unit should be mounted so that all wiring is covered. The operation is dependent on the period of time that SCR1 is turned on. The firing point of SCR1 is determined by the setting of potentiometer R3.

SECTION 2 *Control and Sensing* 97

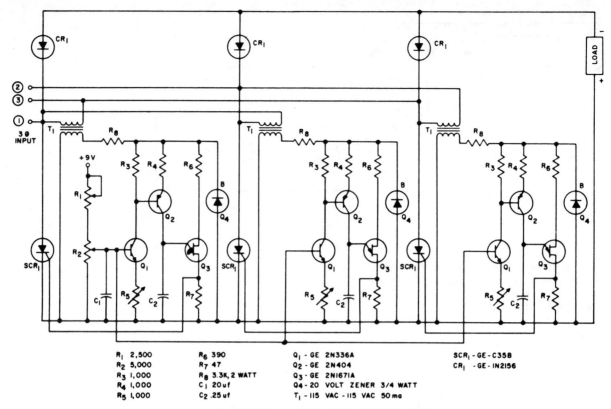

R_1 2,500	R_6 390	Q_1 - GE 2N336A	SCR_1 - GE-C35B
R_2 5,000	R_7 47	Q_2 - GE 2N404	CR_1 - GE-IN2156
R_3 1,000	R_8 3.3K, 2 WATT	Q_3 - GE 2N1671A	
R_4 1,000	C_1 20uf	Q_4 - 20 VOLT ZENER 3/4 WATT	
R_5 1,000	C_2 .25 uf	T_1 - 115 VAC - 115 VAC 50 ma	

ALL RESISTORS 1/2 WATT UNLESS OTHERWISE NOTED

2.3.31 GE full-range stepless 3-phase dc control circuit is composed of three conventional rectifiers and three SCRs. Each SCR is fired independently by its own unijunction transistor firing circuit. The firing angle of each phase is controlled by the collector current of Q2, which acts as a constant-current charging source for the UJT emitter capacitor. The collector current of Q2 is controlled by transistor Q1. Since both Q1 and Q2 are operating as emitter followers, there is no resultant voltage gain and the change in firing angle is directly proportional to the change in the dc control voltage at the base of Q1. The tracking accuracy of the firing angle, which is controlled by potentiometer R5 in the respective emitters of Q1, is plus or minus one electrical degree at 60 Hz.

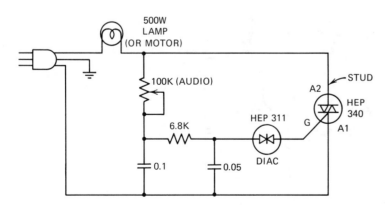

2.3.32 Motorola full-range light dimmer controls up to 500W and can be constructed in a box small enough for mounting in place of a wall switch. Do not allow triac to come into contact with the metal utility box. This circuit can also be used to control the speed of universal-type motors such as those typically found on blenders and other kitchen appliances. The HEP 311 device shown here as cathode-connected series diodes is a diac, which fires at 32V.

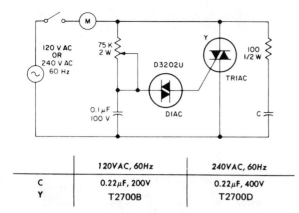

	120VAC, 60Hz	240VAC, 60Hz
C	0.22 µF, 200V	0.22 µF, 400V
Y	T2700B	T2700D

2.3.33 RCA induction motor control for proportional speed adjustment. This type of circuit is best suited to applications that require speed control in the medium- to full-power range. It is specifically useful in applications such as fans or blower-motor controls, where a small change in motor speed produces a large change in air velocity.

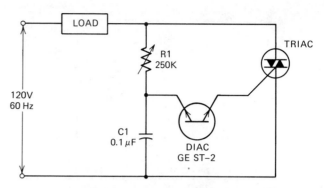

2.3.35 GE inexpensive phase-control lamp dimmer. Adjustable resistor R1 and capacitor C form a single-element phase-shift network. When the voltage across C1 reaches the breakover voltage of the diac, C1 is partially discharged by the diac via the triac gate. This pulse triggers the triac into the conduction mode for the remainder of that half-cycle. This circuit has a limited control range and a large hysteresis effect at the low-input end of the range, but its simplicity makes it attractive and popular.

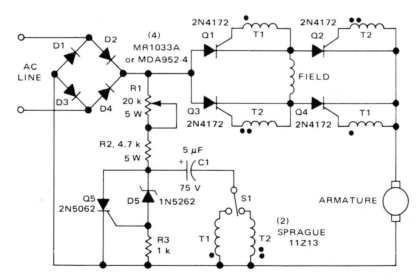

2.3.34 Motorola direction and speed control for series-wound or universal motors. Silicon controlled rectifiers Q1–Q4, connected in a bridge arrangement, are triggered in diagonal pairs. Which pair is turned on is controlled by switch S1, since it connects either coupling transformer T1 or coupling transformer T2 to a pulsing circuit. The current in the field can be reversed by selecting either SCRs Q1 and Q4 or Q2 and Q3 for conduction. Since the armature current is always in the same direction, the field current reverses in relation to the armature current, thus reversing the direction of rotation of the motor.

The speed of the motor is controlled by potentiometer R1. The larger the resistance in the circuit, the longer the time required to charge C1 to the breakdown voltage of zener D5. This determines the conduction angle of either Q1 and Q4, or Q2 and Q3, thus setting the average motor voltage and thereby the speed.

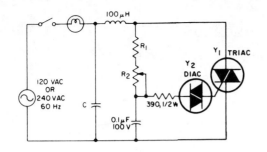

	120VAC, 60Hz	240VAC, 60Hz
R_1	3.3kΩ, ½W	4.7kΩ, ½W
R_2	200kΩ, ½W	250kΩ, 1W
C	0.1μF, 200V	0.1μF, 400V
Y_1	T2800B	T2800D
Y_2	D3202U	D3202U

2.3.36 RCA lamp dimmer with series gate resistor for hysteresis reduction. The series resistor slows down the discharge of the capacitor through the trigger diode. Consequently, the capacitor does not lose as much charge while triggering the triac. As a result of the slower capacitor discharge through the trigger diode, however, the peak magnitude of the gate trigger current pulse is reduced. The size of the trigger capacitor may have to be increased to compensate for the reduction of the gate trigger current pulse.

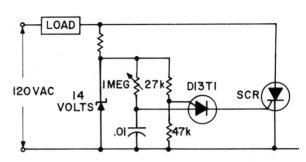

2.3.38 GE low-gain phase-control circuit uses a programmable unijunction transistor (PUT) as a conventional unijunction. Notice the low value of capacitance (0.01 μF) used. The total power consumption of the control circuit is considerably less than that using standard unijunctions.

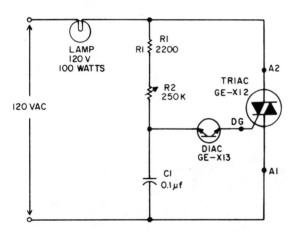

2.3.37 GE lamp dimmer and speed control. When variable resistor R2 is made very small, capacitor C1 charges rapidly at the beginning of each half-cycle of the ac voltage wave. When the voltage across C1 reaches the breakover voltage of the diac (about 32V), the capacitor is discharged into the gate of the triac. This triggers the triac on early in each half-cycle, and it continues to conduct current from the time it is triggered to the end of each half-cycle. The lamp will therefore have current flowing through it for most of each half-cycle and will produce full brightness.

As R2 is increased in resistance, the length of time required to charge C1 to the diac's breakdown point increases. This causes the triggering of the triac to occur later in each half-cycle, shortening the length of time that current is flowing through the lamp. Thus R2 controls the lamp brightness.

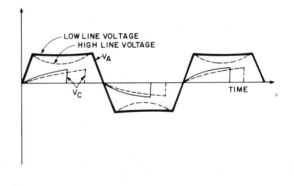

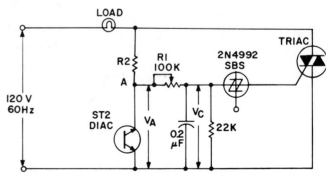

2.3.39 GE line-voltage-compensating dimmer. If a lamp dimmer has been set for a low light level and the line voltage drops for a fraction of a second, the lamps may be extinguished. To eliminate this effect, a compensating circuit such as this one may be used. R2 must give a 1 to 2V droop in the waveform at normal line voltages. When the line voltage drops, the droop decreases.

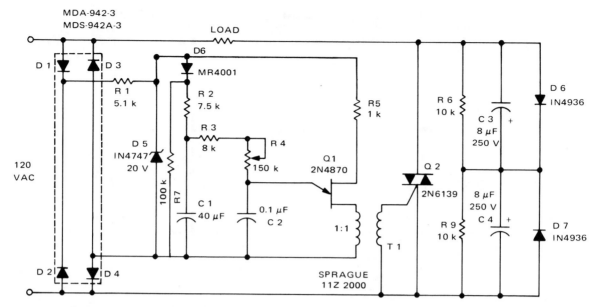

2.3.40 Motorola light dimmer with "soft start" operation. Soft starting is desirable because of the very low resistance of a cold filament compared to its hot resistance. This low resistance causes very high inrush currents when a lamp is first turned on; this leads to short lamp life. Failures caused by high inrush currents are eliminated by the soft start feature, which applies current to the bulb slowly enough to eliminate high surges. Accidental turn-on, which could nullify this advantage, is prevented by a special dv/dt network.

Operation of this circuit begins when voltage is applied to the diode bridge, which rectifies the input and applies a dc voltage to resistor R1 and zener D5. The zener provides a constant 20V to unijunction transistor Q1 except at the end of each alternation when the line voltage drops to zero. Initially, the voltage across capacitor C1 is zero and capacitor C2 cannot charge to trigger Q1. C1 will begin to charge, but because the voltage is low, C2 will have adequate voltage to trigger C1 only near the end of the half-cycle. Although the lamp resistance is low at this time, the voltage applied to the lamp is low and the inrush current is small. Then the voltage on C1 rises, allowing C2 to trigger Q1 earlier in the cycle. At the same time the lamp is being heated by the slowly increasing applied voltage and by the time the peak voltage applied to the lamp has reached its maximum value, the bulb has been heated sufficiently so that the peak inrush current is kept to a reasonable value. Resistor R4 controls the charging rate of C2 and provides the means to dim the lamp. Power to the load can be adjusted manually by varying the resistance of R4.

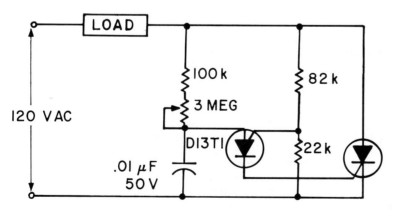

2.3.41 In this GE phase-control circuit the interbase voltage of the GE D13T PUT is derived directly from the ac line. The resistor divider attached to the gate (22K and 82K) is arranged such that the 40V rating on the D13T1 is not exceeded.

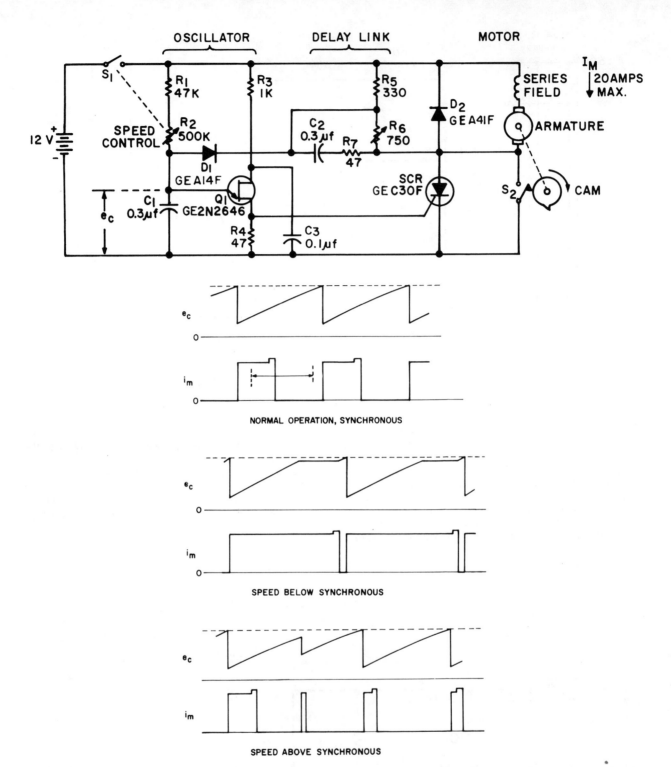

2.3.42 In this GE synchronous control for series motors, unijunction transistor Q1 is used in a relaxation oscillator circuit to produce periodic pulses that trigger the SCR. Since this operates from a dc source, the SCR will conduct current through the series field and the armature until it is turned off. At a certain point in the rotation of the motor shaft, a cam causes switch S2 to close, thus momentarily bypassing the SCR. This action permits the SCR to turn off (commutate) so that no current can flow after switch S2 opens. Thus the oscillator turns the power on and the motor turns the power off.

Rectifier D2 is a free-wheeling diode tht prevents high voltage from appearing across the SCR or the switch by providing a circulating current path for energy stored in the motor inductance. This eliminates most of the arcing normally encountered in the operation of switch S2.

Waveforms show the voltage across capacitor C1 and the emitter of unijunction transistor Q1, and the waveforms of motor current. In normal operation e_c has the normal sawtooth waveform of a relaxation oscillator. The motor current starts at the time Q1 discharges C1 and ends when S2 closes and reopens. Since the voltage drop across the switch is less than that of the SCR, the current is seen to be slightly higher during the time S2 is closed. The time between on and off will depend upon the loading on the motor.

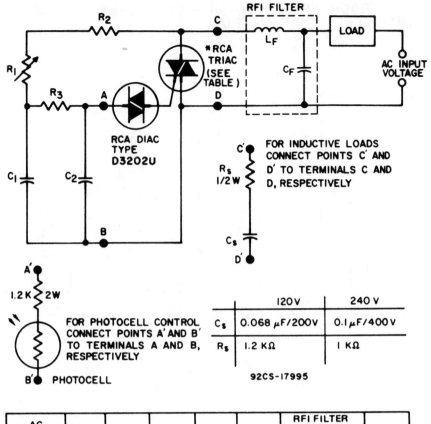

2.3.43 RCA phase-control circuit for lamp dimming, heat control, and universal motor speed adjustment. For photocell control, connect points A and B of circuit to A' and B' of the series cell-resistor network (inset). The table shows capacitance and resistance values for various line voltages at 50 and 60 Hz.

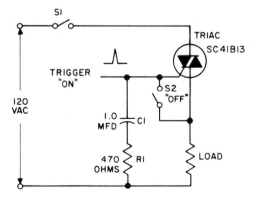

2.3.44 GE triac control circuit requires two switches, one for on and one for off. A trigger pulse turns on triac after S1 is closed.

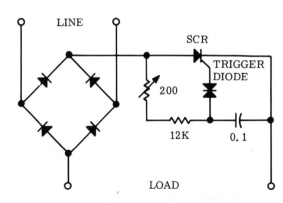

2.3.45 Motorola dc power control circuit capable of controlling a 600W load.

SECTION 2 *Control and Sensing* **103**

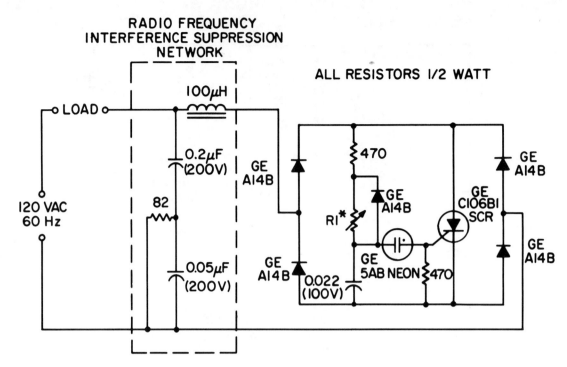

2.3.46 GE low-cost variable voltage control adjusts the speed of small 120V ac shaded-pole fan motors with current ratings up to 1.5A. The speed range ratio is approximately 3:1. With slight modifications the circuit may also be used for dimming high-intensity low-voltage lamps.

So that the single C106B SCR can exert control over both half-cycles of the ac supply, a diode bridge converts alternating current into full-wave pulsating direct current, which is applied directly to the SCR. The ac load (motor or primary winding of the lamp stepdown transformer) is connected in series with one leg of the ac supply. Because the C106 SCR can be triggered with less than 0.5 mA gate current, the inexpensive neon glow lamp (type 5AB) makes an ideal trigger device. Voltage adjustment potentiometer R1 and the 0.22 μF timing capacitor in conjunction with the glow lamp form a relaxation oscillator whose frequency determines the SCR trigger point in each half-cycle. The A14B rectifier connected across the potentiometer resets the timing capacitor to 0V each time the SCR triggers and insures that the output waveform is symmetrical from one half-cycle to the next. For use as a lamp dimmer, R1 should be 250K; for speed control, R1 should be 100K. The radio-frequency-interference filter should not be omitted because, in addition to acting as a noise filter, it prevents fast-rising transients from inadvertently triggering the SCR.

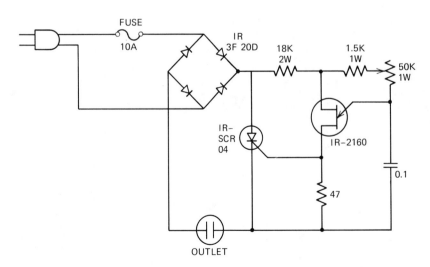

2.3.47 Workman motor control circuit varies speeds of universal motors of 5A and can control light dimming at a full kilowatt.

104 DISCRETE / TRANSISTOR CIRCUIT SOURCEMASTER

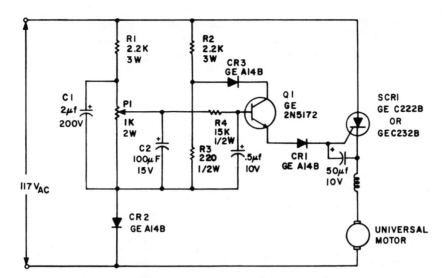

2.3.48 GE motor control for applications requiring low speeds and wide speed ranges uses an intermediate amplifier stage between the reference voltage of P1 and the gate of SCR1. Emitter follower Q1 and its associated resistors and diodes produce a low output impedance from the reference signal and therefore give improved speed performance at low-speed settings. Q1 provides isolation and current gain between the reference voltage and the gate of the SCR.

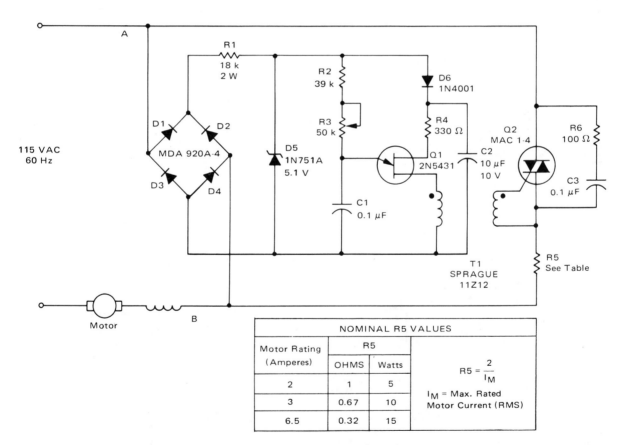

Motor Rating (Amperes)	R5 OHMS	R5 Watts
2	1	5
3	0.67	10
6.5	0.32	15

$$R5 = \frac{2}{I_M}$$

I_M = Max. Rated Motor Current (RMS)

2.3.49 Motorola motor speed control and feedback. When the triac conducts, the normal line voltage, less the drop across the triac and resistor R5, is applied to the motor. By delaying the firing of the triac until a later portion of the cycle, the voltage applied to the motor is reduced and its speed is reduced proportionally. The use of feedback maintains torque at reduced speeds. Nominal values for R5 can be obtained from the table or they can be calculated from the equation given. Exact values for R5 depend somewhat on the motor characteristics, so use an adjustable wirewound resistor that can be calibrated in terms of motor current. If value of R5 is too high, feedback will be excessive and surging or loss of torque will result. Motor current flows through R5, and its wattage must be determined accordingly.

SECTION 2 *Control and Sensing*

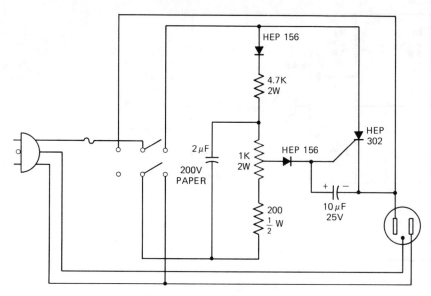

2.3.50 Motorola motor speed control for portable shop tools, kitchen appliances, or any device that uses a dc motor not requiring more than 3.5A (about 400W). The heatsink for the SCR can be a piece of aluminum or copper, 4½ × 2 × ⅛ in. (11.43 × 5.08 × 0.32 cm).

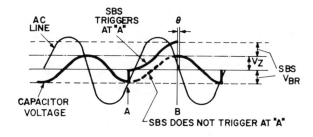

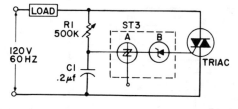

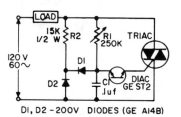

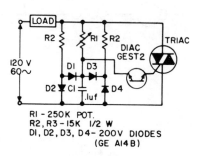

2.3.51 GE no-hysteresis dimmer has component economy. By replacing the diac with the silicon bilateral switch and a zener, a dissymmetry is introduced into the trigger circuit. The silicon bilateral switch and zener combination is designated ST3. The voltage swing of the capacitor per half-cycle, before the SBS triggers, is twice the switching voltage of the SBS. When the SBS switches at A (see waveforms), it allows the capacitor to discharge to approximately zero, since the SBS is basically a thyristor structure. When the triac commutates, the capacitor is again charged to twice the switching voltage. (If the zener is shorted, then θ would be 90°.) With the zener increasing the switching voltage for one polarity, the switching voltage is not reached until much later in the cycle; hence, the snap-on effect is greatly reduced.

2.3.52 GE no-hysteresis dimmer. The hysteresis effect is eliminated entirely by using one of the circuits shown, which resets the timing capacitor to the same level after each positive half-cycle, providing a uniform initial condition for the timing capacitor. If symmetry is required, the right-hand circuit may be used. This circuit is similar in operation to that at the left except that it resets on each half-cycle.

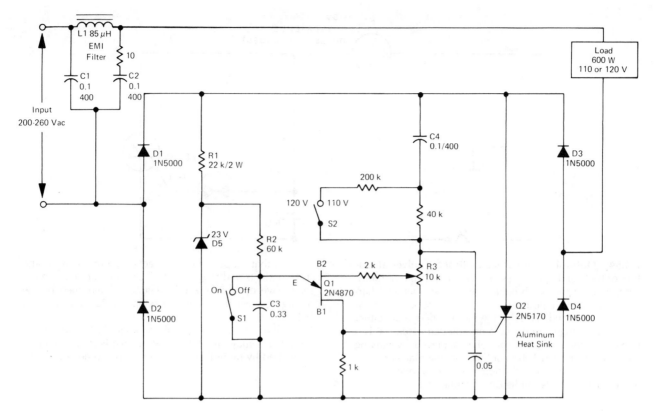

2.3.53 Motorola open-loop compensator for small conduction angles provides an output of 110 or 120V ±2.5V at 600W for an input voltage of 200 to 260V.

A full-wave bridge (D1 through D4) and a single SCR (Q2) are used to obtain full-wave control. A unijunction transistor (Q1) is the trigger device. Basic triggering frequency is determined by the charging and discharging of capacitor C3 through resistor R2. The supply for this circuit is regulated by zener D5.

Root-mean-square compensation is provided by capacitor C4 and the associated circuitry in the B2 circuit of Q1. As the input voltage increases, raising the interbase voltage of Q1, the required trigger voltage also increases, retarding the firing point of the SCR, since it takes longer for timing capacitor C3 to charge to this higher value. Potentiometer R3 allows some adjustment to compensate for differences in individual unijunction transistors so that the output can be set to the desired level.

Switch S1 can be used to turn the circuit on or off. A small switch here can control a relatively large load.

Switch S2 permits a selection of 110 or 12V output.

Choke L1 and capacitors C1 and C2 provide filtering to reduce electromagnetic interference (EMI).

This type of circuit is suitable primarily for applications requiring a conduction angle less than 90° and output voltages of about half the input. At greater angles premature firing of the unijunction might occur. Since at the beginning of each half-cycle the unijunction interbase voltage is zero, a very low trigger voltage will latch the unijunction on at the beginning of the half-cycle. To avoid this latchup, the charging rate of C3 must be sufficiently slow to allow buildup of the interbase voltage before triggering occurs. This requirement on the time constant of the charging circuit limits this type of circuit to small conduction angles. To operate over larger conduction angles, some type of delay on the charging would be required at the beginning of the half-cycle while the trigger voltage was still low.

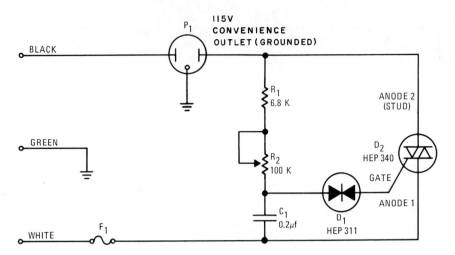

2.3.54 Motorola motor speed control for universal motors of tools and appliances. (If you are not sure of motor type, look for the brush holders or brushes that are characteristic of this type of motor. Also, if the label on the motor housing indicates that the motor will operate either from alternating or direct current, you can assume it is a universal motor.) Power control is achieved by varying the percentage (0 to 100%) of time that the triac conducts on each half-cycle. Conduction time is controlled by the trigger pulse that is applied to the triac gate. When the voltage that is applied to D1 reaches the breakover point, D1 conducts and triggers D2. The particular point on the half-wave cycle where this occurs is determined by the charging time of C1 and the setting of R2 in series with R1.

Long or constant use of the unit under high load-current conditions may cause the triac to overheat; this can be remedied by mounting D2 on an adequate heatsink (HEP 501).

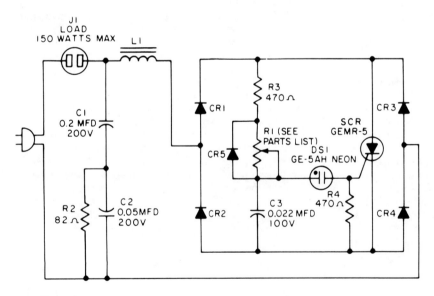

CR1 through
  CR5   GE-504A diode
L1   RF choke: 65 turns No. 18 varnished wire on ¼ in. ferrite rod (such as antenna loopstick rod)

2.3.55 GE dimmer and fan control is a multipurpose circuit that can be used to dim a high-intensity lamp or control the speed of a small fan (1.5A maximum nameplate rating). Operating on the phase-control principle, the circuit can give a 3:1 speed-control range with a shaded-pole fan motor or provide complete on and off for a high intensity lamp.

108   DISCRETE / TRANSISTOR CIRCUIT SOURCEMASTER

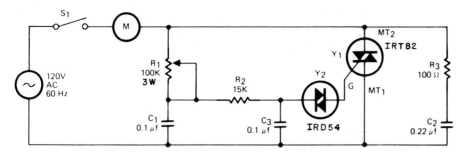

2.3.56 Workman motor speed controller provides the delay necessary to trigger a triac at very low conduction angles and provides practically full power to the load at the minimum-resistance position of the control potentiometer. When this type of control circuit is used, instant range of motor speeds can be obtained from very low- to full-power applications. Triac Y1 is gate-controlled by diac Y2. The voltage across Y2 is adjusted by potentiometer R1, which is the only control in the system. Resistor R2 controls the surge across Y1, and capacitors C1 and C3 maintain proper current. The snubber system (R3 and C2) protects the triac and the controlling system.

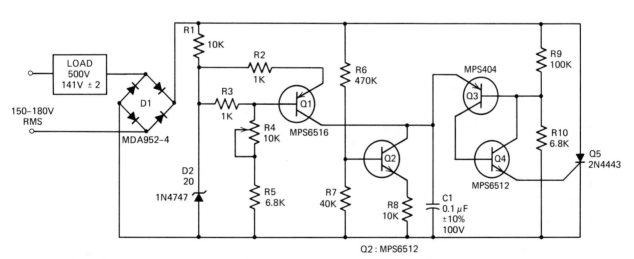

Q2 : MPS6512

2.3.57 Motorola open-loop compensator for large conduction angles provides an output of 500W at 141V rms (±2V) with an input of 150 to 182V. Transistors Q3 and Q4 and resistors R9 and R10 are connected to form a composite device with characteristics similar to a unijunction transistor. The trigger voltage is equal to the voltage drop across R10 plus the emitter–base voltage drop of Q3. As soon as Q3 turns on, Q4 also turns on, providing a discharge path between the capacitor and the gate of the SCR. The resulting discharge current pulse triggers the thyristor.

Latchup is prevented by delaying the turnon of current-source transistor Q1 until the trigger voltage has reached a sufficiently high value. Prior to the conduction of zener D2, the emitter–base voltage of transistor Q1 is determined by the resistive divider comprised of R1, R3, R4, and R5. Since it takes approximately 0.6V across the base–emitter junction to turn a transistor on, the line voltage will be approximately 14V before Q1 begins conduction and charging C1. This delay provides sufficient time to raise the trigger voltage above a value that would cause latchup.

SECTION **2** Control and Sensing

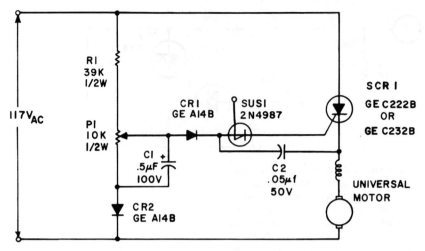

2.3.58 GE universal motor speed control with feedback triggered by silicon unilateral switch (SUS), has good performance at low and high speeds. Because of the higher impedance offered by R1 and P1, the ramp charge on capacitor C1 provides a stable triggering point for the SUS. Since the triggering of SCR1 is accomplished by a pulse rather than continuous current through CR1, this circuit does not require different component values for R1 and P1 with different SCRs and no restriction is necessary on the gate triggering current of the SCR1.

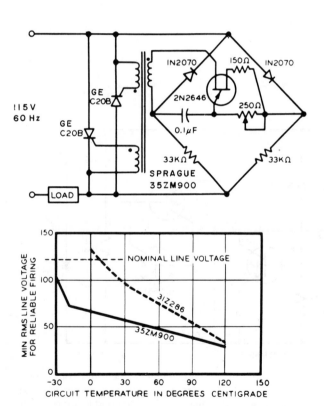

2.3.59 Sprague phase-control circuit employing SCRs and a trigger transformer. The inset chart shows minimum line voltage for reliable firing of the SCRs as a function of circuit temperature in degrees Celsius. The transformer is a Sprague 35ZM900, but the 31Z286 can also be used; the chart compares the performance of the two. Note that the latter transformer cannot trigger the SCRs at very low temperatures.

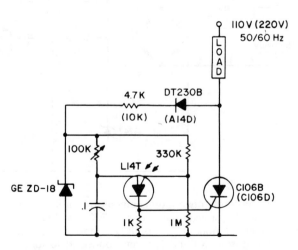

2.3.60 GE phase-control circuit provides 100% isolation by using a light-sensitive element. This half-wave arrangement affords control over the 0 to 45° portion of the 115V cycle. The 100K potentiometer is set so that the L14T triggers at 170 to 180°. Then, as the light increases, the delay angle can be shortened to 10°, thereby varying the power from 0 to 45%.

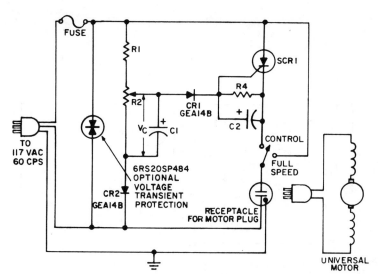

	LOW UP TO 1 AMP NAMEPLATE	MEDIUM UP TO 3 AMPS NAMEPLATE	HIGH UP TO 5 AMPS NAMEPLATE
R2	10K, 1W	1K, 2W	1K, 2W
R1	47K, 1/2W	3.3K, 2W	3.3K, 2W
R4	1K, 1/2W	150, 1/2W OPTIONAL	150, 1/2W OPTIONAL
C1	.5μf, 10V	10μf, 10V	10μf, 10V
C2	1μf, 10V	.1μf, 10V OPTIONAL	.1μf, 10V OPTIONAL
SCR1	GE C106B	GE C22B12 OR C122B12	GE C33B

2.3.61 GE plug-in speed control for standard portable tools and appliances uses a simple half-wave SCR phaser. The main advantage of this circuit over others lies in the fact that no rewiring of the motor is necessary. The circuit uses the counter emf of the motor armature as a feedback signal of motor speed to maintain essentially constant speed with varying torque requirements. There will be some variation in the effectiveness of speed control from one motor to another, depending on the magnitude of the residual field for the particular motor.

During the positive half-cycle of the supply voltage, a reference voltage is established on the arm of potentiometer R2, which is compared with the counter emf of the motor through the gate of the SCR. When the pot voltage rises above the counter emf, current flows through CR1 into the gate of the SCR, triggering SCR, and thus applying the remainder of that half-cycle of supply voltage to the motor. If load is applied to the motor, its speed tends to decrease, thus decreasing counter emf in proportion to speed. The pot reference voltage thus causes current to flow into the SCR gate earlier in the cycle. The SCR triggers earlier in the cycle, and additional voltage is applied to the armature to compensate for the increased load and to maintain the preset speed. The particular speed at which the motor operates can be selected by R2. Stable operation is possible over approximately a 10:1 speed range. Stability at very low speeds can be improved by reducing the value of C1 at the expense of feedback gain.

Normal operation at maximum speed can be achieved by closing switch S1, thus bypassing the SCR. Rectifier CR1 prevents excessive reverse voltage on the gate of SCR. CR2 prevents the inductive field current in the motor from free-wheeling in the SCR gate circuit and is necessary to establish an effective reference voltage. R4 also improves stability by bypassing the commutator hash around the gate of the SCR and by reducing thermal effects on the triggering sensitivity of the SCR. The doublethrow action on the control switch prevents voltage transients from being applied to SCR when the switch breaks motor current in the full-speed position.

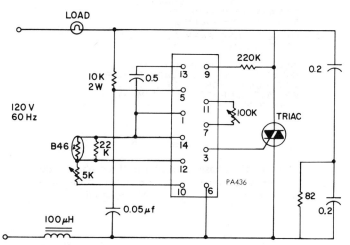

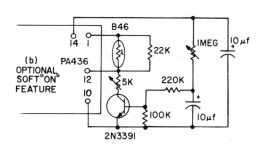

2.3.62 GE high-performance feedback lamp control incorporates GE's PA436 integrated phase-control circuit. In this circuit a simple light feedback network with a level set is shown, but other sensors may be used. The inset shows a soft-on circuit for this system.

SECTION 2 *Control and Sensing* **111**

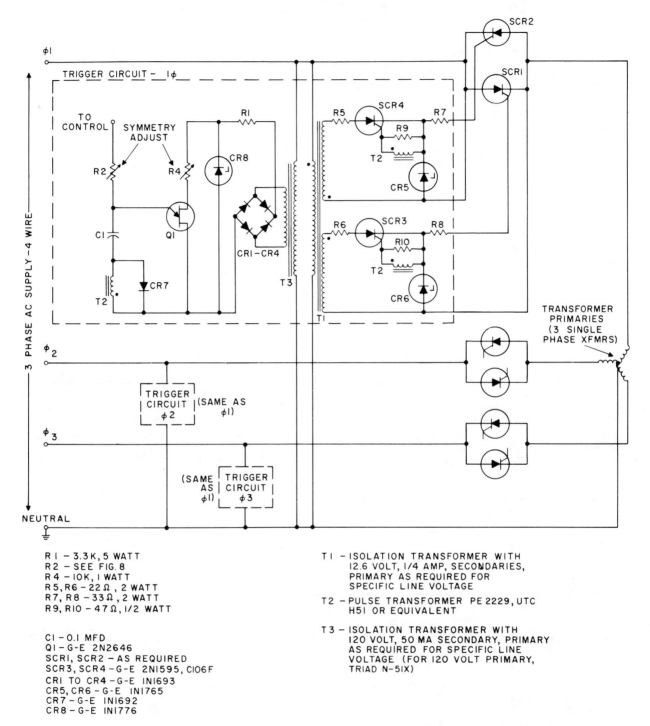

R1 – 3.3K, 5 WATT
R2 – SEE FIG. 8
R4 – 10K, 1 WATT
R5, R6 – 22Ω, 2 WATT
R7, R8 – 33Ω, 2 WATT
R9, R10 – 47Ω, 1/2 WATT

C1 – 0.1 MFD
Q1 – G-E 2N2646
SCR1, SCR2 – AS REQUIRED
SCR3, SCR4 – G-E 2N1595, C106F
CR1 TO CR4 – G-E 1N1693
CR5, CR6 – G-E 1N1765
CR7 – G-E 1N1692
CR8 – G-E 1N1776

T1 – ISOLATION TRANSFORMER WITH 12.6 VOLT, 1/4 AMP, SECONDARIES, PRIMARY AS REQUIRED FOR SPECIFIC LINE VOLTAGE

T2 – PULSE TRANSFORMER PE 2229, UTC H51 OR EQUIVALENT

T3 – ISOLATION TRANSFORMER WITH 120 VOLT, 50 MA SECONDARY, PRIMARY AS REQUIRED FOR SPECIFIC LINE VOLTAGE (FOR 120 VOLT PRIMARY, TRIAD N-51X)

2.3.63 GE phase control for three-phase, four-wire power system feeding wye-connected transformers. Trigger circuits shown as insets (A) through (E).

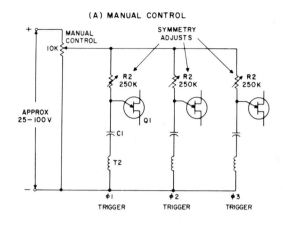
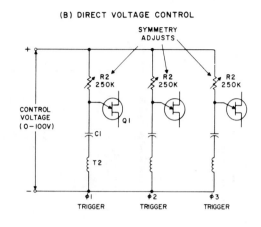
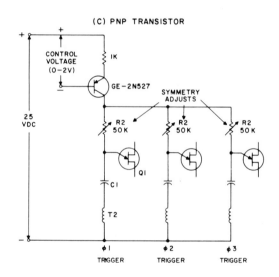
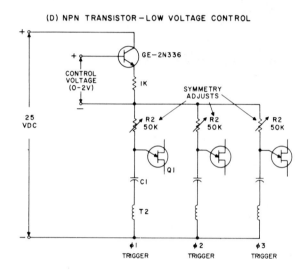
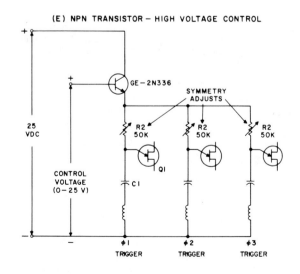

SECTION 2 Control and Sensing

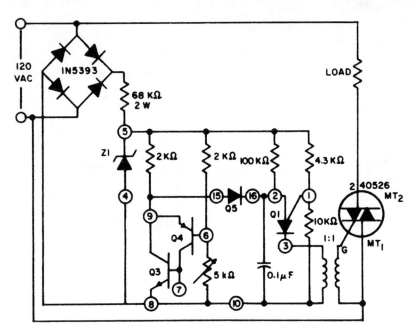

2.3.64 RCA phase-control circuit provides motor speed control, lamp dimming, etc. Pin numbers are CA3097E. To use Q5 as a diode, short terminal 14 to 15.

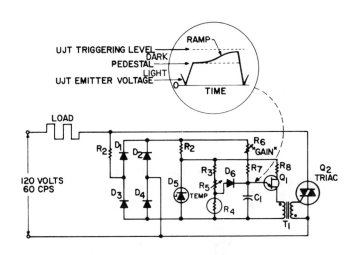

R_1, R_2 = 2200 OHMS, 2 WATTS
R_3 = 2200 OHMS, 1/2 W
R_4 = PHOTO CELL, APPROX 5000 OHMS AT NORMAL OPERATING LEVEL
R_5 = 10,000 OHMS W.W. POTENTIOMETER
R_6 = 5 MEGOHM POTENTIOMETER
R_7 = 100 KΩ, 1/2 W.
R_8 = 1000 OHMS, 1/2 W.

Q_1 = 2N2646
Q_2 = AS REQ'D (CHOOSE FOR CURRENT REQ)
T_1 = SPRAGUE 11Z12 OR EQUIVALENT
D_{1-4} = GE A14B
D_5 = GE Z4XL22
D_6 = GE A14A
C_1 = 0.1 μfd, 30 V.

2.3.65 GE precision proportional lighting control. In systems where feedback and symmetry are of great importance, this circuit should do the job. The diode bridge (D1–D4) rectifies the ac line so that a unijunction ramp and pedestal relaxation oscillator may be used as the trigger circuit. A photocell (R4) and a control pot (R5) form a resistive pedestal control, which varies as the light output. When the light level increases, the pedestal decreases, decreasing the power to the load and hence regulating the light output. Since the trigger circuit is unilateral (all the trigger pulses are positive on the primary of the pulse transformer), the pulse transformer serves to invert the trigger pulse to the more sensitive triac trigger modes and also provides isolation between the trigger circuit and triac.

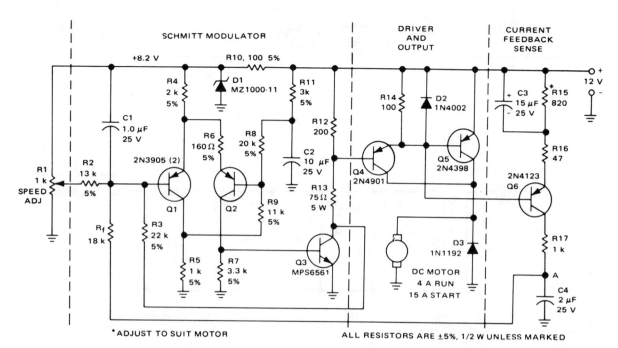

2.3.66 Motorola pulse-width-modulated dc motor speed control incorporating voltage-sensing feedback operates from a 12V source and is capable of driving motors with inrush currents up to 10A. The maximum allowable running current will obviously be less than 10A and will depend to a great extent on the heatsink provided for Q5.

The modulated waveform is provided by the Schmitt trigger consisting of Q1 and Q2, phase-inversion stage Q3, and the delayed feedback through R3 and C1. The output is a variable-width, variable-frequency pulse whose duty cycle and frequency are a function of the dc input. The dc input is the summation of the current through R2, which is connected to speed-adjust potentiometer R1, and the current through R_f, the overall feedback resistor.

The output of the modulator is fed to a Darlington-connected power-amplifier stage consisting of Q4 and Q5, which drives the dc motor. Free-wheeling diode D3 suppresses the inductive kickback of the motor. Diode D2 protects the base–emitter junction of Q5 against reverse breakdown due to voltage transients generated by the motor.

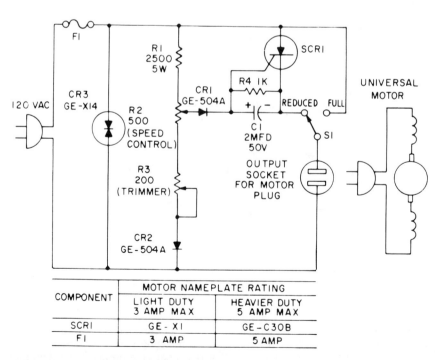

COMPONENT	MOTOR NAMEPLATE RATING	
	LIGHT DUTY 3 AMP MAX	HEAVIER DUTY 5 AMP MAX
SCR1	GE-X1	GE-C30B
F1	3 AMP	5 AMP

2.3.67 GE plug-in speed control for tools or appliances, such as drills, sewing machines, saber saws, food mixers and blenders, movie projectors, fans, etc.

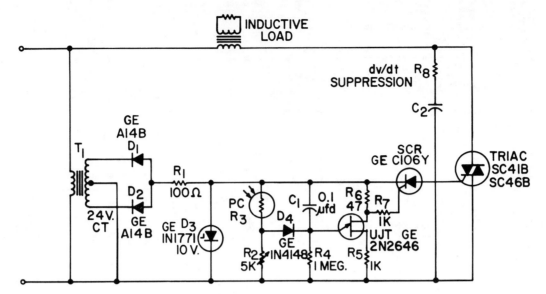

2.3.68 GE ramp-and-pedestal phase control for inductive loads. In this circuit a rectified supply voltage is used with a small pilot SCR to give a continuous drive to the gate following the triggering time. This holds the triac on until the current in the load is high enough for the triac to latch. The phase reference for the firing circuit is the line voltage, not the voltage across the triac. For heavily inductive loads, taking the timing reference from the triac can result in a dc component in the load. For further discussion on this and related problems, see General Electric application note 200.31.

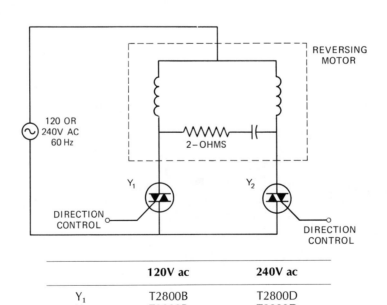

2.3.69 RCA reversing motor control uses two triacs. The reversing switch can be either a manual or an electronic switch used with some type of sensor to reverse the direction of the motor. A resistance is added in series with the capacitor to limit capacitor discharge current to a safe value whenever both triacs are conducting simultaneously. Simultaneous conduction can easily occur because the triggered triac remains in conduction after the gate is disconnected *until the current reduces to zero*. In the meantime, the nonconducting triac gate circuit can be energized so that both triacs are on and large loop currents are set up in the triacs by the discharge of the capacitor.

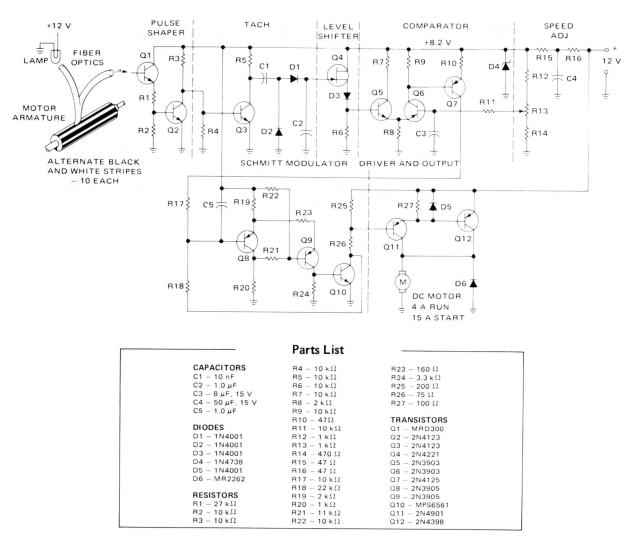

2.3.70 Motorola regulated dc motor control with feedback from optical pickoff.

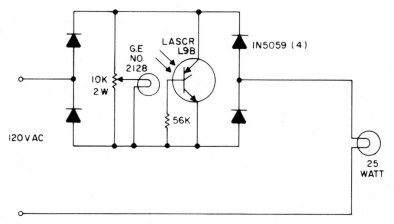

2.3.71 World's smallest phase control, developed by General Electric, provides ideal controllable power variation for lamps, soldering irons, and small heating elements. The GE 2128 miniature lamp has a low-mass filament with a short delay time compared with most lamps. With a low applied voltage, the time to reach the LASCR firing level for the lamp is about three cycles. As the applied voltage is increased, the time is reduced, reaching about 1 ms when directly across the LASCR terminals (thus providing phase control of the LASCR). The lamp voltage is removed when the LASCR fires, protecting the lamp and resetting it for the next half-cycle. The lamp and LASCR should be in direct contact.

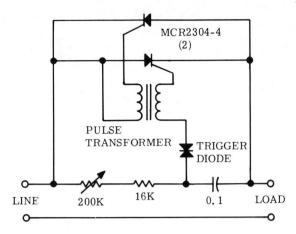

2.3.72 Motorola simple full-wave power control circuit uses the bidirectional three-layer trigger (diac) and the built-in gate shunt resistor of the MCR2304 series of SCRs, which allow triggering in both half-cycles.

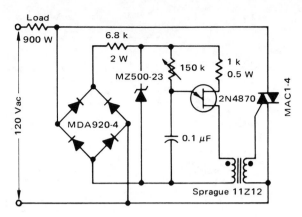

2.3.74 Motorola full-wave trigger circuit for a 900W resistive load.

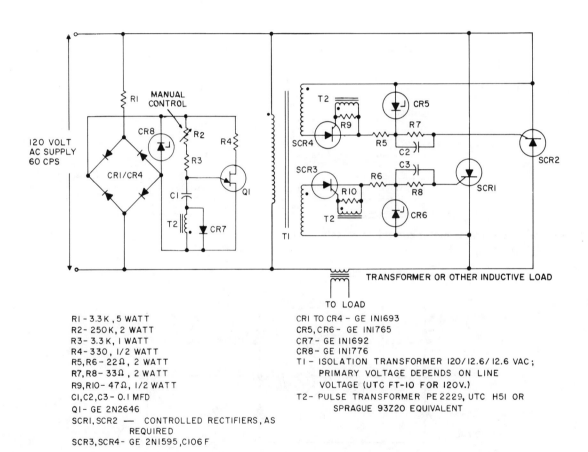

R1 - 3.3K, 5 WATT
R2 - 250K, 2 WATT
R3 - 3.3K, 1 WATT
R4 - 330, 1/2 WATT
R5, R6 - 22Ω, 2 WATT
R7, R8 - 33Ω, 2 WATT
R9, R10 - 47Ω, 1/2 WATT
C1, C2, C3 - 0.1 MFD
Q1 - GE 2N2646
SCR1, SCR2 — CONTROLLED RECTIFIERS, AS REQUIRED
SCR3, SCR4 - GE 2N1595, C106 F

CR1 TO CR4 - GE 1N1693
CR5, CR6 - GE 1N1765
CR7 - GE 1N1692
CR8 - GE 1N1776
T1 - ISOLATION TRANSFORMER 120/12.6/12.6 VAC; PRIMARY VOLTAGE DEPENDS ON LINE VOLTAGE (UTC FT-10 FOR 120V.)
T2 - PULSE TRANSFORMER PE 2229, UTC H51 OR SPRAGUE 93Z20 EQUIVALENT

2.3.73 GE semiconductor nonmagnetic trigger circuit has characteristics similar to the magnetic trigger, but operation can be achieved with less space and weight and lower cost. This circuit is adequate for triggering the largest SCRs available.

Unijunction transistor Q1 is connected across the ac supply line by means of the bridge, thus permitting Q1 to trigger on both halves of the ac cycle. The time constant of R2 in conjunction with C1 determines the delay angle at which the unijunction delivers its first pulse to the primary of pulse transformer T2 during each half-cycle. These pulses are coupled directly to the gates of SCR3 and SCR4. Whichever of these SCRs has a positive anode voltage during that specific half-cycle triggers and delivers voltage to its respective main SCR, firing it. The low-voltage ac supply for the pilot SCRs (SCR3 and SCR4) is derived from a filament transformer. Zener diodes CR5 and CR6, in conjunction with resistors R5 and R6, clip the ac gate voltage to prevent excessive power dissipation in the gates of the main SCRs. The RC networks (R7–C2 and R8–C3) also limit gate dissipation in the main SCRs while delivering a momentarily higher gate pulse at the beginning of the conduction period to accelerate the switching action in the main SCRs.

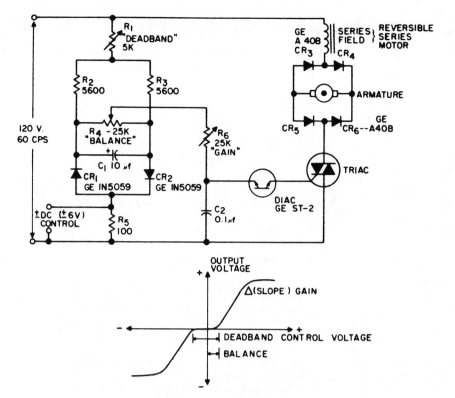

2.3.75 GE reversing motor drive with dc control signal is a positioning servo drive featuring adjustment of balance, gain, and deadband. Mechanical input can be fed into the balance control, or that control can be replaced by a pair of resistance transducers for control by light or by temperature.

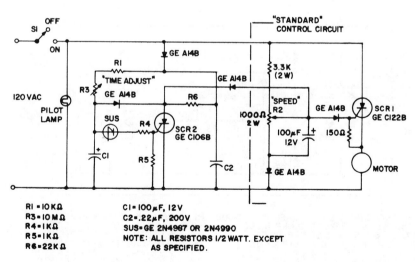

2.3.76 GE self-timing blender motor speed control shuts off the motor at the end of a timed cycle. The automatic shutoff feature is provided by the circuitry shown to the left of the dashed line. When switch S1 is closed, timing capacitor C1 starts charging at a variable rate depending on the set value of R3. When the voltage on the capacitor reaches the forward breakover voltage of the SUS, the latter switches on, applying a current pulse to the gate of SCR2, which brings it into conduction. The R6–C2 network provides SCR2 with enough latching current to keep it on thereafter. SCR2 then accomplishes two functions; it shunts the gate current away from the power SCR1, thus turning it off, which deenergizes the load. It also resets capacitor C1 so that in case of rapid successive load operations it will provide essentially the same time-delayed turnoff. The pilot light remains on until the main power switch is turned off, providing an added safety feature. The maximum time delay obtainable with this circuit is in excess of 1 min.

SECTION **2** Control and Sensing   **119**

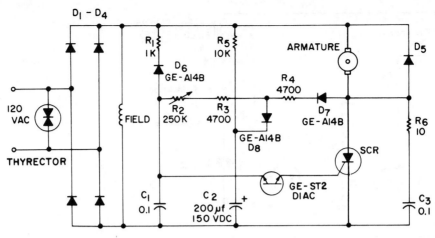

NOTE: MOTOR SIZE DETERMINES TYPES SCR, D_1-D_5, & THYRECTOR

2.3.77 GE shunt motor speed control provides a soft-start function by the charging of C2 when the circuit is initially energized. The charging current for C2 passing through D7, R4, and D8 causes the voltage applied to the charging circuit for C1 to increase slowly. As a result, the SCR triggering phase angle starts at about 170° after a brief delay and then advances to its normal setting over a period of several cycles. Resistor R5 completes the charging of C2 to the average value of the bridge to prevent any further interference with the action of the circuit after the starting period. Resistor R5 also discharges capacitor C2 when the circuit is deenergized.

The thyrector connected across the input power lines suppresses voltage transients that could damage the semiconductors of the circuit. In addition, resistor R6 and capacitor C3 are connected in parallel with the SCR to limit the rate at which the voltage can appear across the SCR after it turns off. If this voltage appears too rapidly, the SCR may not have sufficient time to completely turn off; it will fail to commutate and will produce full power on the armature. At high motor speeds the counter emf of the motor subtracting from the rectifier output voltage increases the time available for the SCR to turn off. If R6 and C3 are not used or are inadequate for the particular motor, a low-speed setting may result in abrupt speed fluctuations.

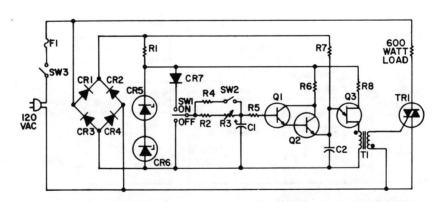

CR1 THRU CR4 : G-E (IN5059) RECTIFIER DIODE	R1 : 3.3 K OHM, 2 WATT RESISTOR
CR5, CR6 : G-E Z4XL7.5 ZENER DIODE	R2, R4: 4.7K OHM, 1/2 WATT RESISTOR
CR7 : G-E (IN5059) RECTIFIER DIODE	R3 : 5 MEGOHM, 1 WATT POTENTIOMETER
C1 : 100 μf, 15 WVDC ELECTROLYTIC CAPACITOR (G-E QT1-22)	R5, R7: 1 MEGOHM, 1/2 WATT RESISTOR
C2 : 0.1 μf, 15 WVDC CAPACITOR	R6 : 2.2 K OHM, 1/2 WATT RESISTOR
Q1, Q2: G-E (2N5172) n-p-n TRANSISTOR	R8 : 470 OHM, 1/2 WATT RESISTOR
Q3 : G-E 2N2647 UNIJUNCTION TRANSISTOR	SW1 : SPDT SWITCH
TR1 : TRIAC	SW2 : SPST SWITCH
F1 : AS REQUIRED	T1 : SPRAGUE 11Z12 PULSE TRANSFORMER

2.3.78 GE slow-on, slow-off, time-dependent phase-control circuit for applications where it is necessary to turn lamps on or off slowly. The circuit is a full-wave ramp-and-pedestal control. The ramp is obtained from resistor R7, and the pedestal height is controlled by the dc voltage across C1 through a Darlington follower. The high input impedance of the Darlington connection permits very long charging and discharging time for the capacitor, which produces a very slow turnon and turnoff of the lamp load. At maximum time setting, approximately 20 min is required to make the full transition from on to off.

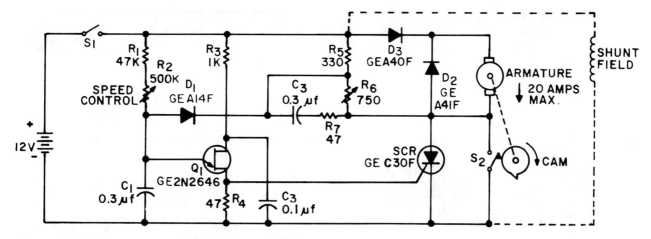

2.3.79 GE synchronous control for shunt or permanent-magnet motors. Rectifier D3 decouples the counter emf of the armature from the SCR during the off condition to prevent interference with the delay link. In the off state, the voltage across the SCR is the supply voltage instead of the supply minus the counter emf.

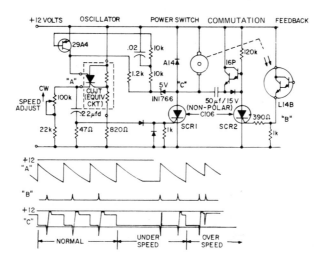

2.3.80 GE synchronous dc motor speed control for portable home and auto entertainment products such as phonographs and tape recorders. The stability should be good in the normal outdoor temperature range. The circuit combines the excellent stability of the CUJT (complementary unijunction transistor) with the good performance of the "synchronous dc motor speed control" presented in General Electric application note 200.43. Here the speed is synchronized to the speed of a free-running relaxation oscillator by means of turnoff feedback pulses from the motor shaft. In this case, light reflection off a disc on the motor is sensed by the L14B. When the 0.22 μF capacitor discharges, it triggers SCR1. This applies dc motor power. As the shaft rotates, light picked up causes SCR2 to turn on, commutating SCR1 off via the 50 μF cross-coupling capacitor. Meanwhile, the 0.22 μF capacitor has been charging toward the peak point and again fires, turning on SCR1. The process repeats itself.

Hence the motor is synchronous with the oscillator. The gate terminal of the CUJT provides the over/under speed regulation. If the motor is running underspeed, it will not have been commutated off before the timing capacitor reaches the peak point, V_P, so that the base current will flow out of the PNP base, thereby raising the peak point current to 20 mA. When the motor is finally commutated, the 0.22 μF capacitor is allowed to discharge, turning SCR1 on immediately. This continues until the motor is up to synchronous speed. In overspeed, feedback pulses arrive before the 0.22 μF is at F_P. In this case the commutation pulse pretriggers the CUJT via the gate current through the 1.2K gate resistor. The diodes at SCR1 gate prevent the negative commutation pulses from SCR2 to be coupled back through to the 0.22 μF capacitor. The 5V zener is used to decouple the motor's residual voltage. Speed is adjustable with the 100K pot.

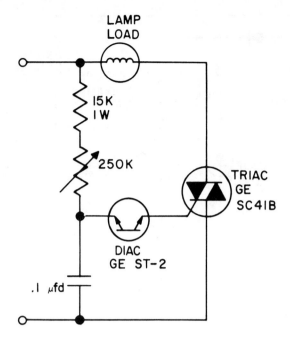

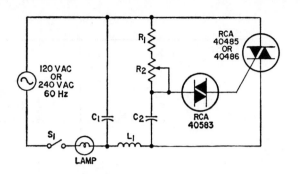

2.3.81 GE table-lamp dimmer with reduced hysteresis can be installed in a very small space. This circuit takes advantage of the availability of both sides of the lamp load. The hysteresis effect is reduced substantially as a result of the charging current that flows into the capacitor after triggering. This residual current acts to replace the pulse charge so that the initial firing point remains close to the lamp turnoff point.

PARTS LIST

Part	Value for 120V, 60 Hz	Value for 240V, 50/60 Hz
C1	0.1 μF, 200V	0.1 μF, 400V
C2	0.1 μF, 200V	0.05 μF, 400V
L1	100 μH	200 μH
R1	3.3K, ½W	4.7K, ½W
R2	250K, ½W	250K, 1W

2.3.83 RCA triac light dimmer provides full-wave control of incandescent lamps. For 120V operation, the 40485 triac is recommended; for 240V operation, the higher-power 40486 triac should be used. A 40583 diac, together with associated resistance–capacitance time-constant networks, is used to develop the gate current pulses that trigger the triac into conduction. When space is at a premium, the triac and diac may be replaced by the 40431 (the 40432 for 240V).

During the beginning of each half-cycle of the input, the triac is off. As a result, the entire line voltage appears across the triac, and the lamp is not lighted. The entire line voltage, however, is also across the RC network that parallels the triac, and this voltage charges the capacitor(s). When the voltage across C2 rises to the diac breakover voltage, the diac conducts and the capacitor discharges through the diac and the triac gate to trigger the triac. Potentiometer R2 is adjusted to control the brightness of the incandescent lamp. If the resistance of the potentiometer is decreased, the trigger capacitor charges more rapidly and the breakover voltage of the diac is reached earlier in the cycle; power is thus applied to the lamp, increasing the intensity of light.

Capacitor C1 and inductor L1 form an RFI suppression network, which minimizes the high-frequency transients generated by the rapid on-and-off switching of the triac so that these transients do not produce noise interference in nearby electrical equipment.

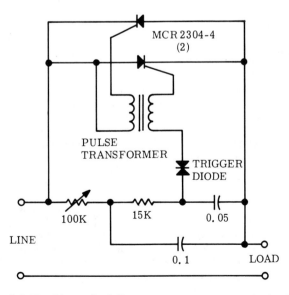

2.3.82 Motorola full-range ac power control circuit uses a double phase-shift network to obtain reliable triggering at conduction angles as low as 5°.

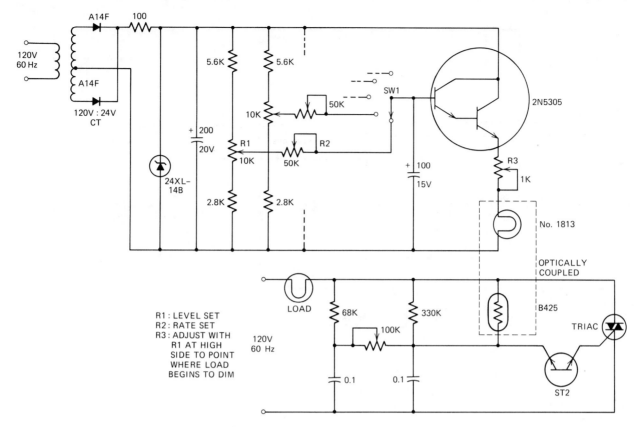

2.3.84 GE theatrical lighting dimmer employs low-voltage remote control with independent presets and fade rate controls. In theatrical lighting it is necessary to return at will to a given light level. This circuit offers as many presets as desired, each with its own control to adjust the rate at which the light level changes. This circuit has a low-voltage remote-control section that can be assembled in a box positioned several hundred feet from the actual power control circuit. This is accomplished by means of an emitter follower that is controlled by the preset. The load of the follower is an 1813 lamp that is optically coupled to a photocell. The photocell replaces the potentiometer in an extended range phase control.

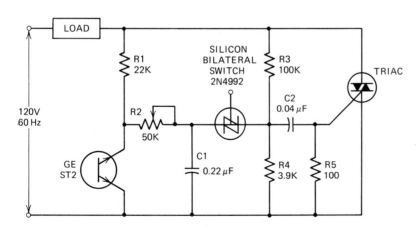

2.3.85 GE line-voltage-compensated phase-control system. In this circuit the ST2 is used as a clamp to keep the charging rate on C1 constant. The feedback is from resistor divider R3–R4. As the line voltage increases, the voltage of the capacitor C1 is offset by its voltage across R4 to delay the firing of the 2N4992. C2 provides low-frequency decoupling of the gate from the resistor divider.

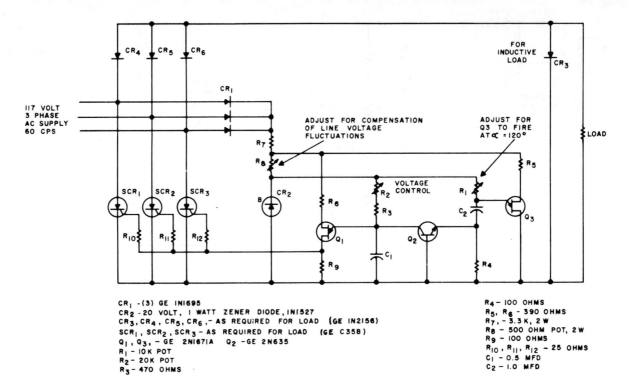

2.3.86 GE three-phase control circuit for variable dc output retains the simplicity of the single-phase firing circuit by preventing any gate signal from being applied to the SCRs when angle α is greater than 120°. Thus, if the control signal calls for a voltage less than 25% of maximum, the dc output voltage drops to zero. When the dc voltage is raised from zero, it jumps abruptly from zero to 25% and then rises without steps above that value.

CR1 supplies positive line voltage to the control circuit whenever the anode voltage on an SCR swings positive with respect to the positive dc bus. This voltage is clipped at 20V by zener CR2 and supplies a conventional unijunction transistor relaxation oscillator firing circuit. R2 controls the firing angle of Q1 by regulating the charging rate to C1. The pulse developed across R9 as unijunction Q1 discharges C1 is coupled to the gates of SCR1, SCR2, and SCR3, through R10, R11, and R12. Whichever SCR has the most positive anode voltage at the instant of the gate pulse starts conduction at that point.

Transistors Q2 and Q3 prevent Q1 from firing at any angle greater than 120°. Q3 is an independent unijunction oscillator that initiates its timing cycle at the same instant as Q1. R1 is set at a fixed value so that Q3 fires at an angle slightly less than 120°. If Q1 triggers before 120°, it fires the SCR whose positive anode voltage is providing the interbase bias for the unijunctions through CR1. Firing this SCR short-circuits this control circuit supply voltage. The interbase bias voltage of Q3 drops to zero, causing Q3 to fire and discharge C2 in preparation for the next cycle. This is the mode of operation when Q1 is controlling the dc output voltage between 25 and 100% maximum. In this mode, Q2 and Q3 have no effect on the functioning of the bridge. If Q1 is delayed beyond 120°, Q3 fires; discharging C2 through the base–emitter junction of Q2, saturating this device; and discharging C1 through Q2. This alternate mode of discharging C1 does not impose a pulse on the SCR gates; the dc output voltage is therefore zero in this mode.

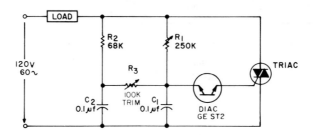

2.3.87 GE widened-range dimmer/phase control. The addition of a second RC phase-shift network extends the range of control and reduces hysteresis to a negligible region. This GE circuit will control from 5 to 95% power in the load but is subject to supply-voltage variations. When R1 is large, C1 is charged through R3 from the phase-shifted voltage appearing across C2. This action not only provides an additional range of phase shift across C1 but enables C2 to partially recharge C1 after the diac has triggered, thus reducing hysteresis. R3 should be adjusted so that the circuit just drops out of conduction when R1 is brought to maximum resistance.

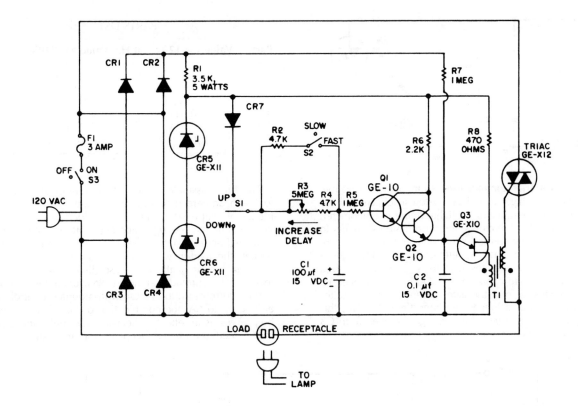

Parts List

C1	100 μF 15V electrolytic capacitor
C2	0.1 μF 15V capacitor
CR1–CR4	GE-504A rectifier diode
CR5, CR6	GE-X11 zener diode
CR7	GE-504A rectifier diode
F1	3A fuse
Q1, Q2	GE-10 transistor
Q3	GE-X10 unijunction transistor
R1	3500-ohm 5W resistor
R2, R4	4700-ohm ½W resistor
R3	5M ½W resistor
R5, R7	1M ½W resistor
R6	2200-ohm ½W resistor
R8	470-ohm ½W resistor
S1	SPDT toggle switch
S2, S3	SPST toggle switch
Triac	GE-X12
T1	Pulse transformer—available from GE distributors as ETRS-4898, or from General Electric Co., Dept. B, 3800 N. Milwaukee Ave., Chicago, Ill. 60641.

2.3.88 GE time-dependent lamp dimmer gives "soft-on," "soft-off" operation that automatically increases or decreases the brightness of a lamp over an adjustable period of time. The trigger circuit is designed so that a time-dependent output is obtained after initially energizing the circuit. Slow turnon or turnoff is obtained after the position of switch S1 is changed. When the switch is placed in the up position, capacitor C1 begins to charge through R2 and R3. For time periods shortly after switching, the capacitor voltage is low. This holds the base of Q1 down and thus the emitter of Q2 is held at a low voltage below the peak-point voltage on the unijunction transistor. Simultaneously, C2 is charged during each half-cycle through R7. The time constant of R7–C2 is rather long compared to a half-cycle of the line voltage. This time constant is selected so that the capacitor voltage just barely reaches the peak-point voltage at the end of the half-cycle with zero voltage on C1. As the voltage on C1 rises, the voltage on C2 also rises and the R7–C2 charging curve starts from a slightly higher voltage at each cycle. This means that the voltage on C2 reaches the peak-point voltage of the unijunction transistor slightly earlier during each cycle, thus gently increasing the output. The double emitter-follower configuration (Q1–Q2) provides an extremely high impedance so that the charging and discharge currents to C1 are not shunted away from it.

When switch S1 is moved to the down position, capacitor C1 discharges through R2 and R3. The operation proceeds as before but in reverse.

SECTION **2** *Control and Sensing*

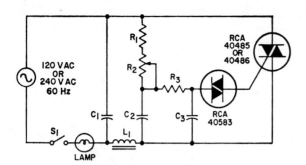

PARTS LIST

Part	Value for 120V, 60 Hz	Value for 240V, 60 Hz
C1	0.1 µF, 200V	0.1 µF, 400V
C2	0.1 µF, 200V	0.05 µF, 400V
C3	0.1 µF, 100V	0.1 µF, 100V
L1	100 µH	100 µH
R1	1K, ½W	7.5K, 2W
R2	100K, ½W	200K, 1W

2.3.89 RCA triac dimmer uses a circuit with a double time constant trigger to reduce hysteresis effects and extend the effective range of the light-control potentiometer. As applied to light dimmers, the term *hysteresis* refers to a difference in the control-potentiometer setting at which the lamp turns on and the setting at which the light is extinguished. C2 reduces hysteresis by charging to a higher voltage than capacitor C3. During gate triggering, C3 discharges to form the gate current pulse. Capacitor C2, however, has a longer discharge time constant; this capacitor restores some of the charge removed from C3 by the gate current pulse.

The triac dissipates power at the rate of about a watt for each ampere of controlled current; therefore, some means of heat removal must be provided to keep the device within its safe operating-temperature range. On a small light-control circuit, such as one built into a lamp socket, the lead-in wire serves as an effective heatsink. The attachment of the triac case directly to one of the lead-in wires provides sufficient heat dissipation for operating currents up to 2A. On wall-mounted controls operating up to 6A, the combination of faceplate and wall box serves as an effective heatsink. (Do not connect the triac directly to a metal faceplate; use an insulator.)

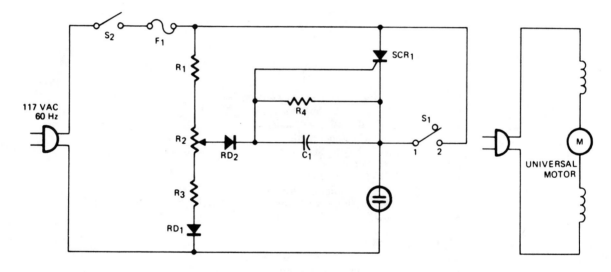

MOTOR SPEED CONTROL PARTS LIST

	Up to 2A Load	Up to 5A Load
R_1	10K, 2W	5K, 2W
F_1	3A	10A
SCR_1	IR122B	SCR-04

RD_1, RD_2	Rectifiers (IR-5A4D)
R_3	200 Ohm, 1 Watt
R_2	1.5K Ohm Potentiometer, 2W
R_4	1000 Ohm, 1 Watt
C_1	2 µF, 50 Volt Electrolytic Capacitor
	Plug and Socket

2.3.90 Workman universal motor speed control applicable to food mixers and blenders, small power shop equipment, and home movie projectors. The speed variation is controlled from zero to approximately 80% of full speed by varying R2. For 100% full speed, switch S1 is turned to position 2, circumventing the SCR control. Two sets of components are specified—one for up to a 2A load, the second for up to a 5A load. Check the label of the equipment for the correct settings. If there is a question, use the 5A system; it will operate 2A without any trouble at all. This is a 120V circuit.

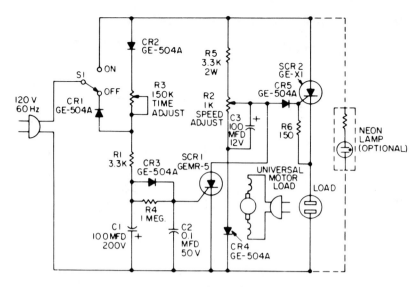

2.3.91 GE universal motor control with built-in self-timer. Constructed from miniature components, this circuit should prove ideal for kitchen blenders.

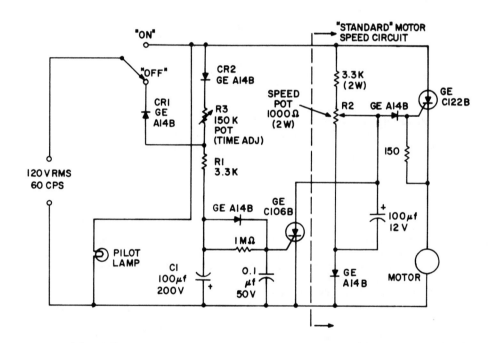

2.3.92 GE universal motor speed control has an electronic timing feature added to shut the motor off automatically at the end of a predetermined time cycle. Both motor speed and time delay are adjustable independently. The motor circuit is provided with speed-dependent feedback that gives excellent torque characteristics to the motor, especially at low rotational speeds where conventional rheostat or variable transformer controls are ineffective. Speed is adjustable via potentiometer R2. With the components shown, the control can handle motors with nameplate rating of up to 2A; by substituting the C32B SCR, heavier duty appliances can be accommodated.

The automatic shutoff feature is derived from circuitry associated with the C106B SCR. The timer SCR diverts gate current from the main speed control SCR on completion of a time cycle. With the main power switch off, the motor is disconnected from the ac line, but capacitor C1 charges to the peak of the line through rectifier CR1 and resistor R1. When the motor is turned on by the power switch, the pilot light comes on and capacitor C1 starts to discharge through resistors R3 and R1 and rectifier CR2. Since the time constant associated with this RC network is numerically long and discharge current can only flow for a short period each cycle, C1 takes many complete cycles of alternating current to discharge. When C1 is finally discharged and starts to charge in the opposition direction, the timer SCR triggers and diverts the gate current from the main SCR. At this point the motor stops. For safety, the pilot light remains on until the main power switch is turned off. The maximum time delay obtainable with this circuit is in excess of 30s.

SECTION 2 Control and Sensing

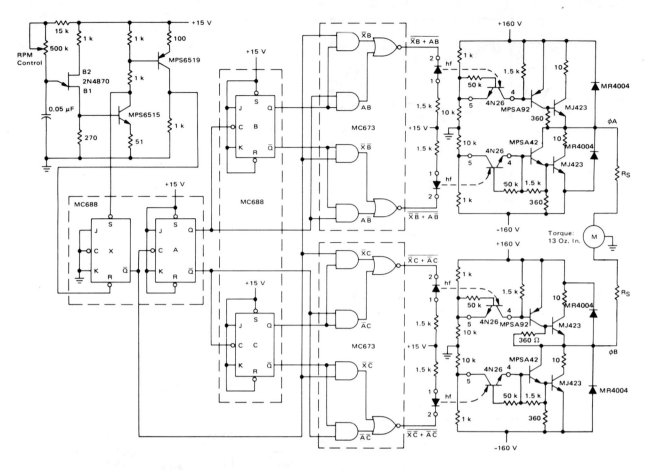

2.3.93 Motorola variable-speed control system for induction motors. A permanent-split-capacitor induction motor requires two drive signals 90° apart. A capacitor is normally used in series with one winding to obtain the necessary phase shift when the motor is operated from a single-phase source. Since the capacitive reactance is inversely proportional to the frequency, the capacitor cannot maintain the proper phase shift when operation over a range of frequencies is desired. A prerequisite, then, is to eliminate the need for such a capacitor. A 2N4870 UJT is used as a free-running oscillator. The frequency range of this oscillator is 40 to 1200 Hz. Because the logic that follows is in a divide-by-4 configuration, the actual drive frequency range is 10 to 300 Hz. This implies a speed range (for an induction motor with two pairs of poles) of 300 to 9000 rpm. In practice, however, the speed range will be limited by a variety of losses.

The resistor in the base 1 lead of the UJT controls the width of the oscillator output pulse. This pulse is shaped and used to control the duty cycle of the LED drive signals.

The MPS6515 and MPS6519 pulse amplifiers translate the oscillator signal to MHTL levels to drive the set and reset inputs of the X flip-flop. Since this flip-flop is configured in the R-S mode, its operation is dependent upon the input levels and duration, exclusive of the input rise and fall times.

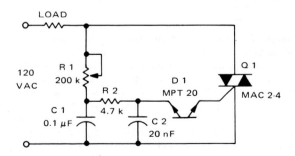

2.3.94 Motorola wide-range light dimmer using very few parts operates from a 120V 60 Hz ac source and can control up to 1000W of power to incandescent bulbs. The power can be varied by controlling the conduction angle of triac Q1.

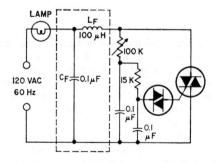

2.3.95 RCA triac control circuit that includes electromagnetic-radiation suppression for the purpose of minimizing high-frequency interference. The values indicated are typical of those used in lamp-dimmer circuits. Select triac (and mating diac) according to the maximum power to be controlled.

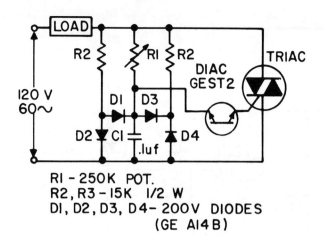

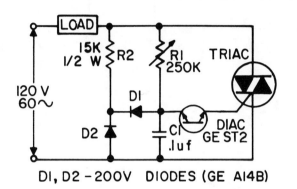

2.3.96 GE wide-range dimmers in which the hysteresis effect is eliminated entirely. The left circuit resets the timing capacitor to the same level after each positive half-cycle, providing a uniform initial condition for the timing capacitor. This is only useful for resistive loads because the firing angle is not symmetrical throughout the range. If symmetrical firing is required, the circuit at the right may be used.

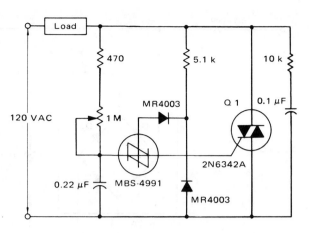

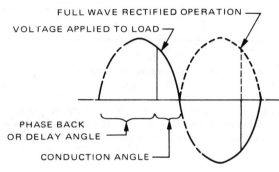

2.3.98 Motorola wide-range light dimmer operates from a 120V 60 Hz ac source and can control up to 1000W of power to incandescent bulbs. Power is varied by controlling the conduction angle of triac Q1. Many circuits can be used for phase control, but the single RC circuit used is the simplest. The control circuit for this triac must create a delay between the time the voltage is applied to the circuit (broken line of the sine wave) and the time it is applied to the load. The triac is triggered after this delay and conducts current through the load for the remaining part of each alternation. This circuit can control the conduction angle from 0° to about 170° and provides better than 97% of full-power control.

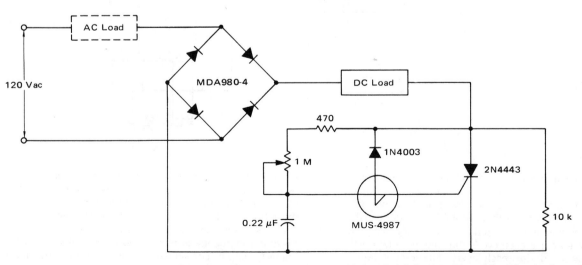

2.3.97 Motorola full-range power controller. A load requiring pulsating direct current may be connected between the bridge rectifier and the SCR. If the load requires an alternating voltage, it may be connected in series with either side of the ac power line. The diode and 10K resistor across the SCR will guarantee the discharge of the capacitor near the end of each half-cycle.

SECTION 2 *Control and Sensing* 129

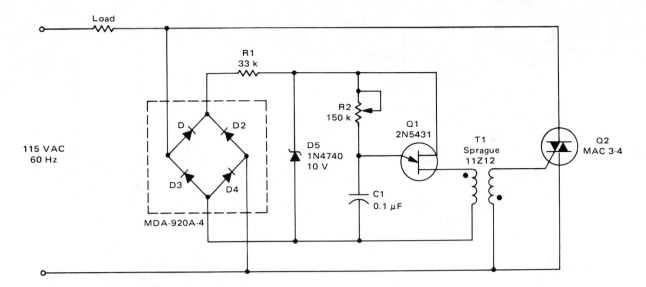

2.3.99 Motorola wide-range light dimmer using a unijunction transistor and a pulse transformer to provide phase control for a triac. The circuit operates from a 115V 60 Hz source and can control up to 800W of power to incandescent lights. The power to the lights is controlled by varying the conduction angle of the triac from 0 to about 170°. The power available at 170° conduction is better than 97% of that at the full 180°.

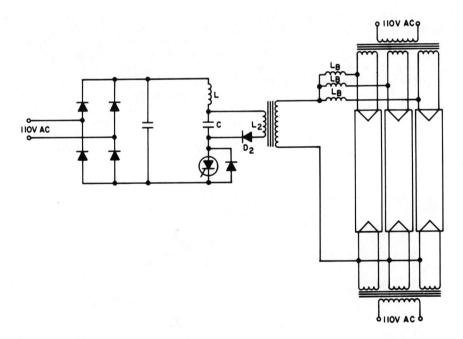

2.3.101 GE fluorescent lamp dimmer overcomes problems typically associated with such light sources. Fluorescent lamps are peculiar loads in that they take some time to ignite; thus they require an inverter capable of operating into an open circuit. The gas in the lamp has to be ionized twice each cycle, thus requiring two high-voltage pulses during each cycle. Fluorescent lamps, therefore, cannot be smoothly dimmed by merely lowering the applied voltage. Because the lamps act as negative resistance, an impedance (the ballast) must be connected in series with each lamp.

This variable-frequency inverter makes an ideal fluorescent lamp dimmer because the output pulse is of constant amplitude regardless of repetition rate, thus meeting the requirements for ionization. Diode D2 prevents the resonance between C and L2 that would occur when the triggering frequency is reduced to a low value for low-intensity output. For similar reasons, only inductive ballasts L_B are used.

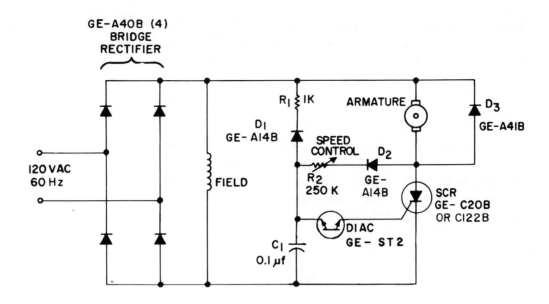

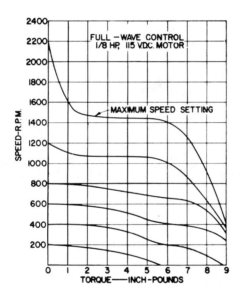

2.3.101 GE low-cost speed control for shunt-wound dc motors. The field winding is permanently connected across the dc output of the bridge rectifier. The armature voltage is supplied through the SCR and is controlled by turning the SCR on at various points in each half-cycle, the SCR turning off only at the end of each half-cycle. Rectifier D3 provides a circulating current path for energy stored in the inductance in the armature at the time the SCR turns off.

At the beginning of each half-cycle the SCR is off and C1 starts charging. When the voltage across C1 reaches the diac breakover voltage, a pulse is applied to the SCR gate, turning the SCR on and applying power to the armature for the remainder of that half-cycle. At the end of each half-cycle, C1 is discharged by the current through rectifier D1, resistor R1, and the field winding. The time required for C1 to reach the diac voltage breakover governs the phase angle at which the SCR is turned on and is controlled by the magnitude of resistor R2 and the voltage across the SCR. Since the voltage across the SCR is the output of the bridge rectifier minus the counter emf across the armature, the charging of C1 is partially dependent upon this counter emf, hence upon the speed of the motor. If the motor runs at a slower speed, the counter emf will be lower and the voltage applied to the charging circuit will be higher. This decreases the time required to trigger the SCR and increases the power supplied to the armature, thereby compensating for the loading on the motor.

Resistor R1 is chosen to limit the discharge current of C1 to a value less than the current through the field winding. If this discharge current is higher than the field, the excess may be diverted through the SCR and can result in failure of the SCR to turn off. If R1 is too large, the voltage on C1 may not be reset at the end of each half-cycle and irregular operation will occur at low-speed settings.

2.4 Temperature Regulation Devices

There are basically two methods to control temperature. The most common approach is to apply a fixed voltage to a heating element for a specified time, depending on the ultimate temperature desired. Another method involves applying a variable voltage to a resistive element, the resultant temperature then being a function of the power dissipated in the element.

Both techniques are represented by the circuits in this subsection. Also included are overtemperature detection and alarm circuits. Although these circuits are intended primarily for thermal control, they are not the only circuits that can be adapted to temperature control applications. Many of the high-power dimmers of Section 2.3 will prove themselves nicely in temperature control, as will some of the on-off switches in the remaining subsections.

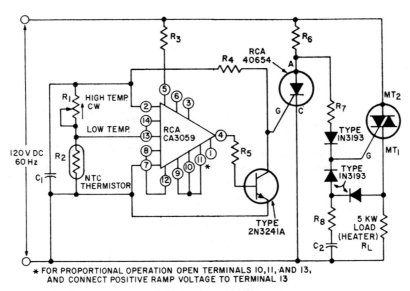

* FOR PROPORTIONAL OPERATION OPEN TERMINALS 10, 11, AND 13, AND CONNECT POSITIVE RAMP VOLTAGE TO TERMINAL 13

Parts List

C_1 = 100 μF, electrolytic 15 V
C_2 = 0.5 μF, 200 V
R_1 = Temperature-control potentiometer
R_2 = Negative-temperature-coefficient thermistor
R_3 = 5000 ohms, 5 watts
R_4 = 1500 ohms, 0.5 watt
R_5 = 10000 ohms, 0.5 watt
R_6 = 2200 ohms, 5 watts
R_7 = 1000 ohms, 0.5 watt
R_8 = 1000 ohms, 2 watts

2.4.1 RCA integral-cycle temperature controller employs CA3059 integrated-circuit zero-voltage switch, a 2N3241A transistor, a 40654 SCR, and a 2N5444 triac to control the ac power applied to an electric heating element. This circuit is completely devoid of half-cycling and hysteresis effects and includes a fail-safe feature that causes power to be removed from the load if the temperature sensor should be accidentally opened or shorted.

The sensor used with the controller is a negative-temperature-coefficient (NTC) thermistor connected between pins 7 and 13 of the CA3059. When the temperature being controlled is low, the resistance of the thermistor is high; the CA3059 produces a positive-voltage output at terminal 4. The 2N3241A inverts this voltage, and the SCR is not turned on. The triac is then triggered directly from the line on positive alternations. When the triac is triggered, power is applied to the heating element (R1), and C2 is charged to the ac input-voltage peak. When the line voltage swings negative, the capacitor discharges to trigger the triac. The diode-resistor-capacitor "slaving" network triggers the triac on negative alternations after it has been triggered on positive alternations to provide only integral cycles of ac power to the load.

When the temperature being controlled rises to the desired level, the resistance of the thermistor decreases and a zero-voltage output is obtained at pin 4 of the CA3059. The positive voltage at the collector of the 2N3241A is applied to the gate of the SCR, which then starts to conduct at the beginning of the positive alternation so that the trigger current is shunted away from the triac gate. The cycle is repeated when the SCR is again turned off by the reversal of the polarity of the ac input voltage.

This circuit can be converted into a proportional integral-cycle temperature controller by application of a positive-going ramp voltage to pin 9 of the CA3059 (with pins 10 and 11 open).

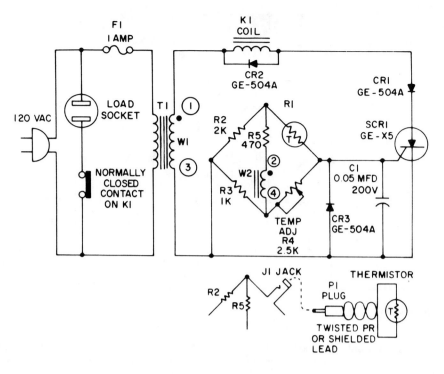

K1   Relay, DPDT, 5A contacts, 6V dc coil (Potter and Brumfield GP11 or equivalent)

R1   GE 1D303 thermistor, 0.3 in. diam, 1K at approximately 70°F

T1   Power transformer: 120V ac primary; two 12.6V secondaries W1 and W2 (UTC F10 or equivalent)

2.4.2 GE automatic thermostatically controlled outlet is powered when more heat is required; it shuts down when desired temperature is reached, which is ideal for hot pots, solder irons, or for improving cheap space heaters. Locating the thermistor on the soldering iron, in the bath water, or in any other zone that must be temperature-controlled will provide the necessary feedback information. If the thermistor is to control a cooling system such as a fan or air conditioner rather than a heating system, opposite action can be secured by either connecting the load to a normally open contact on the relay or by reversing the leads on the secondary winding W1. (For loads greater than 4A, use a "load socket" to switch heavy-duty relay coils. Use relay contacts to switch a heavy load.)

This circuit will control the temperature at thermistor R1 within approximately 1° over the temperature range from 20 to 150°F. For the most precise temperature control in this and other ranges, SCR1 should be kept at a stable temperature. If other temperature ranges are desired, another thermistor can be used. Thermistor R1 should have approximately 1000 ohms resistance in the center of the desired control range.

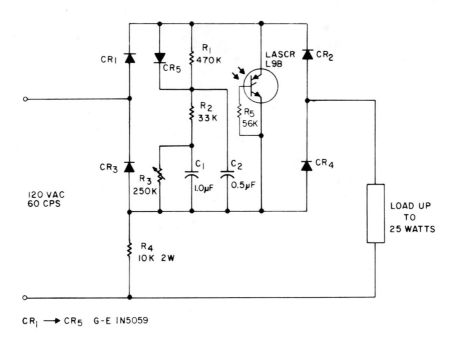

CR₁ → CR₅ G-E 1N5059

2.4.3 GE impulse-actuated, variable on-time switch. A random impulse of light fires the LASCR, applying current to the load. Capacitors C1 and C2 discharge through R2 and R3, and through R1 and the LASCR. As long as this capacitor-discharge current is higher than the holding current, the LASCR cannot commutate, thus applying full-wave alternating current to the load. When the discharge current drops, the LASCR will turn off at the next succeeding current zero, assisted by R4 for inductive loads.

Decreasing R3 reduces the time the switch remains on.

This switch can turn on at any phase angle but will turn off only at a current zero. During conduction, the full sine wave is applied to the load, with virtually no harmonic distortion. Radio noise is therefore negligible. This circuit is useful for the operation of solenoids, contactors, small motors, lamps, etc., particularly in conjunction with an optical programmer.

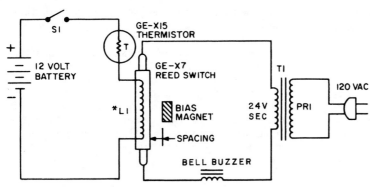

NOTE: BATTERY CIRCUIT CURRENT APPROXIMATELY 10 MILLIAMPERES. SELECT SUITABLE BATTERY

*L1 7000 turns of No. 38 wire (440 ohms). This GE C-1 coil is available at GE distributors.

2.4.4 GE adjustable temperature alarm will ring a bell, actuate a buzzer, or turn on a light when any predetermined temperature is reached. Any temperature from 115 to 165°F may be detected as the alarm sensing point. The approximate spacings of magnet to coil are given in the following table for various temperatures:

Spacing, in.	Switch Point, °F
1½	117
1	129
¾	138
⅝	149
½	164

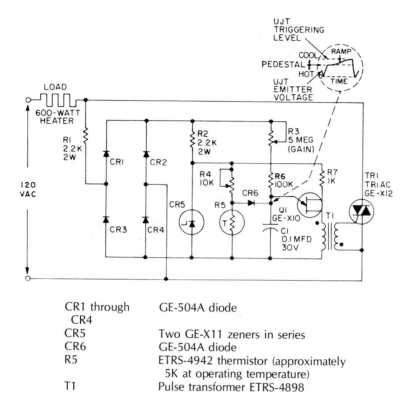

CR1 through CR4	GE-504A diode
CR5	Two GE-X11 zeners in series
CR6	GE-504A diode
R5	ETRS-4942 thermistor (approximately 5K at operating temperature)
T1	Pulse transformer ETRS-4898

2.4.5 GE precision temperature regulator for controlling ovens, hot plates, fluids, air, and gases. A thermistor (temperature-sensitive resistor) probe is used as a temperature detector in a UJT-controlled regulating circuit for a power switching triac. This temperature regulator has a fast response, adjustable gain (bandwidth), adjustable temperature, and built-in protection against transient voltages. It can control 600W of power over the full range with as little as a 2°C change in the thermistor temperature. The triac is triggered by a unijunction transistor and voltage-divider network that is sensitive to changes in the thermistor's resistance. The resistance of the thermistor, of course, is a function of its ambient temperature.

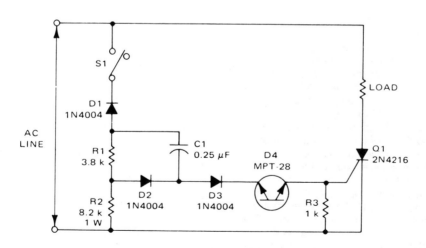

2.4.6 Motorola zero-point switch controls sine-wave power in such a way that either complete cycles or half-cycles of the power supply voltage are applied to the load. This type of switching is primarily used to control power to resistive loads such as heaters; it can also be used for controlling the speed of motors if the duty cycle is modulated (by having short bursts of power applied to the load) and the load characteristic is primarily inertial rather than frictional. Modulation can be on a random basis with an on-off control, or on a proportioning basis with the proper type of proportioning control.

In this circuit a pulse is generated before the zero crossing and provides a small amount of gate current when the line voltage starts to go positive. This circuit is primarily for sensitive-gate SCRs. Less-sensitive SCRs, with their higher gate currents, normally require smaller values for R1 and R2, and the result can be high power dissipation in these resistors.

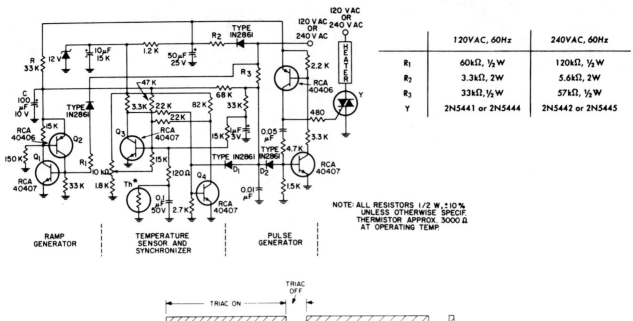

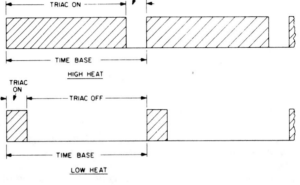

2.4.7 RCA proportional integral-cycle heat control. The disadvantage of thermal overshoot is overcome and RFI is minimized by use of the concept of integral-cycle proportional control with synchronous switching. In this system, a timebase is selected and the on time of the triac is varied within the timebase. The ratio of on to off time of the triac within this interval depends upon the heating power required to maintain the desired temperature.

A timebase ramp voltage is generated by charging capacitor C through resistor R for approximately 2s for the values shown. The length of the ramp is determined by the voltage magnitude required to trigger the regenerative switch consisting of Q1 and Q2. The temperature sensor consisting of Q3 and Q4, together with the controlling thermistor, establishes a voltage level at the base of Q3 that depends upon the resistance value of the thermistor. Q3 and Q4 form a bistable multivibrator. The state of the multivibrator depends upon the base bias of Q3. When Q3 is conducting, Q4 is cut off. The pulse generator is energized and generates pulses to trigger the triac. The output of the pulse generator is synchronized to the line voltage on the negative half-cycle by D2 and R3 and on the positive half-cycle by D1 and R3. The pulses are generated at the zero-voltage crossings and trigger the triacs into conduction at only these points.

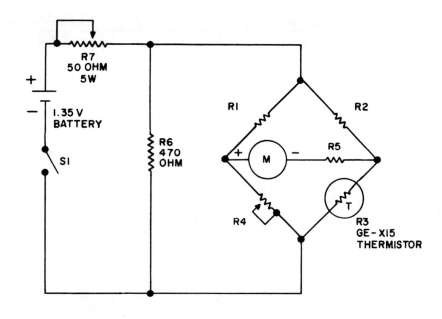

High temperature range (+32 to +122 F)
R1 — 1000-ohm, 1/4-watt resistor
R2 — 1000-ohm, 1/4-watt resistor
R3 — GE-X15 thermistor
R4 — 5000-ohm, 5-watt potentiometer
R5 — 9500-ohm, 1/4-watt resistor
R6 — 470-ohm, 1/4-watt resistor
R7 — 50-ohm, 5-watt potentiometer
S1 — SPST toggle switch
M — 50-microampere d-c meter (G-E Type DW-91), or equivalent
Battery — 1.35-volt mercury cell

Low temperature range (-40 to +32 F)
R1 — 7300-ohm, 1/4-watt resistor
R2 — 7300-ohm, 1/4-watt resistor
R3 — GE-X15 thermistor
R4 — 50,000-ohm, 5-watt potentiometer
R5 — 4850-ohm, 1/4-watt resistor
R6 — 470-ohm, 1/4-watt resistor
R7 — 50-ohm, 5-watt potentiometer
S1 — SPST toggle switch
M — 50-microampere d-c meter (G-E Type DW-91), or equivalent
Battery — 1.35-volt mercury cell

2.4.8 GE thermistor thermometer and temperature alarm will take the outside temperature remotely by the turn of a switch and a glance at a meter instantly from inside the house. When the temperature increases, the thermistor resistance decreases to produce an imbalance in the simple bridge circuit, which causes meter deflection. This deflection must be calibrated to indicate the temperature in degrees. The device will measure temperature in two ranges: −40 to 32°F and +32 to 122°F. The parts list shows values for each range. Each range is determined by the values of the resistance used in each leg of the bridge.

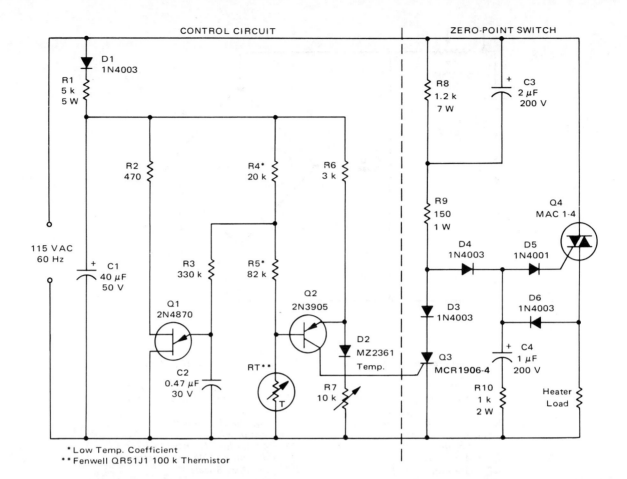

*Low Temp. Coefficient
**Fenwell QR51J1 100 k Thermistor

2.4.9 Motorola modulated triac zero-point switching circuit controls ac heater loads operating from 115V. The circuit at the right is the zero-point switch; at left is the proportional control for the zero-point switch. Diode D1, resistor R1, capacitor C1, and the load on C1 establish the dc supply voltage for the control circuit. The temperature is sensed by thermistor R_T, which is part of the bridge circuit consisting of R4, R5, R6, R7, D2, and R_T. The detector for the bridge is transistor Q2. R7 is set so that the bridge is in balance at the desired temperature. As the temperature increases, R_T decreases, and Q2 turns on and provides a gate drive to SCR Q3. Q3 turns on and shunts the gate signal away from the triac Q4; Q4 shuts off and removes power to the load. Now, as the temperature drops, R_T increases and Q2 turns off, SCR Q3 turns off, and full-wave power is applied to the load. Normally, the circuit would continue to cycle randomly, providing groups of full power to the heater load. However, modulation is applied to proportion the load power in response to small changes in R_T. The modulation is achieved by superimposing a sawtooth voltage on one arm of the bridge through R3. The period of the sawtooth is set to equal 12 cycles of the line frequency. From 1 to all 12 cycles can be applied to the load, thus allowing the load power to modulate in 8% steps from 0 to 100% duty cycle. The sawtooth voltage is generated by the unijunction transistor relaxation oscillator consisting of R2, R3, R4, C2, and Q1. The sawtooth wave modulates the bridge voltage so that over a portion of the 12-cycle group the bridge voltage will be above the null point, and over the other portion it will be below the null point. This action divides each 12-cycle group into an on-portion and an off-portion, the proportioning depending upon the amount R_T has varied from the nominal value.

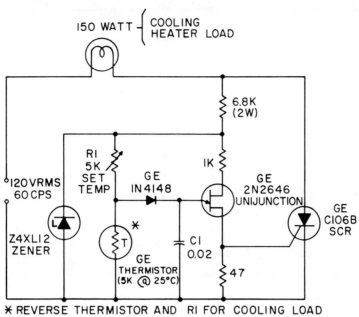

* REVERSE THERMISTOR AND R1 FOR COOLING LOAD
NOTE: ALL RESISTORS 1/2 WATT EXCEPT AS NOTED

2.4.10 GE precision temperature controller for photo developing and similar applications. In this circuit, a GE C106B SCR steplessly adjusts the voltage applied to a heater so that heat supplied the liquid bath load exactly balances the heat losses to the outside. In this way, the bath temperature is held constant. The earlier the SCR turns on during each cycle of the ac supply voltage, the greater the voltage applied to the load. Conversely, the later the SCR firing angle, the less the voltage applied.

SCR firing angle is controlled by a simple RC timed unijunction transistor oscillator. For a given potentiometer setting and capacitor value, unijunction frequency is determined uniquely by the resistance value of the load-monitoring thermistor. As the load temperature drops, the thermistor resistance rises, the UJT frequency rises, and the SCR fires earlier in the cycle. In this way the feedback loop is closed and more voltage is applied to the heater to compensate for the heat loss. By reversing the positions of R1 and the thermistor, the circuit will respond to a cooling (refrigerator-type) load.

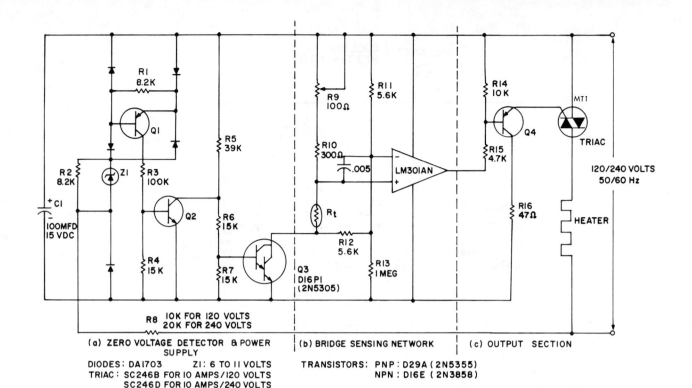

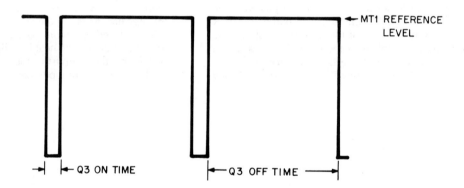

2.4.11 GE low-resistance-sensor, zero-voltage-switching temperature control. Since the sensing element is connected directly across the amplifier input, there is no gain loss. Sensor self-heating and amplifier and power supply loading are also eliminated. This is achieved by *sampling* the temperature being sensed. Sampling time is approximately 100 μs, and it will occur every time the input line voltage goes through zero or every half-cycle. For 60 Hz distribution systems the sampling duty cycle is 1.2%. The bridge sensing network is connected to the collector of Q3 so that it is energized only during transistor Q3 "on" periods. The collector waveshape voltage of Q3 is as shown; this voltage is referenced to the triac MT-1 terminal.

In the bridge sensing network, resistor R13 is connected from the inverting input of the operational amplifier to the negative terminal of the power supply capacitor C1 to insure that the voltage of the noninverting input of the op-amp is higher than the voltage of the inverting input when Q3 is off. For this condition the operational amplifier output is at the triac MT-1 reference level keeping transistor Q4 off. With Q4 off, the current will not flow out of the triac gate.

Resistors R11 and R12 are used to set the reference level of the operational amplifier. Resistors R9, R10, and sensor R_t set the operating point of the temperature controller. With R_t lower in resistance than the combination of R9 plus R10, every time transistor Q3 turns on, the noninverting input of the operational amplifier will be lower than the inverting input. The operational amplifier will turn transistor Q4 on, which triggers the triac at the beginning of every half cycle. With sensor resistance R_t equal to or higher than R9 + R10, the voltage of the noninverting input of the op-amp is higher than the voltage of the inverting input. Under this condition the operational amplifier output is at the triac MT-1 reference level, keeping transistor Q4 and the triac off. Resistor R16 at the collector of Q4 limits the triac gate current to 100 mA.

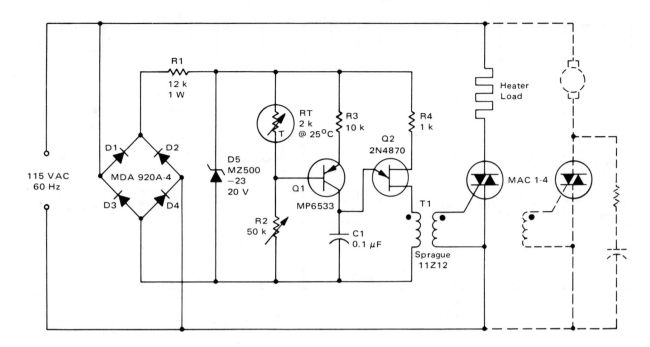

2.4.12 Motorola temperature-sensitive heater control regulates room temperature. This circuit eliminates several of the disadvantages of the mechanical control: large size, high price, unreliability, and poor power regulation. The mechanical control is an on-off arrangement and is not capable of regulating the power. By using phase control, the circuit is able to reduce the power to the load as the desired temperature is reached, thus eliminating much of the overshoot inherent in mechanical controls.

The line voltage is rectified by bridge D1–D4. The output of the bridge is applied to the control circuit through resistor R1 and clamped to 20V by zener D5. Thermistor R_T and variable resistor R2 control the base current for transistor Q1. *R2 is adjusted so that Q1 is off at the desired temperature.* When Q1 is off, no current can flow to capacitor C1, and C1 cannot charge to the firing voltage of unijunction Q2. (Q2 cannot fire the triac.) *If the temperature decreases, the resistance of R_T increases,* Q1 is turned on, and current flows to C1. C1 charges to the firing voltage of Q2, and Q2 fires and turns the triac on through pulse transformer T1. *If the temperature continues to decrease,* the resistance of R_T increases more and Q1 is turned on more. C1 charges faster and the triac is triggered earlier, delivering more power to the load. As the temperature increases, the resistance of R_T decreases and Q1 will conduct less. C1 takes longer to charge and the triac is triggered later in the cycle. *When the desired temperature is reached, Q1 is off and the triac is off.*

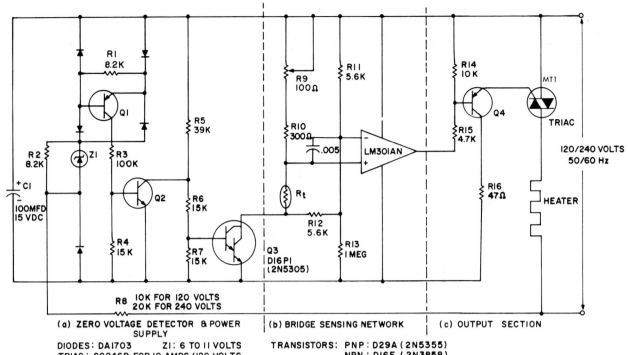

2.4.13 GE zero-voltage-switching temperature control operates with low-resistance wire sensors (which have a positive temperature coefficient) for high sensitivity without cluttered circuitry. Since the sensing element is connected directly across the amplifier input, there is no gain loss. Sensor self-heating and amplifier and power supply loading are eliminated. This is achieved by *sampling* the temperature being sensed. Sampling time is approximately 100 μs, and it will occur every time the input line voltage goes through zero or every half-cycle. For 60 Hz distribution systems the sampling duty cycle is 1.2%. Assume that the sensor output is 10 ohms. The dc power supply is approximately 8V. For good common-mode operation it is advisable to set the operating point at half the power supply voltage, in this case 4V. The sensor power dissipation is 19.2 mW, and power supply loading is 4.8 mA average.

It must be noted that the instantaneous peak power dissipated in the sensor is 1.6W and the peak current through the sensor is 400 mA.

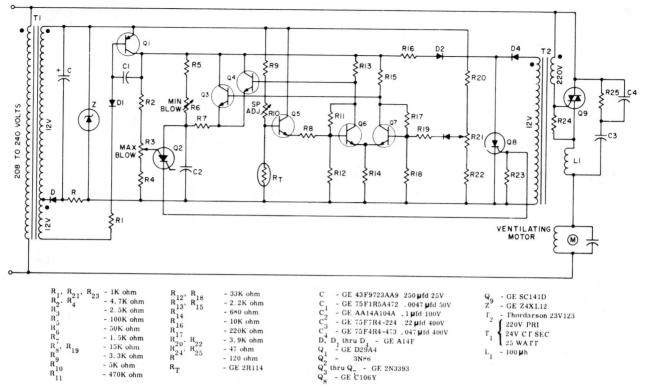

R_1, R_{21}, R_{23}	- 1K ohm	R_{12}, R_{18}	- 33K ohm	C	- GE 43F9723AA9 250 µfd 25V	Q_9	- GE SC141D
R_2, R_4	- 4.7K ohm	R_{13}, R_{15}	- 2.2K ohm	C_1	- GE 75F1R5A472 .0047 µfd 50V	Z	- GE Z4XL12
R_3	- 2.5K ohm	R_{14}	- 680 ohm	C_2	- GE AA14A104A .1 µfd 100V	T_2	- Thordarson 23V123
R_5	- 100K ohm	R_{16}	- 10K ohm	C_3	- GE 75F7R4-224 .22 µfd 400V	T_1	{ 220V PRI, 24V CT SEC, 25 WATT }
R_6	- 50K ohm	R_{17}	- 220K ohm	C_4	- GE 75F4R4-473 .047 µfd 400V		
R_7	- 1.5K ohm	R_{20}, R_{22}	- 3.9K ohm	D_1 thru D_4	- GE A14F		
R_8, R_{19}	- 15K ohm	R_{24}, R_{25}	- 47 ohm	Q_1	- GE D29A4	L_1	- 100 µh
R_9	- 3.3K ohm		- 120 ohm	Q_2	- 3N×6		
R_{10}	- 5K ohm	R_T	- GE 2R114	Q_3 thru Q_7	- GE 2N3393		
R_{11}	- 470K ohm			Q_8	- GE C106Y		

NOTE: ALL RESISTORS 1/2 WATT ± 10% UNLESS OTHERWISE NOTED

2.4.14 GE proportional motor speed control especially designed for fan-and-coil water systems. A fan-and-coil water system is a room temperature regulator capable of heating or cooling by means of passing hot or cold water from a central supply through heat-exchanger coils in each room, where a blower helps transfer the heat from or to the room air. The purpose of this solid-state control is to improve room-temperature regulation by proportionately controlling the blower speed in accordance to room-temperature demands, as indicated by the room thermostat.

Since heating or cooling is accomplished by passing hot or cold water through one coil, the control must be capable of increasing the blower speed for either an over- or under-temperature deviation. The triggering circuit is shown within the dotted block and is synchronized to the power triac, Q1. This type of synchronization is acceptable for the majority of the motors used in this application.

Q1 is triggered on with a pulse supplied by unijunction Q4, through T1. R11 and C4 determine the minimum speed. When a unijunction firing signal is supplied through D7 later in the cycle than that supplied through D8, as a result of C4's charging through R11, the blower rotates at the minimum speed dictated by R11. When the central water control determines that cooling is required and provides water colder than normal room temperature, water-sensing thermistor T_W causes Q2 to be off and Q3 on. This requires C3 to charge through R4 and the room temperature thermistor T_A. If the relative resistance of T_A is such as to "request" room cooling, C3 will charge in a ramp-and-pedestal fashion to the unijunction firing voltage ahead of C4, increasing the blower speed and decreasing the room temperature. As more cooling is required, the blower speed will be modulated up to its maximum. In the event that the central water control determines that heating is required and provides water warmer than nominal room temperature, T_W forces Q3 off and Q2 on. Now C3 charges through R7, R6, and R5. Consequently, when the relative resistance of T_A is such as to request room heating, C3 will again charge to the unijunction firing voltage ahead of C4, increasing the blower speed and room temperature.

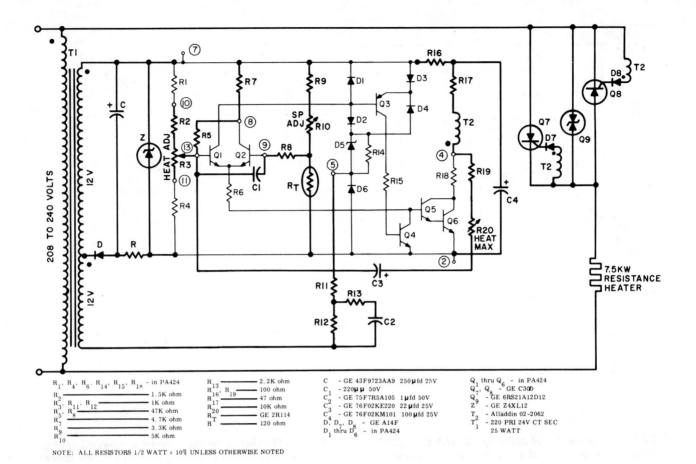

2.4.15 GE ventilating blower control for heating and cooling is designed to be operated from one thermostat located within the room. When neither room heating nor cooling is required, the control operates the blower at minimum speed. As the room temperature decreases, the blower speed is proportionally increased. When the heating source increases the room temperature, the control will proportionally reduce the blower speed. Similarly, when the room temperature increases, the blower speed is proportionally increased and then proportionally decreased when the cooling source lowers the room temperature. The control has RFI and dv/dt suppression and line-voltage synchronization with a continuous triac gate signal for good motor performance and thermostat isolation; it is capable of controlling 6A at 240V. Higher currents can be controlled by using a larger-power triac.

This control is capable of being operated in conjunction with the *proportional heating control* (2.4.16), and it uses common components of T1, D, C, R, Z, and the thermostat (consisting of R9, R10, and thermistor R_T).

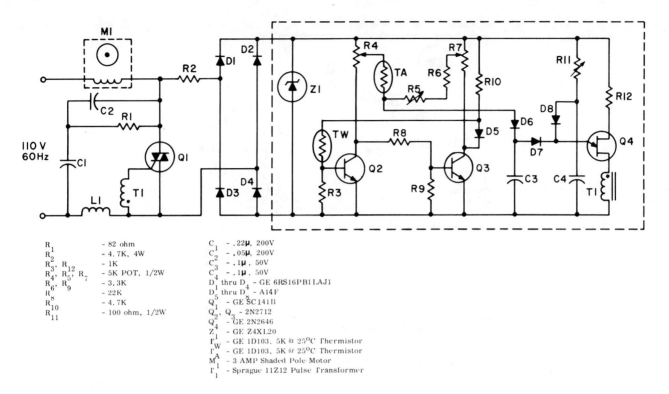

R_1	– 82 ohm
R_2	– 4.7K, 4W
R_3, R_{12}	– 1K
R_4, R_5, R_7	– 5K POT, 1/2W
R_6, R_9	– 3.3K
R_8	– 22K
R_{10}	– 4.7K
R_{11}	– 100 ohm, 1/2W
C_1	– .22μ, 200V
C_2	– .05μ, 200V
C_3	– .1μ, 50V
C_4	– .1μ, 50V
D_1 thru D_4	– GE 6RS16PB1LAJ1
D_5 thru D_8	– A14F
Q_1	– GE SC141B
Q_2, Q_3	– 2N2712
Q_4	– GE 2N2646
Z_1	– GE Z4X1.20
T_W	– GE 1D103, 5K @ 25°C Thermistor
T_A	– GE 1D103, 5K @ 25°C Thermistor
M_1	– 3 AMP Shaded Pole Motor
T_1	– Sprague 11Z12 Pulse Transformer

NOTE: ALL RESISTORS 1/2W ± 10% UNLESS OTHERWISE SPECIFIED

2.4.16 GE proportional heating control is capable of controlling a 7.5 kW resistance heater operating from a 220V power line. The power switching components are the inverse parallel SCR combination of Q7 and Q8, with the triggering circuit being those components connected to the low-voltage side of transformer T1. Transformer T1 allows low-voltage triggering components in addition to electrically isolating the thermostat from the high-voltage power line. The resistance heater is energized and the room temperature is increased when SCRs Q7 and Q8 are switched at zero voltage by the triggering circuit, primarily consisting of the PA424 IC. This circuit accepts a signal from the thermostat components R9, R10, and thermistor R_T, and triggers Q7 and Q8 at zero voltage. The dc power supply does not use diode D6 of the PA424 but consists of C, Z, R, and D. C1 is used to prevent erratic SCR triggering due to noise, and R8 isolates the thermostat from Q2 interference.

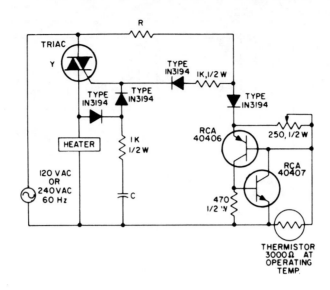

	120VAC, 60Hz	240VAC, 60Hz
R	2.2kΩ, 5W	3.9kΩ, 5W
C	0.5μF, 200V	0.5μF, 400V
Y	T4700B	T4700D

2.4.17 RCA on-off circuit for the control of resistance-type heating elements provides synchronous switching close to the beginning of the zero-voltage crossing of the input waveform to minimize RFI. The thermistor controls the operation of the two-transistor regenerative switch, which controls the operation of the triac. When the temperature being controlled is low, the resistance of the thermistor is high and the regenerative switch is off. The triac is then triggered directly from the line on positive half-cycles of the input voltage. When the triac triggers and applies voltage to the load, the capacitor is charged to the peak value of the input voltage. The capacitor discharges through the triac gate to trigger the triac on the opposite half-cycle.

When the temperature being controlled reaches the desired value as determined by the thermistor, the transistor regenerative switch conducts at the beginning of the positive input-voltage cycle to shunt the trigger current away from the triac gate. The triac does not conduct as long as the resistance of the thermistor is low enough to make the transistor regenerative switch turn on before the triac can be triggered.

2.5 Value, Rate, and Process Monitors

The types of circuits that appear in this subsection may be used to detect, monitor, and control a wide variety of processes. Although but a handful of circuits are given here, they're generally quite easy to adapt to similar but unrelated functions. The first circuit, for example, is a relay-triggering arrangement that can be coupled to a number of different sensing elements—a thermistor, a photocell, a wire grid, and a timing RC network are but a few of the devices that can be used for triggering.

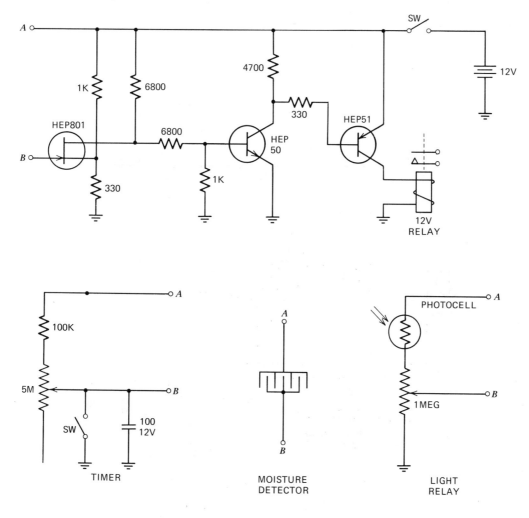

2.5.1 Motorola sensitive relay for use as *moisture detector, light-activated switch, timer, or liquid-level alarm or switch*. The basic circuit can be triggered by any of a number of sensors connected at **A** and **B**. The inset timer circuit provides a range of 5s to 1 min; change R and C to adjust range ($R \times C$ = time in seconds). For the moisture detector, use a network of close-spaced wires deposited on a printed-circuit board or the equivalent (moisture will bridge contacts to trigger the relay). For photocell applications, use the *eight-relay* circuit at the **A–B** input. As a liquid-level indicator, connect a thermistor across **A–B**.

SECTION 2 Control and Sensing

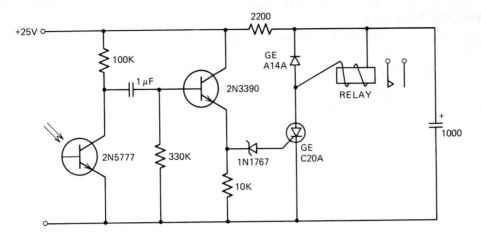

2.5.2 GE go/no-go rate sensor detects the frequency at which an object interrupts a beam of light and will operate a relay when the rate exceeds a predetermined value. This circuit, employing a 2N5777 phototransistor, is particularly valuable for sensing moderately fast changes of light while ignoring the slow drifting rate that results from dust, dirt, lamp aging, atmospheric conditions, and other such interferences.

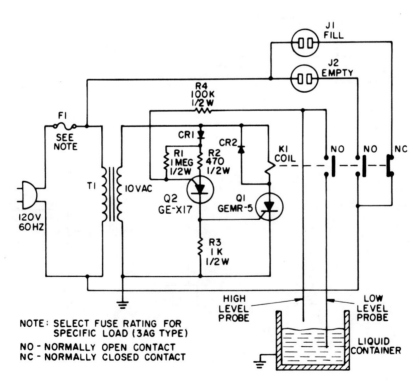

| CR1, CR2 | GE-504A diodes |
| K1 | 12V ac relay with DPDT contacts |

2.5.3 GE automatic liquid-level control for sump pumps, water storage tanks, swimming pools, animal drinking troughs, or other liquid containers when it is desirable to keep the fluid level between two fixed points. Two load sockets, marked FILL and EMPTY, are selected for the proper control when the liquid is either entering or leaving the container. These loads can be either electric motors or solenoid-operated valves operating from 120V ac power. Liquid-level detection is accomplished by two metal probes, one measuring the high level and the other the low. Both probes can be single wires or conductive rods insulated from the mounting supports.

148   DISCRETE / TRANSISTOR CIRCUIT SOURCEMASTER

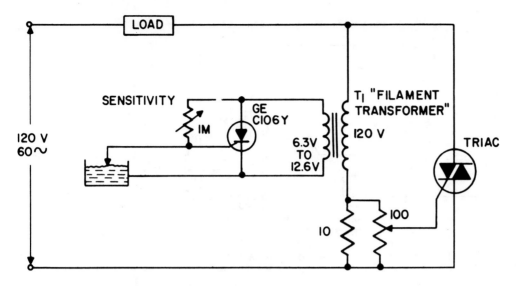

2.5.4 GE water-level-sensing control circuit applies power to the load until the water conducts through the probe and bypasses gate current from the low-current SCR. This gives an isolated low-voltage probe to satisfy safety requirements.

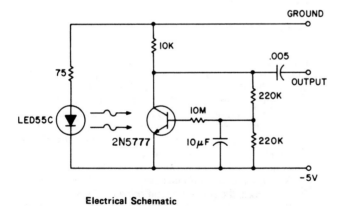

Electrical Schematic

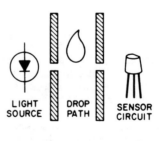

Mechanical Schematic

2.5.5 GE dripping fluid "drop" detector. The photo-Darlington is bias-stabilized by feedback from the collector, compensating for different photo-Darlington gains and LED outputs. The 10 μF capacitor integrates the collector voltage feedback, and the 10M resistor provides a high base impedance to give minimal effect on optical performance. The fluid drop causes a momentary loss in light reaching the chip, which causes the collector voltage to rise momentarily, generating an output signal. The initial light bias is small because of output power constraints on the light-emitting diode and mechanical spacing system. The change in light level is a fraction of this initial bias due to stray light paths and drop translucence.

SECTION **2** *Control and Sensing*

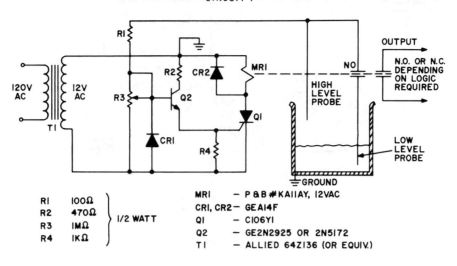

CIRCUIT 1

R1	100Ω	
R2	470Ω	1/2 WATT
R3	1MΩ	
R4	1kΩ	

MR1 — P&B #KA11AY, 12VAC
CR1, CR2 — GEA14F
Q1 — C106Y1
Q2 — GE2N2925 OR 2N5172
T1 — ALLIED 64Z136 (OR EQUIV.)

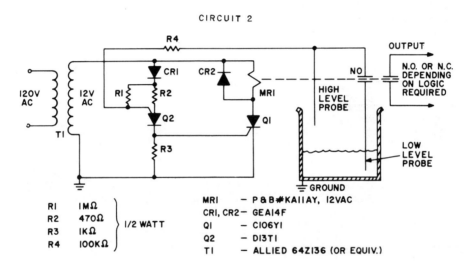

CIRCUIT 2

R1	1MΩ	
R2	470Ω	1/2 WATT
R3	1kΩ	
R4	100kΩ	

MR1 — P&B #KA11AY, 12VAC
CR1, CR2 — GEA14F
Q1 — C106Y1
Q2 — D13T1
T1 — ALLIED 64Z136 (OR EQUIV.)

2.5.6 GE liquid-level controls use hybrid techniques to maintain the liquid content of any reasonably well grounded container between two previously established limits. The controls are ideally suited for use with drink-vending machines, sump pumps, swimming pools, water storage cisterns, cattle drinking troughs, and many other similar applications.

Liquid-level detection in both circuits is accomplished by two metal probes, one measuring the high level, the other the low level. All probes are of single-wire construction.

The leftmost circuit consists of the two level-detector probes, energized with alternating current to eliminate electrolytic corrosion, a transistor that amplifies the probe signals sufficiently to trigger an SCR, and an SCR power switch that drives an output relay. The whole circuit is powered from a low-voltage filament transformer for safety reasons.

With the liquid level just below the low-level probe, no base current flows to Q1, the SCR is off, and the output relay is deenergized. Should the liquid level rise above the low-level probe, the circuit state remains unchanged, since there is a normally open relay contact in series with this probe. When the liquid level reaches the high-level probe, however, the base current flows to Q1 (via the ground, liquid, and probe), Q1 triggers SCR1, and the output function is energized (or deenergized) via MR1. Once MR1 is energized, the normally open contact in series with the low-level probe closes and latches the circuit on until the liquid level drops below the low-level probe (at which time the output function is deenergized). Diode CR1 removes the reverse voltage from Q1's base–emitter junction during negative half-cycles of the ac supply and insures that ac flows through the probes to prevent electrolytic corrosion.

Except that a dc component of current is found flowing in the probes, the second circuit provides the same action as the first. The transistor of circuit 1 has been replaced by the D13T1 programmable unijunction transistor (PUT). When the high-level probe circuit is closed by an impedance of less than 20M, the drop across the 1M resistor is great enough to trigger the PUT, which in turn triggers the C106Y1 SCR. The relay then closes and remains closed until the liquid clears the low-level probe, at which time the SCR returns to a nonconducting state.

Circuit 1 should be used where accumulation of lime deposits on the probes could be a problem, whereas circuit 2 could be used for higher impedance fluids.

The output function (pump, solenoid, etc.) is taken directly from the second set of relay contacts, which can be either normally open or normally closed, depending on the logic required.

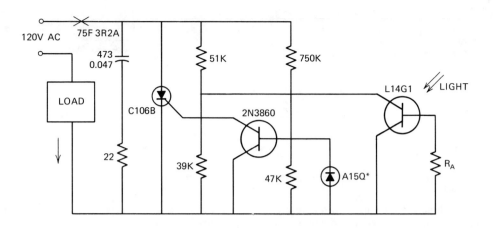

*The A15Q may be replaced by 100 pF shunting a DHD 800 diode. Select R_A from the following table:

ILLUMINATION

Holdoff light level, ≈footcandles	≈20	≈40	≈80	200	≈400
R_A, incandescent light	NA	1500	270	68	33
R_A, flame light	220K	75K	30K	12K	6.2K
R_A, fluorescent light	NA	NA	2.2M	180K	68K

2.5.7 GE flame monitor. The monitoring of a flame and direct switching of a 120V load is easily accomplished through use of GE's L14G1 for "point sources" of light. For light sources that subtend over 10° of arc, the L14H1 should be used and the illumination levels raised by a factor of 5. This circuit provides zero-voltage switching to eliminate phase controlling.

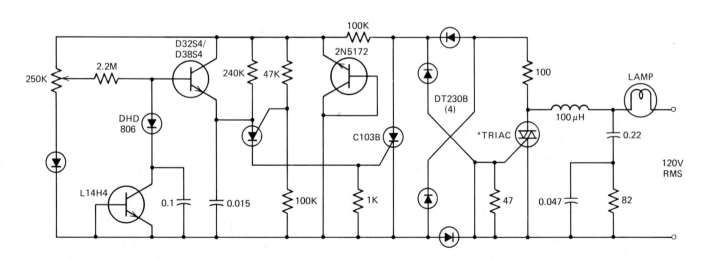

*GE triac	Lamp
SC151D	750W
SC146D	550W
SC136D	100W

2.5.8 GE automatic brightness control maintains a lamp at a constant brightness over a wide range of supply voltages. This circuit utilizes the consistency of photodiode response to control the phase angle of the power-line voltage applied to the lamp. This provides a candlepower range from 100% to less than 10% of the nominal lamp output. The 100 μH filter and shunt RC network eliminate RFI problems.

SECTION 2 *Control and Sensing* **151**

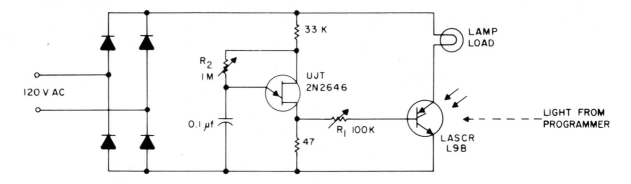

2.5.9 GE lamp switching circuit cuts bulb failure and lowers maintenance cost. For programmed control of lamps in which the operation is repetitive for a large number of cycles, thermal stresses on lamps and on control are severe. This unijunction-transistor control circuit can provide preheating of the lamp by triggering the LASCR late in each half-cycle. The setting of R2 will determine the minimum lamp current required to maintain filament temperature just below the visible level. Gate resistor R1 may be adjusted to control sensitivity of the LASCR to light. One unijunction circuit may be used in conjunction with several LASCR and lamp circuits by using a separate gate resistor for each LASCR.

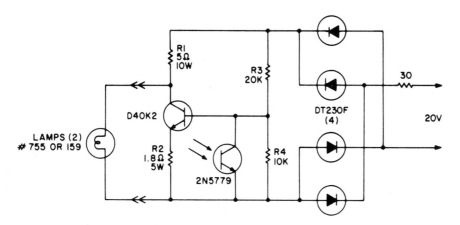

2.5.10 GE brightness control lowers the illumination level of lighted displays as the room ambient light drops to avoid undesirable visual effects. The bias resistors are optimized for the 20V source and must be recalculated for other sources. The GE 2N5779 must be placed to receive the same ambient illumination as the display and must be shielded from the display lamps.

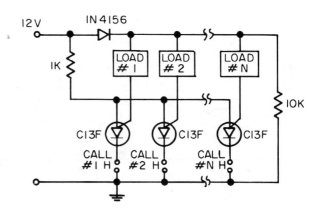

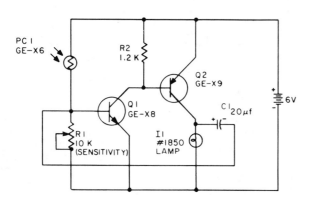

2.5.11 GE light target. The lamp flashes when a spot of light falls on the photocell. Use a 2W pot for R1.

2.5.12 GE anticoincidence circuit provides ground only to the one load selected first. If other loads are selected, the gates on the associated GE CSCRs are reverse-biased until the first load is released. Note that two loads cannot be driven at one time, since any driven load deprives all the others of triggering voltage. Any number of loads may be operated with this circuit at supply voltages up to 50V and load currents to 50 mA. The 1N4156 stabistor diode and 10K resistor provide gate trigger voltage for the CSCR, while the 1K resistor provides the holding current.

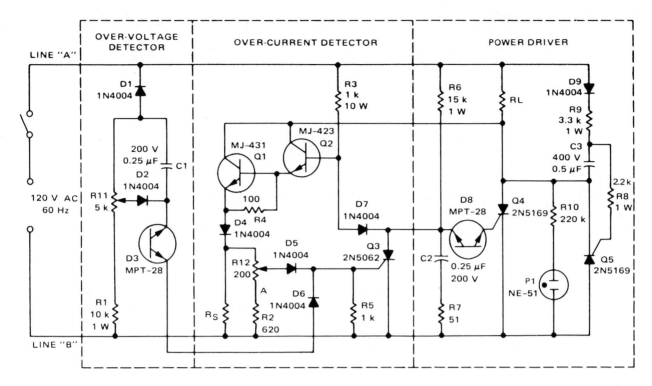

$*R_S \approx 0.1\ R_{LOAD}$

2.5.13 Motorola ac overvoltage and overcurrent protection with automatic reset for a 10A power supply operating from conventional household current (115V, 60 Hz).

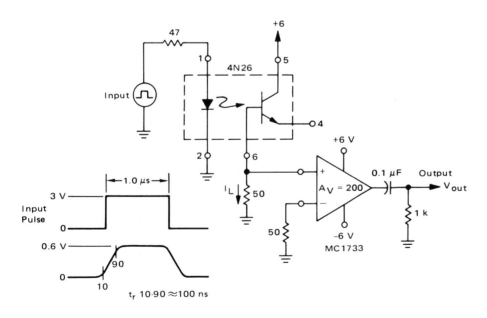

2.5.14 Motorola diode coupler. In this isolated coupling device, the output is taken from the collector–base diode. The emitter is left open, the load resistor is connected between the base and ground, and the collector is tied to the positive voltage supply. Using the coupler in this way reduces the switching time from 2 or 3 μs to 100 ns.

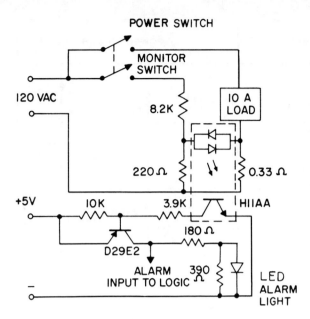

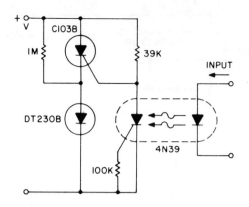

2.5.17 GE normally closed zero-voltage-switching circuit employs GE's relatively low-cost 4N39 SCR optocoupler.

2.5.15 GE load monitor and alarm. In many computer-controlled systems where ac power is controlled, the load dropout due to filament burnout, fusing, etc.—or the opposite situation: load power when uncalled for due to switch failure—can cause serious systems or safety problems. This circuit provides a simple ac power monitor that lights an alarm lamp and provides a "1" input to the computer control in either of these situations while maintaining complete electrical isolation between the logic and the power system. For other than resistive loads, phase-angle correction of the monitoring voltage divider is required.

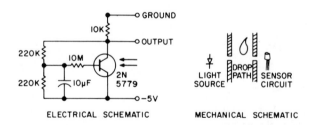

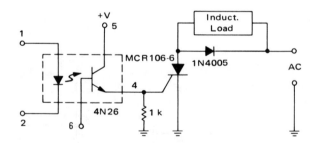

2.5.16 GE low-light-level drop detector uses LED and photo-Darlington. The output may be connected to a simple ac amplifier. The photo-Darlington is dc-bias-stabilized by feedback from the collector, compensating for different photo-Darlington gains and LED outputs. The 10 μF capacitor integrates the collector voltage feedback, while the 10M resistor provides a high base–source impedance to give a minimal effect on optical performance. The detected drop causes a momentary loss in light reaching the chip, which causes the collector voltage to rise momentarily, generating an output signal. The initial light bias is small due to output power constraints on the LED and mechanical spacing system constraints. The change in light level is a fraction of this initial bias due to stray light paths and drop translucence.

2.5.18 Motorola opto coupler driving a silicon controlled rectifier. The SCR is used to control an inductive load, and the SCR is driven by a coupler. The SCR used is a sensitive-gate device that requires only 1 mA of gate current, and the coupler has a minimum current transfer ratio of 0.2 so that the input current to the coupler need be only 5 mA. The 1K resistor connected to the gate of the SCR is used to hold off the SCR. The 1N4005 diode is used to suppress the self-induced voltage when the SCR turns off.

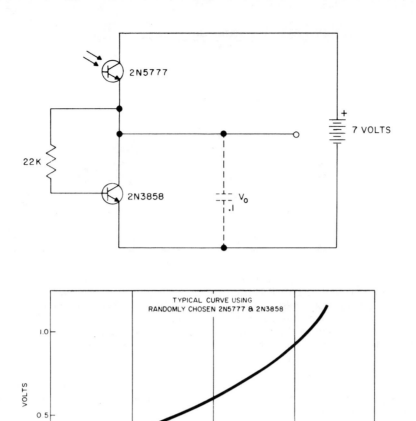

2.5.19 GE logarithmic light and infrared radiation detector will give an output voltage proportional to the amount of received light over at least three decades (1000:1). This is accomplished by connecting a transistor to act as a logarithmic resistor in series with the phototransistor.

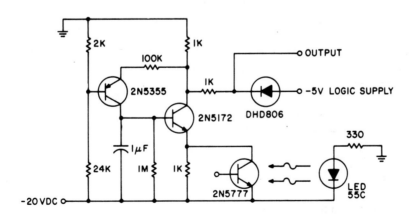

2.5.20 GE paper-tape reader works with logic signal levels. With a nominal −1V at the output dropping to 0.6V below the logic supply, this circuit reflects the requirements of a paper-tape optical reader system. The circuit operates at rates of up to 1000 bits per second. It also must operate at tape translucency such that 50% of the incident light is transmitted to the sensor and provides a fixed threshold signal to the logic circuit. Photo-Darlington speed is enhanced by cascode constant-voltage biasing. The output threshold and tape translucency requirements are provided for by sensing the output voltage and providing negative feedback to adjust the cascode transistor bias point.

SECTION 2 Control and Sensing

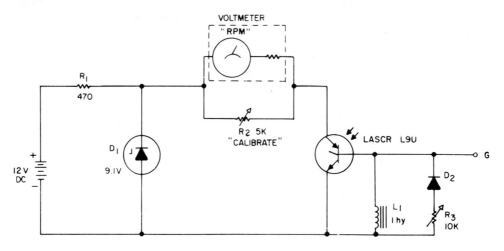

2.5.21 GE optical tachometer is an example of how the average value of anode current in the GE LASCR is directly related to the repetition rate of trigger pulses, either light or electrical. The voltmeter is calibrated in rpm and essentially measures the average current in the LASCR. Since the pulse width is determined by anode current, peak amplitude, and gate inductance, the simplest method of calibrating this tachometer is with a variable resistor (R2), which adjusts the pulse width by the anode current peak. Damping resistor R3 should be set for the maximum value that will permit one-shot operation of the circuit.

If R3 is too high, this circuit oscillates; if R3 is too low, the current through L1 will not decay rapidly and there will be a significant period after turnoff when the LASCR cannot be retriggered. Since this circuit will operate over a fairly wide range of anode currents, a multiple-range tachometer is easily obtained by using various values of R2. Electrical signals may also be used to drive this tachometer by coupling pulses into the gate terminal through sufficiently high impedance to prevent interference with the pulse width of conduction.

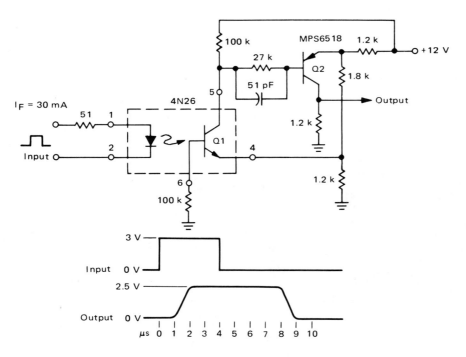

2.5.22 Motorola opto-coupled Schmitt trigger. One of the Schmitt trigger transistors is the output transistor of a coupler. The input to the Schmitt trigger is the LED of the coupler. When the base voltage of the coupler's transistor exceeds $V_e + V_{be}$, the output transistor of the coupler will switch on. This will cause Q2 to conduct and the output will be in a high state. When the input to the LED is removed, the coupler's output transistor will shut off and the output voltage will be in a low state.

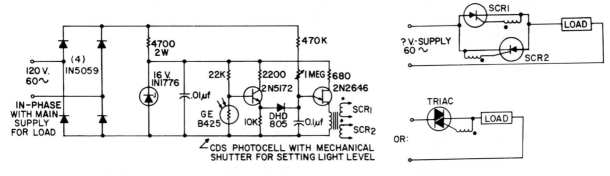

2.5.23 GE photoelectric load-control system uses a ramp-and-pedestal arrangement in which a cadmium sulfide photocell drives a transistor follower to control the pedestal level charging on the capacitor. Since the input impedance of the follower is high, this circuit will accommodate a very wide range of impedance levels in the photocell (or any other resistance-type transducer). This circuit may be used to regulate the output of a lamp or the illumination on a surface. The use of a controllable shutter will permit proportional control of the load in response to mechanical-position input information. In this manner, the load can be controlled by the temperature of a bimetal, the pressure in a bellows, the humidity in a hygrometer, etc.

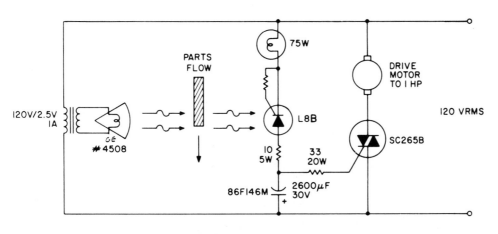

NORMAL FLOW

2.5.24 GE production-line "log jam" control. In many production lines the flow of parts is controlled by a drive motor, which should be turned off if a jam occurs and parts are no longer being removed from the output of the driven conveyance. This simple circuit provides direct control of the drive with a lamp providing a visual indication of normal flow. Snubber networks have not been illustrated to simplify the schematic. The No. 4508 sealed-beam lamp and 75W indicator lamp are run at about two-thirds the rated voltage to provide long service and reliable operation. Light blockages of up to 250 ms are ignored by this circuit (except for the indicator's blinking off) and, if the blockage time extends, the drive motor is turned off. When the light blockage is removed, the drive motor will automatically turn on again.

SECTION **2** *Control and Sensing*

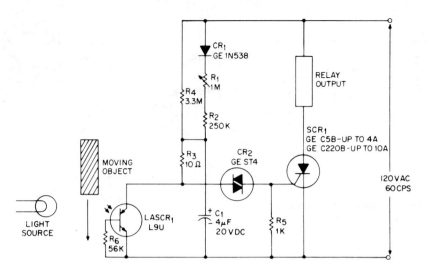

2.5.25 GE production-line monitor "watches" the smooth flow of small components down a high-speed conveyor. It has the capability of overlooking small self-clearing pileups, but it will shut the line down rapidly in the event of an impending catastrophic jam.

The SCR, in series with a relay load, is supplied from the 120V line. SCR1 is normally off, being energized only when a fault condition occurs. With the light on, the LASCR conducts and prevents the voltage from building up across C1. Each time a passing component momentarily interrupts the light beam, LASCR1 is briefly com- mutated by the ac line. During these off periods C1 starts to charge toward the peak line voltage but is shorted to zero once more as light is restored. If the light path to LASCR1 is blocked for more than a few milliseconds, however, capacitor C1 will continue to charge unimpeded by LASCR1, and at some time determined by the time constant $C_1(R_1R_2)$ will exceed the avalanche voltage of CR2 and fire SCR1. SCR1 then activates the load. Reset is automatic when light is restored to LASCR1. Circuit delay time can be adjusted from a few milliseconds to several seconds by adjustment of R1.

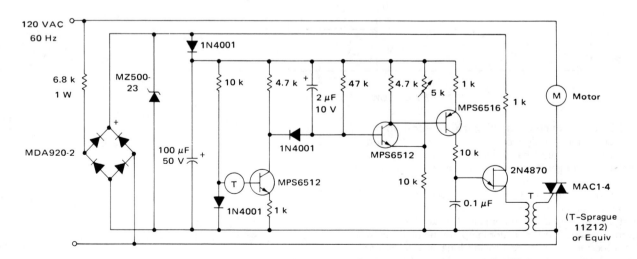

2.5.26 Motorola tachometer speed control. Since the voltage and temperature characteristics closely match those of the transistor base-to-emitter junction, this circuit needs no initial adjustments.

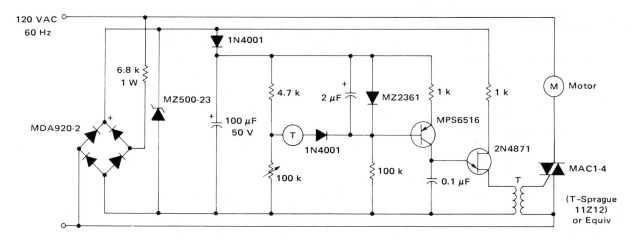

2.5.27 Motorola tachometer speed control in which the rectified and filtered tachometer voltage is added to the output voltage of the voltage divider formed by R1 and R2. If the sum of the two voltages is less than $V_1 - V_{BEQ1}$ (where $V_{Be\ Q1}$ is the base–emitter voltage of Q1), Q1 will conduct a current proportional to $V_1 - V_{BE\ Q1}$, charging capacitor C. If the sum of the two voltages is greater than $V_1 - V_{BE\ Q1}$, Q1 will be cut off and no current will flow into the capacitor.

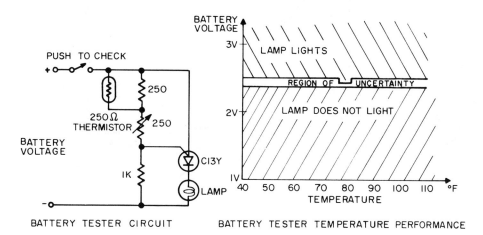

2.5.28 GE temperature-compensated battery voltage monitor employs complementary SCR. If the battery voltage is low, the trigger voltage is not applied to the GE C13Y gate by the voltage divider; the lamp stays off. The thermistor provides temperature stabilization, giving performance as illustrated.

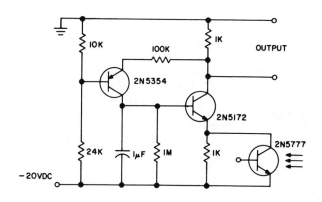

2.5.29 GE high-speed optical paper-tape reader system operates at rates of up to 1000 bits per second. Photo-Darlington speed is enhanced by cascode constant-voltage biasing. The output threshold requirements are provided for by sensing the output voltage and providing negative feedback to adjust the cascode transistor bias point. Circuit tests confirmed operation to 2000 bits per second at ambient light levels equal to signal levels.

SECTION **2** *Control and Sensing*

2.6 Intrusion and Hazard Detectors

The circuits in this subsection are designed to trigger an alarm when a preestablished condition or event occurs that is considered a hazard; included are burglar alarms, smoke and gas detection devices, and an audio alarm that can be interconnected with a variety of TTL-output sensors.

If the "unsafe condition" you want to guard against is not covered by the circuits in this subsection, you should be able to adapt some of those shown here or in the preceding subsection so that they can be operated with the transducer most applicable to your requirement.

Alarm systems are perhaps the easiest of all circuit types to design because they typically require a go/no-go indication rather than an analog of some voltage or frequency. The design process is typically reduced to a systematic selection of the most appropriate sensor or transducer followed by an appraisal of the modular circuit elements already existing that can operate within the framework of the chosen sensor's minimum and maximum values.

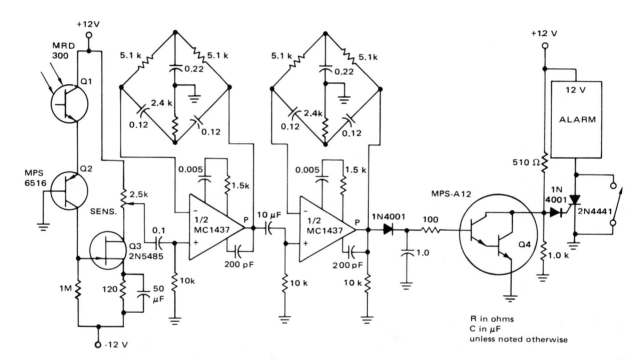

2.6.1 Motorola phototransistor intrusion alarm uses modulated light and is frequency-selective. An MRD300 phototransistor with a common-base speedup stage drives FET preamplifier Q3. The preamp output is fed to a pair of series-connected bandpass amplifiers that use twin-tee feedback networks for frequency selectivity. The filter–amplifier output is fed to a peak detector circuit that holds Q4 in saturation. This deprives the SCR of trigger voltage.

When the input to the phototransistor is interrupted, the SCR is triggered on and the alarm sounds.

The light source is modulated to match the pass frequency of the receiver. Small lamps (No. 327, for example) can be electrically modulated up to several hundred hertz.

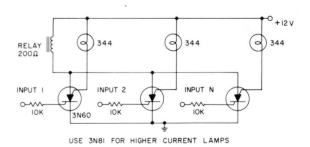

2.6.2 GE multiple-entry intrusion alarm circuit. Any of several inputs pulls in a common alarm relay with lamps giving a visual indication of triggering input. Low-resistance lamps decrease the input sensitivity. Inputs can be provided by actuated (open) normally closed switches. Opening a door or window releases the switch to trigger the alarm.

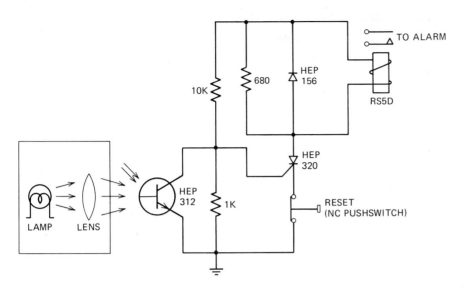

2.6.3 Motorola photoelectric intrusion alarm will close a relay when a light beam is broken. Use a good light source and lens to focus the light beam across the doorway or window to be protected. This circuit requires a 6V dc relay with a coil resistance of 300 to 400 ohms, such as Potter & Brumfield RS5D. The reset switch should be concealed to prevent an intruder from deactivating the system. The power source can be any 9 to 12V battery; the power requirement is minimal.

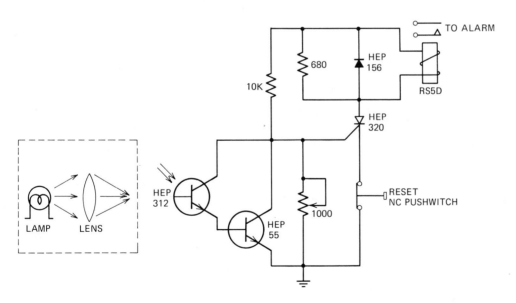

2.6.4 Motorola sensitive photoelectric intrusion alarm has a sensitivity adjustment to allow for varying light-source brightness. This circuit is ideal for stores with wide entrances that are normally unattended. Enclose the light source in a protected box and focus the beam so that it strikes the phototransistor squarely; the light box must be mounted securely to prevent inadvertent defocusing and misalignment of the light beam.

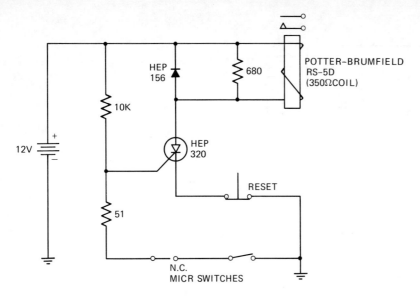

2.6.5 Motorola automotive burglar alarm is triggered when an opened, normally closed microswitch releases the SCR gate from its grounded condition even for an instant. When the SCR begins to conduct, it stays on regardless of whether the gate on the SCR is provided with trigger voltage. To reset the alarm to the armed state, the anode–cathode circuit of the SCR must be broken momentarily.

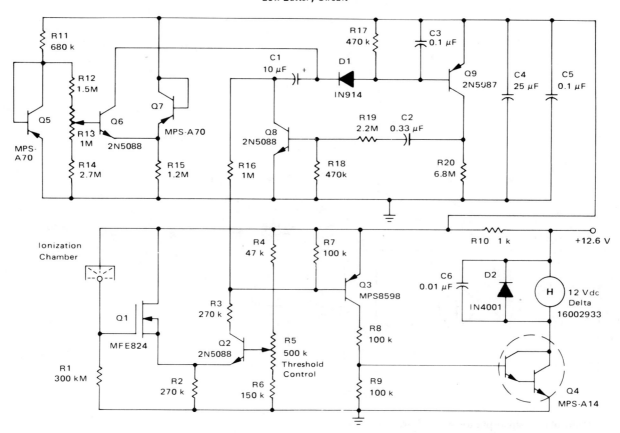

2.6.6 Motorola ionization chamber smoke detector for use when the smoke alarm signal is a continuous one rather than a pulsating one. The standby current for this discrete smoke detector is approximately 70 μA. When Q3 turns on, it supplies the 100 μA base current to the Darlington Q4 (MPS-A14). The worst-case low-temperature h_{FE} (approximately 5000) of this transistor insures that the transistor is fully saturated during the peak-current horn startup condition. The horn will be powered continuously for as long as the smoke content exceeds the detector threshold setting.

162 DISCRETE / TRANSISTOR CIRCUIT SOURCEMASTER

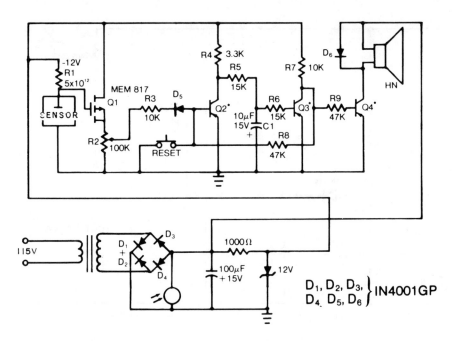

2.6.7 Line-operated smoke detector. General Instrument *Superectifiers* are used in the bridge power supply circuit (D1 through D4). Because of their high reliability and low cost, they are also used as circuit diodes in positions D5 and D6. The sensor is a single ionization chamber, typical of popular smoke detectors.

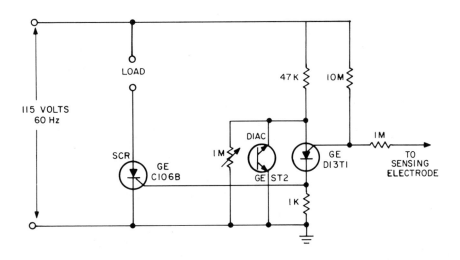

2.6.8 GE proximity detector is actuated by an increase in capacitance between a sensing electrode and the ground side of the line. The sensitivity can be adjusted to switch when a human body is within inches of the insulated plate used as the sensing electrode. Thus this circuit can be used as an electrically isolated touch switch or as a proximity detector in alarm circuits.

The GE D13T1 programmable unijunction transistor (PUT) will switch on when the anode voltage exceeds the gate voltage by 0.5V (the trigger voltage). This anode voltage is clamped at the on voltage of the diac. As the capacitance between the sensing electrode and ground increases (due to an approaching body), the angle of phase lag between the anode and gate voltages of the D13T increases until the voltage differential at some time is large enough to fire this PUT. Because the anode voltage is clamped, it is larger only at the beginning of the cycle; hence switching must occur early in the cycle, minimizing RFI.

Sensitivity is adjusted with the 1M potentiometer, which determines the anode voltage level prior to clamping. This sensitivity will be proportional to the area of the surfaces opposing each other.

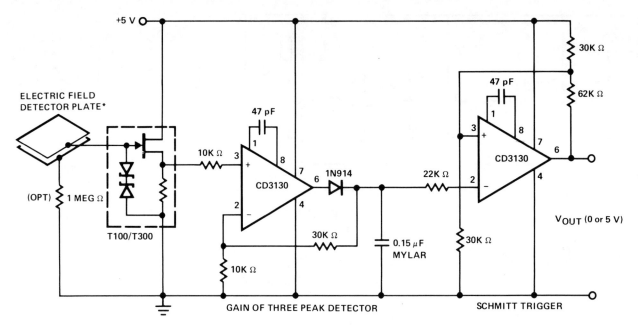

2.6.9 Siliconix self-biased proximity detector works on detected changing field. The sensor plate can be any insulated metallic piece, such as a double-clad PC board. The sensor connects to a Siliconix T300 JFET impedance converter, which drives a gain-of-3 peak detector followed by a Schmitt trigger. The output will be logic 0 (0V) or 1 (5V).

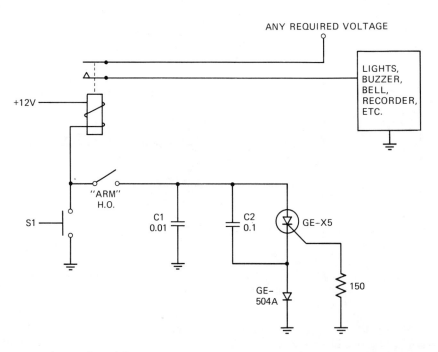

2.6.10 General-purpose alarm adapted from a GE automotive theft alarm. In a car the momentary-contact switch can be the horn button, since this already provides a grounding function. For other applications this switch will serve as the alarm trigger. In a car, any voltage drop at the battery will cause the horn to blow when the ARM switch is on. In other applications, S1 causes triggering when it is activated. S1 can be a normally closed momentary switch that is held open by a shut door or closed window. Releasing the switch grounds the normally 12V line to pull in the relay and trigger the alarm.

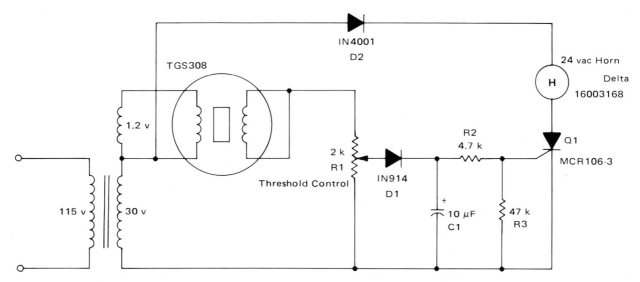

2.6.11 Motorola simple gas/smoke detector using the TGS308 semiconductor sensor and an SCR for half-wave control of an ac horn. The TGS sensor (available from JEMA, 7500 N. Kolmar Ave., Skokie, Ill. 60076) in series with the 2K threshold control potentiometer R1, is powered by the prescribed 1.2V rms heater voltage and 30V rms circuit voltage (manufacturer's specified range: 5 to 30V). When no gas is present, the output voltage (across R1) is approximately 3V. When the sensor detects gas or smoke, the output voltage will rise due to the decrease in resistance of the sensor to some value proportional to the gas concentration, typically 20V in large concentration. This signal is tapped off the threshold control potentiometer (to vary sensitivity), and rectified and filtered to provide dc gate control of the SCR. The fired SCR half-wave powers the ac horn with 21V rms ($e_{rms} = e_{peak}/2$). The SCR remains fired as long as the gate is biased on. Once the gas or smoke has cleared the sensor, the SCR will commutate off when its holding current is reduced to zero.

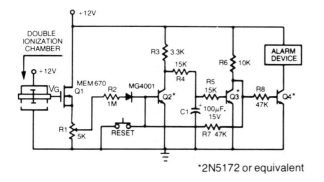

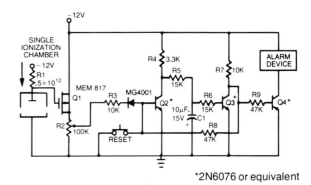

2.6.12 Smoke detector circuit uses General Instrument's MEM 670 N-channel depletion-mode MOSFET operating as a source follower. The sensor shown in the diagram is a double ionization chamber, electrically equivalent to two very large resistors connected in series, with the gate of the MOSFET connected at midpoint. The quiescent voltage at the output of the ionization chamber ranges from approximately +6V (no smoke present) to +11V (smoke present); in the "no smoke present" condition the effective resistances of both chambers are approximately equal, and for the "smoke present" condition the effective resistance of the open chamber is greatly increased by the action of smoke particles on ionized air molecules.

2.6.13 General Instrument smoke detector using a MEM 817 P-channel enhancement-mode MOSFET as the buffer amplifier employs a sensor that consists of a single ionization chamber with a large-value resistor connected in series. Quiescent voltage values at the output of the chamber vary from about −4 to −6V, and detection of smoke will result in an excursion of about −4V.

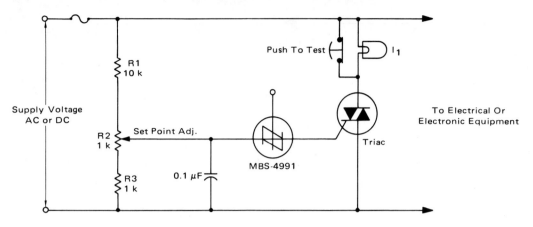

2.6.14 Motorola electronic crowbar, for applications in which it is economically desirable to shut down equipment rather than allow it to operate on excessive supply voltage. Since the triac and SBS are both bilateral devices, the circuit is equally useful on ac or dc supply lines. With the values shown for R1, R2, and R3, the crowbar operating point can be adjusted over the range of 60 to 120V dc or 42 to 84V ac. The resistor values can be changed to cover a different range of supply voltages. The voltage rating of the triac must be greater than the highest operating point as set by R2. A low-power incandescent lamp (I1) with a voltage rating equal to the supply voltage may be used to check the set point and operation of the unit by opening the test switch and adjusting the input or set point to fire the SBS. An alarm unit such as the Mallory Sonalert may be connected across the fuse to provide an audible indication of crowbar action. *This circuit may not act on short, infrequent power-line transients.*

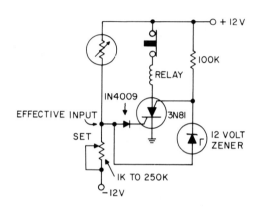

2.6.16 Safety alarm using GE semiconductors triggers relay unless the input is within a volt of ground.

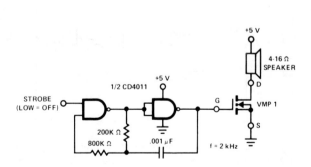

2.6.15 Inexpensive audio alarm system. The Siliconix 34011 gates provide a 2 kHz 50% duty cycle pulse train to the VMP 1, which directly drives an 8-ohm speaker. The VMP 1 can drive nearly any load of 2A or less (3A under pulsed conditions), but some care must be taken to insure proper gate enhancement to support the required current. When V_{GS} is 5V, a maximum of 650 mA will flow regardless of the drain-to-source voltage. For a drain current of 1A, at least 6.25V must be applied to the gate. To allow for worst-case conditions, however, a minimum of 10V should be applied.

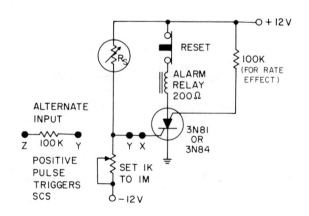

2.6.17 GE universal alarm may be used with a variety of resistive elements. Temperature-, light-, or radiation-sensitive resistors up to 1 megohm readily trigger the alarm when they drop below the value of a preset potentiometer. Alternately, 0.75V at the input (across the 100K resistor) triggers the alarm. Connecting the SCS between ground and −12V permits triggering on a negative input. When R_S decreases sufficiently to forward-bias the SCS, the alarm is activated. Interchanging R_S and the potentiometer triggers the SCS when R_S increases.

2.7 Phone-Line Sensors

The two current-carrying wires of a telephone system convey digital and analog signals that we can use for a wide variety of purposes.

Until fairly recently it was illegal for any equipment to be interconnected with a telephone without the sanction of the phone company; this sanction was almost impossible to get without use of phone-company equipment installed by phone-company personnel. The well known Carter case, in which a company engaged in the manufacture and sale of a telephone-to-radio interfacing system was given the go-ahead against the protest of AT&T, opened the door to experimenters and speculative manufacturers throughout the country, who have devised myriad ingenious systems that depend on phone-line interconnection for their marketability.

It is extremely important to remember that equipment cannot just be connected to a phone line without consideration of the effects such an interconnection might have. Whether we want to operate an automatic phone-answering device or activate external switches when the phone rings, we must provide adequate isolation between the existing phone line and the devices we use or be faced with continual misdialing, excessive phone-line current drain, or simply unsatisfactory operation of the phone system from such pesky problems as hum, noise, feedback, insufficient or excessive signal gain, crosstalk, and a host of other anomalies.

The circuits in this subsection do offer the degree of isolation required for proper phone operation after hookup. Use these circuits as a guide for designs of your own innovation.

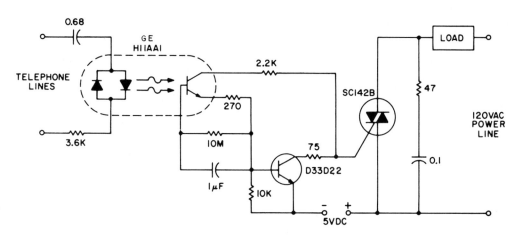

Maximum Load: 500 W Lamp or 800 W Inductive or Resistive

2.7.1 GE telephone ring relay uses +5V from an external source (as a battery) to switch high current power-line loads when the phone rings, while maintaining 100% isolation from the telephone line. The use of GE's isolated-tab triac simplifies heatsinking by removing the constraint of isolating the triac heatsink from the chassis.

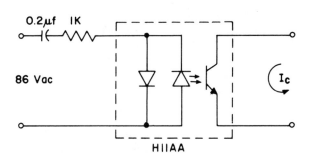

2.7.2 GE telephone ring detector detects the presence of a ring signal in a phone system without any direct electrical contact between the alarm and the line. When the 86V ac ring signal is applied, the output transistor of the GE H11AA is turned on, indicating the presence of a ring signal.

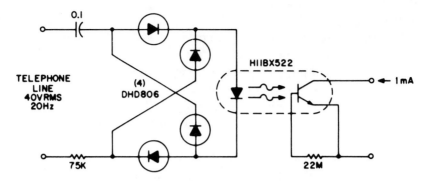

2.7.3 Low-current-loading telephone ring detector results from using the GE H11BX522 photo-Darlington optocoupler, which provides a 1 mA output from 0.5 mA input throughout the −25 to +50°C temperature range.

The circuit allows ring detection down to a 40V ring signal while providing 60 Hz rejection to about 20V rms. Zero-cross filtering may be accomplished either at the input bridge rectifier or at the output.

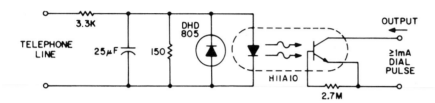

2.7.4 GE telephone-dial pulse indicator senses the switching on and off of the phone-line voltage and transmits the pulses to logic circuitry. A GE H11A10 threshold coupler, with capacitor filtering, rejects high levels of induced 60 Hz noise. The DHD 805 diode provides reverse-bias protection for the LED during transient overvoltage situations. The capacitive filtering removes less than 10 ms of the leading edge of a 60V dial pulse while providing rejection of up to 25V at 60 Hz.

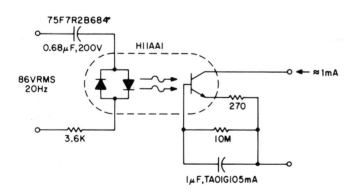

2.7.5 GE telephone ring detector uses the 20 Hz ring signal on the telephone line and initiates action in an electrically isolated circuit. Typical applications would include automatic answering equipment, interconnect/interface, and key systems. The ring detector shown provides about a 1 mA signal for a 7 mA line loading 100 ms after the start of the ring signal. The time-delay capacitor provides a degree of dial click suppression and filters out the zero crossing of the 20 Hz wave.

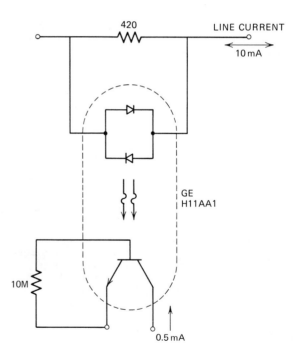

2.7.6 GE phone-line current detector. Detecting line current flow and indicating the flow to an electrically remote point are required in line-status monitoring at a variety of points in the telephone system and auxiliary systems. The line should be minimally unbalanced or loaded by the monitor circuit, and relatively high levels of 60 Hz induced voltages must be ignored. The GE H11AA1 allows line currents of either polarity to be sensed without discrimination and will ignore noise up to approximately 2.5 mA.

2.8 Remote Control and Servos

Remote control can mean different things to different people. To the RC modeler it might mean servos and mechanical linkages; to the communications worker it might mean turning circuits on and off by radio or by frequency-selective circuits in response to precise tones. This subsection is a potpourri of "remote control" circuit ideas that fall into a variety of technological categories.

If servo control is your primary interest, don't forget to also check the audio amplifier complement (Section 7), since many of those shown will be equally adaptable to servo utilization.

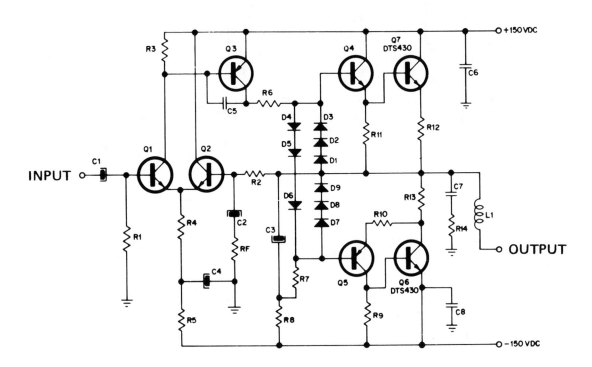

TRANSISTORS

Q_1, Q_2, Q_4 — 2N3440
Q_3, Q_5 — 2N5416
Q_6, Q_7 — DTS-430

RESISTORS

R_1, R_2 — 39k
R_3 — 390
R_4 — 47k
R_5 — 27k
R_6 — 680
R_7, R_8 — 10k
R_9, R_{11} — 100
R_{10} — 5.6
R_{12}, R_{13} — 1.8, 4W
R_{14} — 22, 1W
R_F — 560

CAPACITORS

C_1 — 5µF, 6V
C_2 — 50µF, 6V
C_3 — 100µF, 150V
C_4 — 5µF, 100V
C_5 — 100pF, 300V
C_6, C_7, C_8 — 0.1µF, 150V

DIODES

D_1 to D_9 — 1N3253

INDUCTORS

L_1 — 10µH

2.8.1 Delco direct-coupled 100W servo, industrial control, public address amplifier with a quasi-complementary output stage. Flat frequency response and low output impedance (75 ohms) are its outstanding attributes. A minimum number of circuit components are required.

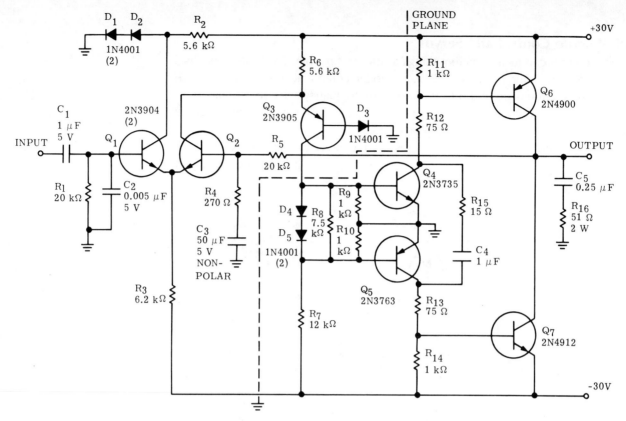

2.8.2 Motorola complementary-transistor servo amplifier eliminates transformers, using direct-coupling throughout. Whereas the transformer-coupled servo amplifier can drive only ac motors, the complementary amplifier can be used with both ac and dc loads.

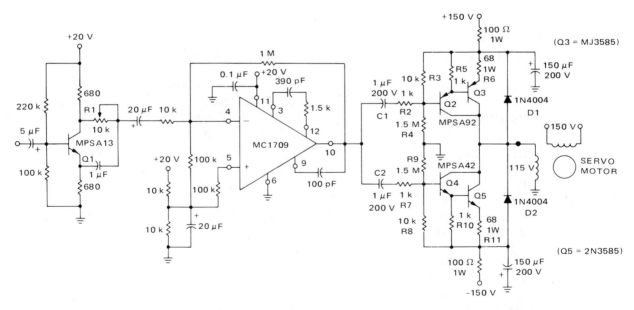

2.8.3 Motorola line-operated servo amplifier consists of three blocks: the push-pull RC phase shifter, the single-ended op-amp preamplifier, and the push-pull class B power amplifier. The power amplifier is of a complementary transistor design using high-voltage (300V) transistors in a common-emitter Darlington configuration.

When using balanced plus-and-minus power supplies, the amplifier output can be directly coupled to ground-referenced servomotor control phase. Thus the coupling capacitor can be eliminated, since the amplifier quiescent voltage is zero.

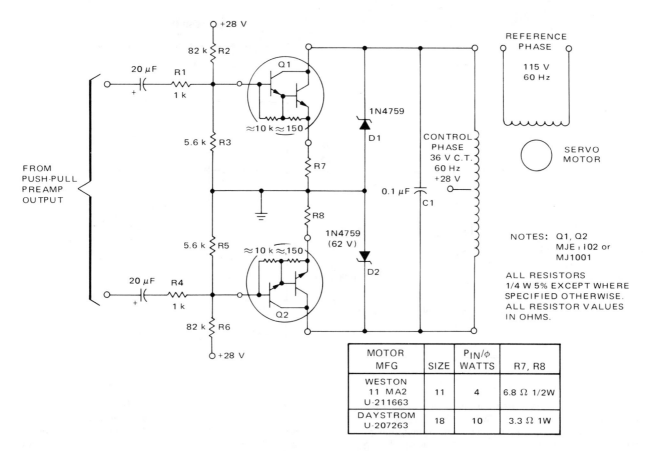

2.8.4 Motorola push-pull power amplifier for low-voltage servo uses power Darlingtons for high current gains and input impedances. The input is driven push-pull and is capacitively coupled to eliminate the input transformer; thus the servo amplifier is completely transformerless. The Darlington collectors are connected directly to the control phase, which results in a collector/collector voltage swing of approximately 4 times the 24V supply voltage less the transistor saturation voltages and emitter resistor voltage drops. This ×4 factor is the result of the autotransformer action of the centertapped control phase. The control phase voltage therefore is 36V rms.

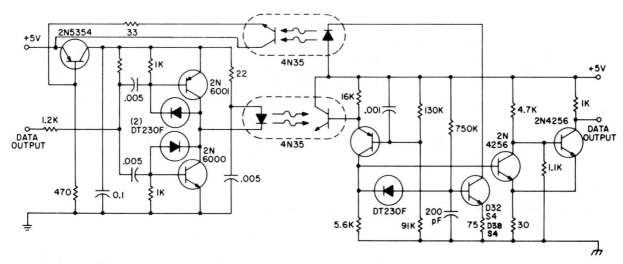

2.8.5 High-speed data transmission isolator/amplifier uses GE's 4N35 as the isolated data link. The circuit provides the data transmission amplifier, the data transmission link, the data receiver amplifier, and bias-current feedback, which allows very high-speed operation of the coupler. By changing the bias resistor values to maintain the threshold and current levels, power supplies of other than 5V can be used. The push-pull driver is used on the data transmission isolator in conjunction with the 0.005 µF capacitor to allow the rapid injection and removal of the IRED's charge. As the effect of incomplete removal of this stored charge is a dc component of light output at high frequency, the feedback network sets feedback current in the data transmitting isolator to minimize this. Hysteresis in the output amplifier sharpens the rise and fall times of the output while maintaining pulse-width relationships. Resistors should be 5%, ½W. Use high-frequency wiring techniques.

SECTION 2 Control and Sensing

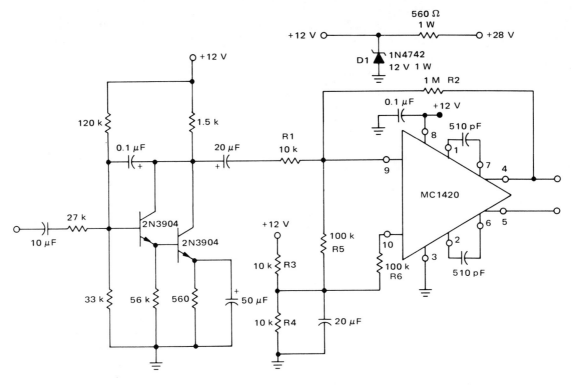

2.8.6 Motorola servo preamplifier uses the 90° operational integrator driving an MC1420 differential-input, differential-output op-amp connected in an inverting configuration. With a single-ended input, a push-pull output results that can drive Darlington power amplifiers. The voltage gain is set by resistors R1-10K and R2-1M, resulting in a gain of approximately 30 dB. The op-amp has excellent operating point stability due to the high dc feedback. The bandwidth is kept relatively narrow (approximately 4 kHz) for servo system stability by the 510 pF compensation capacitors. This op-amp is specified for dual supplies of ±8V maximum.

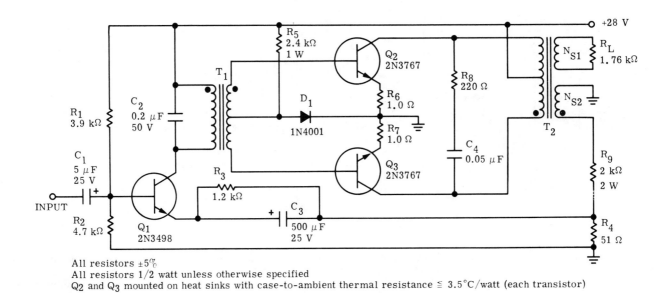

All resistors ±5%
All resistors 1/2 watt unless otherwise specified
Q_2 and Q_3 mounted on heat sinks with case-to-ambient thermal resistance ≤ 3.5°C/watt (each transistor)

2.8.7 Motorola transformer-coupled servo amplifier.
Because of the excellent impedance matching that can be achieved with transformers, little power is lost between stages. Only three transistors are required to provide a stable voltage gain of 100.

172 DISCRETE / TRANSISTOR CIRCUIT SOURCEMASTER

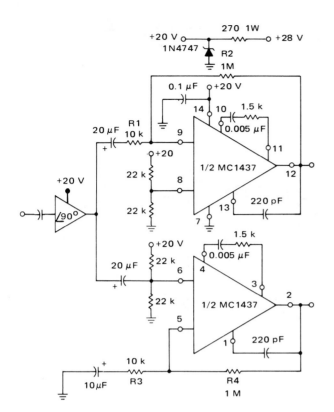

2.8.8 In this Motorola servo preamp, two op-amps are connected in parallel from a common input, one in an inverting configuration and the other noninverting. The respective outputs are complementary. A dual op-amp MC1437 is powered by a single +20V zener-regulated supply. This device can sustain a dual supply of ±18V max. The voltage gain, 40 dB, is set by 10K resistor R1, 1M resistor R2 in the inverting configuration. Direct-current stability is excellent by virtue of the high dc feedback. Single-supply operation is obtained by the 22K resistor networks connected to each amplifier input. The bandwidth is limited to approximately 6 kHz by the input lag network of a 0.005 μF capacitor, a 1.5K resistor, and an output lag capacitor of 220 pF.

2.8.10 Workman solar-operated device is designed as a **remote control for small models**. The motor should be a 1.5 to 3V low-friction motor, such as IR's EP50. The circuit is controlled by applying a light beam, such as from a flashlight, to the solar cell. The resulting voltage is applied as the base bias of transistor Q1. This amplified signal is applied to transistor Q2. When the solar cell is dark, no bias current flows. As light is applied, more and more power is applied to the motor. A heatsink should be used for transistor Q2.

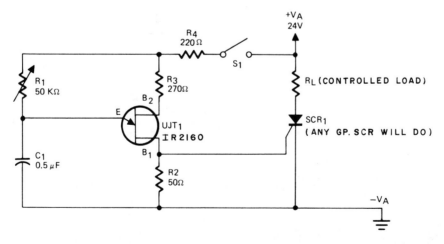

2.8.9 Workman unijunction transistor oscillator circuit applies positive pulses to the gate of the SCR. The value and rating of the load and of SCR1 may be varied depending on application. When the switch is closed, capacitor C1 starts to charge through resistors R4 and R1 until the applied voltage reaches the firing voltage of the unijunction transistors. As the transistor fires, the capacitor discharges through the emitter into base 1, developing the pulse to trigger the SCR on. Increasing or decreasing the value of R1 varies the charging time of C1 and the firing time of the SCR.

SECTION 2 Control and Sensing

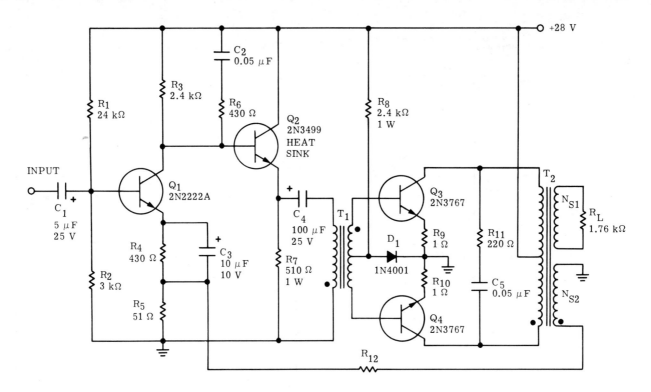

All resistors ±5%
All resistors 1/2 watt unless otherwise specified
Q_3 and Q_4 mounted on heat sinks with case-to-ambient thermal resistance $\leq 3.5°C/watt$ (each transistor)
Q_2 mounted on heat sink with case-to-ambient thermal resistance $\leq 25°C/watt$

2.8.11 Motorola transformer-coupled servo amplifier.

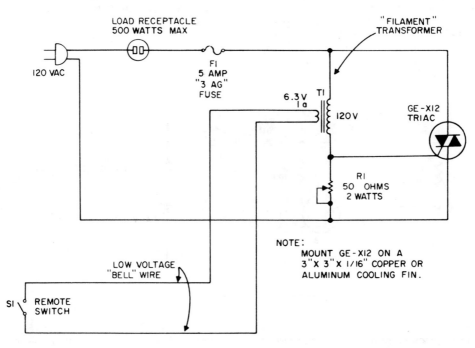

2.8.12 GE remote control for lamp or appliance can switch up to 500W. The primary current of a small 6.3V filament transformer actuates a triac and energizes the load. When switch S1 in the secondary of the transformer is open, a small current flows through the primary wind- ing; since this may be large enough to trigger the triac, a shunting resistor, R1, is required to prevent it. Adjust it for the highest resistance that will not cause the triac to trigger with S1 open.

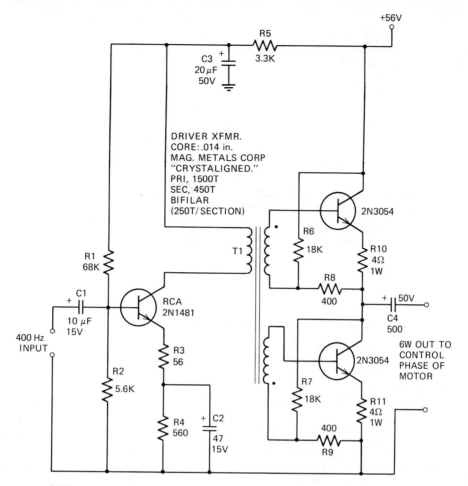

2.8.13 RCA servo amplifier can supply up to 6W of power to the drive motor of a servo system. The amplifier is driven by a 400 Hz ac signal and is operated from a dc supply voltage of 56V. A pair of RCA 2N3054 power transistors are used in a class AB push-pull output stage to develop the required output power.

A 2N1481 common-emitter input stage amplifies the 400 Hz input to the level required to drive the output transistors. The amplified 400 Hz signal at the collector of the 2N1481 transistor is coupled to the base of each 2N3054 output transistor by transformer T1. The secondary of T1 is split to form two identical windings oriented so that the inputs to the output transistors are equal in amplitude and 180° out of phase.

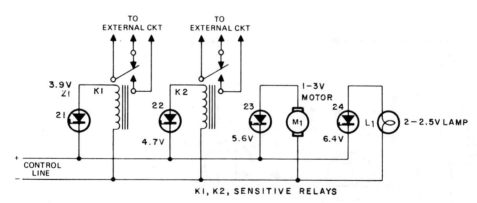

2.8.14 Workman zener remote control circuit is a unique demonstration unit as well as a highly practical control unit. The zener diodes are used to control a variety of discrete control functions over a single pair of control lines. By using a variable voltage input, you can control the functions of the devices shown in the circuit (or any number of functions) as long as power is available. Zeners are chosen in accordance with the sequence of operation. For example, let us assume that Z1 is rated at 3.9V. All of the devices would remain inactive as long as the line voltage remained less than 3.9V. If the voltage exceeds 3.9V but is less then 4.7V, Z1 would conduct, permitting the relay to close and actuating the relay's system. If the control voltage is raised still further to over 4.7V but below 5.6V, the second zener would conduct, its relay closing and its circuit going into operation. Of course, Z1 would also be conducting, as would all control devices operating from voltages less than the zener being keyed. The circuits connected to the diodes must be able to handle the total voltage of the control line—in our example, 9V. Your example may be different. Use Ohm's law to determine the wattage required for the zeners.

SECTION **2** *Control and Sensing*

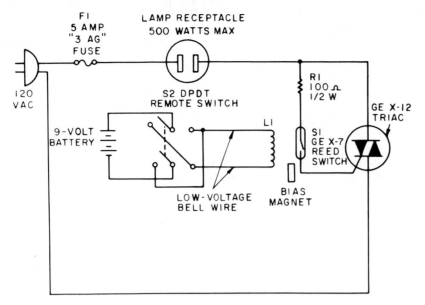

2.8.15 GE transformerless remote control of 500W circuit. When switch S1 (single-pole, double-throw) is closed in the on position, the magnetic field of the reed switch coil, L1, is reinforced by the bias magnet field and the reed switch closes. This triggers the triac and energizes the load. When the remote switch is released to the open position, the force of the bias magnet keeps the reed switch closed and the load remains energized. Depressing the switch to off and then returning to open reverses the coil field, which opposes the bias magnet field, and the reed switch opens. This turns off the triac and the load.

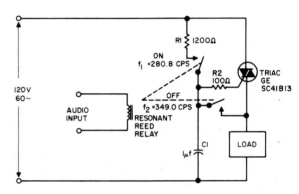

2.8.17 GE radio system remote control circuitry for power on-off functions in tone-responsive circuitry. Here two individual tones control on and off functions.

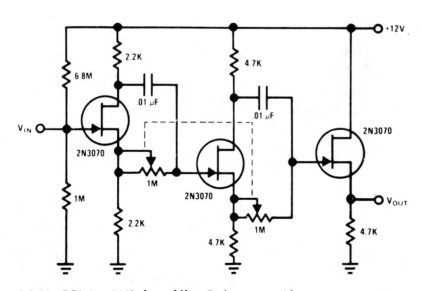

2.8.16 RCA 0 to 360° phase shifter. Each stage provides 0 to 180° shift. By ganging the two stages, a 0 to 360° phase shift is achieved. The 2N3070 JFETs are ideal, since they do not load the networks.

Section 3
TIMING CIRCUITS

The circuits in this section depend on timing networks for their operation—the categories are flashers, ring counters, and timers. Flashers are circuits that either turn a single lamp on and off at predetermined intervals or turn on two alternating lamps for a prescribed switching period; the flasher is thus a very slow multivibrator with power-switching capability. The ring counters featured here are extensions of the flasher in that power switching on a timed basis is employed, but typically more than two switched lines are involved; the counter turns on and off an array of lamps (or other loads) sequentially. It is no coincidence that most of the power-switching circuits of the first two subsections incorporate thyristors at the gated current carrier; such devices require but small trigger currents and are capable of handling heavy electrical loads.

The final subsection contains time-delay circuits, delayed-dropout devices, and adjustable-period timers for short, medium, and long durations.

3.1 Flashers

A flasher can be an important tool for the motorist who finds it necessary to stop along the narrow shoulder of a heavily trafficked roadway, but it is also the ideal device for attracting attention in darkness, whether for advertising or for marking hazards. The single most necessary ingredient in a flasher is reliability—the flasher usually has to operate with no "operator" in attendance, and the individual who positions a flasher for his own safety certainly cannot afford to have the device malfunction during use. Except for the obvious low-current "toys," these circuits have all been built and tested to insure continuous and predictable operation under a variety of environmental conditions.

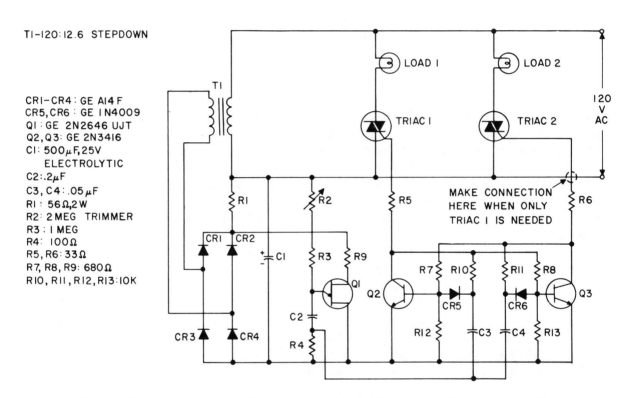

3.1.1 GE ac flasher with no moving parts is built around a free-running unijunction oscillator triggering a transistor flip-flop that fires two triacs capable of handling 1 kW load each. If a single lamp output with only on-off performance rather than two alternately flashing lamps is desired, triac 2 can be left out, but making the proper connection as shown. The operation of the circuit is as follows: Transformer T1, diodes CR1 through CR4, resistor R1, and capacitor C1 provide the dc supply to oscillator Q1 and to the flip-flop of Q2, Q3. Because of the ripple on base 2 of unijunction Q1, C2 can reach the peak-point voltage of Q1 only at the beginning of the half-cycles, thus firing Q1 early in the half-cycle. The frequency of oscillation of Q1 is determined by the setting of R2.

Resistors R5 and R6 form a divider network with R1 supplying about 6V to the flip-flop. When Q2 is on and Q3 is off, the collector of Q2 will be at a negative potential with respect to the gate and terminal 1 of triac 1, while the collector of Q3 will be at the same potential as the gate and terminal 1 of triac 2. The negative potential seen at the gate will cause the current to flow out of the gate from the positive side of the dc supply, through terminal 1 and the gate of triac 1, R5, the collector of Q2, and then out of the emitter of Q1 to the negative side of the dc supply. Gate current flow of triac 1 will cause it to conduct, energizing load 1. Since the gate and terminal 1 of triac 2 do not see a different potential, there will be no current flow to or from the gate so that triac 2 will remain off.

Timing capacitor C2 charges through R2 and R3, and when the voltage across it reaches the peak-point voltage of the unijunction, it discharges to produce a negative-going pulse across resistor R4. This will change the state of the flip-flop, turning Q2 off and Q3 on, causing triac 1 to stop conducting and triac 2 to conduct.

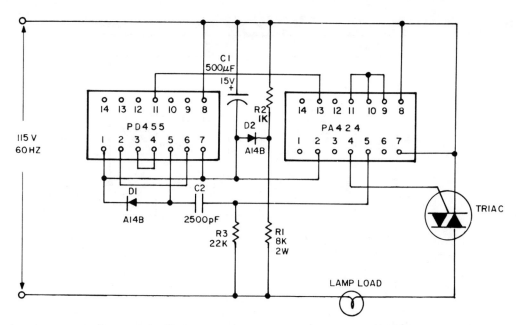

3.1.2 GE ac flasher minimizes the space needed for trigger circuitry. The two IC packages, two capacitors, two diodes, and three resistors make up the entire circuitry. The resistor divider, along with D2 and C1, form a dc supply to power the two integrated circuits. The PD455 is a six-stage frequency divider. The stages are connected in such a series that the input of 60 Hz line frequency is divided by 2 six times. This gives a final output period of $64/60$ or just over a second. (If five stages were used, the period would be $32/60$ second, etc.). The PA424, a zero-voltage switch, triggers the triac every half-cycle while the output of the PD455 is on. When the PD455 output switches to the other state, the PA424 does not trigger, and the triac is off. For different frequencies, move the connection from pin 11 of the PA455 and connect to any pin, from 9 to 14; each is a different frequency.

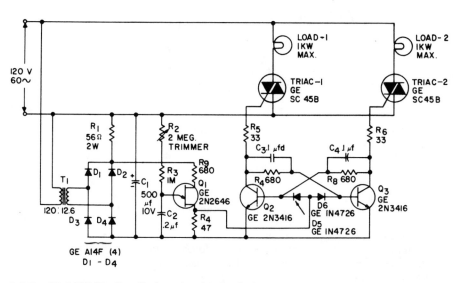

3.1.3 GE 1 kW flip-flop flasher circuit. The flashing rate can be adjusted from about 100 ms to a 10 s cycle time.

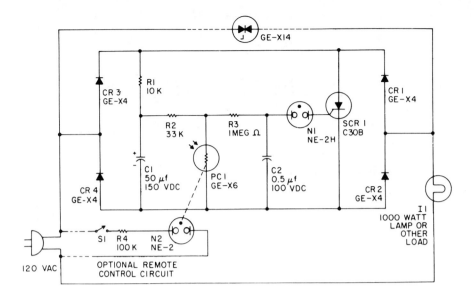

3.1.4 GE 1 kw flasher with photoelectric control can be used to actuate warning lights on towers, piers, or construction hazards. The controlled devices may consist of motors, sirens, neon signs, or incandescent lamps, up to a total power load of 1000W. Operation of the control may be by its photoelectric cell, which could start the lamp (or other load) flashing after sunset and turn it off at dawn. In addition, a highly sensitive remote control of the flasher is available by merely adding a neon lamp to actuate the photoelectric cell from an isolated source.

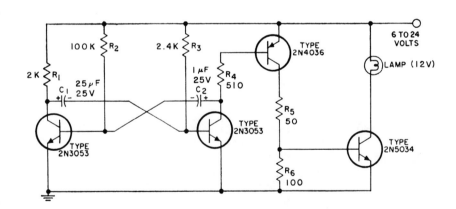

3.1.5 In this RCA lamp flasher circuit, a free-running multivibrator is used to gate the operation of a two-stage amplifier. An incandescent lamp or other load may be connected in series with the collector of the output transistor, and each time the transistor conducts, voltage will be applied across the load. The input power may be any dc voltage from 6 to 24V.

The rectangular wave developed at the collector of the second transistor is resistively coupled to the base of the 2N4036, gating it on and off. This stage in turn gates the operation of the 2N5034 used in the output stage. A lamp is connected from the positive side of the power supply to the collector of the output transistor. The lamp flashes at the frequency of the multivibrator.

The repetition rate ($\approx$ 2 pps) may be changed by altering the values of C1 and C2. The on time changes proportionally with C2, and the off time changes proportionally with C1.

The dissipation requirements of the resistors are proportional to the square of the supply voltage; for 12V operation the dissipation rating for R5 is only 2.5W. Bulbs and resistive loads up to 2A may be used; however, if the flasher circuit is used to switch loads that have inductive components, diode protection must be provided for the 2N5034 output transistor.

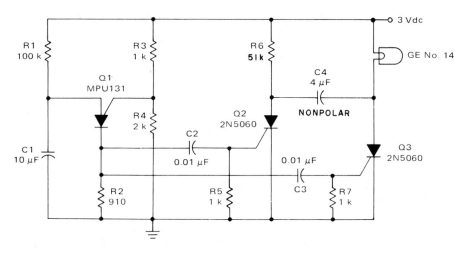

3.1.6 Motorola low-voltage lamp flasher is composed of a relaxation oscillator formed by Q1 and an SCR flip-flop formed by Q2 and Q3. With the supply voltage applied to the circuit, timing capacitor C1 charges to the firing point of the PUT (2V plus a diode drop). The output of the PUT is coupled through two 0.01 μF capacitors to the gates of Q2 and Q3. Note that C4 is nonpolarized.

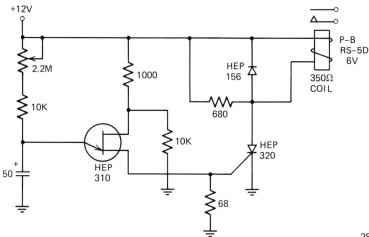

3.1.7 Motorola photographic timer is adjustable from 2s to nearly 5 min. The relay pulls in at the end of the timed period, which begins as soon as voltage is applied to the power input terminal. The 2.2M pot may be a screwdriver trimmer mounted on a PC board or a linear-taper panel-mounting pot. The unijunction transistor fires after the 50 μF capacitor charges sufficiently.

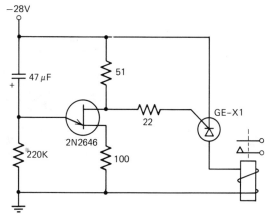

3.1.8 Sessions 15s interval timer is designed for operation with negative 28V repeater control systems. The relay pulls in 15s after −28V is applied at power terminal. Since the 220K resistor controls the timed period, and since the relation between this resistor's value and the period is linear, simply increase the value of the resistor to lengthen the period.

SECTION 3 *Timing Circuits* **181**

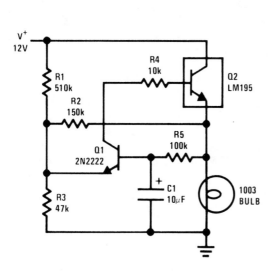
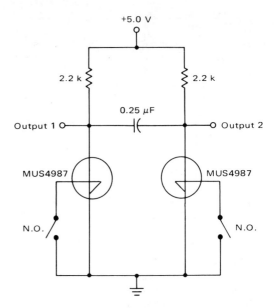

3.1.9 National power lamp flasher is designed to flash a 12V bulb at about a once-per-second rate. The reverse base current of Q2 provides biasing for Q1, eliminating the need for a resistor. Typically, a cold bulb can draw 8 times its normal operating current. Since the LM195 is current limited, high peak currents to the bulb are not experienced during turnon. This prolongs the bulb life as well as easing the load on the power supply.

3.1.11 Motorola bistable switch with memory. Initially, both SUSs are in the off state, since the supply voltage is less than V_s of the devices. A momentary contact in either of the normally open switches will gate its SUS on. That device will remain in the on state and its anode voltage less than +1V until the other device is gated on. The negative anode swing of the SUS turning on is coupled through the commutating capacitor and reverse-biases the other device, turning it off. The mechanical switches may be replaced, of course, by other solid-state devices or a negative-going pulse referenced to the supply line.

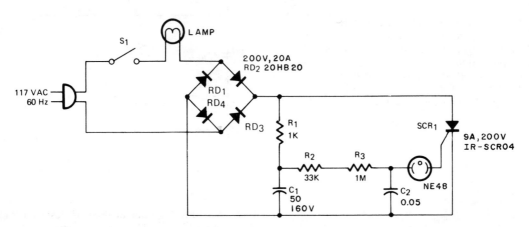

3.1.10 In this Workman 115V flasher system the two SCRs operate somewhat like a flip-flop circuit, where they alternately conduct and then turn off. Note that the lamp is between the input voltage and the rectifier bridge. The rest of the circuit operates on pulsating direct current supplied through the bridge. Resistor R1 and capacitors C1 and C2 form a charging network. When capacitor C2 exceeds the breakdown voltage of the neon bulb, the capacitors discharge into the gate of SCR1 and fire on it. The SCR is turned off as the line voltage drops to zero.

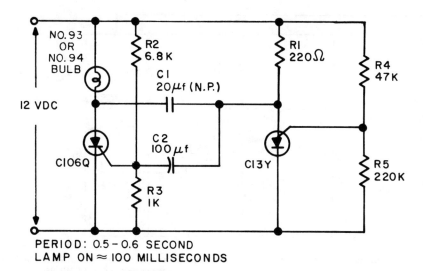

PERIOD: 0.5-0.6 SECOND
LAMP ON ≈ 100 MILLISECONDS

3.1.12 GE dc flasher has only two active devices: a C106Q SCR and a C13Y complementary SCR. Upon energizing the circuit, the C106Q must turn on, since C2 is discharged. C1 and C2 charge through R1 until the anode potential of the C13Y exceeds the gate potential by a diode drop. The C13Y turns on and commutates off the C106Q by means of C1. With the C13Y on and the C106Q off, capacitor C1 charges through the load. When the C13Y turns on, the charge of C2 drives the gate of the C106Q negative with respect to the cathode; this reverse bias aids in turning off the SCR.

Capacitor C2 discharges through the R2–R3 divider until the gate potential has risen high enough to trigger the SCR. The SCR turns on and commutates off the CSCR by means of C1.

The on-off ratio can be changed by changing the resistor ratios. Changing the R2–R3 divider alters the off time; changing the R4–R5 divider on the value of R1 changes the on time.

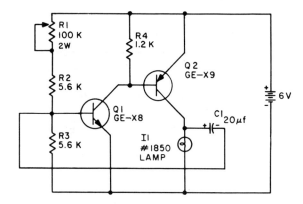

3.1.14 GE automatic lamp flasher is a two-stage, direct-coupled transistor amplifier connected as a free-running multivibrator. Both flash duration and flash interval can be changed by turning potentiometer R1.

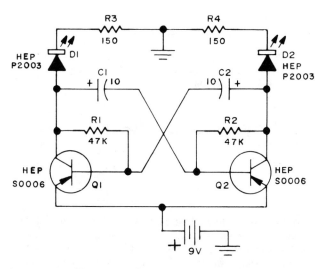

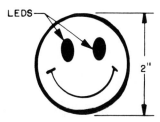

3.1.13 In this Motorola novelty blinker circuit, the "eyes" of the smiling face blink alternately. The circuit is also ideal for model railroad wigwag applications.

SECTION **3** Timing Circuits **183**

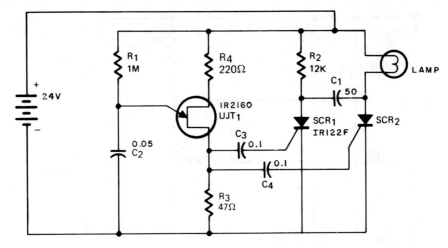

3.1.15 Workman SCR light flasher circuit is a battery-powered system using a unijunction transistor in the firing circuit. This circuit uses a neon light as a breakover unit to fire the SCR. The lamp used should be for flasher service, since it will get repeated heavy-duty use. The flasher unit operates from a unijunction transistor oscillator through capacitor C4. SCR2 produces a high-energy flash in the lamp. SCR1 turns itself off automatically because of the high resistance of R2, preventing sufficient current through SCR1 to maintain the conducting state.

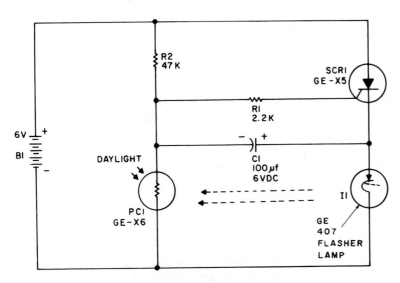

3.1.16 GE automatic flashing light is ideal for marking buoys and identifying road construction sites at night. The circuit operates only in the darkness. Use half-watt resistors.

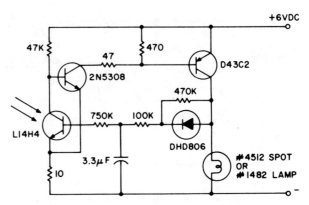

3.1.17 GE flashing light of high brightness and short duty cycle provides maximum visibility and battery life. This necessitates use of an output transistor that can supply the cold filament surge current of the lamp while maintaining a low saturation voltage, the addition of dynamic feedback, and the use of a phototransistor sensor to minimize sensitivity variation.

184 DISCRETE / TRANSISTOR CIRCUIT SOURCEMASTER

3.2 Ring Counters

The ring counters in this section differ from logic-circuit ring counters only in their power-handling capacity; these circuits are designed to switch relatively heavy loads on and off in a predictable sequence. They're used as *chasers*—strings of lights that are individually activated in rapid sequence to simulate the effect of movement—directional signals for cars and traffic-control arrows, and for a variety of home-entertainment effects.

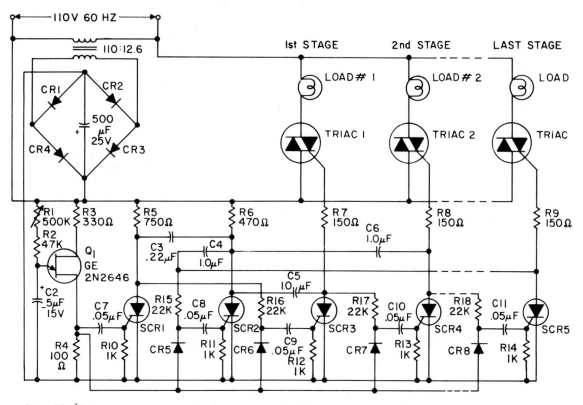

ALL SCR'S GE C106Y   ALL RESISTORS 1/2 W
ALL DIODES GE A14F
ADDITIONAL STAGES MAY BE ADDED BETWEEN DOTTED LINES

3.2.1 GE ac chaser turns on a string of lamps sequentially to produce the effect of moving. When power is applied, all the SCRs will be off. The free-running unijunction oscillator receives its power from the bridge rectifiers and filter capacitor C1. At the end of the R1 time delay the UJT will fire and the pulse at base 1 will only turn on SCR1. When the UJT fires again, SCR2 will turn on, thus firing triac 1. The next two pulses will turn on SCR4 and SCR5, in that order, firing triacs 2 and 3. The following pulse will fire SCR2, which will commutate off SCR1, SCR3, SCR4, and SCR5, removing the gate drive to all the triacs. The next pulse will start the cycle again, turning SCR1 on and SCR2 off. With this arrangement, the off time takes two pulses.

SECTION 3 Timing Circuits

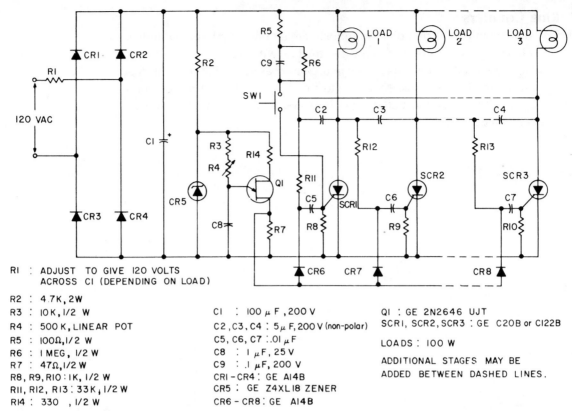

```
R1  : ADJUST TO GIVE 120 VOLTS
      ACROSS C1 (DEPENDING ON LOAD)
R2  : 4.7K, 2W
R3  : 10K, 1/2 W
R4  : 500K, LINEAR POT
R5  : 100Ω, 1/2 W
R6  : 1 MEG, 1/2 W
R7  : 47Ω, 1/2 W
R8, R9, R10 : 1K, 1/2 W
R11, R12, R13 : 33K, 1/2 W
R14 : 330 , 1/2 W

C1  : 100 μF, 200 V
C2, C3, C4 : 5 μF, 200 V (non-polar)
C5, C6, C7 : .01 μF
C8  : 1 μF, 25 V
C9  : .1 μF, 200 V
CR1–CR4 : GE A14B
CR5 : GE Z4XL18 ZENER
CR6–CR8 : GE A14B

Q1 : GE 2N2646 UJT
SCR1, SCR2, SCR3 : GE C20B or C122B

LOADS : 100 W

ADDITIONAL STAGES MAY BE
ADDED BETWEEN DASHED LINES.
```

3.2.2 GE 3-stage ring counter switches 100W. When power is first applied, diodes CR1–CR4 and capacitor C1 supply 120V dc to the anodes of the SCRs through the loads, causing diodes CR6, CR7, and CR8 to be reverse-biased. Out of this dc supply, zener CR5 provides 18V to the free-running unijunction oscillator circuit. The trigger pulses from the UJT cannot turn any SCR on due to the reverse bias on the gate diodes. To start the circuit, switch SW1 must be closed momentarily to provide a pulse to the gate of SCR1. When SCR1 turns on, the reverse bias on CR7 is removed; the next time the UJT supplies a pulse, SCR2 turns on, causing SCR1 to turn off because of the commutating C3. Similarly, when SCR2 turns on, the reverse bias on CR8 is removed; the next time the UJT fires, SCR3 turns on and causes SCR2 to turn off. Every time a pulse appears at the common shift line, the power to the load transfers sequentially from one stage to the next, always in the same direction.

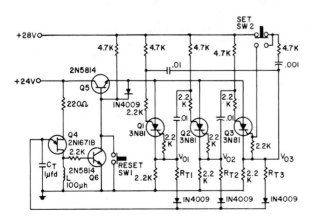

3.2.3 GE ring counter has variable timing capability. Shift pulses are generated by the unijunction transistors; the intervals between pulses are controlled by C_T and R_T. A different R_T can be selected for each stage of the counter.

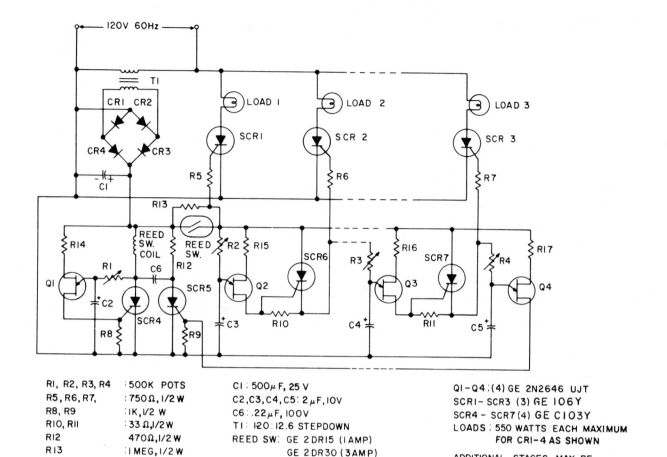

R1, R2, R3, R4	: 500K POTS	C1 : 500 μF, 25 V	Q1–Q4 : (4) GE 2N2646 UJT
R5, R6, R7,	: 750 Ω, 1/2 W	C2, C3, C4, C5 : 2 μF, 10V	SCR1–SCR3 (3) GE 106Y
R8, R9	: 1K, 1/2 W	C6 : .22 μF, 100V	SCR4 – SCR7 (4) GE C103Y
R10, R11	: 33 Ω, 1/2 W	T1 : 120 : 12.6 STEPDOWN	LOADS : 550 WATTS EACH MAXIMUM
R12	: 470 Ω, 1/2 W	REED SW: GE 2DR15 (1 AMP)	FOR CR1-4 AS SHOWN
R13	: 1 MEG, 1/2 W	GE 2DR30 (3 AMP)	
R14, R15, R16, R17	: 330 Ω, 1/2 W	REED SW COIL: 10,000 T #39 WIRE 825 Ω	ADDITIONAL STAGES MAY BE
		CR1–CR4 : (4) GE A14A	ADDED BETWEEN DASHED LINES.

3.2.4 GE chaser circuit sequentially fires ac loads at the half-wave rate. When power is applied, all SCRs are non-conducting and Q1 starts timing. At the end of the time delay (set by R1) Q1 fires SCR4, which energizes the reed switch coil. The reed switch contact then connects the dc supply to the remaining portion of the circuit and at the same time applies a dc drive to the gate of SCR1. The closure of the reed contact starts Q2 timing. At the end of its time delay (as set by R2) it fires SCR6, which fires SCR2. This sequential firing continues until Q4 fires, which causes SCR5 to turn on and SCR4 to turn off. The reed switch coil deenergizes, causing all the SCRs (except SCR5) to turn off, resetting the circuit. At the same time that SCR4 is turned off, Q1 starts timing and the cycle repeats. The 1M resistor across the reed switch contact is to prevent SCR6 and SCR7 from triggering because of the rate of rise of the voltage when the contact closes. The anodes of these two SCRs essentially see the dc supply voltage even when the reed switch opens, but these SCRs turn off because the SCR current does go below the holding current level.

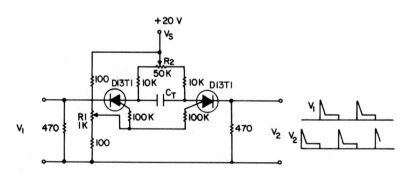

3.2.5 GE flip-flop trigger circuit uses GE PUTs connected as relaxation oscillators. When one of the two trigger devices is in the on state, the other is always in the off state. Turning on one device will instantaneously produce a negative voltage on the other due to the presence of capacitor C_T. This will shift it to the off state. The frequency is adjusted by R1, and the symmetry is trimmed by R2.

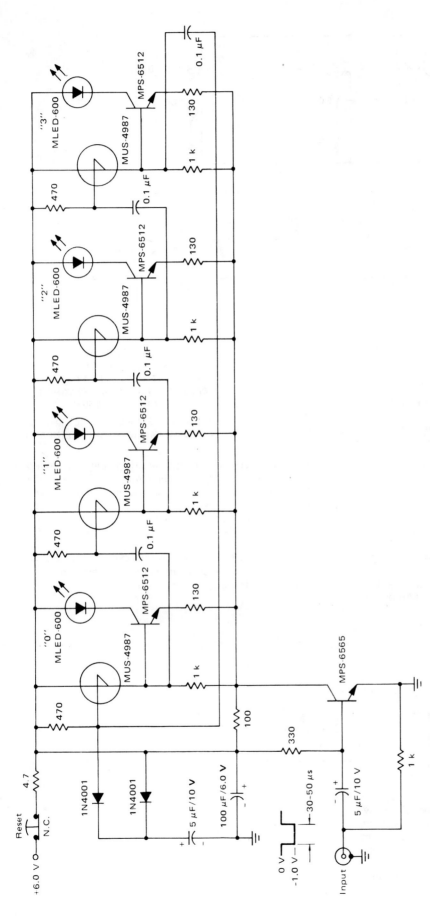

3.2.6 Motorola 4-stage ring counter. When power is applied, the SUS in stage "0" will be switched on due to the surge of gate current required to charge the 5 μF capacitor connected to it. The MPS 6565 transistor is saturated and its collector near ground. Therefore, most of the supply voltage appears across the 1K cathode resistor in stage "0." This voltage determines the current level through MPS 6512 and the light-emitting diode is on, indicating a count of "0." Each stage is connected to the next through a 0.1 μF capacitor. The capacitor connecting stage "0" to stage "1" will be essentially discharged, since the voltage at each side is near the supply. Each of the other three coupling capacitors is charged to approximately supply voltage. An event to be counted must be shaped to a 1V negative square wave with a duration of 30 to 50 μs and fed to the counter input. Each input pulse will turn off the MPS 6565 transistor, and the 100-ohm resistor will allow its collector to rise to the supply voltage. During this interval there is no voltage across the SUS in stage "0" and it will return to the off state. When the MPS 6565 is driven into saturation following the first input pulse, only the coupling capacitor between stages "0" and "1" will require charging, and most of the required current must flow out of the SUS gate in stage "1," turning it on with the MLED 600 in that stage, indicating a count of "1." Each succeeding input pulse is able to advance the count around the ring a stage at a time. Opening the normally closed reset switch will cause any stage to return to the off state, and only stage "0" will be gated on when it is closed. The 4.7 ohm resistor and 100 μF capacitor limit the dv/dt of the supply line to prevent stages other than the first from switching on when the reset switch is closed. Any number of stages may be cascaded.

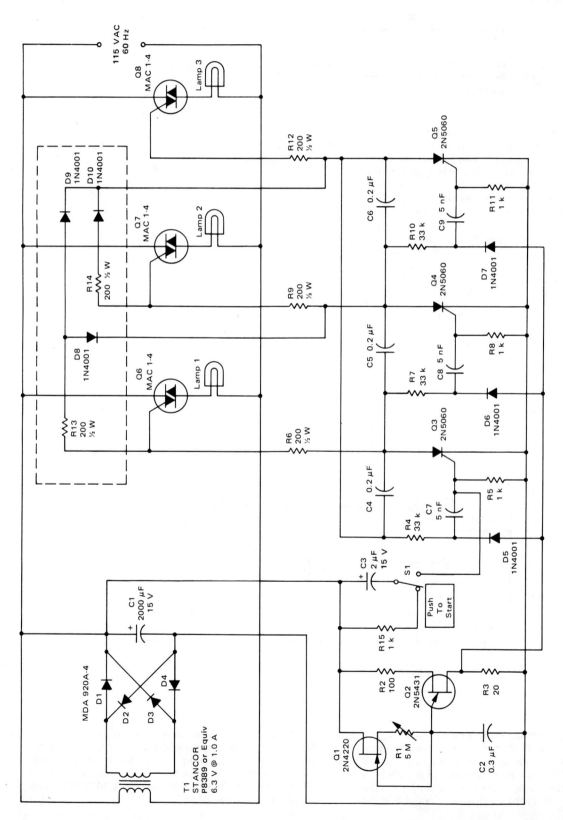

3.2.7 Motorola sequential ac flasher. Transformer T1, the diode bridge consisting of D1–D4, and capacitor C1 form a 9V dc supply for the ring counter and pulse circuit. The pulse circuit for the ring counter is composed of unijunction transistor Q2, field-effect transistor Q1, variable resistor R1, and capacitor C2. C2 charges through Q1 and R1 to the breakover voltage of Q2. Q2 fires and C2 discharges through the emitter of Q2 and resistor R3. This pulse is applied to the ring counter. Q1 is used in conjunction with R1 to obtain a wide range of pulse rates. The pulse rate can be adjusted from approximately one every 0.1 second to one every 8s.

SECTION 3 *Timing Circuits* 189

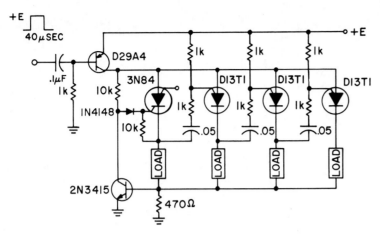

3.2.8 GE common-anode SCS ring counter. The object of a *ring counter* is to have power applied to one load at a time, and by applying input pulses, to advance the count from one stage to the next. When power is applied to the circuit, the PNP transistor immediately turns on and applies power to all the SCSs. Since none of the stages turns on, no current passes through the base of the 2N3415 transistor. Hence its collector voltage begins to rise to V+ and turns on the first SCS (3N84). With the application of the next pulse at the input, the D29A4 is momentarily turned off, thereby turning off the first SCS. When power is reapplied, the first GE 13T1 is turned on via the 0.05 µF coupling capacitor. Similar action takes place from the next counter pulse. The first 13T1 is turned off, and upon reapplication of power, the second 13T1 turns on.

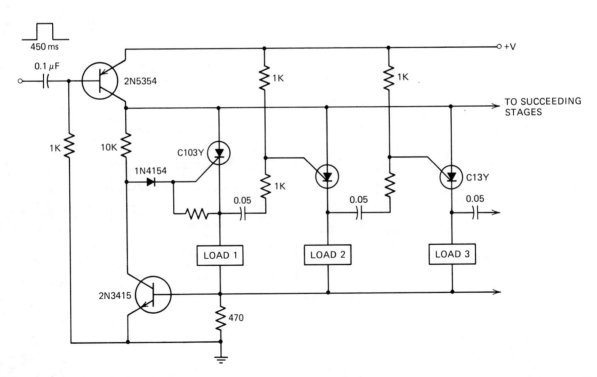

3.2.9 GE complementary-SCR ring counter. When power is applied, the PNP transistor turns on, applying power to the CSCR anode bus. Since none of the stages is on, no current passes through the base of the 2N3415 so that the collector voltage rises to +E and turns on the first SCR (C103Y). With the application of a pulse at the input, the PNP is momentarily turned off, turning off the SCR. When power is reapplied, the first C13Y is turned on via the 0.05 µF coupling capacitor. Similar action takes place for each pulse from the counter until the last C13Y is turned off, at which time the C103Y SCR is retriggered by the 2N3415.

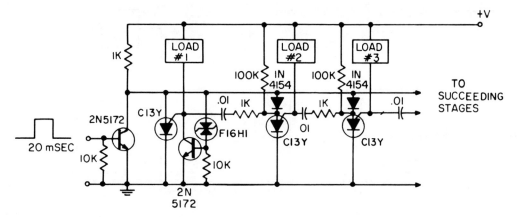

3.2.10 GE CSCR ring counter employs a gate-driven load. The first stage is triggered by the transistor–zener combination when the anode line reaches 7V. When the C13Y is turned on, the gate–cathode circuit becomes a low impedance and power is delivered to the load. If a pulse is present, the input transistor removes anode current from the conducting CSCR and turns it off; the gate becomes a high impedance. When the pulse is removed from the input, the succeeding stage is turned on by the 0.01 μF coupling capacitor. The long discharge time constant of this capacitor allows the use of low-value capacitors and long shift pulses. When the GE C13Y is used the load current should be limited to 50 mA.

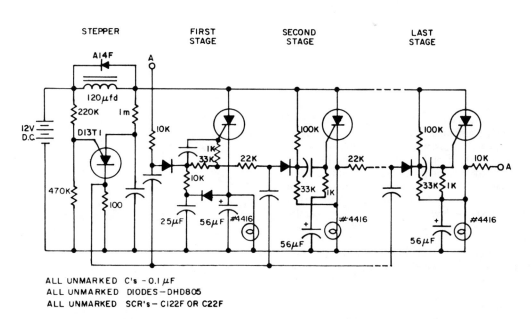

3.2.11 GE LC commutating ring counter uses electrolytics to cut cost. When the system is energized, all capacitors are uncharged, but the gate coupling capacitor charges through the 100K–33K divider to reverse-bias the logic diodes (except in the first stage). In the first stage the 25 μF capacitor provides this bias, so that on the first stepper pulse the first stage diode is forward-biased and triggers the first SCR. With this SCR on, the current flows through the choke and the lamp. The 56 μF capacitor charges to the lamp voltage, the second-stage biasing diode becomes forward-biased, and the current through the choke stabilizes. When the next step pulse occurs, the second SCR is triggered on and presents the impedance of the capacitor to the choke. The anode line voltage at once collapses to the forward drop of the SCR, the first SCR becomes reverse-biased by the charge on the 56 μF capacitor, while the 56 μF capacitor of the second stage charges.

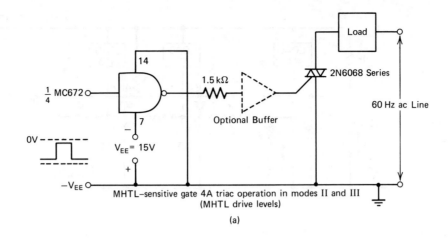

(a)

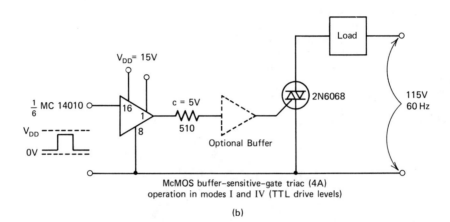

(b)

3.2.12 Motorola sensitive-gate triac serves as a *direct trigger for power loads with HTL, TTL, MOS, or integrated-circuit operational amplifiers*. The circuits shown in A, C, and D show preferred systems, which involve triac operation in quadrants II and III (avoiding quadrant IV, where gate-current requirements are normally greater). Other advantages of triac operation in quadrants II and III: IC power dissipation is minimized through the use of devices having active low outputs, and a comparably higher di/dt capability for the triac is yielded for a given IC output.

Similar systems using the low logic side as the common reference can be implemented with ease but are better using devices with active pullups and the more sensitive triacs; a simple pullup resistor can be added for devices with less sensitive gates.

The circuit in B shows the common grounding system and incorporates a device from Motorola's CMOS (called McMOS) line exemplifying the advantage of Motorola's sensitive-gate triacs. Operational amplifiers, as shown in E, can be used in comparator, amplifier, and many more operational circuits.

A typical transistor buffer, such as the one in F, can be added to any of the circuits where extra drive capability is required. This may be necessary for temperature extremes or for driving triacs of greater current ratings.

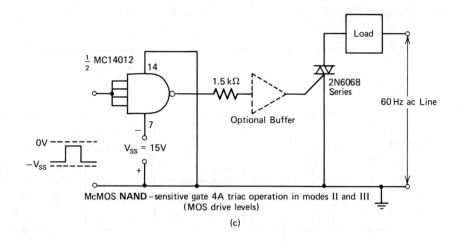

McMOS NAND-sensitive gate 4A triac operation in modes II and III
(MOS drive levels)
(c)

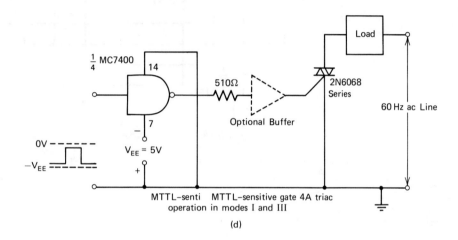

MTTL-senti   MTTL-sensitive gate 4A triac
operation in modes I and III
(d)

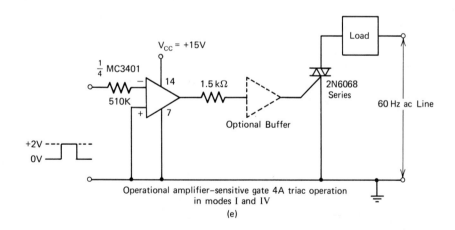

Operational amplifier-sensitive gate 4A triac operation
in modes I and IV
(e)

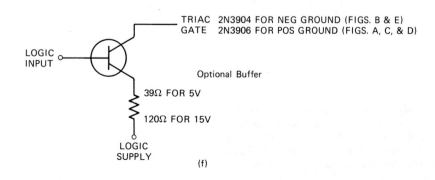

(f)

SECTION 3 *Timing Circuits*

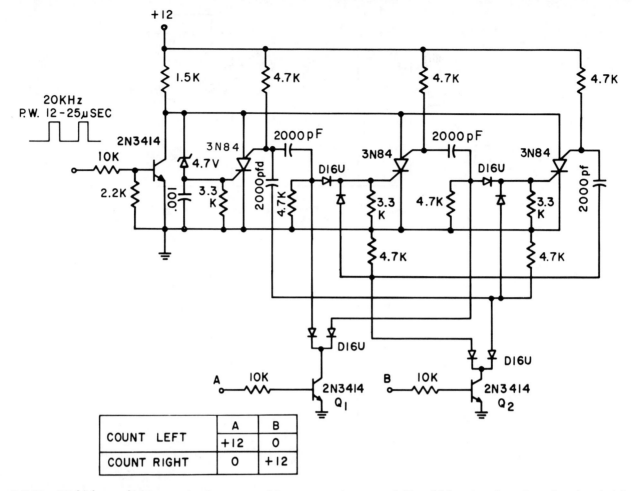

3.2.13 GE high-speed ring counter is connected in a "reversible" arrangement. This counter is capable of count rates of 20 kHz and higher. When Q1 is in conduction, the count shifts from right to left. The trigger current that normally would trigger the stage to the right is now shunted through Q1. The advantages of this circuit are that it takes only the turnon time of the transistor to inhibit a trigger pulse from a preceding stage; also, there is no possibility of false triggering when changing the direction of the count. The latter advantage removes the need for a resistor in series with the inhibit, allowing higher speed operation.

The output of this circuit is taken at the anode gate of the SCS. The voltage at this point is at about +1V when the stage is in the 1 state (SCS on) and at +12V when the stage is in the 0 state.

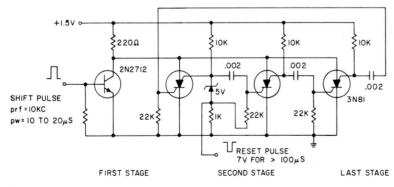

3.2.14 GE extremely low-power ring counter (less than 6 mW) operates from 1.0 to 6.0V. The reset pulse turns on the first stage with its trailing edge. The maximum shift pulse width increases with voltage and approaches 70 μs for a 6V supply. The minimum pulse width is 10 μs.

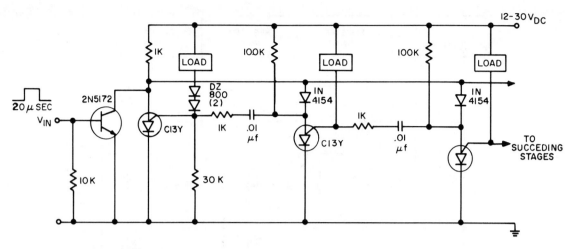

3.2.15 GE ring counter with extremely broad tolerances puts the load in the gate and triggers at the anode of GE complementary SCRs. When power is applied, the first stage is triggered due to the diode drop in the gate. The first pulse turns off the first stage, and when power is restored the second device is triggered by the charge on the coupling capacitor. The long discharge time constant of this capacitor permits the use of low-cost capacitors and long counting pulses. Succeeding stages are similarly triggered. Speed is primarily determined by the trigger time of the first stage.

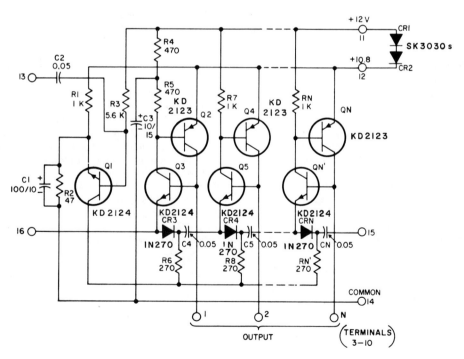

3.2.16 In this RCA shift register, the successive outputs from the various stages are delayed (or shifted) from those of the preceding stages by a controlled time interval (the duration between input trigger pulses). These outputs can be used to operate lamps or can be coupled through OR gates to establish a timing sequence for various operations of other equipment. If terminal 15 of the circuit is connected to terminal 16, the register becomes regenerative and can be used as a ring counter.

In a ring counter, each stage follows in sequence as if placed around a circle or ring. Each pulse input to the ring advances the counter one stage; when the last stage is reached, the next pulse activates the first stage. The cycle repeats until input pulses are stopped. The shift register may incorporate as many stages as desired.

SECTION 3 *Timing Circuits* 195

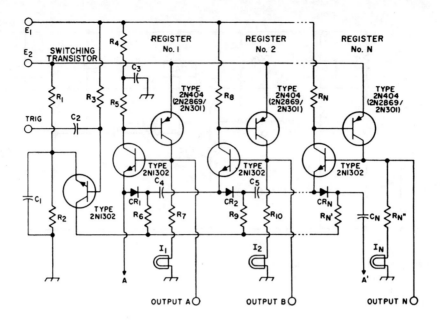

C1	= 100 µF, 6V
C2, 4, 5, CN	= 0.05 µF (or 0.1 µF), 50V ceramic
C3	= 1µF (or 25 µF), 25V
CR1, 2, CRN	= 1 N270 or equiv.
I1, 2, IN	= No. 49 2V, 60 mA (or No. 1488 14V, 150 mA)
R1	= 1K, ½W (or 680 ohms, 1W)
R2	= 27 ohms, ½W (or 12 ohms, 1W)
R3	= 1K, ½W
R4	= 1K, ½W (or 330 ohms, ½W)
R6, 9, RN'	= 660 ohms, ½W (or 180 ohms, 1W)
R7, 10, RN"	= 150 ohms, 1W (or 82 ohms, 2W)
R5, 8, RN	= 2.2K, ½W (or 680 ohms, ½W)

3.2.17 In this RCA basic shift register, the successive outputs from the various stages are delayed (or shifted) from those of the preceding stages by the duration between input trigger pulses. These outputs are coupled through OR gates (not shown) and may be used to program the timing sequence for various digital switching operations. If point **A'** on the circuit is connected to point **A**, the register becomes regenerative and may be used as a ring counter.

Voltages E_1 and E_2 are obtained from separate taps on a resistive voltage divider. With these voltages applied, the switching transistor is immediately triggered into conduction by the positive voltage applied to its base through R3. One of the register stages must be triggered simultaneously to provide a complete path for the current through the switching transistor. The register can be reset so that the operation starts with the first stage at any time by discharging capacitor C3.

The shift register may use as many stages as desired and may be made regenerative by connecting points **A** and **A'**. In addition, the basic circuit can be adapted for operation at many different output-current levels. The circuit as shown is designed for an output-current level of 40 mA ($E_1 = 12V$; $E_2 = 9V$). Transistor types and component values shown in parentheses indicate the changes necessary for operation at an output-current level of 3A ($E_1 = 27V$; $E_2 = 24V$). Voltages E_1 and E_2 should be obtained from a well-regulated dc power supply.

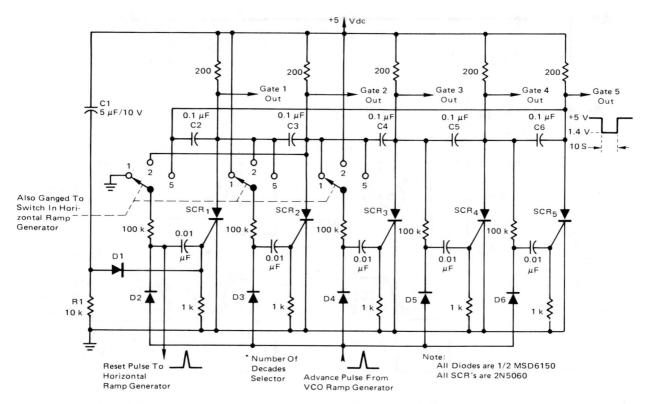

3.2.18 Motorola SCR ring counter. A ring counter generally may be considered as a circuit that sequentially transfers voltage from one load to another when a number of loads are connected to form a closed loop. Transfer around the loop proceeds always in the same direction and is initiated by pulsing a common shift line. The ring counter is actually an extension of the basic flip-flop circuit. In logic functions the ring counter is analogous to a single-bit register. The advance pulse for the ring counter is derived from the reset transition of the VCO sweep ramp generator and is synchronized in time at the interim between each frequency decade.

A selector switch permits the user to select the number of decades that the counter rings through.

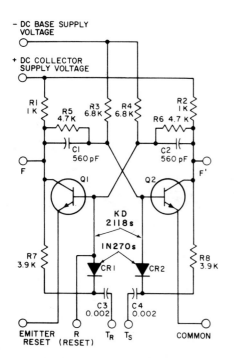

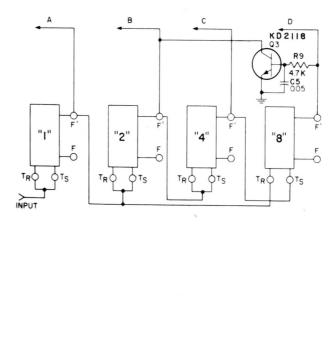

3.2.19 RCA basic flip-flop circuit as used in discrete-device dividers and counters. The inset shows the decade counter that produces binary-coded decimal output.

SECTION 3 Timing Circuits

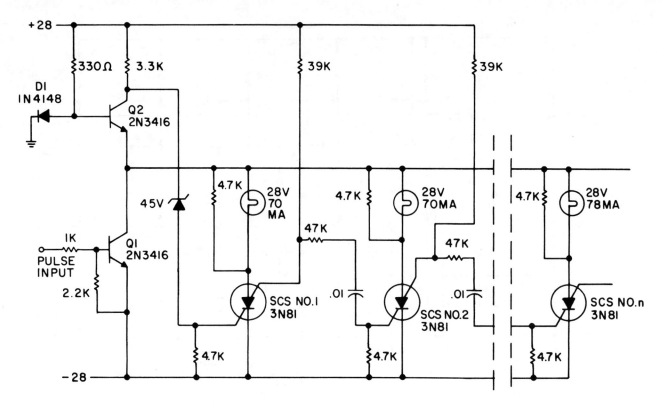

3.2.20 GE SCS incandescent-lamp ring counter with automatic reset capability. The coupling from the last stage to the first is through the breakdown diode. The first stage turns on only if all other stages are off. Transistor Q2 takes care of the variation in load current if a bulb is missing. This transistor is saturated if a stage without a bulb is on. If there is a bulb in the circuit that is on, the additional current flows through the transistor base. If all stages of the counter are off, the collector voltage rises to a point where the breakdown diode starts to conduct. This turns the first stage on. If more than one stage is on, the first stage will not turn on until sufficient input pulses occur to turn off all other stages.

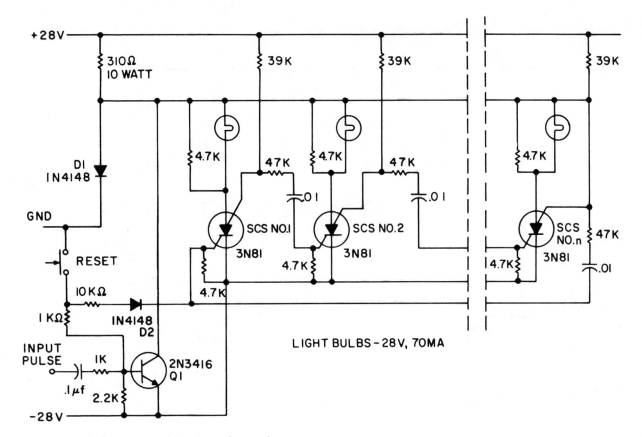

3.2.21 GE SCS ring counter drives incandescent lamps. SCS2 is switched on and SCS1 is switched off by turning on Q1 for between 20 and 2000 μs. This turns SCS1 off. The anode gate voltage of SCS1 rises slowly at a rate determined by the 0.01 μF capacitor. The 47K resistor allows the anode gate voltage to rise immediately above the anode voltage so that the SCS does not turn on due to anode gate current after the end of the input pulse. When Q1 turns off, the current through this capacitor turns SCS2 on. Every time a pulse occurs at the input to Q1, this sequence repeats and the stage that is on moves right one position.

The 4.7K resistor in parallel with the lamp provides a holding current to the SCS so that the circuit operates with the lamp removed or with the lamp burned out. Diode D1 holds the trigger line at about ground if a stage without a lamp is the one that is on.

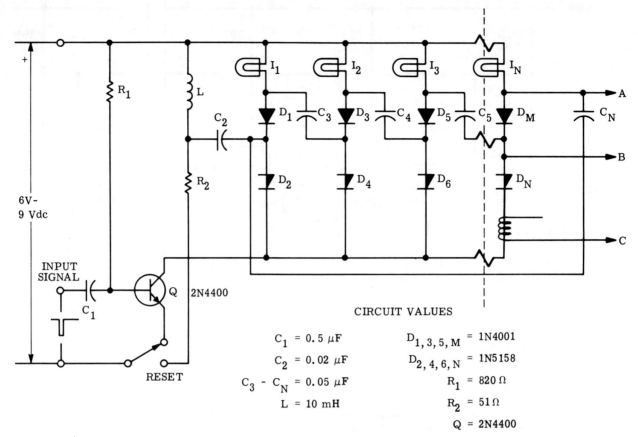

CIRCUIT VALUES

$C_1 = 0.5\ \mu F$	$D_{1,3,5,M} = 1N4001$
$C_2 = 0.02\ \mu F$	$D_{2,4,6,N} = 1N5158$
$C_3 - C_N = 0.05\ \mu F$	$R_1 = 820\ \Omega$
$L = 10\ mH$	$R_2 = 51\ \Omega$
	$Q = 2N4400$

3.2.22 Motorola ring counter. The first pulse triggers I1, the second pulse then turns off I1 and turns on I2, the third pulse turns off I2 and turns on I3, etc. Likewise, the nth pulse turns on I_n and the $n + 1$ pulse turns off I_n and returns to I_1, turning it on, and the cycle is repeated. Any number of stages may be employed; 10 stages form a decade counter.

3.3 Timers

The timer is one of the most versatile tools available to the system designer. If you're enlarging photos in the darkroom, you need a circuit that can be triggered on at the touch of a button but which will shut down automatically at a specified delay following turn-on—this is instant-on, delayed-off in response to an *on* command. If it's a long walk between the garage and the house, you might want an outdoor light that can be turned on in a conventional manner but turned off only after a specified period following actuation of the switch. This is instant-on, delayed-off in response to an *off* command, or *delayed dropout*. Delayed-on circuits apply power to a load only after a set period following activation of the *start* switch or signal; circuits of this genre might be used to apply B+ to a critical vacuum tube such as a mercury vapor rectifier after the filament has had a sufficient warmup time.

The applications for timers are virtually endless; and the ones shown here are easy to adapt to periods of your choice.

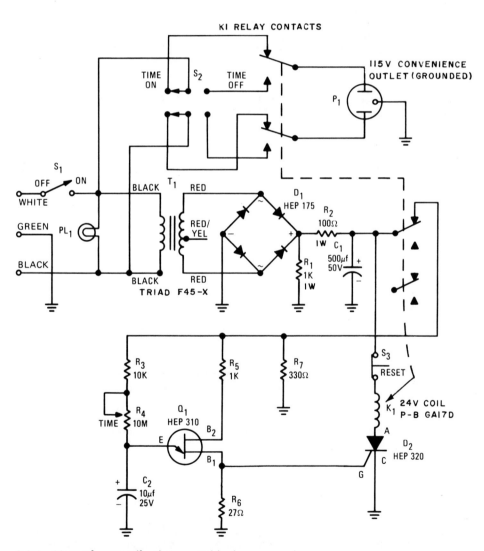

3.3.1 Motorola versatile timer suitable for any application requiring either a "time on" or "time off" sequence. The actual time on or off is variable from 0.5s to approximately 3 min.

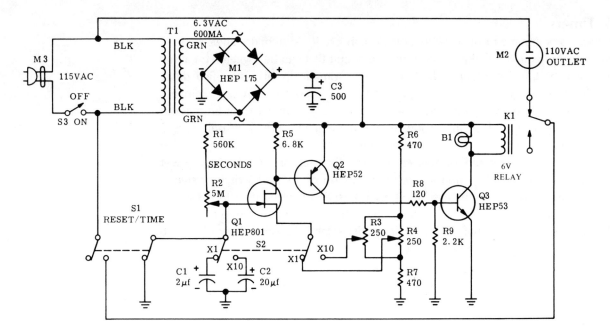

3.3.2 Motorola universal timer is adaptable to almost any electrically controlled switch, shutter, or light circuit. It will take the guesswork out of photographic projection printing and is especially useful for exacting color photo work.

The range of setting is 600 ms to 6s for the ×1 scale and 6 to 60s for the ×10 scale. Shorter or longer times may be obtained by changing the values of the timing capacitors. If this change is made, you must consider two factors:

1. The shorter time will be limited by the relay operation time.
2. The longer time will be limited by the leakage rate of the timing capacitor.

If the latter condition is encountered, some improvement can be obtained by using a tantalum capacitor, which has a lower leakage rate.

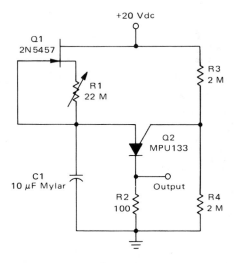

3.3.3 Motorola long-duration timer circuit provides a delay of up to 20 min. The circuit is a standard relaxation oscillator with a FET current source in which resistor R1 is used to provide reverse-bias on the gate-to-source circuit of the JFET. This turns the JFET off and increases the charging time of C1. C1 should be a low-leakage Mylar capacitor.

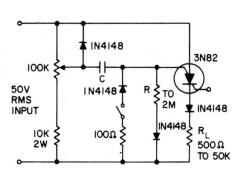

3.3.4 GE ac time delay. The switch is normally closed, charging C and causing the SCS to block. The delay is initiated by opening the switch and discharging C through R. Since R is connected for only half of each cycle, the delay is lengthened beyond the RC time constant. The delay is varied by R, C, and the setting of the potentiometer. Following the delay the SCS conducts alternate half-cycles.

202   DISCRETE / TRANSISTOR CIRCUIT SOURCEMASTER

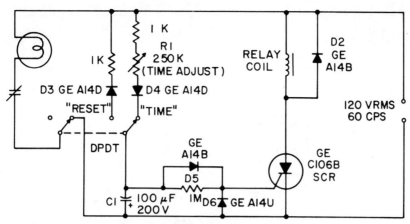

NOTE: ALL RESISTORS 1/2 WATT

3.3.5 GE hybrid time delay relay will switch an electrical circuit on, then shut it off after a timed period from 10 ms to about 1 min. The SCR functions as a very sensitive relay in this circuit; its purpose is to supply sufficient current to energize the low-cost output relay coil while it is being triggered by only the few microamperes of output current available from the timing network.

With the switch in the *reset* position, capacitor C1 quickly charges to the peak negative value of the supply voltage (approximately 150V). In this position, the lamp load is off. When the switch is at *time*, the lamp comes on and C1 starts to discharge toward zero at a rate determined by the setting of R1. Since the R1–C1 time constant is numerically long and the discharge current only flows for a short period during each cycle, this process takes many complete cycles of the ac supply. When the voltage across C1 finally drops to zero, reverses, and reaches about 1V, the SCR triggers. At this point, the lamp load is deenergized through the relay. With a positive gate voltage and negative anode voltage, a large leakage current could flow through the SCR due to remote-base transistor action. The diode D1 limits the reverse current to prevent excessive dissipation in the SCR.

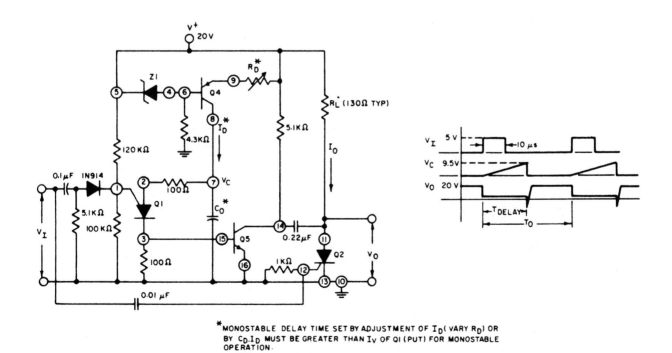

*MONOSTABLE DELAY TIME SET BY ADJUSTMENT OF I_D (VARY R_D) OR BY C_D. I_D MUST BE GREATER THAN I_V OF Q1 (PUT) FOR MONOSTABLE OPERATION.

Q2 (SCR) SWITCHING TIMES:
GATE-CONTROLLED TURN-ON TIME (t_{gt}) = 50 ns (TYP)
CIRCUIT-COMMUTATED TURN-OFF TIME (t_q) = 10 μs (TYP)

3.3.6 RCA monostable (one-shot) multivibrator with variable delay. Numbers are pin designators of CA3097E.

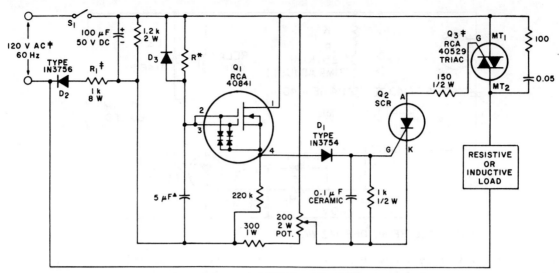

▲ Cornell-Dubilier Electronics—Type MMW or equivalent.
∗ R controls duration of time delay. At R = 60 MΩ up to 5-minute delay (IRC resistor, Type CGH or equivalent)

TIMING CIRCUIT CHARACTERISTICS

T_A = −25°C to +60°C
Accuracy: ±10% (over temperature)
Repeatability: ±3% (at 25°C)
Reset Time: Less than 150 ms

Q2: V_{DRM} = 60V
I_{GT} = 200μA
I_T = 0.8A

D3: I_R = 1 nA
V_R = 60V

3.3.7 RCA timing circuit for industrial control applications using 40841 in a single-gate configuration can also be used at supply voltage of 240 and 24V ac by changing the values of R1 and Q3.

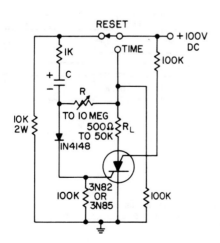

3.3.8 GE universal timer. Power is delivered to the load one-half time constant after the switch is set to *time* position. The time constant is a product of R and C (in ohms and farads).

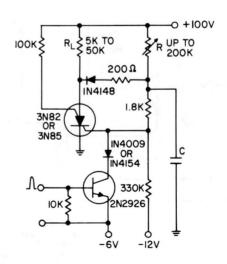

3.3.9 In this GE timer, an input pulse turns off SCS, which triggers after a delay of one time constant.

204 DISCRETE / TRANSISTOR CIRCUIT SOURCEMASTER

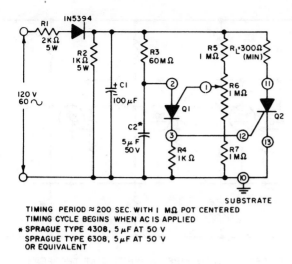
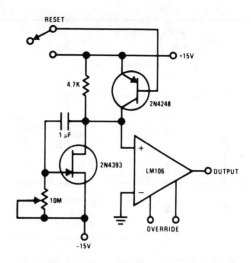

3.3.10 RCA ac line-operated one-shot timer.

3.3.12 In this National long-time comparator circuit, the 2N4393 is operated as a Miller integrator. The high Y_{fs} of the 2N4393 (over 12,000 microsiemens at 5 mA) yields a stage gain of about 60. Since the equivalent capacitance looking into the gate is C times gain and the gate source resistance can be as high as 10M, time constants as long as a minute can be achieved.

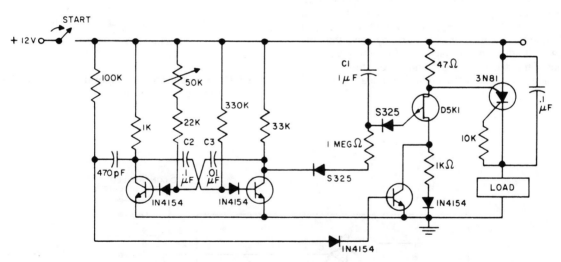

3.3.11 GE 100s timer, in which a free-running multivibrator of very low duty cycle is used to apply a timing voltage to a 1 megohm, 1 µF RC timing combination. This produces a chopped exponential waveform that appears as a very fine exponential staircase function. At the end of each multivibrator charging cycle, a negative pulse is applied to the base of the sampling transistor through a 470 pF capacitor. This allows about a 10 µs sample period. Note the additional 1N4154 diode used to help in the compensation of the two external S325 low-leakage diodes.

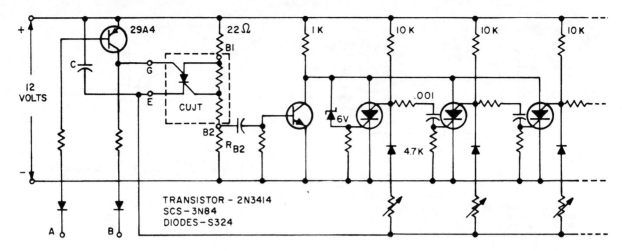

A OR B INHIBIT C FROM FIRING. EITHER INPUT MUST BE ZERO VOLTS TO FIRE. A INHIBITS AT THE PEAK POINT, B INHIBITS IN THE VALLEY.

3.3.13 GE sequential timer with inhibits. Here an RC timer is driven from an SCS ring counter. Only one SCS is on at a time, and successive output pulses at B2 advance the "on" stage to the right. The time intervals are adjustable for each stage. The GE CUJT can be inhibited from firing in one of two ways. It is inhibited when either A or B is down to zero voltage. When at +12V they are uninhibiting and the timer operates in the normal fashion. When lead A is pulled down, the PNP transistor is turned on so that there is an effective gate-to-base-1 short on the CUJT. Hence, when the capacitor charges to V_P there will be insufficient current available from the timing resistor to put it over the peak point (which is now 20 mA). So the timer will stop until input A is raised to +12V, at which point the capacitor will discharge and the ring counter will advance one stage. Input B draws gate current out of the gate of the CUJT and inhibits the capacitor at its valley voltage. When input B is released, normal timing takes place again.

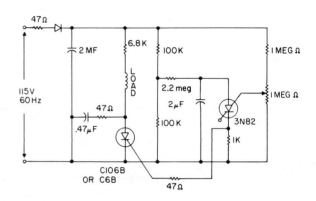

3.3.15 GE adjustable ac timer uses GE SCR. The period is variable from 200 ms to 10s.

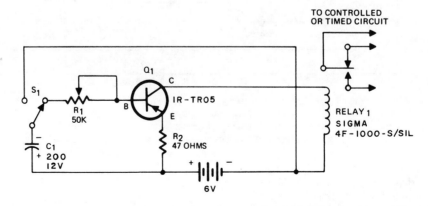

3.3.14 Workman simple timing circuit using a minimum of components is readily applicable to photo enlargers or any other short-term precision timing requirements.

The circuit operates a relay that controls the external equipment. Capacitor C1 is charged by the battery when switch S1 is pressed. Base bias to transistor Q1 is controlled by resistor R1. Resistor R2 limits the base-to-emitter current and increases the input impedance. The collector current operates the relay.

The discharge rate of capacitor C1 (the length of time the relay remains closed) is controlled by adjustment of resistor R1.

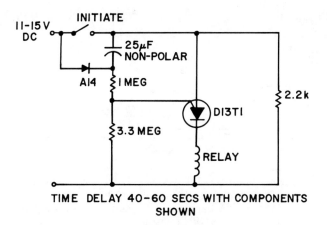

TIME DELAY 40-60 SECS WITH COMPONENTS SHOWN

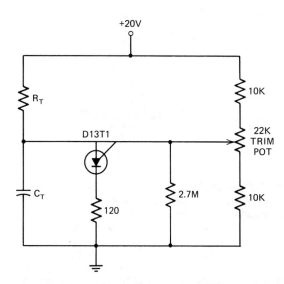

3.3.16 GE variable 60s delayed-on timer incorporates GE D13T1 PUT, which doubles as a timing threshold and load driver. Power is applied to the circuit with the *initiate* switch open-circuited. The 25 μF capacitor charges through the GE A14 diode and 2.2K resistor to a full supply voltage. When the switch is closed, the low side of the capacitor is suddenly raised to +12V, which raises the diode side of the capacitor to approximately +24V. The capacitor immediately begins discharging through the 1M resistor in series with 3.3M. The PUT gate becomes forward-biased and the device turns on, thereby applying power to the relay. The delay is virtually independent of supply voltage.

3.3.18 GE precision timer can be fabricated with low-tolerance RC components if a cheap trimmer is used for adjustment and calibration. Pick R_T and C_T according to delay time required: delay time in seconds equals product of C_T (farads) and R_T (ohms).

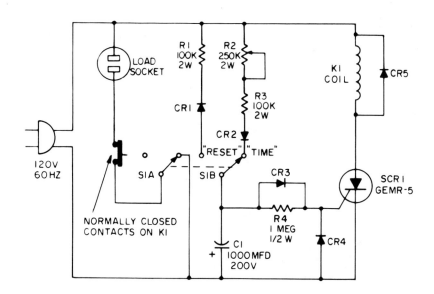

CR1 through — GE 504A diodes
  CR5
K1   —115V ac relay (Potter and Brumfield
            KA11AY or equivalent)

3.3.17 GE delayed-dropout relay is a simple electromechanical timing circuit. A silicon controlled rectifier functions as a sensitive relay to supply sufficient current to energize the low-cost output relay while being triggered by a few microamperes of output current from the timing network, R1–CR1.

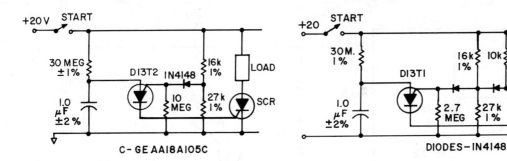

3.3.19 GE precalibrated 30s timers require no adjustment pots. Calibration has been eliminated by using 1% components for the intrinsic standoff resistors and RC timing components. Note the additional use of the compensation diode 1N4148.

When using the GE D13T1, one must either significantly decrease the value of the timing resistor and increase the capacitance or use a sampling scheme as shown at the right. This is precisely the same timer as at the left, but with the addition of the 10K resistor, a diode, and the sampling transistor. A 1 kHz pulse train is applied to the base of the NPN transistor. Each pulse lasts 10 μs and modulates the intrinsic standoff voltage once every millisecond to sample at the capacitor voltage. The 13T1 derives its peak-point current from the capacitor.

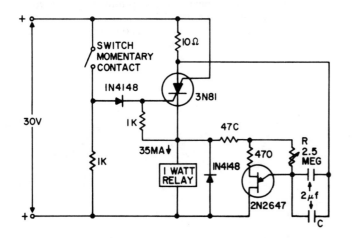

3.3.20 GE 10s timer. A positive pulse to the gate of the SCS triggers it on, supplying power to the relay load and unijunction timing circuit. At the completion of the RC timing interval a negative pulse to the anode turns off the SCS.

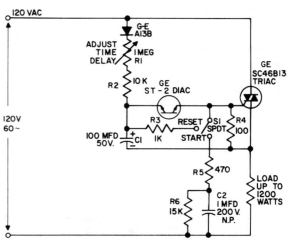

3.3.22 GE delayed-on latching circuit applies power to the load an adjustable period after the start switch is closed.

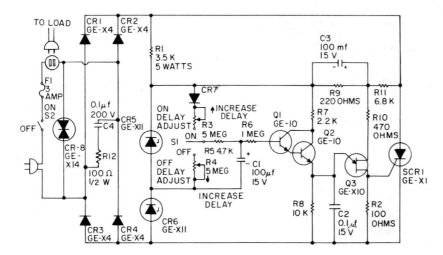

3.3.21 GE long-time-delay power switch provides delayed-on and delayed-off, both fully adjustable.

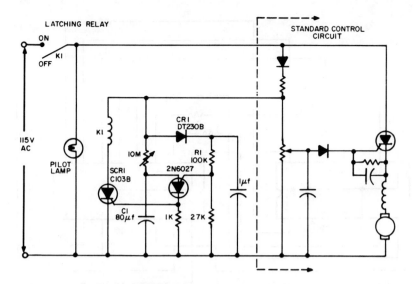

3.3.23 GE PUT timing circuit is ideal for blender control. By varying the setting on a time delay pot, the cook can regulate how long the batter should be mixed when preparing a cake. The timer is initiated by setting the 10-megohm potentiometer. When the voltage on C1 exceeds the PUT gate voltage, the 2N6027 turns on, triggering the C103. This latches the relay R1 off, thereby completely disconnecting the load from line voltage. The timing range will exceed 5 min with the components shown.

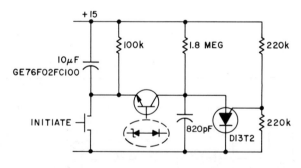

3.3.24 GE "pretimed" oscillator generates a constant frequency lasting for a predetermined time. A 1 kHz oscillator is formed with the GE D13T2 in conjunction with the 1.8-megohm resistor and 820 pF capacitor. The timing section is formed by the 2N2926 transistor (used here as a zener) and the 100K resistor and 10 μF capacitor. When power is supplied, the 13T2 is made to latch by virtue of excess current flowing through the 100K in series with the zener. When the *initiate* switch is pushed momentarily, the capacitor is charged to a full 15V. When it is discharging toward the +15V supply, the circuit oscillates. Eventually, the 10 μF capacitor charges to a point where the voltage across the 2N2926 (when the D13T2 is on) is sufficient to maintain zener breakdown, and the circuit relatches.

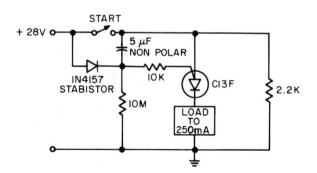

3.3.25 GE 1-min time delay and load driver, in which the high-gate-blocking voltage rating of the GE CSCR allows the 5 μF capacitor to be charged to 25V above the anode when the start switch is closed. The capacitor then discharges through the 10M resistor until the gate is forward-biased and the C13F triggers. The gate bias level eliminates rate-effect problems.

SECTION **3** *Timing Circuits*

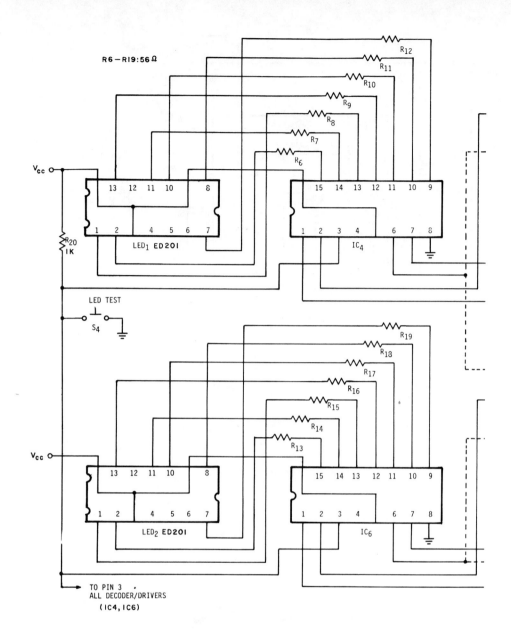

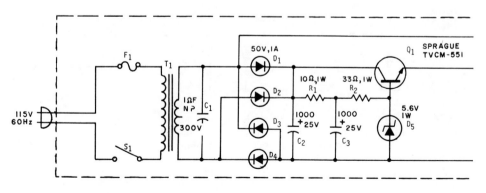

3.3.26 Sprague seconds timer with full LED digital readout, using three digits, operates from the 60 Hz frequency "standard" available at any 115V convenience outlet. The transformer furnishes the 60 Hz voltage to rectifier D6 and transistor Q2. The output of Q2 is a square wave at 60 pps. Closing switch S2 allows Q2 to produce continuous pulses while S3 will allow Q2 to produce pulses only as long as it is held closed.

The output of Q2 is connected to the input of IC1, a decade divider that produces 1 pulse for every 10 input pulses. The output of IC1 is therefore pulsing at the rate of 6 pps.

IC2 produces a pulse for every 6 input pulses. The 60 Hz power line has now been reduced to 1 pps. This signal is now fed into the first counter of the clock chain consisting of IC3, IC4, and LED1.

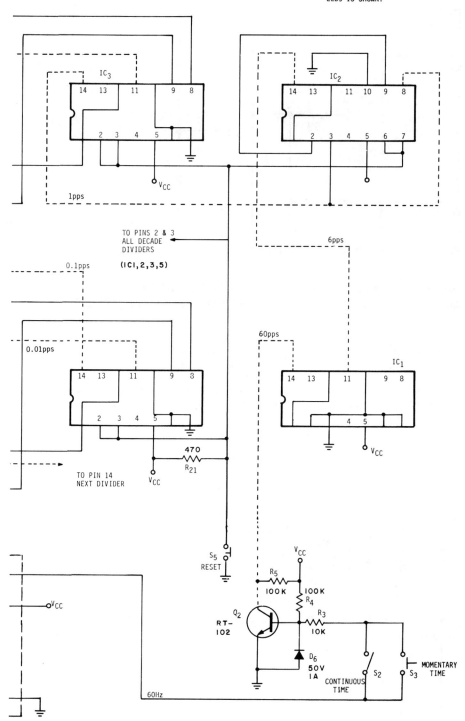

The 1 pps signal is divided by 10 in IC3. At the same time, IC3 converts the 1 pps input into four-line BCD for IC4, which converts BCD into the proper outputs for LED1.

The output at pin 11 of IC3 is a 0.1 pps signal connected to the input of the second counter stage (IC5, IC6, and LED2).

Each time the first stage counts from 9 to 10, an output pulse causes the second stage to count by one. IC5, IC6, and LED2 operates the same as IC3, IC4, and LED1 except that it displays tens of seconds. The system shown counts to 99s and then resets to 00. Additional counter stages can be added to extend the count capability.

SECTION 3 *Timing Circuits*

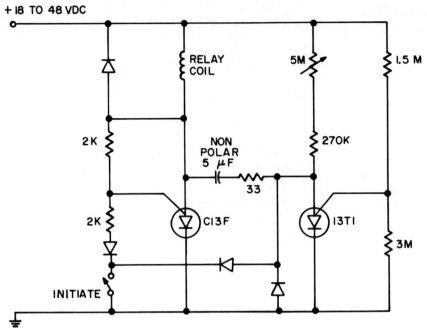

NOTES:
1. A NEW TIME DELAY CYCLE STARTS AT EACH INITIATE SWITCH OPENING.
2. DELAY ADJUSTABLE FROM ≈1 SEC TO ≈30 SEC FOR VALUES INDICATED.
3. ALL DIODES GE TYPE 55322

3.3.27 GE delayed-dropout relay features a controlled delay time regardless of when the initiate time is activated or deactivated. The initiate switch is normally closed; the GE C13F is on and the D13T1 is off. When the *initiate* switch is opened, the 5 μF capacitor charges to the supply voltage. At some voltage set by the GE D13T1 gate-divider network, the D13T1 turns on, turning the C13F off. The relay coil voltage collapses through the diode and the relay drops out. Dropout delay can be adjusted from 1 to 30s by the 5M pot.

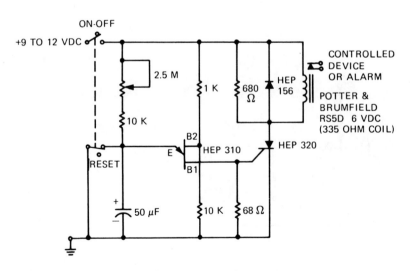

3.3.28 Motorola electronic timer features an adjustable period from 2s to several minutes, for kitchen, darkroom, or household board games that require a timed period for each player. This circuit requires little current and will operate satisfactorily from a standard 9V "transistor" battery. No tone or switched device is included; use a flashlight bulb, tone oscillator, or other indicating device that is suitable to your application.

SECTION 4
HOBBY

The assignment of the "Hobby" title for this section was an arbitrary decision. Not all photographers are hobbyists, and the circuits in that subsection do indeed apply to the professional and amateur alike. Similarly, the radio circuits included in Section 4.2 are largely applicable to professional communications applications. Generally, the circuits of this section are those which seem to fit best here; circuit shoppers who begin their search in this section should follow up by examining other divisions of this book that might contain additional applicable schematics.

4.1 Photographic Circuits

If it seems peculiar that the schematics included in this subsection all deal with electronic flashes, it is because the myriads of other circuits applicable to photography have been included in other sections of this book. If you don't find what you're looking for here, consult applicable portions of Sections 2 and 3. In Section 2.2 are photoelectric switches that permit control of virtually any function according to the amount of light impinging on a surface. In Section 2.3 are dimmers, which allow adjustment of the radiance emanating from a darkroom safelight. Many of the circuits in Section 2.4 will prove satisfactory for regulating the temperature of photo-processing chemicals. In Section 3.3 are a variety of electronic timers that can be used to control camera and enlarger exposures with precision.

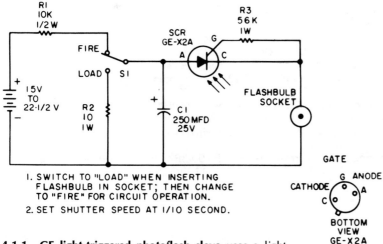

4.1.1 GE light-triggered photoflash slave uses a light-activated SCR as an electronic switch to trigger a second flashbulb whenever the camera's own flashbulb fires. To increase the sensitivity of the LASCR, mount it behind a lens or reflector. If it is mounted at the focal point of a flashlight reflector, the effective operating distance of this photoflash slave is increased by approximately 4 times.

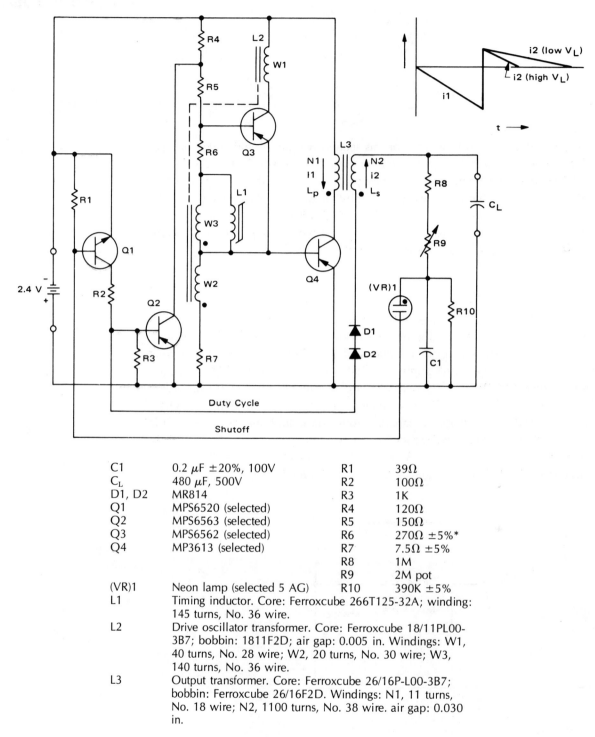

C1	0.2 µF ±20%, 100V	R1	39Ω
C_L	480 µF, 500V	R2	100Ω
D1, D2	MR814	R3	1K
Q1	MPS6520 (selected)	R4	120Ω
Q2	MPS6563 (selected)	R5	150Ω
Q3	MPS6562 (selected)	R6	270Ω ±5%*
Q4	MP3613 (selected)	R7	7.5Ω ±5%
		R8	1M
		R9	2M pot
(VR)1	Neon lamp (selected 5 AG)	R10	390K ±5%
L1	Timing inductor. Core: Ferroxcube 266T125-32A; winding: 145 turns, No. 36 wire.		
L2	Drive oscillator transformer. Core: Ferroxcube 18/11PL00-3B7; bobbin: 1811F2D; air gap: 0.005 in. Windings: W1, 40 turns, No. 28 wire; W2, 20 turns, No. 30 wire; W3, 140 turns, No. 36 wire.		
L3	Output transformer. Core: Ferroxcube 26/16P-L00-3B7; bobbin: Ferroxcube 26/16F2D. Windings: N1, 11 turns, No. 18 wire; N2, 1100 turns, No. 38 wire. air gap: 0.030 in.		

*All other resistors are ±10%, 0.25W.

4.1.2 Motorola photoflash converter circuit stops and starts itself to maintain the voltage on the load capacitor within narrow limits. It is also self-compensating for battery voltage variation and tends to charge the load capacitor in the same length of time regardless of decreasing battery voltage.

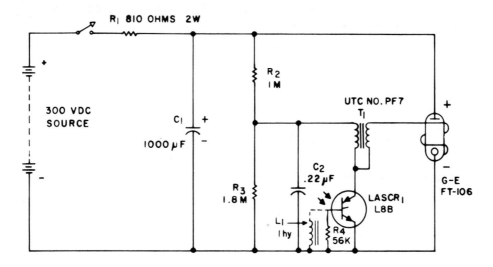

4.1.3 GE "slave" electronic flash results by modifying an industry-standard flashgun circuit with the LASCR. With switch S1 closed, C1 charges to 300V through R1, and C2 charges to 200V through R2 and R3. When the master flashgun fires (triggered by the flash contacts on the camera), its light output triggers LASCR1, which then discharges C2 into the primary winding of T1. Its secondary puts out a high-voltage pulse to trigger the flashtube. The flashtube discharges C1, while the resonant action between C2 and T1 reverse-biases LASCR1 for positive turnoff. With the intense instantaneous light energy available from present-day electronic flash units, the speed of response of the LASCR is in the low microsecond region, leading to perfect synchronization between master and slave.

High levels of ambient light can also trigger the LASCR when a resistor is used between gate and cathode. Although this resistance could be made adjustable to compensate for ambient light, the best solution is to use a 1H inductance, which will appear as a low impedance to ambient light and as a very high impedance to a flash.

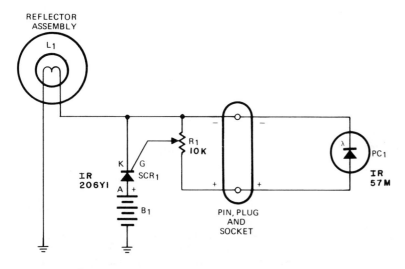

4.1.4 Workman self-firing auxiliary flash unit for photographic applications. You can have a self-firing flash unit for fill, back-lighting, and special photo effects. The entire unit fits inside the barrel of a flash gun. It uses the tube, plug, reflector, and bulb base and assembly to facilitate construction and usefulness.

Mount the solar cell toward the flash on the camera. As the camera-mounted flash fires, the solar cell (PC1) on the auxiliary unit triggers SCR1 on, and the surge of current fires the photo flash bulb (L1). The firing of L1 breaks the circuit and the entire unit turns off until a new bulb is mounted in L1 and the photocell is again exposed to a flash of light.

SECTION 4 *Hobby* 215

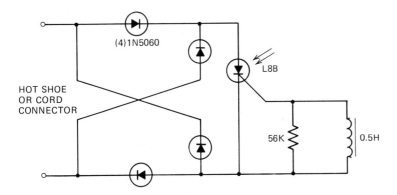

Slave photoflash trigger.

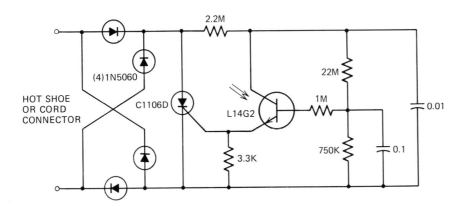

Sensitive directional slave photoflash trigger.

4.1.5 GE slave photographic xenon flash trigger is designed for the trigger cord or "hot shoe" connection of a commercial portable flash unit. This provides remote operation without need for wires or cables between the various units. The flash trigger unit should be connected to the slave flash before turning the flash on, and the L8B should be pointed in the general direction of the camera flash unit. The choice of inductor value will set the sensitivity of the circuit; at 1H no problem was noted with false triggering from fluorescent lights while triggering seemed adequate at 0.1H.

If a longer-range remote trigger unit is desired, the circuit may be modified to use a GE L14G2 lensed phototransistor as the sensor. The lens on this transistor provides a viewing angle of 10° and gives a 10:1 improvement in light sensitivity; it also allows the elimination of the inductor, which may be objectionable due to its bulk or expense. Note that the phototransistor is connected in a self-biasing circuit which is relatively insensitive to slow-changing ambient light and yet discharges the 0.01 μF capacitor into the C106D gate when illuminated by a photoflash.

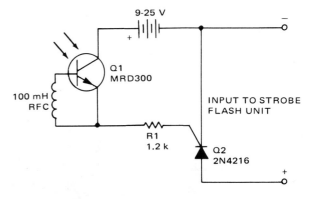

4.1.6 Motorola remote strobe flash slave adapter. At times when using an electronic strobe flash, it is desirable to use a remote or slave flash synchronized with the master. This circuit provides the drive needed to trigger a slave unit and eliminates the necessity for synchronizing wires between the two flash units.

The MRD300 phototransistor used in this circuit is cut off in a V_{CER} mode due to the relatively low dc resistance of RF choke L1 even under high ambient light conditions. When a fast-rising pulse of light strikes the base region of this device, L1 acts as a very high impedance to the ramp and the transistor is biased into conduction by the incoming pulse of light.

When the MRD300 conducts, a signal is applied to the gate of SCR Q2. This triggers Q2, which acts as a solid-state relay and turns on the attached strobe flash unit.

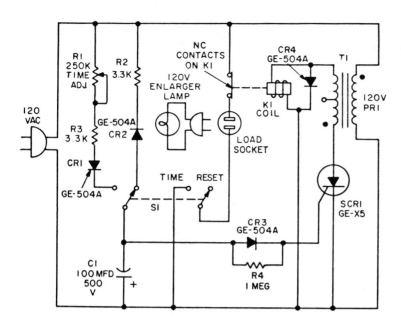

4.1.7 GE enlarger phototimer. Both delayed-off and delayed-on switching functions are interchangeably available by interchanging relay contacts. The circuit can be used as a print-exposure timer; time delays from a fraction of a second to nearly 1 min are attainable with values shown. Use 2W potentiometer, ½W resistors.

4.2 Amateur Radio and CB

Circuits for radio hobbyists include code practice oscillators, electronic keying devices, mike amplifiers, and of course such staples as receivers, transmitters, converters, and RF preamplifiers. This subsection contains those schematics that could not easily be classified under other heads—the tone oscillators, keyers, and microphone amplifiers; receivers, transmitters, and other RF circuits appear in Section 6.

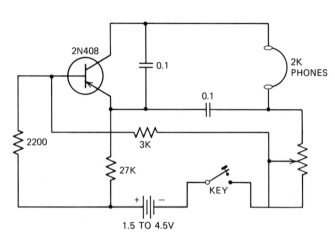

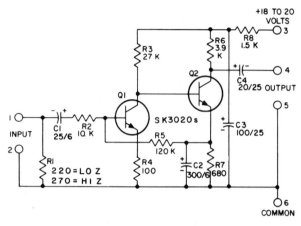

4.2.1 Workman code practice oscillator operates from a dc supply of 1.5 to 4.5V, depending on the amount of output desired. Magnetic headphones provide an audible indication of keying. When the key is closed, the 2N408 transistor supplies energy to the resonant circuit formed by C1, C2, and the inductance of the headphones; this circuit resonates to produce an audio tone in the headphones. Positive feedback to sustain oscillation is coupled from the resonant circuit through C1 and C2 to the emitter of the 2N408. R4 is adjusted to obtain the desired level of sound from the headphones.

4.2.3 RCA microphone preamplifier is capable of boosting the output of a dynamic microphone to 0.5 to 1V level. Frequency response is 20 to 35,000 Hz. When the circuit is in operation, the base bias current for the input transistor Q1 is obtained from the emitter of output transistor Q2 through R5. Q2 obtains its base bias current through the collector resistor of Q1 and R3. This unique bias circuit provides dc feedback for stabilization of the operating points of the transistors.

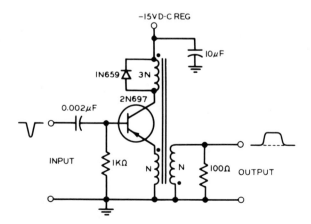

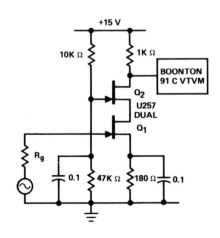

4.2.2 Sprague blocking oscillator delivers 1 μs pulses. The transformer shown is a Sprague 35Z type with a turns ratio of 3:1, which results in a pulse voltage across the primary of about 20V.

4.2.4 Siliconix cascode circuit has applications as a buffer amplifier for use with high-stability oscillators or in low-level power amplifiers mainly due to its low reverse transfer characteristics. FET is a Siliconix dual-unit type.

218 DISCRETE / TRANSISTOR CIRCUIT SOURCEMASTER

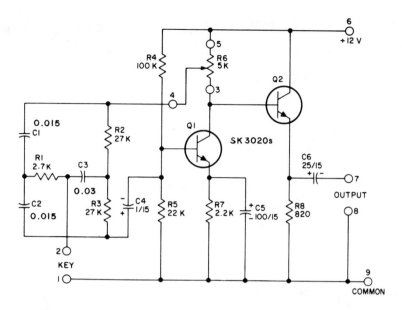

4.2.5 RCA code practice oscillator provides a single-tone sine wave at any frequency from 10 Hz to 175 kHz and can be operated from a wide range of operating voltages. The circuit has an extremely good waveshape, a low output impedance, and excellent keying characteristics. The oscillator is useful in testing high-fidelity audio equipment and can also be used as a generator. When the output signals from two oscillators are mixed, the result is a signal ideal for testing amateur single-sideband transmitters.

Oscillation is produced as some of the output of Q1 is fed back to its base through a twin-tee network consisting of C1, C2, C3, R1, R2, and R3. The frequency characteristic of the feedback network determines the frequency of oscillation of the circuit; the phase shift of the signal through the network is 180°, thus providing the positive feedback required to support oscillation. *Note:* Increasing C1, C2, and C3 will lower the audio pitch.

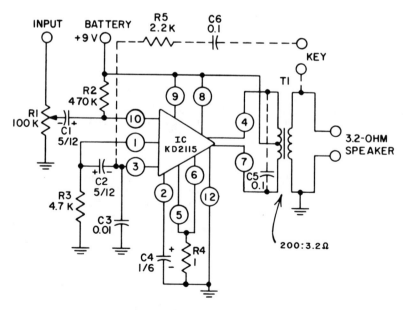

4.2.6 RCA IC audio amplifier/tone oscillator can be used in any application that requires a low-power portable unit. The amplifier requires an input signal of 40 mV and provides an output power of ½W. The audio oscillator can be used with a telegraph key as a code practice oscillator or with an on-off switch to provide a continuous tone.

SECTION 4 *Hobby* **219**

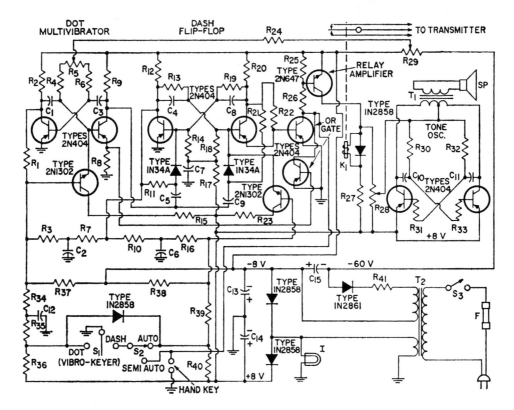

Parts List

C_1, C_3 = 1 μF, paper (or Mylar), 200 V
C_2 = 0.47 μF, ceramic, 25 V
C_4, C_8 = 560 pF, ceramic, 600 V
C_5, C_9 = 330 pF, ceramic, 600 V
C_6, C_7 = 0.01 μF, ceramic, 50 V
C_{10}, C_{11} = 0.02 μF, ceramic, 50 V
C_{12} = 0.1 μF, ceramic, 50 V
C_{13}, C_{14} = 2000 μF, electrolytic, 15 V
C_{15} = 16 μF, electrolytic, 150 V
F = fuse, 1 ampere
I = indicator lamp No. 47
K = dc relay; coil resistance =1350 ohms; Potter & Brumfield RS5D-V or equiv.
R_1 = 39000 ohms, 0.5 watt
R_2, R_9, R_{12}, R_{20} = 3900 ohms, 0.5 watt
R_3, R_{16} = 18000 ohms, 0.5 watt
R_4, R_6 = 51000 ohms, 0.5 watt
R_5, R_{29} = potentiometer, 10000 ohms
R_7, R_{10} = 22000 ohms, 0.5 watt
R_8, R_{22} = 180 ohms, 0.5 watt
R_{11}, R_{21} = 15000 ohms, 0.5 watt
R_{13}, R_{19} = 33000 ohms, 0.5 watt
R_{14}, R_{18}, R_{30}, R_{32} = 27000 ohms, 0.5 watt
R_{15}, R_{23} = 270 ohms, 0.5 watt
R_{17} = 68000 ohms, 0.5 watt
R_{24} = 100000 ohms, 0.5 watt
R_{25} = 68 ohms, 0.5 watt
R_{26} = 560 ohms, 0.5 watt
R_{27} = 620 ohms, 0.5 watt
R_{28} = volume-control potentiometer, 50000 ohms
R_{31}, R_{33} = 10000 ohms, 0.5 watt
R_{34} = 6800 ohms, 0.5 watt
R_{35} = 8200 ohms, 0.5 watt
R_{36}, R_{39}, R_{40} = 15000 ohms, 0.5 watt
R_{37}, R_{38} = 47000 ohms, 0.5 watt
R_{41} = 10000 ohms, 1 watt
S_1 = Vibroplex keyer, or equiv.
S_2 = toggle switch, double-pole, double-throw
S_3 = toggle switch; single-pole, single-throw
T_1 = push-pull output transformer (14000 ohm to V.C.), Stancor No. A3496, or equiv.
T_2 = power transformer, Stancor PS8415, PS8421, or equiv.

4.2.7 RCA compact electronic keyer can be used for automatic keying of a CW transmitter at speeds up to 60 words per minute. Two multivibrator trigger circuits using RCA 2N404 transistors automatically control the dot and dash transmissions. A Vibro-Keyer spring-loaded to the off position selects the type of transmission desired. Unless the Vibro-Keyer is moved to either the dot or the dash position, both multivibrators are held inoperative by the biasing action of 2N1302 clamping circuits.

When Vibro-Keyer S1 is deflected to the dot position, the first 2N1302 clamp transistor becomes inoperative and the dot multivibrator is allowed to operate as a free-running circuit. Feedback circuits in the multivibrator insure continued operation, regardless of whether S1 remains in the dot position long enough to develop the square wave that controls both the duration of the dot and the space that follows it. When S1 is set to the dash position, both clamp transistors become inoperative. The dot multivibrator and the dash flip-flop then operate simultaneously. The dash flip-flop is triggered by the positive pulses from the dot multivibrator. The 1N34A steering diodes prevent triggering of the flip-flop by negative pulses. Because two positive pulses are required to produce one complete cycle of output from the flip-flop, the frequency of this circuit is half that of the dot multivibrator.

The keying speed is determined by the frequency of the dot multivibrator. This frequency is adjustable by means of R29, which varies the amplitude of the negative dc voltage. As the negative voltage is increased to a maximum of 60V, the keying speed is increased to a maximum of 60 words per minute. Potentiometer R5 controls the ratio of on to off time of the dot multivibrator transistors, and thus determines the duration of both dot and dash transmissions and the minimum spacing between successive transmissions.

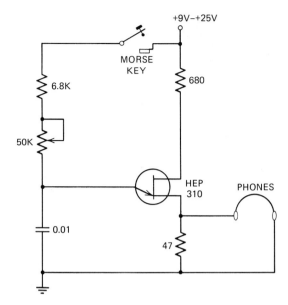

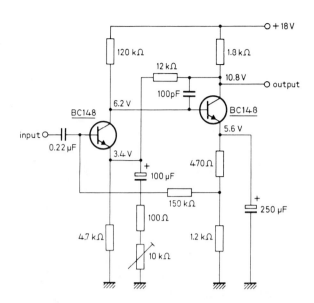

4.2.8 Motorola code practice oscillator with adjustable tone control. A loudspeaker of between 45 and 100 ohms may be substituted for the 47-ohm resistor and headphones. To feed the signal into an external audio amplifier, use a 0.01 μF series capacitor and pick up the signal from unijunction terminal B2 without otherwise making any circuit modifications.

4.2.10 Amperex microphone amplifier with a voltage gain adjustable between 13 and 40 dB. It has only 0.15% distortion for a gain of 13 dB, and 0.75% for a gain of 40 dB, with an output of 2V.

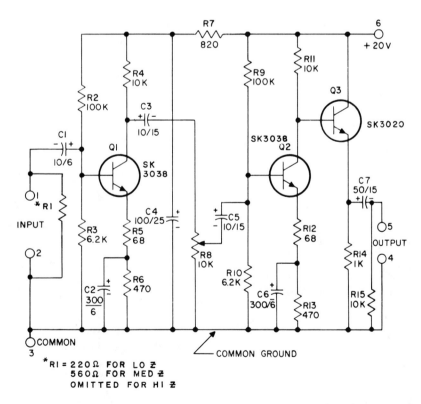

4.2.9 RCA high-dynamic-range microphone preamplifier, intended to be used with low-impedance dynamic microphones, is designed to handle loud passages of music or close talking without adverse effect on the output. The amplifier has a gain of 1500 to 2000 and can provide a maximum undistorted output voltage of 2V rms to a load impedance of 500 ohms or more. The maximum undistorted input is 400 mV rms. The frequency response of the preamplifier is flat from 20 Hz to 30 kHz.

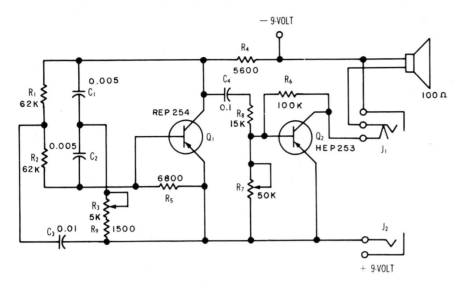

4.2.11 Motorola code practice oscillator can be used for individual instruction with headphones or for group instruction using the built-in speaker. The circuit, a twin-tee oscillator, is a very versatile audio generator because it generates almost perfect sine waves. Also, the twin-tee circuit, with proper parts, can be made to operate at any frequency. By adjusting control R3, the tone frequency can be adjusted between about 600 and 1300 Hz.

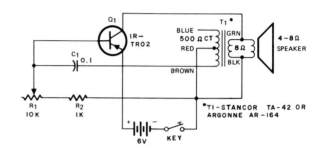

4.2.13 Workman code practice oscillator is a modified audio oscillator employing a small output transformer to provide the positive feedback necessary to start and maintain oscillation. The transistor collector load is the transformer secondary, and the speaker coil is parallel. The feedback signal is applied to the transistor base through a dc blocking capacitor. Transistor base bias is adjustable through resistors R1 and R2.

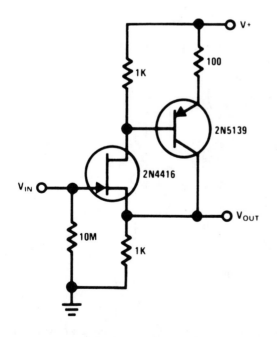

4.2.12 National high-impedance, low-capacitance wideband buffer. The 2N4416 features low input capacitance, which makes this compound-series feedback buffer a wideband unity-gain amplifier.

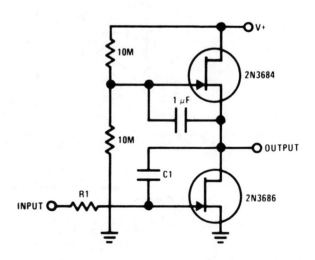

4.2.14 National JFET ac-coupled integrator uses the "μ-amp" technique to achieve very high voltage gain. Using C1 in the circuit as a Miller integrator (capacitance multiplier) allows this simple circuit to handle very long time constants.

222   DISCRETE / TRANSISTOR CIRCUIT SOURCEMASTER

4.3 Communicators

The communicator circuits in this subsection are the types that are not used by CBers, hams, or two-way people—some of them use light as the intelligence-carrying medium, some use wires and thus are intercoms, and others are devices that are typically used to transmit a modulated signal through a local radio receiver.

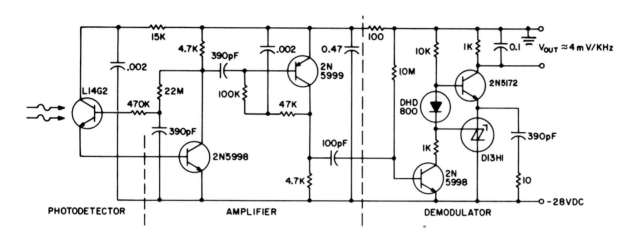

4.3.1 GE optical receiver picks up signals from a companion transmitter. The major constraint on the receiver performance is the signal-to-noise ratio. Shielding, stability, bias points, parts layout, etc., become significant details in the final performance. This receiver circuit consists of a GE L14G2 detector, two stages of gain, and an FM demodulator.

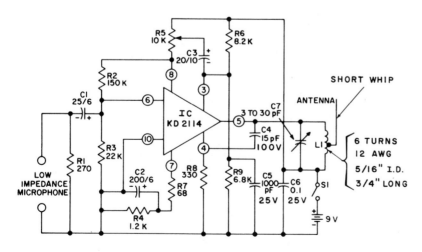

4.3.2 RCA wireless microphone can transmit FM on any frequency within the range of 88 to 108 MHz. The range is 15 to 50m, depending on the location of the receiver antenna relative to the antenna of the wireless microphone.

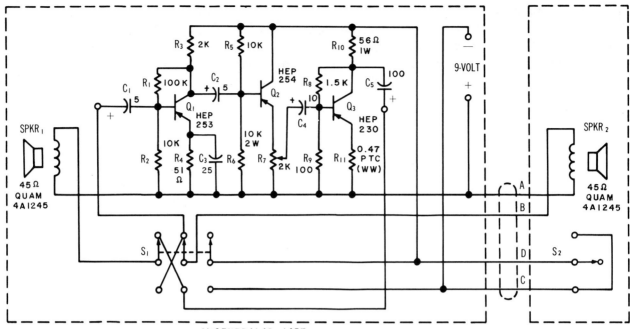

4.3.3 Motorola transistor intercom can be operated from a battery and is particularly useful for communication between a truck cab and camper or between car and house trailer. A special feature is that power is not used while the station is in the standby position. The only time that power is used is when messages are being sent. Thus a 6V lantern-type battery or series-connected flashlight batteries are suitable power sources.

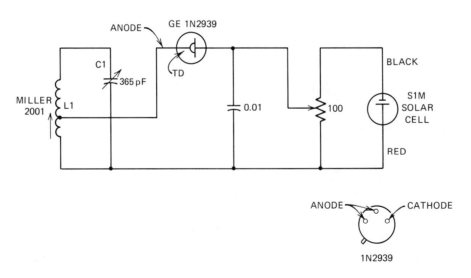

4.3.4 Tunnel-diode radio transmitter will generate a strong signal on the AM standard broadcast band. A single solar cell provides 0.5V, and a small portion of this voltage is applied to the diode through a 100-ohm potentiometer.

Place a portable radio tuned to a weak station near the high end of the band near the coil. Rotate the potentiometer and tuning capacitor at the same time. At one setting you should hear the radio station disappear, indicating oscillation of the tunnel diode. (The GE tunnel diode is no longer being manufactured, but it may be found in abundance in the surplus electronics market.)

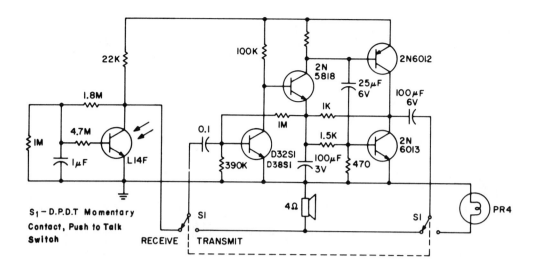

4.3.5 GE light-coupled transceiver designed around 6V flashlights. The lamp current is modulated at an audio rate, which modulates the light beam. The light beam is detected by a photo-Darlington, amplified, and used to drive a small speaker for audio output. Driving the lamp with an ac signal cuts the voltage to about one-third the 6V, requiring use of the 2V PR4 bulb in the 6V circuit. The GE L14F can be mounted on the axis of the beam facing the reflector of the flashlight just above the bulb (the bulb filament is at the focal point of the reflector).

Fidelity is not high due to the low-pass characteristics of the lamp filament, but intelligible conversation at distances up to 100 ft have been reported.

Using an infrared-emitting diode (IRED) for the light source eliminates the response time of the light source as a limit to fidelity but requires the design of a pulse source and applying AM or FM techniques to allow the IRED to generate enough light power to transmit appreciable distances.

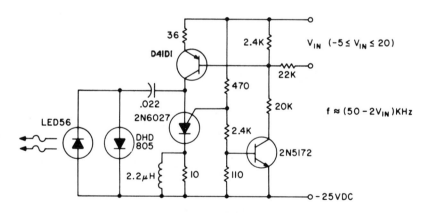

4.3.6 GE FM optical transmitter is designed around a programmable unijunction transistor (PUT) pulse generator. The basic circuit can be operated at 80 kHz and is limited by the PUT/capacitor combination, as higher frequency demands a smaller capacitance, which provides less peak output. The pulse repetition rate is relatively insensitive to temperature and power supply voltage and is a linear function of V_{IN}, the modulating voltage. Using lenses at the light emitter and detector or transmitting the signal through low-cost fiber optics greatly increases the range and minimizes stray-light noise effects. The average power consumption of the transmitter circuit is less than 3W.

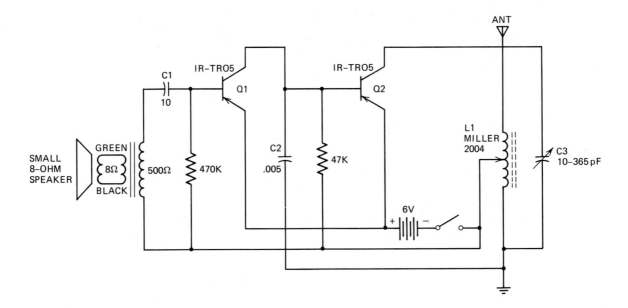

4.3.7 Workman wireless microphone can be used to broadcast over a short distance to standard AM home receivers. The microphone in this project is actually a small loudspeaker. The two transistors and related components amplify the signal to the antenna. The loudspeaker microphone, the matching transformer, and input capacitor C1 may be replaced by a crystal microphone connected directly across R1. The antenna should be a meter or two in length. In operation C3, the variable capacitor across the antenna, should be tuned to a dead spot on the receiver's band. In operation, the signal is picked up by the loudspeaker microphone, stepped up by the transformer T1, and applied through coupling capacitor C1 to the amplifier. Q2 is used as a modified Hartley oscillator with its operating frequency determined by L1, C3 tuned circuit. Feedback for oscillation is furnished by L1.

4.4 Electronic Novelties and Toys

Despite the title, these schematics are not all frivolous. The first few are ideal for serious electronic-music experimenters (who will also be interested in the oscillator circuits of Section 6.1); these are followed by circuits for model railroaders, light-organ builders, and other hobbyists.

The electronic metronomes are full-fledged tone oscillators, and the rhythmic clicking of their output signal can be changed to a harmonic-rich audio tone by simply juggling the values of the RC network's components. The siren circuits are also worthy of being considered something more than toys; they have been successfully employed in civil defense and marine applications by simply feeding the wailing output signal into a high-power audio amplifier.

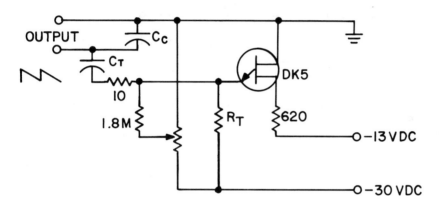

4.4.1 GE sawtooth tone generator. If precautions are not taken to prevent loading the GE unijunction oscillator circuit, frequency stability will suffer. The value of C_c must be 50 to 100 times that of C_t. With a signal stability of better than 500 ppm/°C, the output level is 80 to 160 mV.

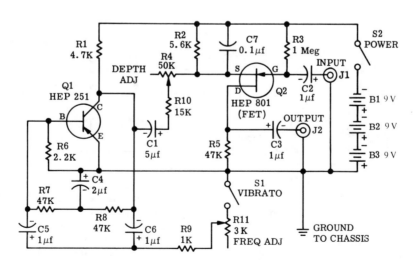

4.4.2 Motorola vibrato circuit is entirely self-contained and can be used with any high-impedance input amplifier and high-impedance pickup or microphone. Q1 is a twin-tee oscillator that produces a 6 Hz sine wave. This output is applied to the source of FET Q2, where it amplitude-modulates any incoming signal applied to the gate (G) through input jack J1. The amplitude of this modulation can be varied by potentiometer R4, thus creating either a light or a deep tremolo effect. The vibrato rate is controlled by R11. During normal vibrato operation, both signals are applied to Q2 and are amplified and presented through C3 to J2 as a combined signal.

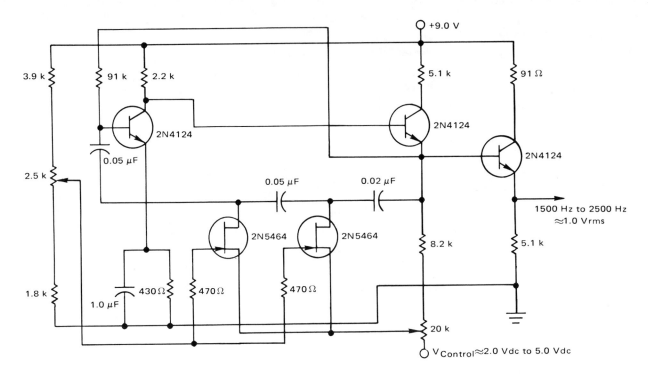

4.4.3 Motorola voltage-controlled oscillator produces a good sine wave and is linear over the range of frequency from 1500 to 2500 Hz.

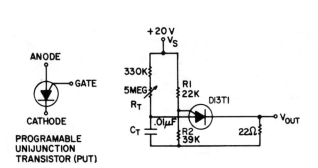

4.4.5 GE relaxation oscillator. If the PUT gate is maintained at a constant potential, the device will remain in its off state until the anode voltage exceeds the gate voltage by one diode forward voltage drop. At this voltage the peak point is achieved and the device turns on. In the circuit shown, the gate voltage of the PUT is maintained from the supply voltage by the resistor divider, R1 and R2. This voltage determines the peak-point voltage V_p. The peak-point current and the valley-point current both depend upon the equivalent impedance of the gate, $R_1R_2/(R_1 + R_2)$ and the source voltage. The frequency is controlled by R_T.

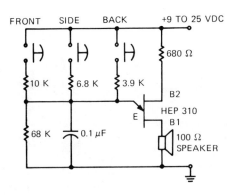

4.4.6 Motorola electronic doorbell uses an oscillator to generate tone. Different series resistors allow one tone frequency for each of several entrances so that you know which door has the visitor by the sound of the tone. If a 100-ohm speaker is unavailable, use two 45-ohm speakers in series. It is a good idea to employ better-quality pushbuttons for this application, since they must endure a variety of changing weather conditions; the best are the types used as auxiliary automotive horn buttons.

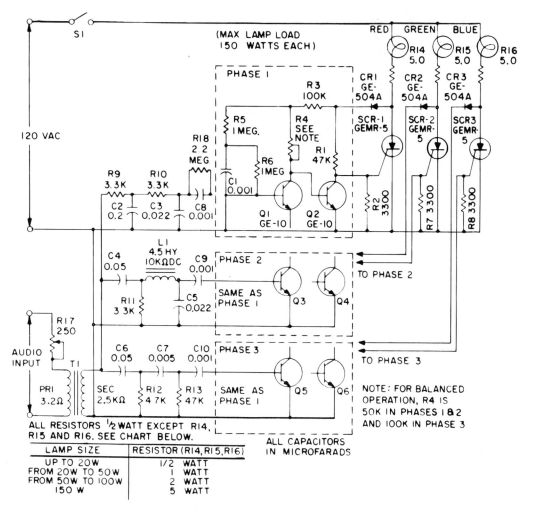

Parts List

C1, C8, C9, C10	0.001 µF 100V capacitor (total 6)
C2	0.2 µF 50V capacitor
C3, C5	0.022 µF 50V capacitor
C4, C6	0.05 µF 50V capacitor
C7	0.005 µF 50V capacitor
CR1–CR3	GE-504A diode (total 3)
L1	4.5H (10K) choke (Triad S-12X or equivalent)
Q1–Q6	GE-10 transistor (total 6)
R1, R13	47K 10% resistor (total 4)
R2, R7, R8, R9, R11	3.3K 10% resistor
R3	100K 10% resistor (total 3)
R4	Two 50K potentiometers and one 100K potentiometer
R5, R6	1 M 10% resistor (total 6)
R10	33K 10% resistor
R12	4.7K 10% resistor
R13	47K 10% resistor
R14–R16	5 Ω
R17	250Ω potentiometer
R18	2.2M 10% resistor
S1	SPST toggle switch, 5A minimum
SCR1–SCR3	GEMR-5
T1	2.5K/3.2Ω audio output transformer (Triad A-3332 or equivalent)
Minibox	6" × 5" × 4" (Bud CU-3007-A, or equivalent)
	Line cord and grommet
	Sockets, (total 3)
	Vectorbord and pins

4.4.7 GE color-light organ takes the sounds coming from the amplifier and separates the low-, medium-, and high-frequency audio tones. These three signals are then amplified and used to drive a combination of colored lamps of up to 150W per channel. The input can be connected directly across a speaker.

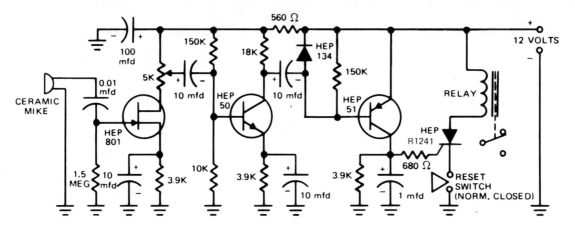

4.4.8 Motorola sound-activated relay, featuring adjustable sensitivity and unlatching by a manual reset, allows control of any circuit with a clap of the hands or some other sharp sound. Use a linear-taper potentiometer, 15V electrolytics, and half-watt resistors with a tolerance of 10%.

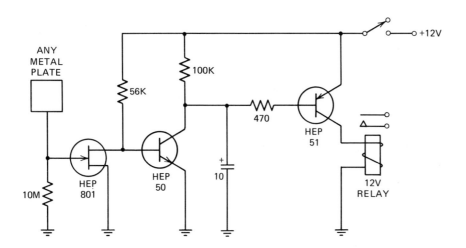

4.4.9 Motorola capacitance touch switch is sensitive enough to be used as a proximity detector. Connect the trigger wire to any metal plate or the knob of a door. The relay closes when the plate is touched. To increase the sensitivity, replace the 10M resistor of the FET gate with a 100M pot.

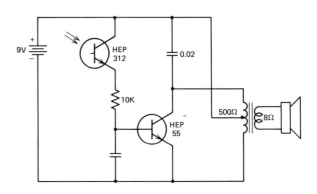

4.4.10 Motorola electronic "rooster" crows with a soft audio tone when the sun comes up. Use for hunting purposes; or position the sensor part of the circuit on the garage to hear when a car enters the driveway at night. This circuit is reported to be sensitive enough to be used as a fire detector. The output transformer has a 500-ohm centertapped primary and an 8-ohm secondary.

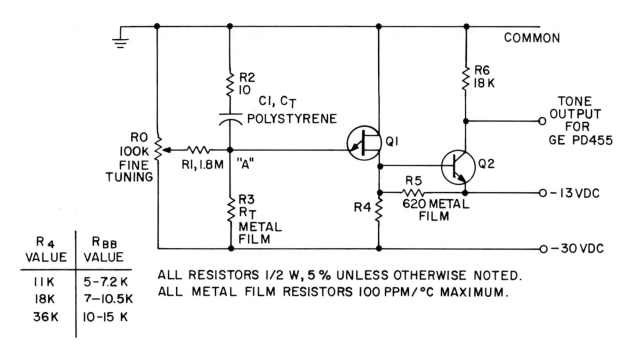

R_4 VALUE	R_{BB} VALUE
11K	5–7.2K
18K	7–10.5K
36K	10–15K

ALL RESISTORS 1/2 W, 5% UNLESS OTHERWISE NOTED.
ALL METAL FILM RESISTORS 100 PPM/°C MAXIMUM.

4.4.11 GE master tone generator circuit for electronic organs is the familiar unijunction relaxation oscillator, its major unique feature being the opposite voltage polarity applied. This polarity is required to take advantage of the characteristics of the GE complementary unijunction transistor (CUJT). The value of R4 optimizes the circuit. R_T and C_T are chosen for the desired tone, with R_O providing the fine-tuning adjustment needed for tolerance and standard value compensation.

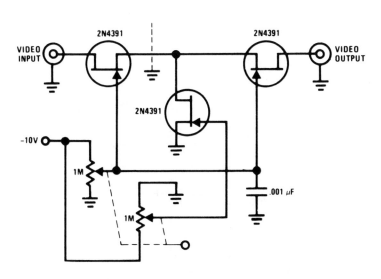

4.4.12 National voltage-controlled variable-gain amplifier. The 2N4391 provides a low $R_{DS(ON)}$ (less than 30 ohms). The tee attenuator provides for optimum linear dynamic range; if complete turnoff is desired, attenuation of greater than 100 dB can be obtained at 10 MHz provided that proper RF construction techniques are employed.

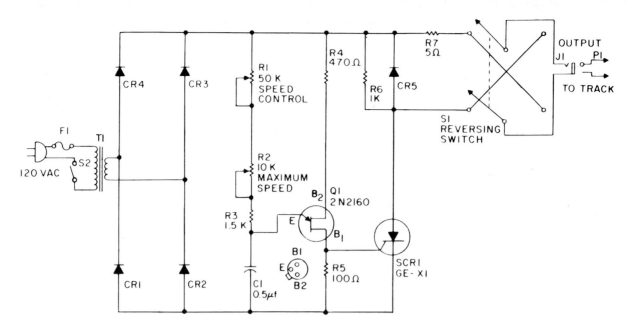

Parts List

C1	0.5 µF 200-volt capacitor
CR1-CR5	GE-504A rectifier diode
F1	0.5A fuse and holder
J1	Output jack
P1	Output plug to track connections
Q1	G-E 2N2160 unijunction transistor
R1	50K 2W potentiometer with SPST switch
R2	10K 2W potentiometer
R3	1.5K resistor
R4	470Ω resistor
R5	100Ω resistor
R6	10K resistor
R7	5Ω 20W resistor (or two 10Ω 10W resistors in parallel)
S1	DPDT switch
S2	SPST switch on (R1)
SCR1	GE-X1 silicon controlled rectifier
T1	Transformer: primary, 120V; secondary, 25V (Stancor P-6469 or equivalent)
Minibox	Aluminum 6" x 5" x 4"

4.4.13 GE scale-speed control for a model railroad.

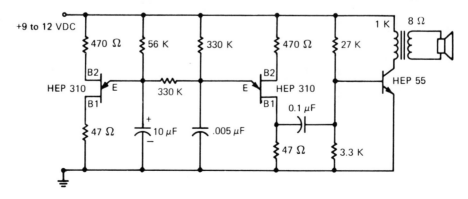

4.4.14 Motorola wailing electronic siren has an automatic rising and falling audio signal and is ideal for burglar alarm applications. For industrial use, where PA amplification is required, substitute a 1200-ohm resistor for the output transformer and feed the audio from the collector of HEP 55 through a 10K series resistor into the amplifier.

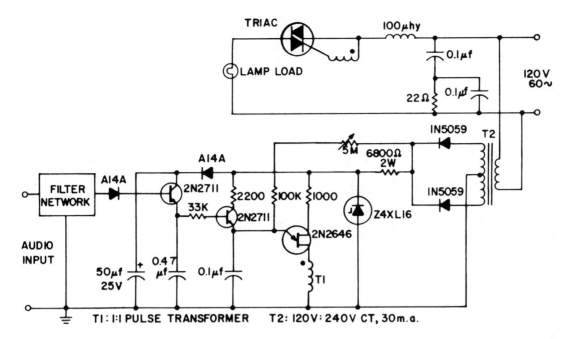

4.4.15 GE color organ "cell." The circuit input can be attached directly to the output of a sound system's amplifier. By combining several similar circuits, a color organ may be constructed. Normally, three systems are used with low-, medium-, and high-frequency bandpass filters used for the three segments. By using red, blue, and green lights for the three channels, all colors of the spectrum can be created. Red lamps overpower equal-wattage blue and green lamps; to compensate, an SCR can be used for the red stage, thereby reducing its applied voltage by 0.707 of the other stages.

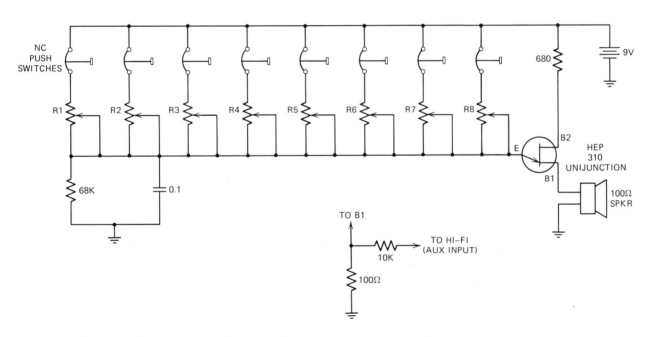

4.4.16 Sessions toy electronic organ plays through an integral 100-ohm speaker. Substitute a resistor for the speaker and feed audio into your hi-fi system through 10K series resistance for louder, richer sound. Potentiometers R1 through R8 are ultraminiature PC types (screwdriver-tunable) of 50K each. Adjust the pots to whatever key is desired (series resistors may be required on two or three of the pots at the end of the audio range; determine this by experiment). A standard 9V "transistor" battery may be used, but disconnect it when the organ is not in use.

SECTION 4 Hobby

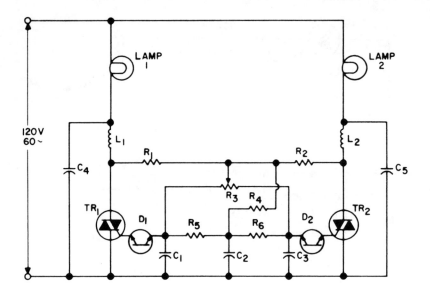

$R_1 = R_2 = 6800 \Omega$, 1 WATT
$R_3 = 150K \Omega$ LINEAR POT. 1W
$R_5 = R_6 = 22K\Omega$, 1/2 W.
$R_4 = 15K\Omega$, 1/2 W.
$TR_1 = TR_2 =$ TRIAC,
$D_1 = D_2 =$ GE ST-2 DIAC

$L_1 = L_2 = 60\mu hy$ (FERRITE CORE)
$C_1 = C_2 = C_3 = 0.1\mu f$ 50V
$C_4 = C_5 = 0.1\mu f$ 200 VOLTS

NOTE: TOTAL LIGHT LEVEL (SUM OF LAMPS 1+2) CONSTANT WITHIN 15%.

4.4.17 GE audio/visual slide fader control. This circuit may be used for fading between two slide projectors (cross fader circuit). As R3 is moved to either side of the center, one triac is fired earlier in each half-cycle, and the other later. The total light output of both lamps stays about the same for any control position.

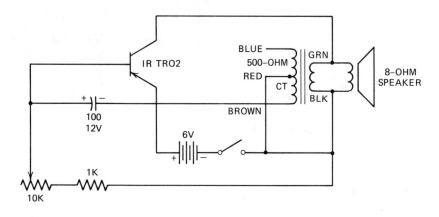

4.4.18 Workman electronic metronome provides adjustable spaced "clicks" for rhythmic reference (beat). The large feedback capacitor causes the oscillator to block at a rate determined by the R1, R2, and C1 time constant, thus producing a series of clicks rather than a continuous tone. Variable resistor R1 serves as a beat rate control.

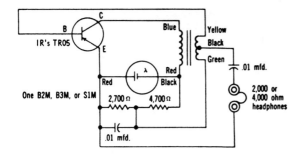

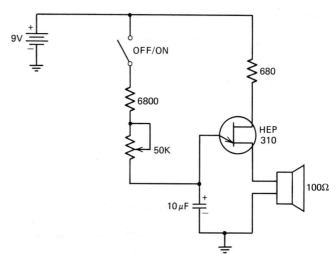

4.4.19 Workman solar-powered 400 Hz audio oscillator. The energy amplified by the transistor is applied to the transformer primary (blue–red). A portion of the energy is fed back to the base of the transistor, where it is again reamplified. Connected in this manner, the circuit current constantly builds up and then breaks down.

The transformer can be any 10K:2K type, centertapped. Although 2000- or 4000-ohm headphones are specified, almost any type can be used. With only a single cell, you will find that the volume is extremely high. You can use the audio oscillator for code practice by inserting a telegraph key in series with the cell.

4.4.21 Motorola metronome for music instruction. The click frequency is controlled by a 50K pot. The battery may be left in the circuit with the switch off as shown, since, the unijunction presents nearly infinite resistance until "triggered" at the emitter.

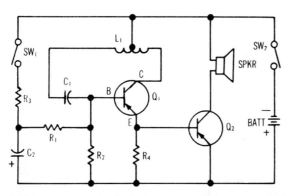

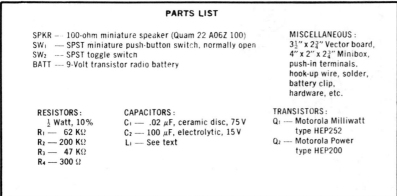

PARTS LIST

SPKR — 100-ohm miniature speaker (Quam 22 A06Z 100)
SW₁ — SPST miniature push-button switch, normally open
SW₂ — SPST toggle switch
BATT — 9-Volt transistor radio battery

MISCELLANEOUS:
3½" x 2¾" Vector board,
4" x 2" x 2¾" Minibox,
push-in terminals,
hook-up wire, solder,
battery clip,
hardware, etc.

RESISTORS:
½ Watt, 10%
R₁ — 62 KΩ
R₂ — 200 KΩ
R₃ — 47 KΩ
R₄ — 300 Ω

CAPACITORS:
C₁ — .02 μF, ceramic disc, 75 V
C₂ — 100 μF, electrolytic, 15 V
L₁ — See text

TRANSISTORS:
Q₁ — Motorola Milliwatt type HEP252
Q₂ — Motorola Power type HEP200

4.4.20 Motorola "panic button" produces a true siren wail, making it suitable for civil defense or emergency-duty applications, especially so when combined with a small PA system.

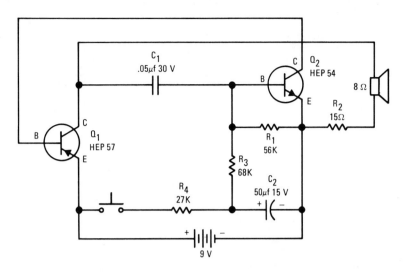

4.4.22 Motorola HEP "panic button" is a fascinating electronic novelty that produces a loud rising and falling tone that sounds very much like a commercial siren. Since it produces a true siren wail, it is ideally suited for civil defense and other special emergency duties when combined with a public address amplifier system.

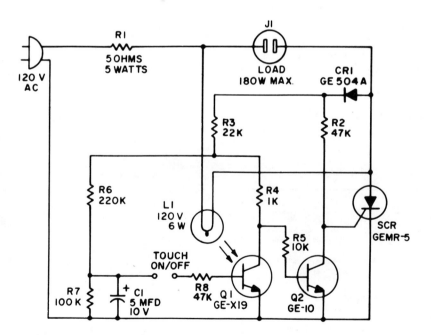

4.4.23 GE touch switch turns on the circuit with one touch and turns it off with the next. This circuit uses an SCR to switch a load of up to 180W. The SCR is triggered by transistor Q2 and resistor R2. When Q2 is conducting, the current from R2 is shunted away from the SCR gate and passes through the collector to the emitter of Q2. The only essential requirement that remains is to turn Q2 on and off with a touch of the finger, and this is accomplished by the rest of the circuit.

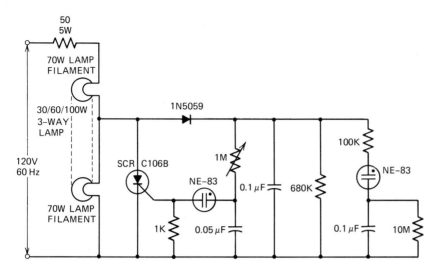

4.4.24 GE flickering "candle" employs a low-cost circuit for synthesizing the flickering effect of a flame. This circuit can be used with a single lamp, but the illusion of a flame is greatly enhanced by the appearance of motion obtained with either two lamps or a dual-filament (3-way) lamp. Various degrees, rates, and the randomness of flickering may be obtained by varying the component values in the circuit. One neon-lamp relaxation oscillator triggers the SCR every other half-cycle at a phase angle that is modulated by a second neon-lamp oscillator operating at a much slower rate. Adding another modulating oscillator that operates at a different rate can improve the effect of random flickering.

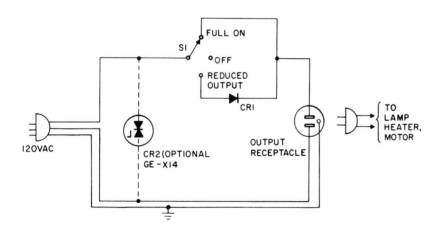

Parts List

CR1 — GE-504A rectifier diode for 130 watts output
— GE-X4 rectifier diode for higher output
CR2 — GE-X14 thyrector diode (optional transient voltage suppressor)
S1 — SPDT, 3-amp, 125-volt a-c switch with center "off" position

4.4.25 GE simple high-low power switch.

SECTION 5
RADIO/TV

Few people build their own radios or television receivers unless they do so according to instructions supplied with one of the popular kits on the market. For the most part, the complete receiver circuits presented in these pages are types than can be easily built and with little expense. The remainder of the circuits in each of the following subsections are functional circuit blocks or modules that may be used either in existing receivers or as a basis for receiver design. For additional radio circuits that might be applicable to your requirements, check the diagrams included in Section 6.

5.1 AM and FM Radio

Most of the circuits presented here might be considered "casual" in that they don't have the performance capability of high-quality commercial equipment. This doesn't mean much when it comes to AM reception because of the limited frequency response of the medium, but it could be significant for experimenters looking for a good FM or FM stereo receiver diagram.

If you're thinking of designing your own receiver and using the circuits in this book as modular circuit blocks, also consult the diagrams for communications receivers (Section 6).

The AM receivers shown here have generally excellent sensitivity but they are not particularly selective—which might make them ideally suited for applications where fidelity is more important than adjacent-channel rejection.

All of the AM receivers in this section will operate nicely in conjunction with the appropriate-frequency transmitter circuits in the Communicator subsection (Section 4.3).

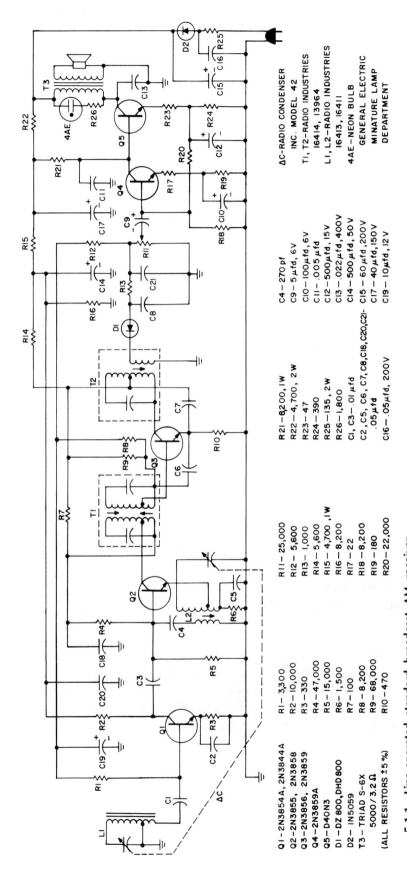

5.1.1 Line-operated standard broadcast AM receiver with untuned RF stage employs five discrete GE transistors. The design has excellent AGC performance on strong signals in spite of the amplifierless RF stage.

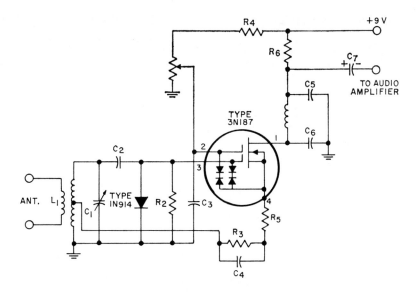

Parts List

C_1 = tuning capacitor, 0 to 100 pF, E.F. Johnson No. R-149-5 or equiv.
C_2 = 47 pF, ceramic
C_3 = 0.05 µF, ceramic
C_4, C_5 = 0.01 µF, ceramic
C_6 = 0.001 µF, ceramic
C_7 = 0.1 µF, electrolytic, 15 V
L_1 = refer to coil-data chart
L_2 = refer to coil-data chart
L_3 = rf choke, 25 mH
R_1 = feedback control, potentiometer, 0.1 megohm, 0.5 watt
R_2, R_3 = 0.1 megohm, 0.5 watt
R_4 = 1800 ohms, 0.5 watt
R_5 = 150 ohms, 0.5 watt
R_6 = 4700 ohms, 0.5 watt

5.1.2 RCA regenerative detector stage can be interconnected with a general-purpose audio amplifier to form a simple CW or AM regenerative receiver. This receiver can extract the audio-signal information from an amplitude-modulated input signal as small as 0.5 µV. By selection of the proper values for the tuned-circuit components, the detector circuit can be adapted for operation over a wide range of frequencies.

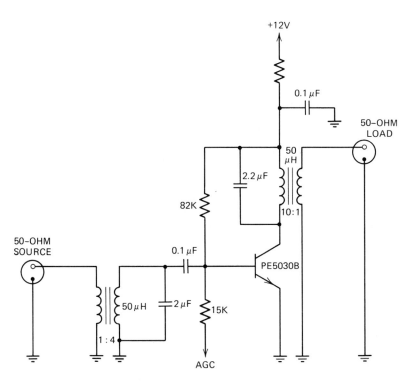

5.1.3 Fairchild 455 kHz AM IF amplifier. The collector current should be 2 mA and the collector–emitter voltage, 6V. At these conditions, a 35 dB gain is obtained.

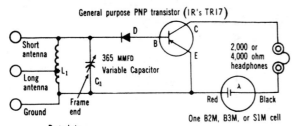

Parts list:
C_1 — 365 mmfd variable capacitor (J. W. Miller #2111 or equivalent)
D — Crystal diode, 1N34, 1N48, etc.
L_1 — Transistor loopstick antenna (J. W. Miller #2001 or equivalent)

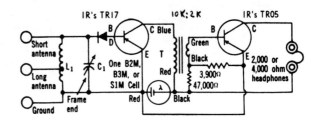

Parts list:
T — 10,000 ohm to 2,000 ohm interstage transformer (Stancor TA-35 or equivalent) (Other parts, same as Fig. 12)

5.1.4 Workman broadcast radio derives its power from the sun. The signal from the radio station is intercepted by the antenna, which is connected to a coil and tuning capacitor. Once the desired signal is selected, the diode detector converts the radio frequency energy to audio frequencies so that they can be heard.

The tiny electrical signal from the detector is passed on to a transistor for amplification. The amplifier signals then energize the headphones, which convert the electrical impulses to sound waves. No battery other than the photocell is required to power the radio; however, a penlight or other small flashlight cell could be used to operate the set at night. The radio is designed to use any antenna length from 3m to 1 km.

One of the L1 wires will be doubled up (two wires in one) and this lead connects to the long antenna. The lead nearest the double wire goes to the frame of the tuning capacitor and to earth ground. The remaining wire goes to the lugs on the side of the tuning capacitor. Connections to the frame of the tuning capacitor can be made by inserting a short screw in one of the front holes and wrapping a wire under the screw head. Be sure the screw does not touch the aluminum plates.

For best performance, connect the receiver to an antenna of 6m or more; the longer the antenna, the greater the volume and number of stations you can receive.

5.1.6 Workman amplified AM solar radio. A transformer is needed to couple the output of one transistor to the input of the next. The transformer may be any interstage type, such as the Stancor TA-35 or a Triad TY56X.

This radio will always work best with the antenna on the long connection. Even so, you may find it has too much volume for clear reception. If this is the case, you can connect a 100 pF capacitor in series with the antenna. For weak stations you can obtain more volume by connecting several cells in series or by using an IR-Workman S3M cell.

If additional cells are used and if you use different transistors or headphones, it may be necessary to vary the value of the 3.9K resistor for best reception.

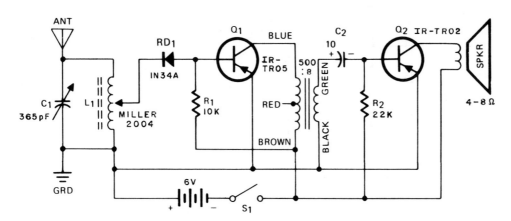

5.1.5 Workman two-stage transistor radio. Signals are picked up by the antenna and selected by a tuned circuit of coil L1 and tuning capacitor C1. The signal is detected by diode RD1 and coupled to transistor Q1, the audio amplifier. Transformer T1 matches the output impedance of Q1 to the input impedance of Q2. Power transistor Q2 in turn drives the loudspeaker.

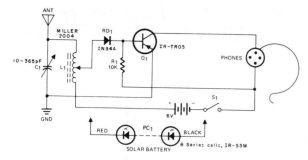

5.1.7 Workman two-stage solar radio includes a power transitor amplifier stage. This radio should work best with the antenna on the long connection. Note that only one solar cell is used. Therefore, this is a low-voltage, low-current unit. If weak stations are not strong enough, more volume is required. Simply add a second solar cell in series with the one indicated.

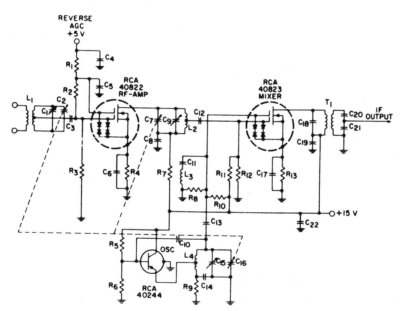

C1, C9, C15 = Trimmer capacitor, 2 to 14 pF
C2, C7, C16 = Ganged tuning capacitors, each section = 6 to 19.5 pF
C3, C6, C14, C17, C22 = 2000 pF, ceramic
C4, C5 = 1000 pF, ceramic disc
C8, C19 = 0.01 μF, ceramic disc
C10 = 3.3 pF, NPO ceramic
C11 = 270 pF
C12 = 500 pF, ceramic disc
C13 = 3 pF, NPO ceramic
C18 = 68 pF, ceramic
C20 = 50 pF, ceramic
C21 = 1200 pF, ceramic
L1 = antenna coil; 4 turns of No. 18 bare copper wire; inner diameter, 9/32 inch; winding length, 3/8 inch; nominal inductance, 0.86 μH; unloaded Q, 120; tapped approximately 1 1/4 turns from ground end; antenna link approximately 1 turn from ground end
L2 = rf interstage coil; same as L1 antenna link
L3 = rf choke, 1 μH

L4 = oscillator coil; 3 1/4 turns of No. 18 bare copper wire; inner diameter, 9/32 inch; winding length, 5/16 inch; nominal inductance, 0.062 μH, unloaded Q, 120; tapped approximately 1 turn from low end
R1, R10 = 0.56 megohm, 0.5 watt
R2 = 0.75 megohm, 0.5 watt
R3 = 0.27 megohm, 0.5 watt
R4, R13 = 270 ohms, 0.5 watt
R5 = 22000 ohms, 0.5 watt
R6 = 56000 ohms, 0.5 watt
R7 = 330 ohms, 0.5 watt
R8, R12 = 0.1 megohm, 0.5 watt
R9 = 4700 ohms, 0.5 watt
R11 = 1.6 megohms, 0.5 watt
T1 = first if (10.7 MHz) transformer; double-tuned with 90 per cent of critical coupling; primary: 15 turns of No. 32 enamel wire, space wound at 60 turns per inch on 0.25-by-0.5-inch slug; secondary: 18 turns of No. 36 enamel wire, close wound on 0.25-by-0.25 inch slug; both coils wound on 9/32-inch coil form.

5.1.8 FM tuner using RCA 40822 and RCA 40823 MOS-FET ICs for the RF amplifier and mixer stages.

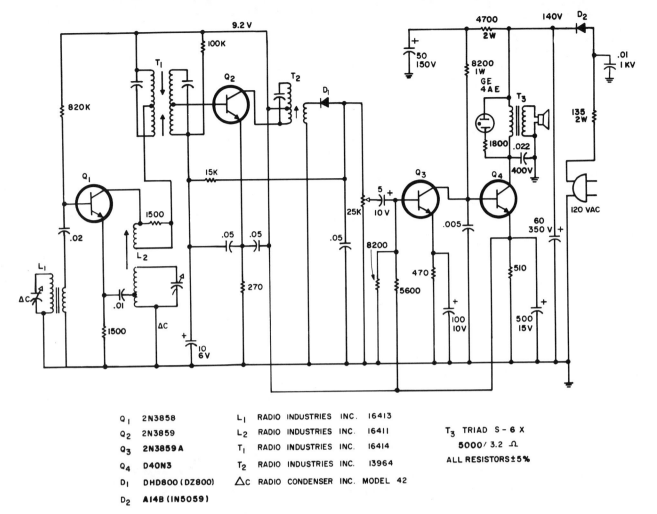

Q_1 2N3858	L_1 RADIO INDUSTRIES INC. 16413
Q_2 2N3859	L_2 RADIO INDUSTRIES INC. 16411
Q_3 2N3859A	T_1 RADIO INDUSTRIES INC. 16414
Q_4 D40N3	T_2 RADIO INDUSTRIES INC. 13964
D_1 DHD800 (DZ800)	$\triangle$C RADIO CONDENSER INC. MODEL 42
D_2 A14B (1N5059)	

T_3 TRIAD S-6X 5000/3.2 Ω
ALL RESISTORS ±5%

5.1.9 GE line-operated four-transistor radio receiver uses D40N3 high-voltage transistor in the audio output stage. Sensitivity runs under 100 μV/m for 100 mW reference, and a power output of 1W or greater with less than 10% distortion is typical.

The power supply is a single rectifier in a half-wave configuration furnishing 140V of B+. This high a B+ voltage is necessary to make the radio economically feasible as it allows operation without a stepdown transformer or an excessively large dropping resistor. With a half-wave circuit, the B+ regulation is such that a class A audio output stage is mandatory. Supply for the audio driver is derived through a simple RC filter from the main 140V supply. The current drain to the remainder of the radio is quite low and is derived from the class A output transistor emitter current. The small capacitor from the rectifier to ground is to prevent damage to the rectifier from line transients that occur during the blocking half-cycle.

The output stage is the key to the entire design. Most ac/dc receivers have an audio output power on the order of 1W. This requires 3W or so of collector dissipation. Some additional dissipation capability is required to allow for operation at an elevated ambient temperature, so that the total dissipation required is about 4W.

Biasing of the stage is ≃124V V_{CE} at 22 mA I_C for a collector dissipation of 2.73W. The theoretical maximum efficiency in class A is 50%, although in actual practice a value of 37% is more usual, so that 1.0W is the power into the speaker. The primary impedance of the output transformer is 5600 ohms, calculated by

$$R_L = \frac{V_{CE}}{I_C} \approx \frac{124V}{0.022A}$$

The voltage gain approaches 60 dB with this high a load impedance. Destructive transient voltages can be generated in the collector circuit, due to the inductance of the output transformer. Stray radiation from electric motors, appliances, spark plugs, power lines, etc., is intercepted by the ferrite rod antenna and amplified through the radio. When this "noise" causes a strong negative pulse to appear at the base of the output stage just at a time when the collector is conducting a large current, the resulting change of current caused by the transistor abruptly switching off will cause an extremely high-voltage transient to be generated. The voltage transient can be prevented by providing an alternative path for the transformer current to flow through. One convenient method is to shunt the transformer with a GE 4AE neon lamp in series with an 1800-ohm resistor. This will limit the collector voltage in the positive direction.

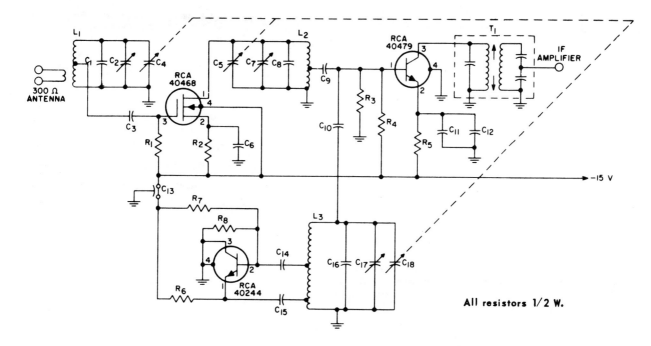

R_1 = 100 KΩ
R_2 = 220 KΩ
R_3, R_4 = 47 KΩ
R_5 = 4.7 KΩ
R_6 = 8.2 KΩ
R_7 = 120 KΩ
R_8 = 22 KΩ

C_1, C_8, C_{16} = 16 pF
C_2, C_7 = 2-12 pF, Trimmer
C_3, C_6 = 0.002 µF
C_4, C_5, C_{18} = 5.5-22.5 pF, ganged tuning capacitor
C_9 = 5000 pF
C_{10} = 2.7 pF
C_{11} = 0.01 µF
C_{12}, C_{14}, C_{15} = 1000 pF
C_{13} = 1000 pF feedthrough type
C_{17} = 2-10 pF, Trimmer

L_1 = #18 bare copper wire, 4 turns, 1/4" I.D., 7/16" winding length, Q_o at 100 MHz = 130.
Tunes with 34 pF capacitance at 100 MHz.
Antenna Link approximately 1 turn from ground end.
Gate Tap approximately 1-1/2 turns from ground end.

L_2 = #18 bare copper wire, 4 turns, 1/4" I.D., 7/16" winding length, Q_o at 100 MHz = 120.
Tunes with 34 pF capacitance at 100 MHz.
Base Tap approximately 3/4-turn.

L_3 = #18 bare copper wire, 4 turns, 7/32" I.D., 7/16" winding length, Q_o at 100 MHz = 120.
Tunes with 34 pF capacitance at 100 MHz.
Emitter Tap approximately 1-1/2 turns from ground end.
Base Tap approximately 2 turns from ground end.

5.1.10 FM receiver front end using an RCA 40468 MOSFET.

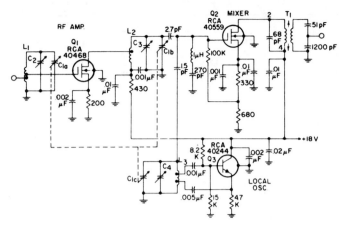

C_{1a}, C_{1b}, C_{1c} - 3-gang tuning capacitor, TRW 5-plate Model V2133 with trimmers stripped off.
C_2, C_3, C_4 - Arco 402 trimmer, maximum value 10 pF
L_1 - No.18 bare copper wire, 5 turns on 19/64" form, coil length 1/2", with IRN .250" x .250" Arnold slug. $Q_0 = 164$. Antenna tap at 0.8 of a turn, output tap at 1.4 turns.
L_2 - No.18 bare copper wire, 5 turns on 15/64" form, coil length 3/8", with 0.181" x 0.375" Arnold slug. $Q_0 = 104$.
L_3 - No.18 bare copper wire, 5 turns, air core with 3/8" O.D., coil length 1/2". Emitter tap on 1-1/2 turns. Feedback tap on 2 turns. $Q = 164$.
T_1 - Double tuned, 90 per cent of critical coupling. Primary unloaded uncoupled $Q = 137$ with 68-pF tuning capacitance, secondary unloaded uncoupled $Q = 76$ with 47-pF tuning capacitance. Secondary has a turns ratio of 26.2 to 1.0. Primary, No.32 enamel wire, 15 turns, space wound at 60 TPI, 0.250" x 0.500" TH slug. Secondary, No.36 enamel wire, 18 turns, close wound, 0.250" x 0.250" TH slug. Both coils on 9/32" form without shield.

5.1.11 FM tuner using MOS transistors for the RF amplifier and mixer stages. RF amplifier transistor RCA 40468 operates in the common-source configuration with a stage gain of 12.7 dB. The mixer transistor, an RCA 40559, also operates in the common-source configuration, with both the RF and local-oscillator signals applied to the gate terminal. The bipolar oscillator transistor (RCA 40244) operates in the common-collector mode. The conversion power gain from the mixer stage is 17.5 dB; the total gain of the tuner is 30.2 dB.

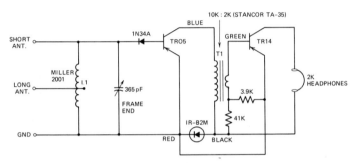

5.1.13 Workman single-stage transistor radio provides more sensitivity than the primary crystal receiver. The antenna should be at least 6m long. This project may be powered by the two single 1.5V cells. Signals are picked up by the antenna network and selected by tuning coil L1 and capacitor C1. The audio from the signal is detected by diode CR1 and amplified by the transistor.

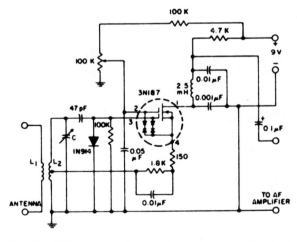

5.1.12 RCA regenerative receiver using protected dual-gate MOSFET IC. The circuit is basically an amplifier with controlled feedback adjusted to the verge of oscillation. Gate 2 provides a convenient means to adjust the amplifier gain to the requisite level. Detection is accomplished in the gate 1 input circuit by the interaction of the diode in parallel with the 100K resistor and the 270 pF capacitor.

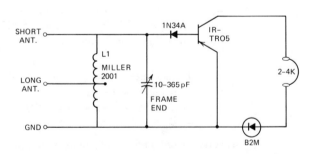

5.1.14 Workman inexpensive solar radio can be mounted in a small box and yet deliver a good signal. The signal is picked up by a 3 to 6m antenna. This circuit also requires a ground.

A small penlight battery could be connected parallel to the photocell for dark operation. When connecting coil L1, you will find three output wires. One of the wires will be doubled; this lead connects to the long antenna. The lead nearest the double wire connects to the frame of the tuning capacitor and to an earth ground. The remaining wire goes to the lugs on the side of the tuning capacitor.

5.2 Video and Deflection

Every electronic equipment item that requires a cathode-ray tube for display—be it an oscilloscope or a TV set, a closed-circuit monitor or an alphanumeric computer peripheral—requires electronic circuitry for deflecting the electron beam vertically and horizontally across the screen. The circuits in this subsection are applicable to all such video devices.

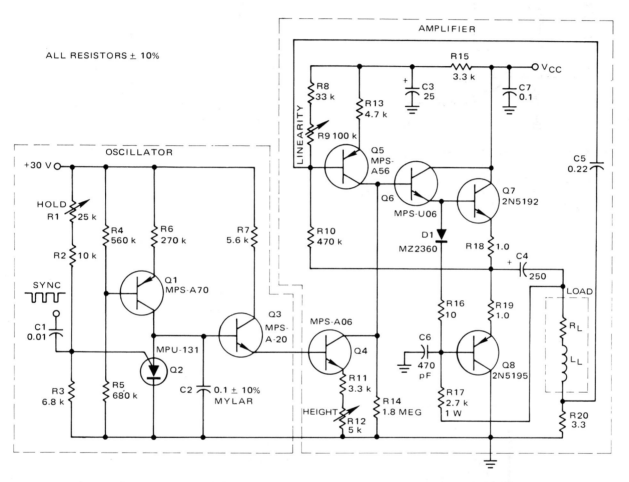

5.2.1 Motorola complementary vertical deflection circuit. In the oscillator, Q1 acts as a constant-current source to charge capacitor C2. The voltage across C2 thus increases linearly with time. Capacitor C2 is discharged by programmable unijunction transistor Q2. The oscillator frequency is controlled by the voltage level at which Q2 turns on. This is accomplished by controlling the gate voltage on Q2 with potentiometer R1. Negative-going sync pulses on the gate lead of Q2 will turn on the PUT if the free-running frequency of the oscillator is close to that of the sync pulses. The lock-in range is determined by the magnitude of the sync pulses. For example, a 1V sync pulse will lock in the oscillator in the 50 to 60 Hz range. Q3 serves as a buffer between the oscillator and the amplifier to prevent nonlinear charging of capacitor C2.

Transistor Q4 serves a voltage-controlled current source. Transistor Q5 acts as a constant-current source (excluding feedback effects) such that transistors Q6 and Q7 are turned on when the oscillator voltage is low. As the oscillator voltage increases, the collector current in Q4 increases, decreasing the drive current to the base of transistor Q6. Q6 begins to turn off, as does Q7. When the oscillator voltage increases further, Q7 cuts off completely and Q8 begins to conduct. When the oscillator voltage reaches its maximum, transistor Q6 and Q7 will be off and Q8 will be on. At this point the oscillator voltage drops to zero, turning off transistor Q4. Transistors Q6 and Q7 then become biased on and Q8 biased off. However, the inductive load will not allow an abrupt change in the collector current. As a result the inductor voltage will increase, thus increasing the voltage at the center point of the output pair. Since Q5 is biased through R10, Q5 and Q6 begin to turn off. Turning off Q6 reduces the voltage across R17, keeping Q8 on until the current in the load goes to zero. When the inductor current reaches zero, the output voltage begins to fall. This causes Q7 to turn on and Q8 to turn off, thus completing the current reversal. The output voltage will remain high until the current builds up to maximum in the inductor.

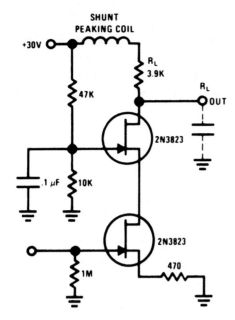

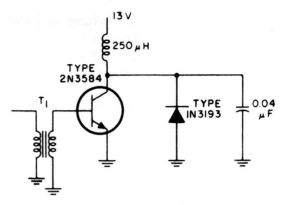

T_1 = C.P. ELECTRONICS X-9370 OR EQUIV.

5.2.2 RCA FET cascode video amplifier features very low input loading and the reduction of feedback to almost zero. The RCA 2N3823 is used because of its low capacitance and high Y_{fs}. The bandwidth of this amplifier is limited by R_L and the load capacitance.

5.2.4. RCA transistor magnetic deflection circuit. The yoke, Celco HD 428-S560 or equivalent, is used to drive a cathode-ray tube for an alphanumeric display with a 36° full-deflection angle and a 12 kV acceleration potential. The yoke inductance is 250 μH and the energy required is 225 microjoules. The sweep time is 50 μs and the retrace time, 10 μs. From this information, the peak collector current I_p of the deflection-circuit transistor is calculated as follows:

$$I_p = \sqrt{\frac{2(225)10^{-6}}{(250)(10^{-6})}} = 1.35 \text{A}$$

The supply voltage V_{cc} required is given by

$$V_{cc} = \frac{2LI_p}{t_s} = \frac{2(250)(10^{-6})(1.35)}{(50)(10^{-6})} = 13.5 \text{V}$$

The tuning-capacitor value C is given by

$$C = \left(\frac{t_r}{\pi}\right)^2 \left(\frac{1}{L}\right) = \frac{(100)(10^{-12})}{(\pi)^2(250)(10^{-6})} = 0.40 \; \mu\text{F}$$

Finally, the maximum collector voltage V_{CE} is given by

$$V_{ce} = 13.5 + (1.35)\frac{\pi}{(10)(10^{-6})}250(10^{-6}) = 118\text{V}$$

The breakdown voltage must therefore be greater than (118)(1.3) = 155V.

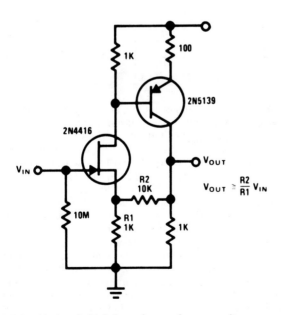

5.2.3 National high-impedance, low-capacitance amplifier. This compound series-feedback circuit provides high input impedance and stable, wideband gain for general-purpose video amplifier applications.

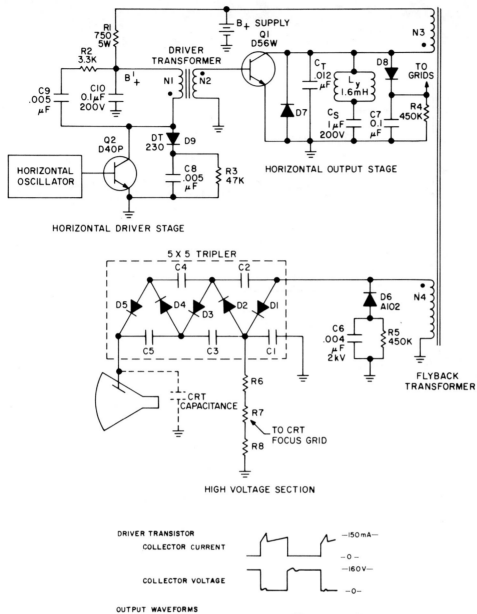
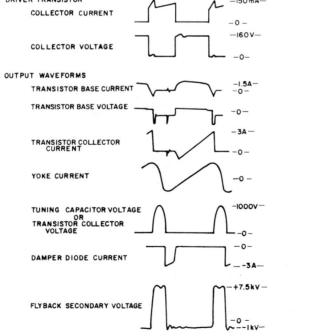

5.2.5 GE TV horizontal sweep circuit, including horizontal driver, horizontal output, and high-voltage circuit. Waveforms for various parts of the circuit are shown.

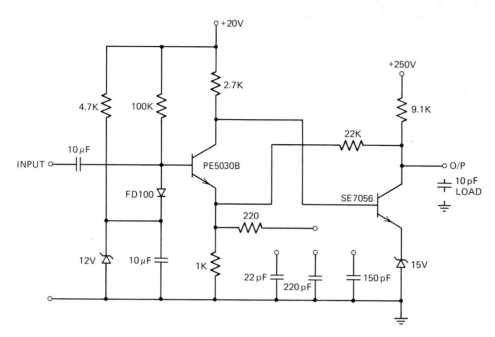

5.2.6 Fairchild video amplifier suitable for camera viewfinders and picture monitors. The amplifier uses a high-voltage video transistor, the SE7056, and is capable of more than 100V peak-to-peak output. The output stage is driven by a PE5030B transistor. The rise time into a 10 pF load is less than 0.1 μs and the bandwidth is over 6 MHz.

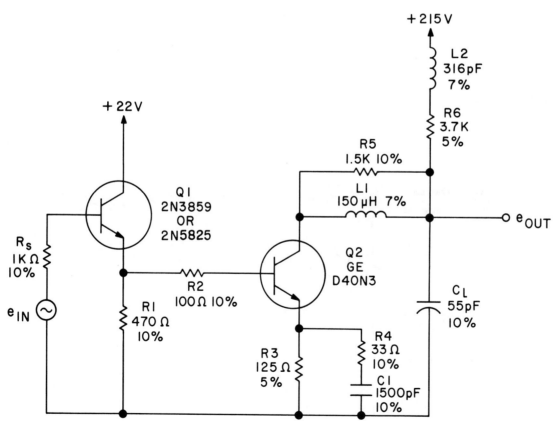

5.2.7 In this GE video amplifier parallel impedances have been combined and the 3.58 MHz trap and zener deleted. A source impedance of 1K is assumed for the input signal.

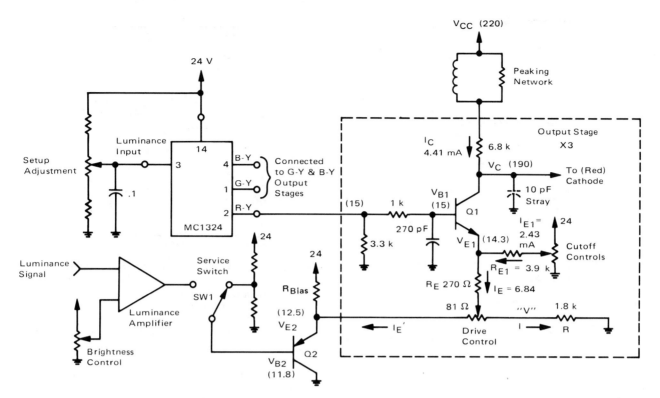

5.2.8 Motorola unitized-gun picture-tube driver requires operation with a 220V B+ and a voltage gain of 24 to supply a 120V drive with 5V of signal from the demodulator. Black-level voltage of 190V with a 40V range on black-level adjustment is required. Complete equations for design and exhaustive design details of this, as well as data covering the complete video system, are available from Motorola in application note AN-761, entitled "Video Amplifier Design: Know Your Picture Tube Requirements," by Steve Tainsky, Applications Engineering.

Any modulation on the 220V supply will appear directly at the cathode; also, any modulation on the 24V supply is coupled to the cathode via the resistor connecting the cutoff adjustment potentiometer to the emitter of the output stage. Even with the low 6.8K collector load, the 3 dB bandwidth of this stage working into the 10 pF load is less than 2.5 MHz; therefore, peaking components must be used to obtain the full 3.5 MHz bandwidth.

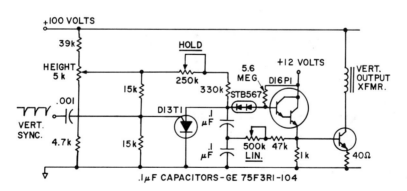

5.2.9 TV vertical deflection oscillator uses GE's D13T1 PUT. The basic oscillator is formed by the two 15K resistors in series (R_{BB}) and the 250K variable resistor in series with the 330K resistor charging the 0.1 μF capacitors. The *hold* pot is used to adjust to the 60 Hz sweep rate of the TV set. Synchronization is accomplished with negative pulses capacitively coupled to the intrinsic standoff point (4V amplitude). Since the frequency of a relaxation oscillator of this sort is relatively independent of the interbase voltage, this makes a convenient means to adjust the height of the resultant sawtooth waveform across the capacitor. This is done with the resistor divider network shown on the left side of the diagram. The GE 16P1 is a low-cost, very high-gain Darlington for driving the output power transistor. The multipellet diode is added for dc level shifting so that the output power transistor will not be cut off at the beginning of each sweep cycle. Linearity is achieved by bootstrap feedback through the 500K variable resistor.

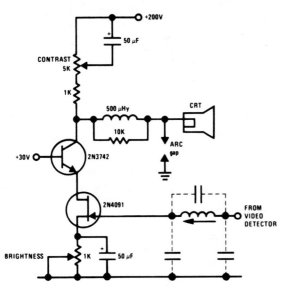

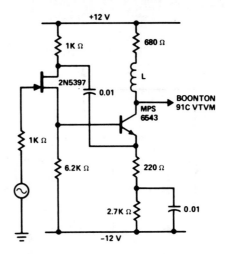

5.2.10 National JFET bipolar cascode circuit will provide full video output for a CRT cathode drive. The gain is about 90. The cascode configuration eliminates Miller-effect problems with the 2N4091 JFET, thus allowing direct drive from the video detector. An *m*-derived filter using stray capacitance and a variable inductor prevents the audio at 4.5 MHz from being amplified by the video amplifier.

5.2.12 Video amplifier. Siliconix FET and bipolar transistor combination makes a good video amplifier because the FET input provides the voltage gain, thus obtaining a superior gain–bandwidth product. The feedback capacitor ac couples the emitter to the drain. The ac voltage at the gate is nearly equal to that at the source. This source voltage is dc-coupled to the base to produce an ac voltage at the emitter whose amplitude is almost equal to that at the base. Thus all three signals are in phase and Miller capacitance is largely eliminated.

The frequency response of this circuit is controlled by the output time constant if f_t of the transistor is much greater than the amplifier bandwidth. In the circuit shown, the ac load is 2.5 pF.

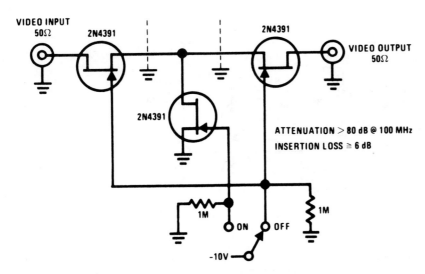

5.2.11 National high-frequency switch. The 2N4391 provides a low *on* resistance of 30 ohms and a high *off* impedance (at less than 2 pF) when off. A proper layout is an absolute must if these performance levels are to be achieved in practice.

252 DISCRETE / TRANSISTOR CIRCUIT SOURCEMASTER

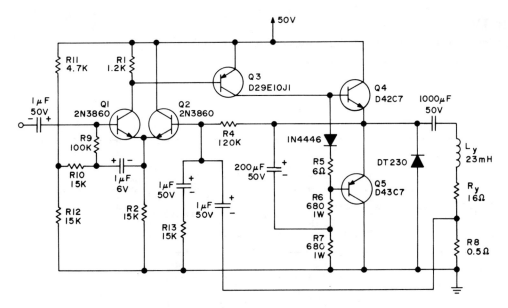

GE TV vertical deflection circuit for capacity-coupled, saddle-yoke load using a single power supply.

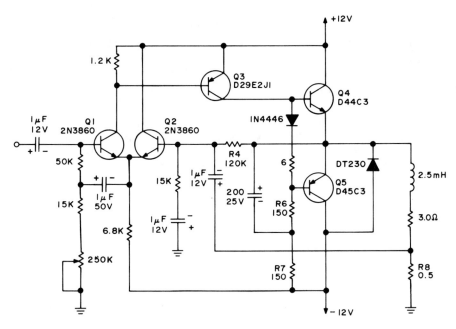

5.2.13 GE TV vertical deflection circuit driving directly a toroidal yoke and using two power supplies. The semiconductors are GE.

5.3 Black-and-White TV

Most of these circuits are intended for black-and-white television receiver applications but some of them—audio stages, for example—are equally applicable to color TV and monitor receivers.

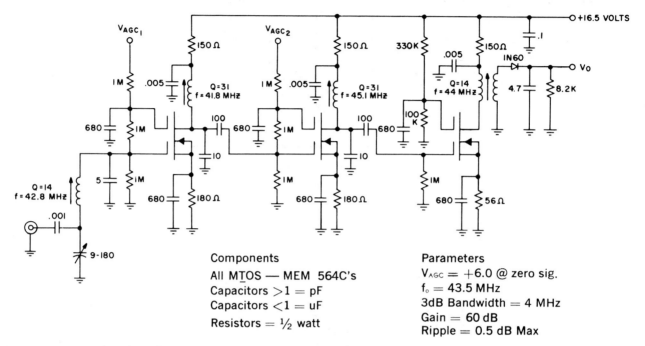

Components
All MTOS — MEM 564C's
Capacitors >1 = pF
Capacitors <1 = uF
Resistors = ½ watt

Parameters
V_{AGC} = +6.0 @ zero sig.
f_o = 43.5 MHz
3dB Bandwidth = 4 MHz
Gain = 60 dB
Ripple = 0.5 dB Max

5.3.1 44 MHz video IF amplifier uses three General Instrument MEM 564C MTOS devices. Sound and picture traps have been omitted for simplicity. The interstage networks comprise a four-pole Butterworth design (maximally flat amplitude response).

Since the stages are stagger-tuned, somewhat less gain can be expected than an equivalent sync-tuned amplifier. An overall gain of 60 dB (voltage gain) was measured, which includes a 6 to 10 dB detector loss.

The design of this IF strip is similar to that of a vacuum-tube type. The drain of a stage is capacitively coupled directly to the gate of the succeeding stage. $R_e(Y_{is})$ is approximately 60 microsiemens and $R_e(Y_{os})$ is approximately 120 microsiemens. Direct coupling results in only a 1 dB loss per stage.

With all loss resistance taken into account, the total shunt resistance R_s at each output node is about 5K.

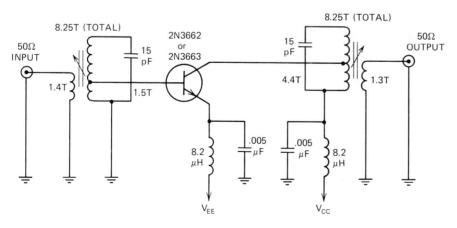

5.3.2 Motorola 45 MHz IF amplifier has an unloaded Q of 80 for the two single-tuned transformers. The overall bandwidth is 4.5 MHz and the stage gain is 20 to 25 dB.

254 DISCRETE / TRANSISTOR CIRCUIT SOURCEMASTER

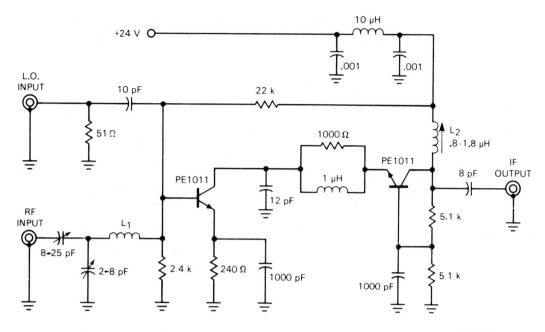

	Channel 6	Channel 13
L_1	330 nH	75 nH

5.3.3 High-performance mixer for VHF TV uses Fairchild's PE1011, an NPN transistor designed for VHF operation. The circuit is a basic cascode mixer with the exception of the network between the two transistors. Devices are biased at a nominal collector current of 6.5 mA, a collector-to-emitter voltage of 10V for the input transistor and a collector-to-base voltage of 12V for the output transistor. In an actual tuner, the 10 pF coupling capacitor would be from 1 to 3 pF to reduce loading of the local oscillator. Actual local-oscillator injection at the base of the first transistor ranges between 20 and 200 mV rms.

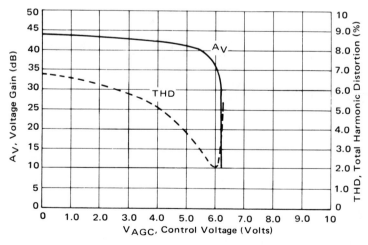

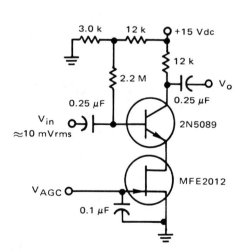

5.3.4 Motorola simple automatic gain control (AGC) circuit. The FET changes the voltage gain over a 30 dB range with only a 1V change in V_{GS}. This circuit suffers from a bit of harmonic distortion due to the unbypassed emitter degeneration.

SECTION 5 Radio/TV 255

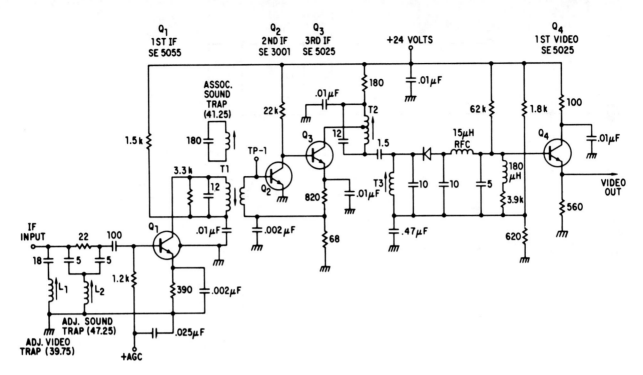

5.3.5 Fairchild monochrome video IF amplifier has the following characteristics:

AGC Bias	Sensitivity of 80% Modulated 45.75 MHz for 1V p-p Detected Video
2V	125 μV ($\pm$4 dB)
5V	12.5 mV (pm6 dB)

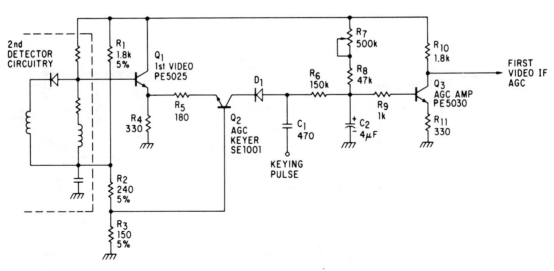

5.3.6 Fairchild second detector, video buffer, and AGC circuitry. Setup control is eliminated by using 5% resistors in a common-divider network feeding both the video buffer and the AGC keyer. Using 2% resistors for R1 and R2 reduces tolerances further for some of the more critical applications. The peak-to-peak video level is determined by the voltage developed across R2 plus a slight detector diode prebias across the detector load. The $V_{BE(on)}$ voltages of the keyer and video buffer are opposing. The worst-case spread in peak-to-peak video caused by resistors would be under 10% (4% using 2% resistors). (The semiconductors are Fairchild types.)

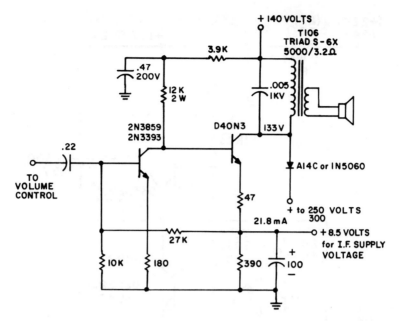

5.3.7 GE audio amplifier stage for a portable TV sound system. The 2N3859 or 2N3393 driver is biased at 8.2 mA with a dc feedback loop from the output emitter to the driver base providing the forward bias for both stages. The feedback loop action gives the audio circuit very good stability and tolerance to large changes in beta and temperature. If the current is suddenly increased in the output, the bias on the driver increases and lowers the forward bias on the output to return it to normal. The driver emitter is unbypassed to provide emitter degeneration for good ac stability and to achieve lower harmonic distortion with less loading on the ratio detector. The class A output also has a small unbypassed emitter resistor (47 ohms), which provides degeneration for good stability, reduces high-voltage transients, and increases the input impedance. The input impedance of the output is essentially the load (R_L) for the driver; the larger the load the greater the voltage gain obtained from the driver. The output stage is biased at 21.8 mA.

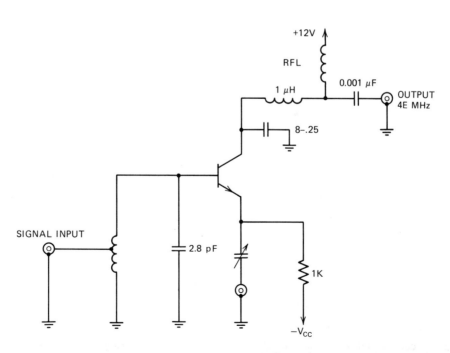

5.3.8 Television mixer stage using the Fairchild PE5030B. The local oscillator, at 258 MHz, is injected into the emitter, and the signal at 213 MHz (channel 13) is applied to the base. The IF is at 45 MHz. The optimum noise source conductance for a mixer is lower than for an amplifier at the same frequency so that a stepup transformer is used to transform the signal generator impedance (50 ohms) to about 200 ohms.

SECTION 5 Radio/TV

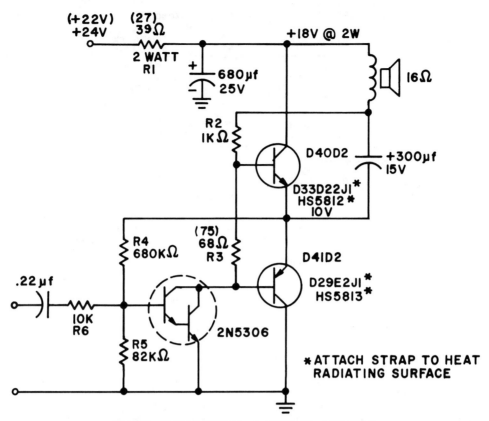

110mV INPUT GIVES 2 WATTS OUTPUT.

5.3.9 GE low-cost 2W TV audio amplifier operates from a ratio detector or from the higher audio level available from the sound processing IC. The output transistors are protected against intermittent load shorts and can be protected for a continuous load short by increasing R1 to 68 ohms (24V supply). This limits the power output to 1W continuous and 2W music power. The GE D40D2 and D41D2 complementary power transistors can be used in this circuit without heatsinks. The other complementary types given must have the aluminum strap attached to a heat-radiating surface equivalent to a 13 cm² aluminum fin for operation up to 60°C ambient temperature. The heatsink design must keep the transistors within their dissipation rating using a thermal resistance rating of $R_{\theta JC}$ = 110°C/W for aluminum strap units.

The power supply decoupling provided by R1 and a 500 to 1000 μF capacitor permits the peak currents required for push-pull operation to be supplied from the electrolytic and prevents modulation of other TV functions using the common low-voltage supply.

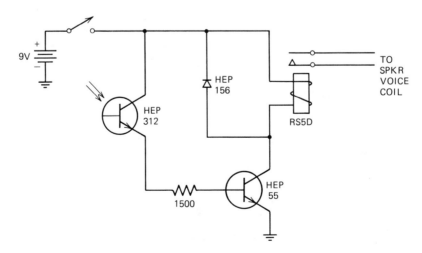

5.3.10 Motorola TV commercial killer may be connected at the set and triggered with a flashlight from across the room; the circuit should be used only with sets that have an audio output transformer, since activation places a short across the speaker terminals. For transformerless transistor circuits, use 1 or 2 ohms in series with relay contact leads. The speaker will be disabled (or lowered in volume if series resistors are used) for as long as light falls on the phototransistor.

258 DISCRETE / TRANSISTOR CIRCUIT SOURCEMASTER

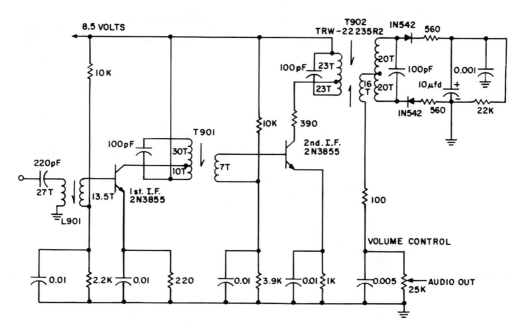

5.3.11 GE sound IF strip for portable TV is a typical high-gain grounded-emitter stage with input and output mismatch to insure stability without neutralization. The audio takeoff, L901, is a single series-wound stepdown transformer used to select the 4.5 MHz sound IF from the video information remaining after the 4.5 MHz detector. The stage is biased typically at 3.6 mA with very little change from high to low h_{FE}. Output transformer T901 is a tapped, single-tuned, bifilar-wound stepdown transformer with unloaded $Q_o = 64$ and loaded $Q_L = 56$. The maximum available gain is approximately 46.5 dB with 21.0 dB total insertion loss, which gives a usable gain of 25.5 dB. The bandwidth of the first IF stage is 80 kHz.

The second sound IF stage is a GE 2N3855 common-emitter amplifier biased at 1.7 mA to obtain stable gain with a higher impedance transformer. The maximum available gain at this current is approximately 47.7 dB, and the usable gain is approximately 27.5 dB. The insertion loss needed to stabilize the stage is 20 dB.

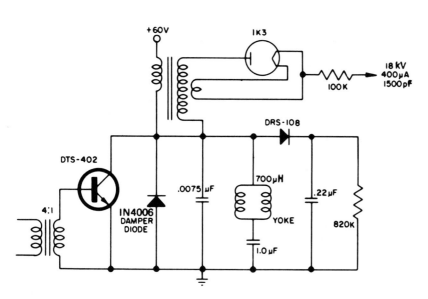

5.3.12 Delco power transistor in the typical *horizontal output circuit* of a black-and-white television receiver.

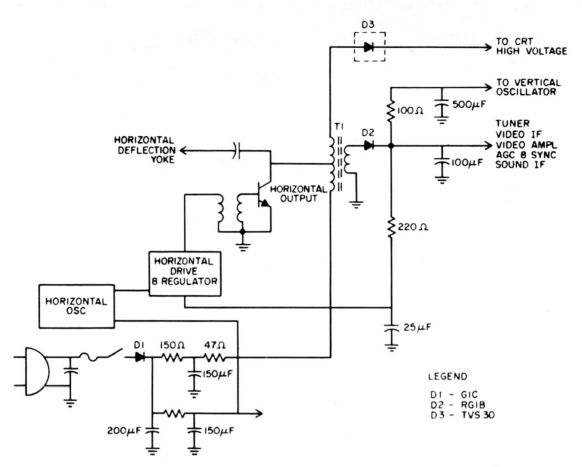

5.3.13 General Instrument TV power supply uses the existing high-frequency supply associated with the horizontal deflection system. In this case, additional low-voltage windings are used in the secondary of the flyback transformer. Rectifier D1 supplies operating voltage for the horizontal oscillator and audio output only. Power for the rest of the set comes from D2, a fast-recovery rectifier operating off the flyback transformer. Regulation is obtained by voltage feedback to the horizontal regulator that controls the horizontal output voltage. High voltage for the picture tube is rectified by D3, a General Instrument TVS-30 high-voltage fast-switching rectifier.

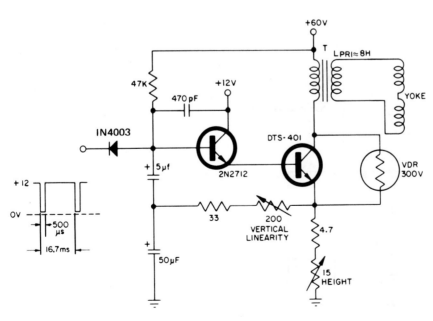

5.3.14 Vertical output circuit for black-and-white TV employs two Delco power transistors. (The circuit is typical.)

260 DISCRETE / TRANSISTOR CIRCUIT SOURCEMASTER

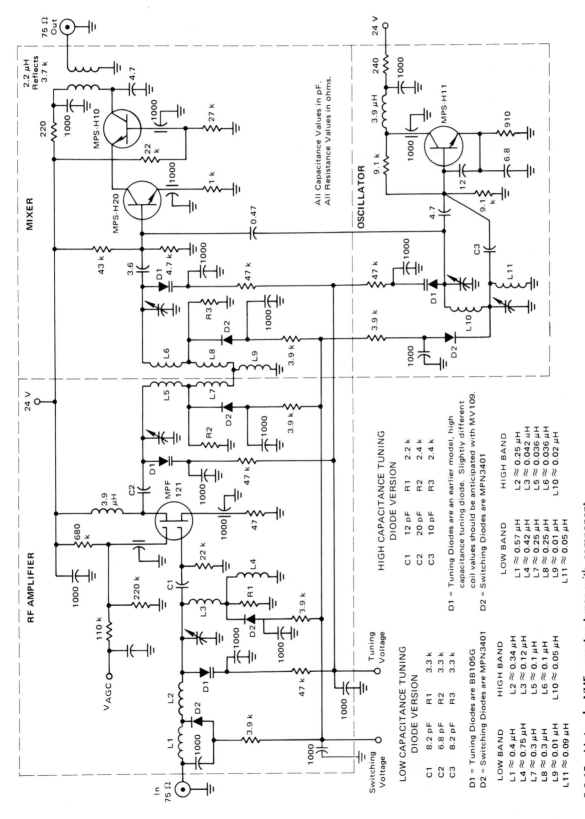

5.3.15 Motorola VHF varactor tuner with component values for both high-capacitance and low-capacitance varactors. Circuit board layout and photographs are available from Motorola (AN-544A).

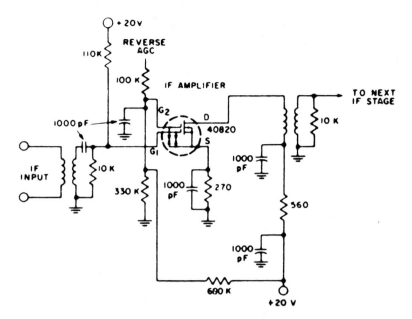

5.3.16 TV IF amplifier stage using RCA 40820 MOSFET IC. Gate 2 is for AGC. The reverse AGC bias applied to gate 2 makes gate 1 move in a positive direction. Evaluations of the relationship between AGC and cross modulation show that it is desirable to allow the voltage between gate 1 and the source to move in a positive direction when gate 2 is reverse-biased. Various circuit arrangements have been used to achieve this action.

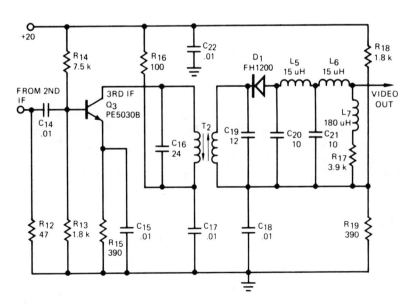

5.3.17 TV video IF amplifier final. Due to the low feedback capacitance of the Fairchild PE5030B it is unnecessary to neutralize the amplifier to achieve high gain. The transistor is operated at a collector current of 10 mA, which enables the transistor to deliver large output currents with little distortion. The detector is a hot-carrier diode.

5.4 Color TV

It will not take more than a glance to see that these are not circuits for complete television receivers; rather, they are circuits for specific stages that are common to color sets.

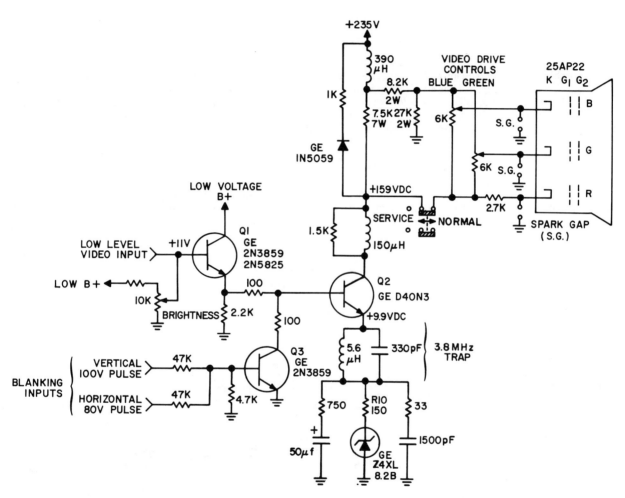

5.4.1 GE improved luminance amplifier. The composite video signal from the low-level amplifier stage is coupled to emitter follower Q1. The signal may be dc-coupled or ac-coupled and dc-restored at this point. A brightness control varies the dc level at the base of Q1, controlling the degree of conduction of Q2 and the resulting scene brightness. An automatic brightness limiting control signal could be injected at the base of Q1 (or into the low-level video stages) to automatically protect the CRT from excessive drive and beam current. The low-impedance drive provided by Q1 minimizes the Miller effect at the base of Q2 and reduces the dependence of the performance of the circuit on the output transistor's beta. Performance data, waveforms, and design hints are available from GE in application note 90.82.

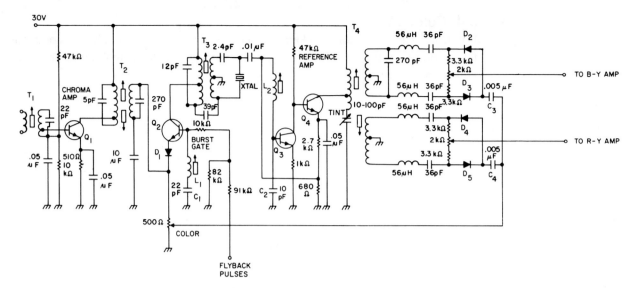

5.4.2 Fairchild chroma processing circuitry for color TV uses crystal filter concept for subcarrier synchronization. It contains three stages of amplification before decoding takes place: The chroma bandpass stage amplifies the chrominance sidebands and the reference burst signal; the burst gate stage selects and amplifies the reference burst signal and drives the crystal filter; and the reference or subcarrier stage amplifies and limits the 3.58 MHz subcarrier.

The last stage, the reference amplifier, has as its load the demodulation transformer, which is designed to provide the current phase relationships of the subcarrier to feed the synchronous detectors and initiate color decoding. The demodulation transformer may have two or three outputs. With only two outputs available, R−Y and B−Y signals, two synchronous detectors and a matrix network are necessary. The matrixing function obtains the G−Y components. A demodulation transformer can be designed to provide R−Y, G−Y, and B−Y components at its secondary windings and have the outputs fed to the respective R−Y, G−Y, and B−Y synchronous detectors and color difference amplifiers. All of the matrixing is done by the demodulation transformer, and thus no further matrixing networks are necessary.

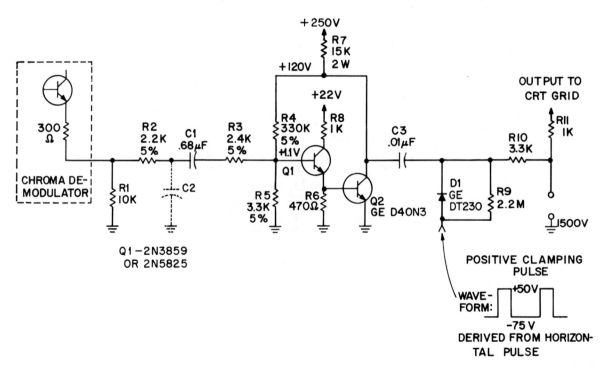

5.4.3 GE color difference amplifier uses overall voltage feedback between the output and the input. Most of the performance characteristics are largely dependent on the feedback network impedances, which include C1, R2, R3, and R4. The use of voltage feedback greatly reduces the output impedance, which (1) makes the frequency response and small-signal transient response independent of the load, (2) facilitates good output clamping, and (3) minimizes the vulnerability of the output to cross coupling and other types of stray signal pickup. Performance details and waveform photos available from GE in application note 90.81.

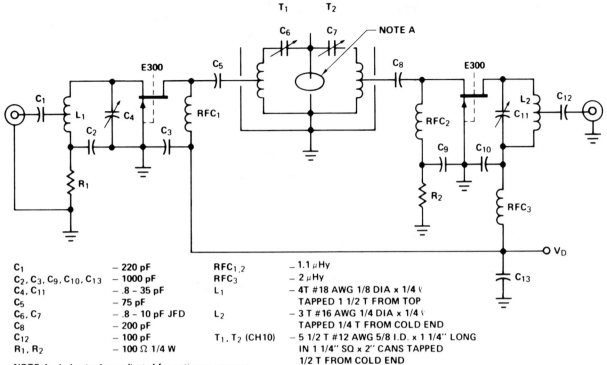

C_1	– 220 pF	$RFC_{1,2}$	– 1.1 µHy
$C_2, C_3, C_9, C_{10}, C_{13}$	– 1000 pF	RFC_3	– 2 µHy
C_4, C_{11}	– .8 – 35 pF	L_1	– 4T #18 AWG 1/8 DIA x 1/4 ℓ TAPPED 1 1/2 T FROM TOP
C_5	– 75 pF		
C_6, C_7	– .8 – 10 pF JFD	L_2	– 3 T #16 AWG 1/4 DIA x 1/4 ℓ TAPPED 1/4 T FROM COLD END
C_8	– 200 pF		
C_{12}	– 100 pF	T_1, T_2 (CH10)	– 5 1/2 T #12 AWG 5/8 I.D. x 1 1/4" LONG IN 1 1/4" SQ x 2" CANS TAPPED 1/2 T FROM COLD END
R_1, R_2	– 100 Ω 1/4 W		

NOTE A: Inductive loop adjusted for optimum response

5.4.4 Selective VHF amplifier for color boosting in weak TV reception areas. Here, Siliconix E300 FETs are used as buffers between a low-impedance input and a near-unity Q ratio resonator. The circuit is a conventional VHF amplifier capacitively coupled to a two-pole helical coil resonator. The compact resonators offer unusually high unloaded Q, often exceeding 1000 in the VHF spectrum. The impedance match between the resonator input and output is achieved by tapping up on the helix. The amplifier is stagger-tuned to provide a 4.5 MHz peaked response and has 15 dB gain with a noise figure of less than 4 dB. Selectivity for fringe-area frequencies is achieved simply by tuning the amplifier to the desired fringe channel frequency.

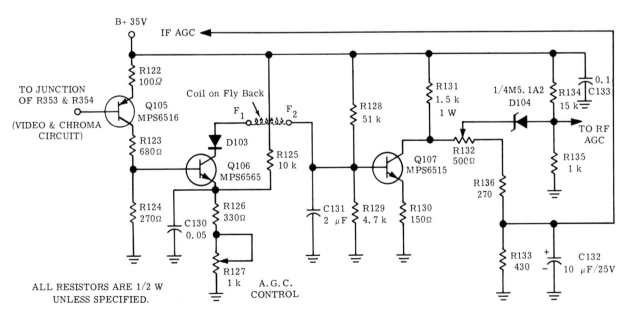

5.4.5 Motorola color TV receiver phase inverter, AGC keyer, and AGC amplifier.

SECTION 5 Radio/TV **265**

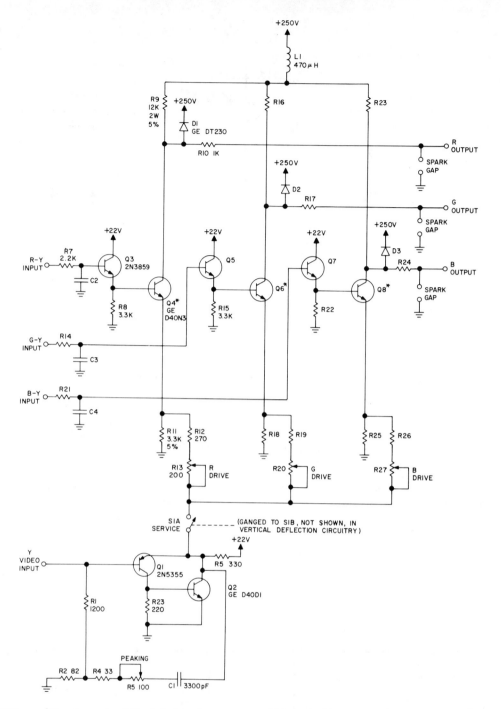

5.4.6 GE RGB amplifier for color TV minimizes interaction between controls and cuts crosstalk. The luminance signal is applied to the base of Q1. It has already undergone certain processing functions in the video preamplifier, as follows:

1. Amplification to provide a maximum luminance amplitude of approximately 3.5V white to black level.
2. Delay of approximately 0.8 μs to match that caused by reduced bandwidth of the chroma information.
3. Application of contrast and brightness functions.
4. Some degree of video peaking (high-frequency enhancement) if desired.
5. 3.58 MHz chroma carrier attenuation if desired.

In addition, the output impedance of the video preamplifier must be high (greater than 10K), as would be provided by the collector of a conventional amplifier stage.

This high driving impedance permits voltage feedback to the base of Q1.

Transistors Q1 and Q2 and their associated circuitry provide a very low-impedance output with the necessary power capability to drive the output stages, and they give increased gain to high frequencies, or video peaking, for enhanced transient response. Emitter followers Q3, Q5, and Q7 provide low-impedance drive to the output stages, Q4, Q6, and Q8. The output stages, with the color difference signals applied to their bases and the luminance signal to their emitters, perform matrixing that results in composite output information to the picture tube, containing both luminance and chroma information.

Q4, Q6, and Q8 require small (Staver type F7-2 or equivalent) heatsinks. If a TO-18 lead arrangement is desired, use a 2N5825 for Q3, Q5, and Q7, and a GE GET3638A for Q1.

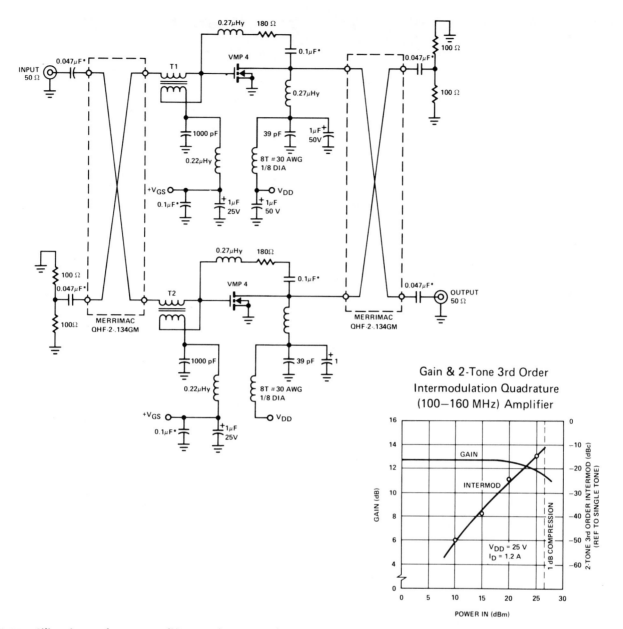

5.4.7 Siliconix quadrature amplifier employs two Siliconix power FETs to achieve performance as shown. Transformer T1 is fabricated by winding four turns of 22-gage solid twisted-pair wire over Indiana General transformer form F625-9Q2. Asterisked capacitors are integral with chip.

SECTION 6
COMMUNICATIONS

This section contains circuits of every description, designed for operation over virtually the entire radio spectrum from the very low frequencies to the ultrahigh and beyond. The first subsection, **Oscillators**, comprises circuits that may be used in both transmitting and receiving systems. The **Multiplier** subsection contains circuits for increasing the frequency of an input signal from an oscillator or other such generator. Complete transmitters, including oscillators, multipliers, modulators, and amplifiers, appear in a separate subsection, followed by a subsection containing circuits for linears, power amplifiers of other classes, and modulators.

With the exception of oscillators, those circuits designed for incorporation into receivers appear in the final two subsections. I have chosen the breakdown given— **RF Preamplifiers and Converters** (Section 6.5) and **Front Ends, RF/IF Amplifiers, and Mixers** (Section 6.6)—to avoid confusion between different circuits that bear the same name: converters.

To some of the longtime veterans of radio, a converter is a mixer stage in a receiver; but to many—particularly ham radio operators—a converter is a modular stage inserted between antenna and receiver, which accepts signals of one frequency and converts them to signals of a lower frequency that is within the tuning range of the receiver with which it is associated. (The fact that a converter is also a dc power supply operated from a dc source only fans the fires of confusion.) The converters in Section 6.5 are the type that connect directly to an antenna and feed into either a broadcast radio or a communications receiver; those of Section 6.6 are of the type that accept two input signals and produce a third as a result of mixing.

6.1 Oscillators

The oscillators included here are built around a number of discrete devices—JFETs, MOSFETs, tunnel diodes, and bipolars—and operate from a variety of supply voltages. There are a great many additional oscillator circuits in this book, included as stages of larger circuits as special-purpose signal sources. For additional oscillators, see **Inverters** (Section 1.3), **Remote Control and Servos** (Section 2.8), **Flashers** (Section 3.1), **Timers** (Section 3.3), **Hobby** (Section 4), the appropriate circuits of Section 5, and other subsections of Section 6.

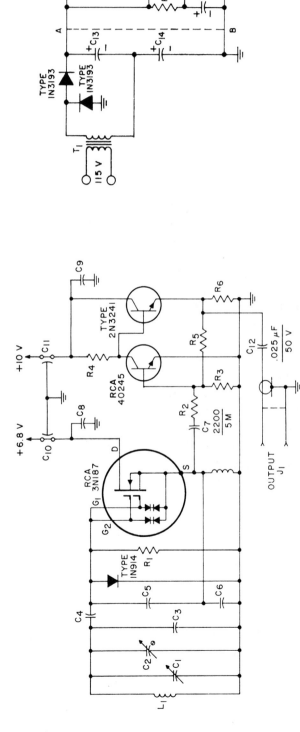

Tuned-Circuit Data

	3.5–4.0 MHz	5.0–5.5 MHz	8.0–9.0 MHz
L_1 No. of turns	17*	14¾*	11½**
Wire size	20	20	18
Turns/inch	16	16	8
Diam., inches	1	1	1
C_1, p.	100	50	50
C_2, pf.	25	25	25
C_4, pf.	100	None	None
C_5, pf.	390	390	270
C_6, pf.	680	680	560
C_9, pf.	680	680	560

 * B & W 3015, AirDux 816T, or equiv.
 ** B & W 3014, AirDux 808T, or equiv.

Parts List

C_1 = Double-bearing variable capacitor, Millen 23100 or 23050 (or equiv.) depending upon frequency range (see Tuned-Circuit Data)
C_2 = Air-type trimmer capacitor, 25 pF maximum, Hammarlund APC-25 or equiv.
C_3, C_4, C_5, C_6 = silver-mica capacitors (see Tuned-Circuit Data for values)
C_7 = 2200 pF, silver mica
C_8 = 0.05 pF, ceramic disc, 50 V.
C_9 = 0.1 pF, ceramic disc, 50 V.
C_{10}, C_{11} = 1500 pF, feed-through
C_{12} = 0.025 μF, ceramic disc, 50 V.
C_{13} = 500 μF, electrolytic, 12 V.; select value for 2-volt peak output level at input to transmitter
C_{14} = 500 μF, electrolytic, 12 V.
C_{15} = 50 μF, electrolytic, 12 V.
CR_1 = Zener diode, 12-volt, 1-watt
CR_2 = ener diode, 6.8 volt, 1-watt
J_1 = Coaxial connector
L_1 = Variable inductor (see Tuned-Circuit Data for details)
L_2 = Miniature rf choke, 2.5 mH, iron core
R_1 = 22000 ohms, 0.5 watt
R_2 = 12000 to 47000 ohms, 0.5 watt; select value for 2-volt peak output level at input to transmitter
R_3 = 12000 ohms, 0.5 watt
R_4 = 820 ohms, 0.5 watt
R_5 = 47000 ohms, 0.5 watt
R_6 = 240 ohms, 0.5 watt
R_7 = 2200 ohms, 0.5 watt
R_8 = 220 ohms, 0.5 watt
R_9 = 180 ohms, 0.5 watt
T_1 = 6.3-volt, 1.2-ampere filament transformer

6.1.1 Stable variable-frequency oscillator uses an RCA 40823 dual-gate-protected MOS transistor and bipolar transistors in a two-stage isolation (output) amplifier to achieve exceptional frequency stability at low dc operating potentials. The MOS oscillator circuit is useful at any frequency up to and including the 144 MHz band. Tuned-circuit data are provided for the standard 3.5 MHz band, for the 5 MHz band for single sideband transmitters, and for the 8 MHz band for 50 and 144 MHz transmitters.

The dc operating potentials for the VFO circuit can be obtained directly from a 12V source. For operation from a 117V 60 Hz ac source, a low-voltage dc supply may be used to supply the required voltage. The 117V ac source voltage is stepped down to 6.3V ac by power transformer T1 and then converted to a dc voltage of 12V by the voltage doubler.

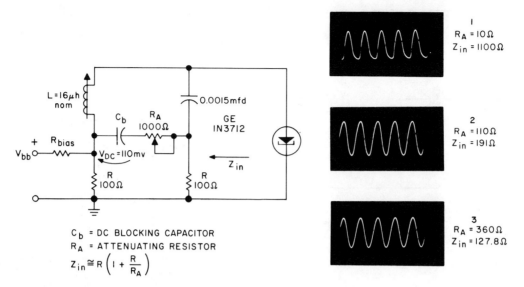

6.1.2 GE variable-amplitude oscillator. Waveform 1 shows the result of $R_A = 10$ ohms. The amplitude is about 340 mV peak-to-peak. Since the tunnel diode is biased at 110 mV, this means that it is swinging from -60 to $+280$ mV. Hence it is not exceeding the valley point but is going well past the peak-point voltage, which is 60 mV for the tunnel diode. Waveform 2 shows a signal of 80 mV (peak-to-peak) amplitude; therefore, the diode swings from $+70$ to $+150$ mV and the resulting improvement in the sine wave can be seen readily ($R_A = 110$ ohms). The last waveform shows a 10:1 reduction in amplitude from the first one (achieved with $R_A = 360$ ohms); hence an amplitude of 34 mV peak-to-peak.

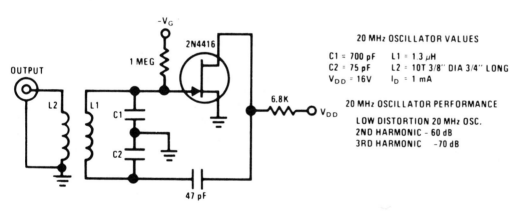

6.1.3 RCA low-distortion oscillator. The 2N4416 JFET is capable of oscillating in a circuit where harmonic distortion is very low. The JFET oscillator is excellent when a low harmonic content is required for a good mixer circuit.

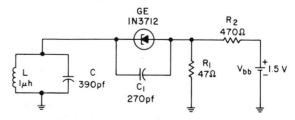

6.1.4 GE tunnel diode oscillator provides signal swing of 300 mV peak-to-peak at 5.46 MHz. The use of a variable inductance for L permits frequency tuning or fine-frequency vernier adjustments.

SECTION **6** Communications

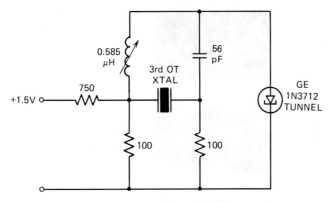

6.1.5 GE CB crystal-controlled oscillator produces output on 27.255 MHz (within the tolerance of the third-overtone crystal itself) over a temperature range of −55 to +85°C and a bias range of 110 to 150 mV. The output power is on the order of 30 μV.

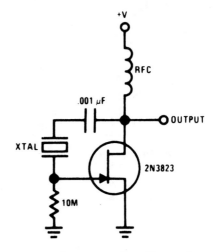

6.1.7 RCA JFET Pierce crystal oscillator allows a wide frequency range of crystals to be used without circuit modification. Since the JFET gate does not load the crystal, good Q is maintained, insuring good frequency stability.

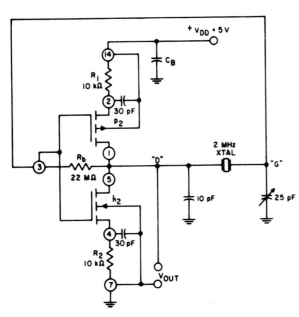

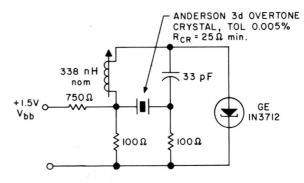

6.1.8 GE VHF oscillator generates 47.100 MHz signal for direct or multiplier applications. The crystal is an Anderson third-overtone type. The oscillator frequency operates within the quartz crystal tolerance over a temperature range of from −55 to +85°C and over a bias range of from 110 to 150 mV.

6.1.6 RCA oscillator circuits using CMOS transistor pairs have been widely used for several years in clock and watch circuits because of their low power consumption and good frequency stability. The design involves the provision of an amplifying section to operate compatibly with an appropriate feedback network. The familiar pi network has been connected between the input and output terminals, points **D** and **G**, to provide the required 180° phase shift for stable oscillator performance. The frequency-determining crystal is an integral part of the pi network's feedback circuit. Resistors R1 and R2 decrease the total power consumption of the oscillator at a particular supply voltage and enhance the frequency stability. Variable frequency oscillators can be built by replacing the crystal with an appropriate inductance and tuning the pi network by conventional means.

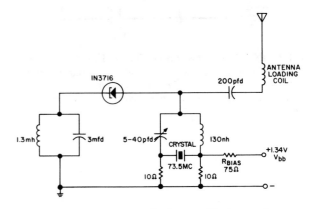

6.1.9 GE VHF tone-producing oscillator produces a powerful signal (150 μW) at 73.5 MHz. The only major caution is to maintain an adequate R/R$_{CR}$ ratio. The diode is operated with 130 mV at about 2.5 mA, and the total battery drain is 5 mA at 1.34V. The oscillator is self-modulated by an audio tank circuit (L1 and C1) and this tone is designed to activate a relay in the receiver.

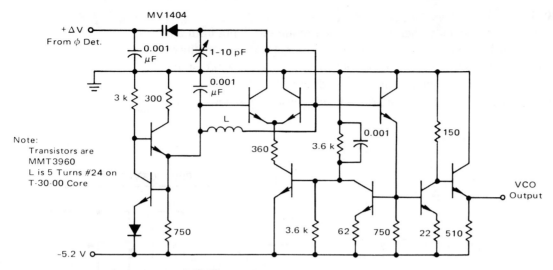

6.1.10 Motorola voltage-controlled oscillator is an ECL tuned-collector type utilizing high-frequency NPN transistors and Motorola's Epicap MV1404 hyperabrupt-junction tuning diode. The tuned circuit elements should have as high a loaded Q as possible to provide good output spectral purity. Good VHF RF layout techniques should be used with all voltage inputs RF decoupled. The sine-wave output of the VCO should be buffered before external use.

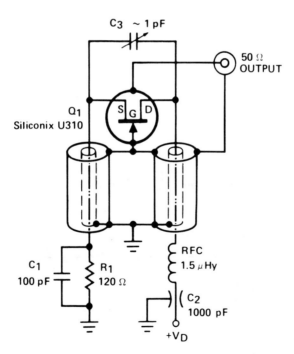

6.1.11 This 700 to 800 MHz oscillator includes a Siliconix U310 FET that has a forward transconductance value greater than 18 millisiemens at zero bias. The oscillator consists of two coaxial resonators, one for the FET source and the other for the drain. Oscillation is achieved by capacity coupling between the two resonators; output coupling is derived from the magnetic coupling that exists at the open ends of the resonators. The best resonator Q is achieved by designing the coaxial resonators for a characteristic impedance of 75 ohms. Measured performance of the oscillator at 25°C is shown in the following table.

V_D (V)	+10	+15	+20	+25
I_D (mA)	15	16.2	18.2	21
P_{out} (dBm)	+6.6	+15.2	+18.3	+20
f (MHz)	725	742.7	754.7	762.9

6.2 Multipliers

Two kinds of multipliers are included here: those which require an input from an oscillator and those which require a relatively high-power signal from an amplified source. Be sure to check the input-signal requirements before selecting a circuit for your own requirement. In general, multipliers that are designed to deliver an unusually high-frequency signal to an antenna are driven with considerable power, and they exhibit substantial loss.

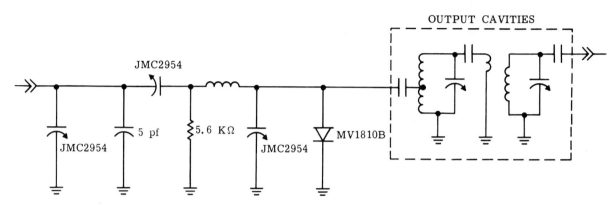

6.2.1 Motorola 500 to 400 MHz one-step multiplier. The input is a lumped pi network with a series-resonant LC arm. The output consists of two cascaded resonant cavities. The most critical aspect of this circuit is the coupling from the varactor to the first cavity. This coupling has to be exceedingly tight to allow some idler current to flow. Photographs and design criteria are available from Motorola (AN-213).

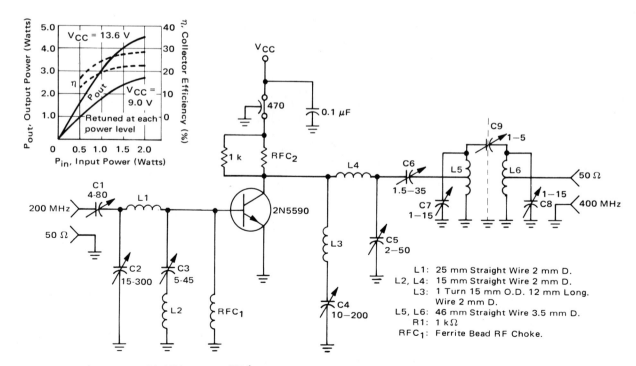

6.2.2 Motorola 200 to 400 MHz power RF frequency doubler uses balanced-emitter transistor.

274   DISCRETE / TRANSISTOR CIRCUIT SOURCEMASTER

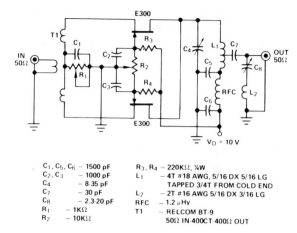

C_1, C_5, C_6 — 1500 pF	R_3, R_4 — 220KΩ, ¼W
C_2, C_3 — 1000 pF	L_1 — 4T #18 AWG, 5/16 DX 5/16 LG TAPPED 3/4T FROM COLD END
C_4 — 8-35 pF	L_2 — 2T #16 AWG 5/16 DX 3/16 LG
C_7 — 30 pF	RFC — 1.2 μHy
C_8 — 2.3-20 pF	T1 — RELCOM BT-9
R_1 — 1KΩ	50Ω IN-400CT-400Ω OUT
R_2 — 10KΩ	

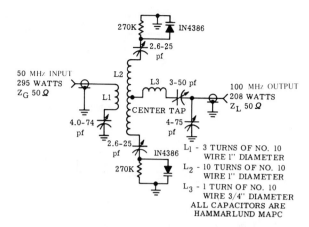

6.2.3 Siliconix FET frequency doubler takes advantage of the transfer curve of the FET, which approximates a square-law response and produces a strong second-order (and negligible higher-order) harmonic output. The FET frequency multiplier circuit requires only one trap for effective harmonic suppression and operates up to 100% efficiency. The circuit uses two matched FETs as common-gate amplifiers in a balanced push-push configuration. The Siliconix E300 FETs in this application exhibit a high forward transconductance of 700 microsiemens and can be well matched in relation to transfer curves. Matched 2N5397s are also suitable, as are dual VHF FETs such as the U257, E420, or the 2N5911. A common-gate configuration is used because the common-gate impedance of the devices closely matches the secondary impedance of available wideband transformers. The Relcom BT-9 input transformer is used because of an excellent bandpass, 8:1 impedance ratio, and isolated secondary.

6.2.5 Motorola 50 to 100 MHz push-push doubler circuit. All components used are lumped elements. The push-push terminology arises from the fact that the varactors are connected in phase opposition to the input signal and parallel to a common load at the even-harmonic output signal. Twice the power handling capability is obtained along with the added benefit of odd-harmonic suppression.

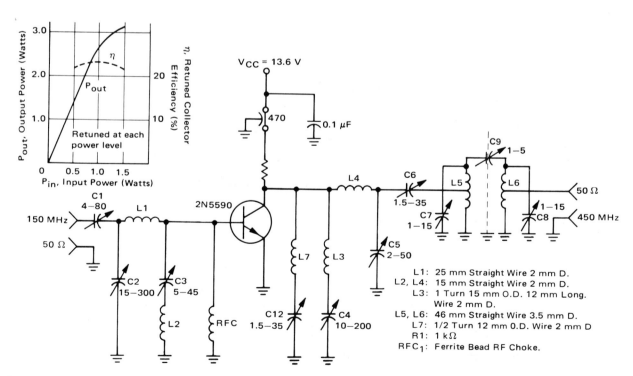

6.2.4 150 to 450 MHz tripler uses a Motorola balanced-emitter transistor for increasing the resistance to "mismatch" burnout.

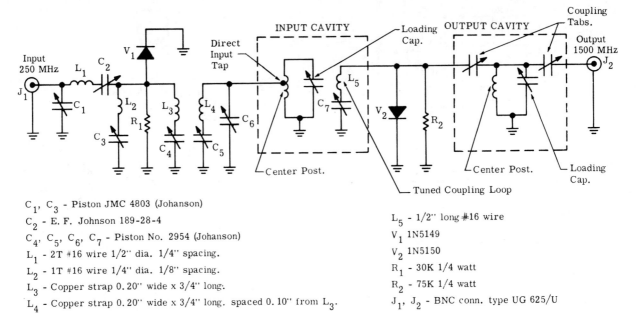

C_1, C_3 - Piston JMC 4803 (Johanson)
C_2 - E. F. Johnson 189-28-4
C_4, C_5, C_6, C_7 - Piston No. 2954 (Johanson)
L_1 - 2T #16 wire 1/2" dia. 1/4" spacing.
L_2 - 1T #16 wire 1/4" dia. 1/8" spacing.
L_3 - Copper strap 0.20" wide x 3/4" long.
L_4 - Copper strap 0.20" wide x 3/4" long. spaced 0.10" from L_3.
L_5 - 1/2" long #16 wire
V_1 1N5149
V_2 1N5150
R_1 - 30K 1/4 watt
R_2 - 75K 1/4 watt
J_1, J_2 - BNC conn. type UG 625/U

6.2.6 Motorola 0.25 to 0.75 GHz tripler consists of lumped-constant components. The 1N5149 varactor is mounted to ground and self-biased with a 30K resistor. The idler circuit is tuned to the second harmonic (0.5 GHz) and shunted across the varactor. The idler coil orientation is 90° to the input series coil to prevent interaction. The output portion of this stage consists of a mutual coupled circuit. This design provides greater spurious rejection than the common series resonant LC circuit. The coupling coils of the output consist of nothing more than copper straps 0.2 in. (0.40 cm) wide and about 0.75 in. (2 cm) long. Straps (rather than wire) provide the needed inductance and coupling.

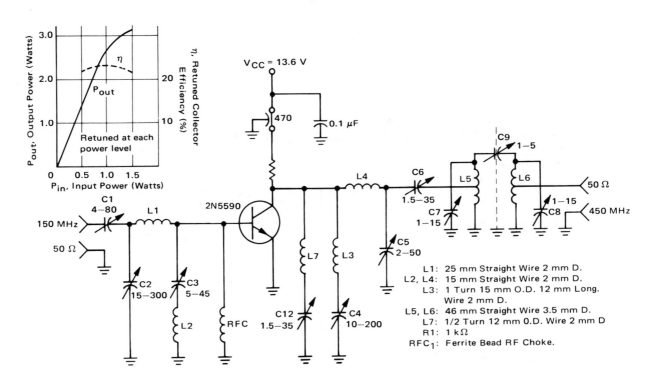

6.2.7 150 to 450 MHz RF tripler circuit uses a Motorola balanced-emitter transistor that can survive short-duration mismatches of considerable magnitude.

276 DISCRETE / TRANSISTOR CIRCUIT SOURCEMASTER

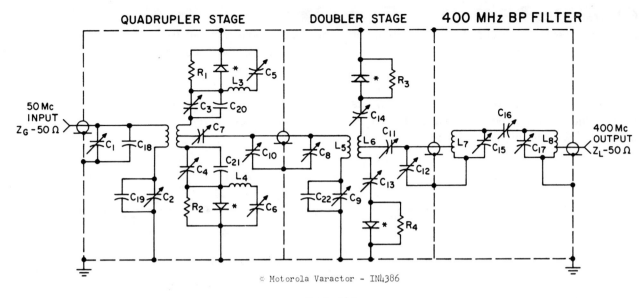

Parts List:

C_1-HammarLund MAPC-75	R_1 - 68KΩ	C_{11}-HammarLund MAPC-25
C_2-HammarLund MAPC-75	R_2 - 68KΩ	C_{12}-Johanson-JMC-1801
C_3-HammarLund MAPC-75	R_3 - 150KΩ	C_{13}-HammarLund MAPC-25
C_4-HammarLund MAPC-75	R_4 - 150KΩ	C_{14}-HammarLund MAPC-25
C_5-HammarLund MAPC-50		C_{15}-Johanson-JMC-1801
C_6-HammarLund MAPC-50		C_{16}-Johanson-JMC-2951
C_7-HammarLund MAPC-25		C_{17}-Johanson-JMC-1801
C_8-E.F. Johnson Co.-Type "M"-160-110		C_{18}-20 pf (Fixed)
C_9-E.F. Johnson Co.-Type "M"-160-110		C_{19}-32 pf (Fixed)
C_{10}-HammarLund MAPC-25		C_{20}-140 pf (Fixed)

C_{21}-140 pf (Fixed)
C_{22}-15 pf (Fixed)

L_1-3 Turns 1" dia. 1/8 Tubing
L_2-6 Turns 5/8" dia. 1/8 Tubing
L_3-2 Turns 13/16" dia. 1/8 Tubing
L_4-2 Turns 13/16" dia. 1/8 Tubing
L_5-1 Turns 5/8" dia. 1/8 Tubing
L_6-2 Turns 5/8" dia. #8 Wire
L_7-3 Turns 1/4" dia. #8 Wire
L_8-3 Turns 1/4" dia. #8 Wire

6.2.8 Motorola varactor multiplier provides 40W RF at 400 MHz when driven with a 50 MHz signal at a power of 140W.

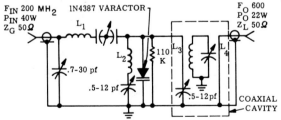

L_1 - 3 TURNS OF 1/16" WIRE 1/2" DIA. x 1" LONG.
L_2 - 1 TURN 1/8" TUBING 3/8" O.D.
L_3 - STRAIGHT COUPLING LOOP 1/8" TUBING 2" LONG SPACED APPROX. 1/8" FROM CENTER CONDUCTOR.
L_4 - STRAIGHT COUPLING LOOP 1/16" WIRE 1-1/2" LONG, SPACED APPROX. 1/16" FROM CENTER CONDUCTOR.

6.2.9 Motorola harmonic tripler loses less than 50% of the input power in converting from 200 to 600 MHz.

6.3 Transmitters

Except for the first two circuits in this subsection, which are simple modulated oscillators capable of delivering their outputs directly to an antenna, the circuits here consist of multistage transmitters. The operating frequency of any of these transmitters may be altered by making the appropriate adjustments in the tuned circuits. To increase the operating frequency, use fewer turns on the coils specified; to lower the operating frequency, either add turns to hand-wound coils or pad ready-made coils with additional capacitance.

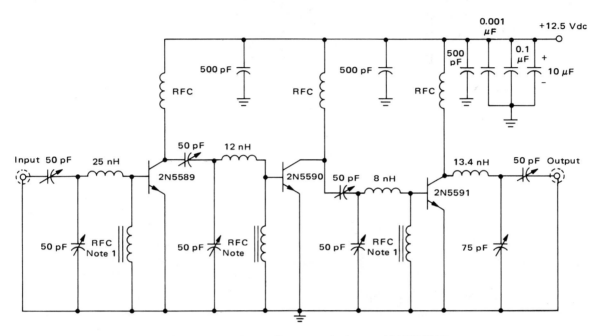

Note   All base chokes are Ferroxcube ferrite type VK-200 19/4B.

6.3.1 175 MHz 25W transmitter for 12.5V operation uses Motorola 2N5589, 2N5590, and 2N5591 balanced-emitter RF power transistors. The predriver, driver, and power-amplifier stages have an overall gain of 21.2 dB. The overall efficiency is 46.5% and all harmonics and spurious outputs are at least 40 dB below the 175 MHz output signal.

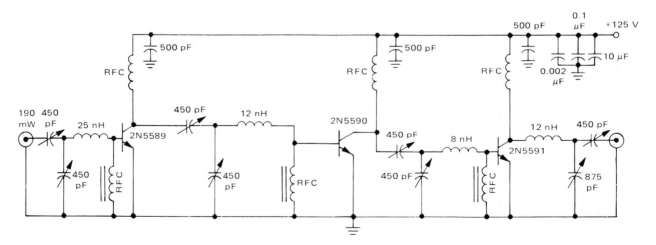

6.3.2 Motorola VHF transmitter for 175 MHz supplying 25W with a 12.5V power supply. Balanced-emitter transistors are used.

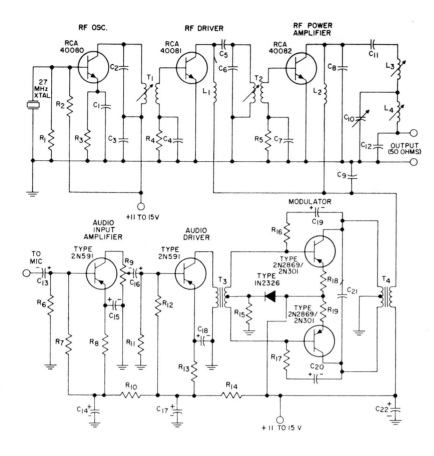

C_1 = 75 pF, ceramic
C_2 = 30 pF, ceramic
C_3, C_9 = 0.01 µF, ceramic
C_4 = 0.001 µF, ceramic
C_5 = 47 pF, ceramic
C_6 = 51 pF, mica
C_7 = 0.002 µF, ceramic
C_8 = 24 pF, mica
C_{10} = variable capacitor, 90 to 400 pF (ARCO 429, or equiv.)
C_{11} = 100 pF, ceramic
C_{12} = 220 pF, ceramic
C_{13} = 5 µF, electrolytic
C_{14}, C_{17} = 50 µF, electrolytic, 25 V
C_{15}, C_{16}, C_{18} = 10 µF, electrolytic, 15 V
C_{19}, C_{20} = 0.2 µF, electrolytic, 15V
C_{21} = 0.1 µF, ceramic

C_{22} = 500 µF, electrolytic, 15V
L_1, L_2 = rf choke, 15 µF, Miller 4624, or equiv.
L_3 = variable inductor (0.75 to 1.2 µH); 11 turns No. 22 wire wound on ¼-inch CTC coil form having a "green dot" core; Q = 120
L_4 = variable inductor (0.5 to 0.9 µH); 7 turns No. 22 wire wound on ¼-inch CTC coil form having a "green dot" core; Q = 140
R_1 = 510 ohms, 0.5 watt
R_2, R_{12} = 5100 ohms, 0.5 watt
R_3 = 51 ohms, 0.5 watt
R_4 = 120 ohms, 0.5 watt
R_5 = 47 ohms, 0.5 watt
R_6 = 0.1 megohm, 0.5 watt
R_7 = 10000 ohms, 0.5 watt

R_8 = 2000 ohms, 0.5 watt
R_9 = potentiometer, 10000 ohms
R_{10} = 3600 ohms, 0.5 watt
R_{11} = 15000 ohms, 0.5 watt
R_{13} = 1000 ohms, 0.5 watt
R_{14} = 1200 ohms, 0.5 watt
R_{15} = 240 ohms, 0.5 watt
R_{16}, R_{17} = 2700 ohms, 0.5 watt
R_{18}, R_{19} = 1.5 ohms, 0.5 watt
T_1 = rf transformer; primary 14 turns, secondary 3 turns of No. 22 wire wound on ¼-inch CTC coil form having a "green dot" core (CTC No. 1542-3 or equiv.); slug-tuned (0.75 to 1.2 µH); Q = 100
T_2 = rf transformer; primary 14 turns, secondary 2¾ turns of No. 22 wire wound on ¼-inch (0.64 cm), CTC coil form having a "green dot" core; slug-tuned (0.75 to 1.2 µH); Q = 100.
T3 = primary: 2500 ohms; secondary: 200 ohms, center-tapped; Microtran SMT 17-SB or equivalent.
T4 = transformer; primary: 100 ohms, center-tapped; secondary: 30 ohms; Stancor TA-12 or equivalent.
XTAL = 27 MHz transmitting crystal, standard third-overtone type.

6.3.3 RCA 5W CB transmitter operates directly from a 12V supply. Its low power drain makes it adaptable to portable use with small storage batteries.

The RF section develops 3.5W of output power at 27 MHz. Both the driver and the power amplifier are modulated to achieve 100% modulation.

The RCA 40080 crystal oscillator provides excellent frequency stability with respect to collector supply voltage and temperature and delivers 100 mW to the driver.

The modulation input is applied to the collector circuit. This stage delivers a minimum of 400 mW to the power amplifier. A heat dissipator should be mounted on the case of the 40081. The 40082 amplifier is modulated through the collector circuit. The double-pi network used as the output resonant circuit provdes a harmonic rejection of 50 dB.

In the modulator section of the transmitter, two 2N591 class A amplifier stages are used to drive a class AB push-pull output stage using two 2N2869/2N301 transistors. This design provides maximum efficiency with low distortion. A 1N2326 compensating diode is used in the biasing network to provide thermal stability. Modulation transformer T4 matches the collector-to-collector impedance of the modulator to that of the driver and power amplifier.

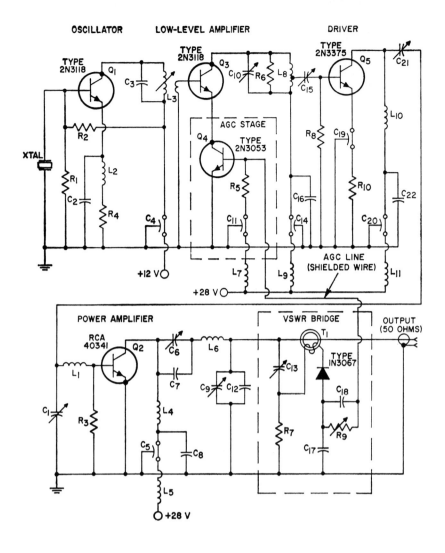

C_{17} = 1000 pF, ceramic
C_{18} = 0.01 μF, ceramic
C_{21} = variable capacitor, 3 to 250 pF, Vitramon No. 464 or equiv.
L_1 = 1 turn of No. 16 wire; inner diameter, 5/16 inch; length, 1/8 inch
L_2 = rf choke, 1 μH
L_3 = oscillator coil; primary, 7 turns; secondary, 1-3/4 turns; wound from No. 22 wire on CTC coil form having "white dot" core
L4 = 5 turns of No. 16 wire: inner diameter, 5/16 in.; length, 1/2 in.
L5, L7, L9, L10, L11 = Rf choke; 7 μH
L6 = 4 turns of B&W No. 3006 coil stock.
L8 = 6 turns of No. 16 wire: inner diameter, 3/8 in.; length 3/4 in.
R1, R6 = 510 ohms, 0.5W
R2 = 3.9K, 0.5W
R3, R8 = 2.2 ohms, wirewound, 0.5W
R4 = 51 ohms, 0.5W
R5 = 2.4K, 0.5W
R7 = 240 ohms, 0.5W
R9 = AGC control, 50K potentiometer
R10 = 5.6 ohms, 1W
T1 = Current transformer (toroid), Arnold No. A4437-125-SF or equivalent
XTAL = 50 MHz transmitting crystal

C_1 = variable capacitor, 90 to 400 pF, Arco No. 429 or equiv.
C_2 = 51 pF, mica
C_3 = 30 pF, ceramic
C_4, C_5, C_{11}, C_{19}, C_{20} = feedthrough capacitor, 1000 pF
C_6 = variable capacitor, 1.5 to 20 pF, Arco No. 402 or equiv.
c_7 = 36 pF, mica
C_8, C_{16}, C_{22} = 0.02 μF, ceramic
C_9, C_{10} = variable capacitor, 8 to 60 pF, Arco No. 404 or equiv.
C_{12} = 91 pF, mica
C_{13} = variable capacitor, 0.9 to 7 pF, Vitramon No. 400 or equiv.
C_{15} = variable capacitor, 14 to 150 pF, Arco No. 426 or equiv.

6.3.4 RCA 40W CW transmitter for 6m uses a VSWR bridge circuit to maintain a steady-state dissipation in the output stage under all conditions of antenna mismatch. This technique makes it possible to realize the full power potential of the RCA 40341 overlay transistor used in the output stage.

The crystal-controlled 2N3118 oscillator stage develops the low-level excitation signal for the transmitter. The 50 MHz output signal from the collector of the oscillator transistor is coupled by L3 to the base of a second 2N3118 used in a predriver stage. The collector circuit of the predriver is tuned to provide maximum signal output at 50 MHz. This signal is coupled from a tap on inductor L8 to the input circuit of the driver stage.

The output of the transmitter is sampled by current transformer (toroid) T1 loosely coupled about the output transmission line. This transformer is the sensor for a VSWR bridge detector used to prevent excessive dissipation in the output stage under conditions of antenna mismatch. If the antenna is disconnected or poorly matched to the transmitter, large standing waves of voltage and current occur on the output transmission line. a portion of this energy is applied by T1 to the 1N3067 diode in the bridge circuit. The rectified current from this diode charges capacitor C18 to a dc voltage proportional to the amplitude of the standing waves. The output of the AGC stage biases the 2N3118 predriver so that its gain changes in inverse proportion to the amplitude of the standing wave. As the amplitude increases, the input drive to the output stage is reduced. This compensating effect maintains a steady-state dissipation in the output transistor regardless of mismatch conditions between the transmitter output circuit and the antenna.

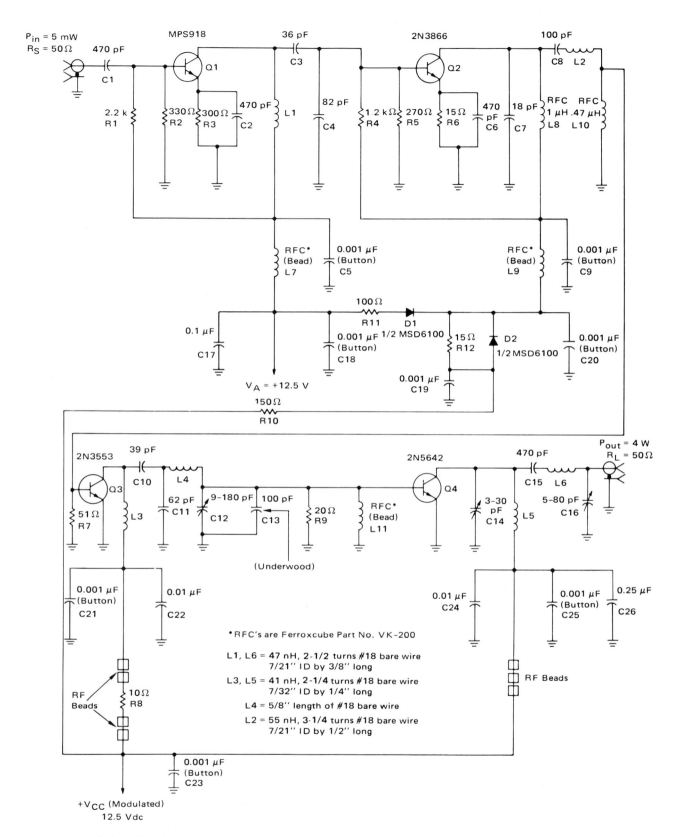

6.3.5 Motorola broadband AM transmitter intended for use in light aircraft. The transmitter has an output power of 4W over a range of 118 to 150 MHz with an input of 5 mW. It requires no tuning in changing frequency. The transmitter operates from a 25V dc supply; the RF circuitry is designed for operation from a 12.5V supply with collector modulation, and the 25V dc requirement is due to the series modulation employed. The modulator requires no modulation transformer, yet provides upward modulation of over 90%. In cases where it is necessary to operate the entire transmitter from a 12.5V source, a conventional modulator using a transformer could be employed.

SECTION 6 Communications

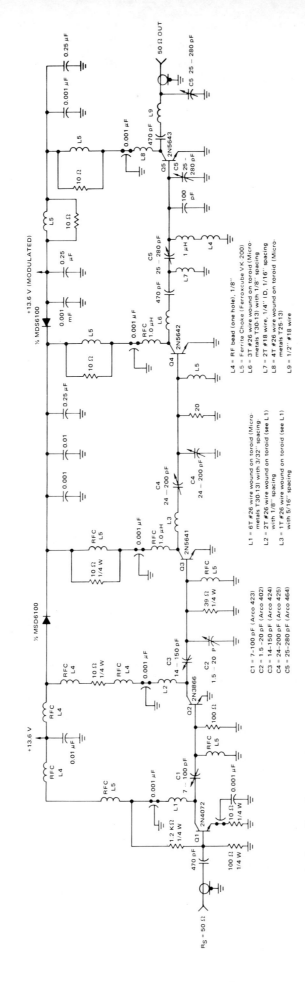

6.3.6 Motorola broadband transmitter intended for commercial aircraft covers the frequency range of 118 to 136 MHz and operates from a 27.2V supply. The RF circuitry is designed for operation from a 13.6V supply with collector modulation; 27.2V is required due to the series modulation method employed. Five amplifying stages are used to generate the 13W carrier.

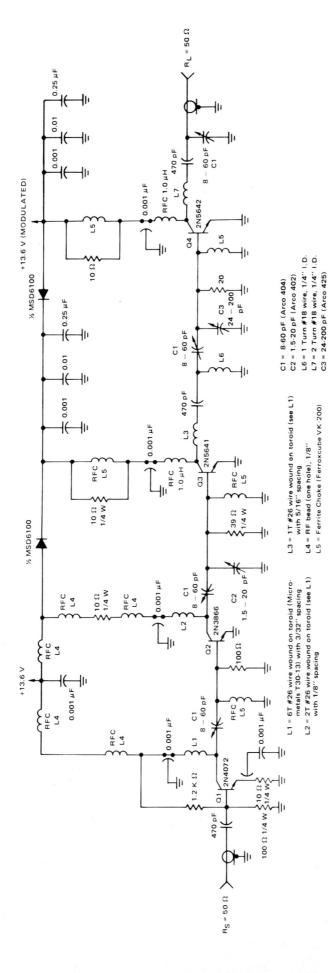

6.3.7 Motorola low-power broadband aircaft transmitter requires low-level signal from oscillator to produce 7W of RF power at the output. Any of the series modulators in this book will prove effective. The frequency range is 118 to 136 MHz.

SECTION 6 *Communications*

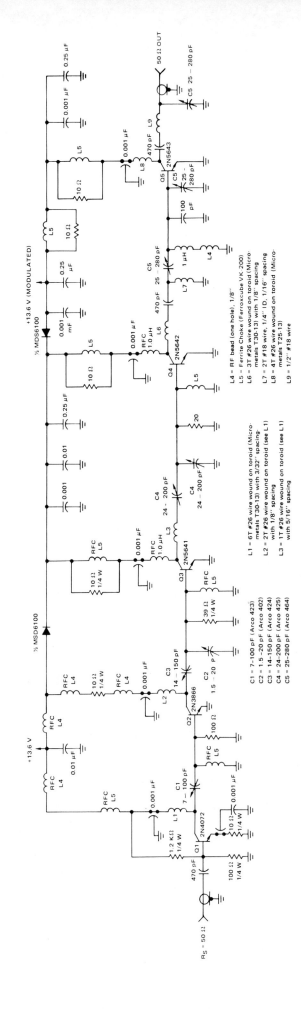

6.3.8 13W broadband aircraft transmitter consists of five stages operated in the common-emitter configuration. The output stage, Q5, uses a Motorola 2N5643. The driver stage, Q, uses a Motorola 2N5642, which is in the same type of package as the 2N5643. The predriver stages use the 2N4072 (Q1), 2N3866 (Q2), and 2N5641 (Q3). Q1 is packaged in a TO-18 case and Q2 in a TO-39 case. The third stage, Q3, uses a Motorola 2N5641 in a package similar to that of the 2N5643, but with smaller leads.

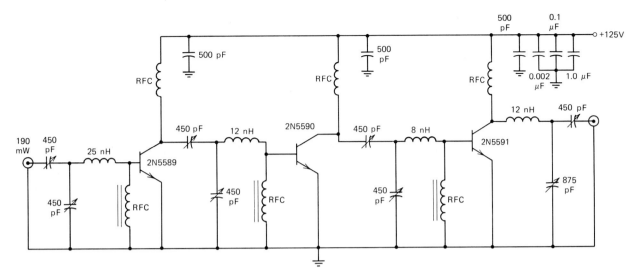

6.3.9 Motorola VHF transmitter for 175 MHz supplying 25W with a 12.5V power supply. Motorola NPN balanced-emitter transistors are used.

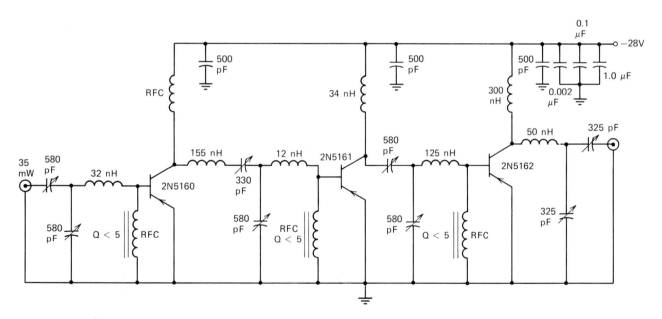

6.3.10 175 MHz transmitter using Motorola PNP balanced-emitter transistors. The output is 30W with a 28V supply.

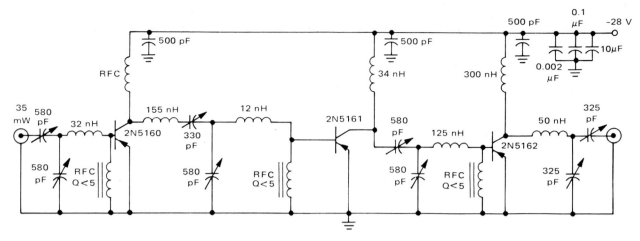

6.3.11 Motorola 175 MHz transmitter using PNP balanced-emitter transistors. The output is 30W with a 28V supply.

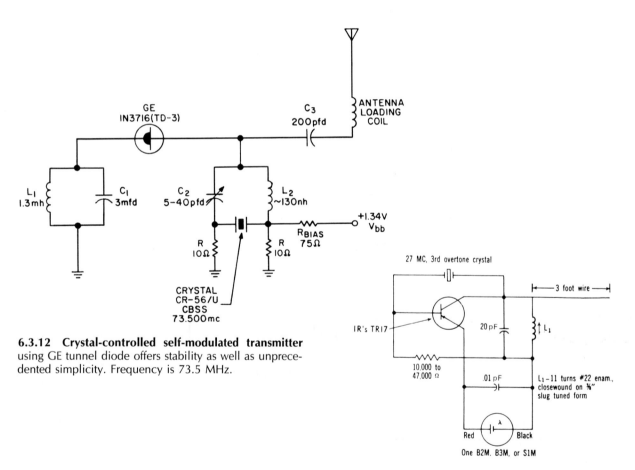

6.3.12 Crystal-controlled self-modulated transmitter using GE tunnel diode offers stability as well as unprecedented simplicity. Frequency is 73.5 MHz.

6.3.13 Workman crystal-controlled sun-powered CB transmitter (27 MHz). The frequency of the transmitter is controlled by a quartz crystal (third-overtone type). The coil and its associated capacitor tunes the transistor to amplify and oscillate at the crystal frequency. The small length of wire serving as an antenna permits the energy to travel several hundred yards.

The transistor may be IR-Workman's TR17. After the circuit is completed, it may be necessary to vary the value of the resistor between 10 and 47K to obtain maximum signal.

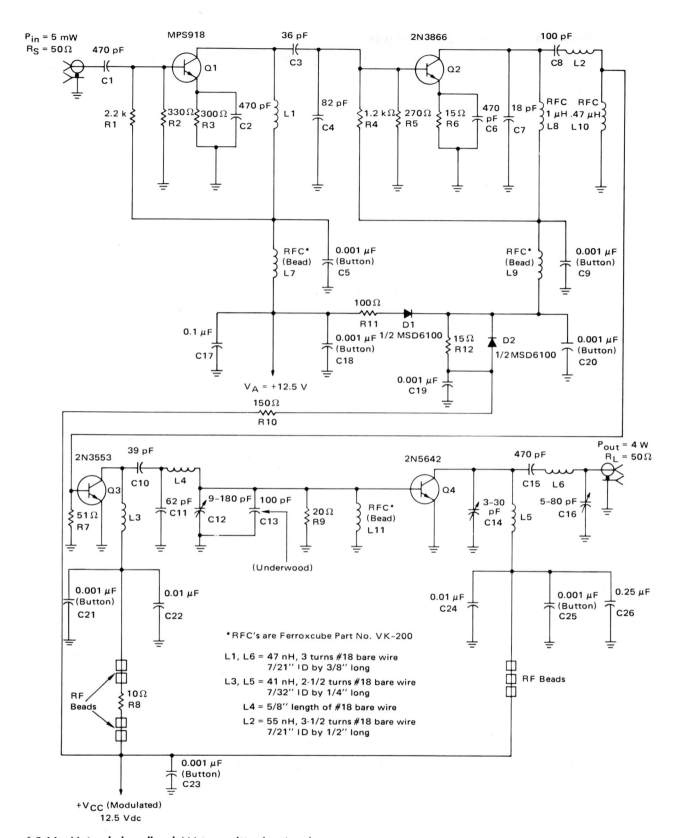

6.3.14 Motorola broadband AM transmitter for aircraft applications. Operation is from 118 to 136 MHz and can be upped to 150 MHz by appropriate shortening of the coils. The circuit will produce 4W of RF with +12V input. (The modulator is shown separately.)

6.4 RF Power Amplifiers and Modulators

The modulators included in this subsection are designed for use with the unmodulated transmitters of Section 6.3. The power amplifiers cover the range from 50 kHz to 500 MHz. As modules, most of these power amplifiers can be joined to appropriate multipliers of Section 6.2.

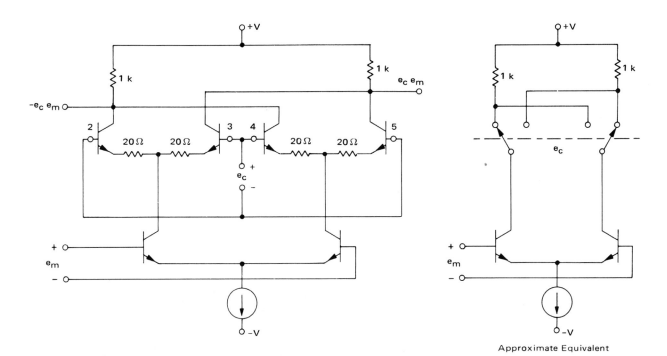

6.4.1 Balanced modulator using Motorola MC1545. The input differential amplifiers have been connected so that their collectors are cross-coupled. If the carrier level is sufficient to completely switch the top differential amplifier pairs, the circuit functions as shown by the approximate equivalent circuit. Here the modulation signal is alternately switched between differential amplifiers at the carrier rate. The result is that the modulation input signal is multipled by a symmetrical switching function which shifts the spectrum of the modulation input and places it symmetrically about the odd harmonics of the carrier.

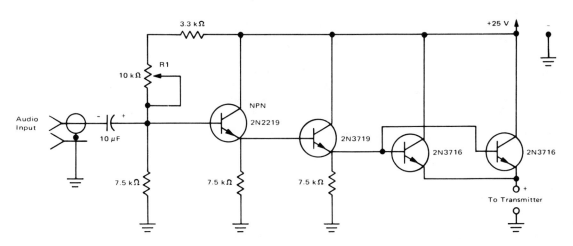

6.4.2 Motorola series modulator for 4W broadband aircraft transmitter. This circuit uses an audio power transistor instead of a transformer secondary. The collector supply to the modulator is 26V. The dc voltage to the transmitter is adjusted with R1. Although this modulator has such disadvantages as thermal drift and distortion, it nevertheless provides a convenient method of modulating a transmitter. A more practical series modulator could include external feedback to reduce the envelope distortion.

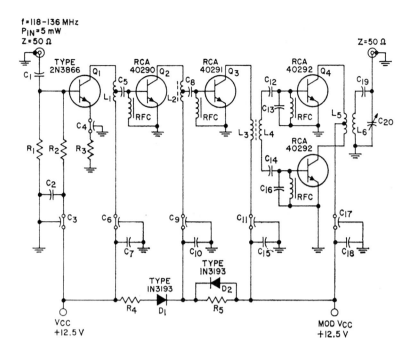

$C_1 = 300$ pF, silver mica
$C_2 = 0.005$ μF, ceramic
$C_3, C_4, C_6, C_9, C_{11}, C_{17} =$ Feedthrough capacitor, 1000 pF
$C_5 = 50$ pF, silver mica
$C_7, C_{10}, C_{15}, C_{18} = 0.5$ μF, ceramic
$C_8, C_{12}, C_{14} = 82$ pF, silver mica
$C_{13}, C_{16}, C_{19} = 150$ pF, silver mica
$C_{20} =$ Variable capacitor, 8 to 60 pF, Arco No. 404 or equiv.
$L_1 = 7$ turns of No. 22 wire, 13/64 inch in diameter, 9/16 inch long, tapped at 1.5 turns
$L_2 = 5.5$ turns of No. 22 wire, 13/64 inch in diameter, closely wound on Cambion IRN-9 (or equiv.) core material, tapped at 2 turns
$L_3 = 6$ turns of No. 22 wire, 13/64 inch in diameter, interwind with L_4 on Cambion IRN-9 (or equiv.) core material
$L_4 = 4$ turns of No. 22 wire, 13/64 inch in diameter, interwind with L_3 on common core
$L_5 = 5$ turns of No. 22 wire, 13/64 inch in diameter, center-tapped; interwind with L_6
$L_6 =$ Same as L_5; interwind with L_5
RFC = 1 turn of No. 28 wire, ferrite bead Ferroxcube No. 56-590-65/4B, or equiv.
$R_1 = 470$ ohms, 0.5 watt
$R_2 = 1500$ ohms, 0.5 watt
$R_3 = 47$ ohms, 0.5 watt
$R_4 = 15$ ohms, 0.5 watt
$R_5 = 33$ ohms, 0.5 watt

6.4.3 RCA 40W aircraft amplifier for 118 to 130 MHz AM transmitters is simple and requires a minimum of adjustments. The amplifier is capable of delivering a peak envelope power of 40W at 95% modulation with a collector voltage of 12.5V. The overall efficiency of the amplifier is 48 to 53% and the envelope distortion is less than 5%.

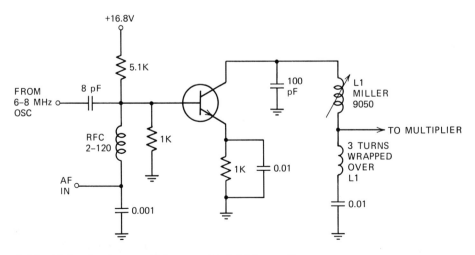

6.4.4 Motorola phase modulator and 7.5 MHz amplifier for low-power FM transmitters requires an audio-frequency signal of about 250 mV for full deviation. Transistor is HEP 55.

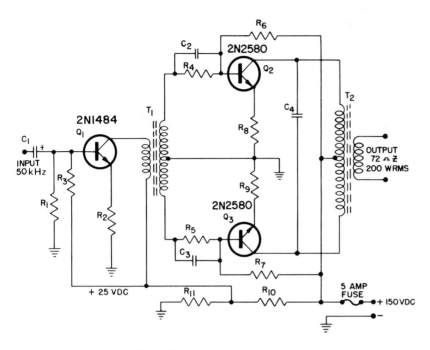

Parts List

C1	6 μF-5V electrolytic
C2, C3	1.0 μF 50V dc paper
C4	0.02 μF 1000V ceramic
R1	1K ½W
R2	33-ohm ½W
R3	6.2K ½W
R4, R5	15-ohm 2W
R6, R7	6K 5W
R8, R9	0.27-ohm 2W
R10	1.2K 25W
R11	200-ohm 5W
Q1	2N1484
Q2, Q3	Delco 2n2580
Heatsink (2)	Delco 7281366 or 7281369
Insulating spacer (4)	Delco 7269634
Mounting kit (2)	Delco 7274633
T1 core	Ferroxcube 3C 206 F 440
primary	40 turns No. 26 enameled wire
secondary	12 turns CT No. 22 enameled wire bifilar wound
T2 core	Allen Bradley U2375C127A ferrite
primary	50 turns CT No. 20 enameled wire bifilar wound
secondary	40 turns No. 20 enameled wire

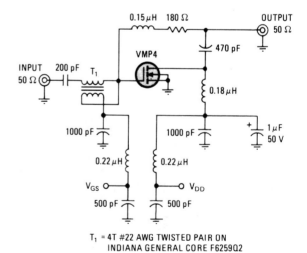

T_1 = 4T #22 AWG TWISTED PAIR ON INDIANA GENERAL CORE F6259Q2

6.4.5 Delco VLF power amplifier delivers 200W of power at 50 kHz to a 72-ohm load. Only 0.5 mW is required for driving power. It is important to provide adequate cooling facilities for the Delco 2N2580 transistors in this application. The heatsinks must have a thermal resistance of 1.0°/W maximum. This value can be obtained by using forced air flow on 7281366 or 7281369 heatsinks. With the specified heatsinks, the amplifier will operate up to 45°C ambient.

6.4.6 In this Siliconix wideband VHF amplifier a single power FET has a power gain of 15 dB flat to within ±1 dB from 40 to 180 MHz—not easily achieved with equivalent bipolar power transistors. The circuit can deliver 10 to 12W into a 50-ohm load, depending on the input. A key feature is the circuit's ability to withstand infinite-load standing wave ratios without any special circuitry power.

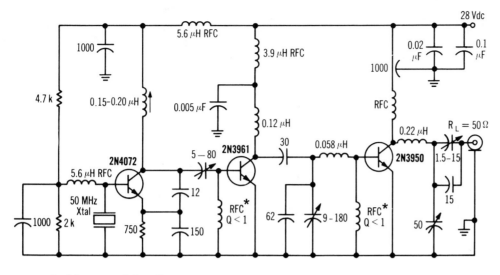

6.4.7 Motorola VHF power amplifier provides 50W continuous output at 50 MHz from a 28V supply with an overall efficiency of 62%. The final transistor, an NPN silicon unit of multiple-emitter design, is mounted in a grounded-emitter TO-60 package. The large package area in contact with RF ground provides a very low-impedance emitter ground path. Design details and performance data are available from Motorola (AN-282).

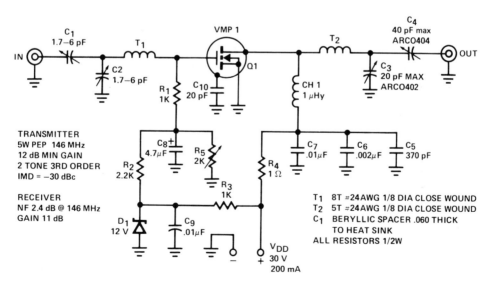

6.4.8 Siliconix 2m amateur-band linear amplifier delivers 5W peak envelope power with a gain of 12 dB from 144 to 150 MHz. The power FET in this circuit is a Siliconix device. The coils (T1 and T2) may be constructed by close-winding 24-gage solid wire into configurations of ⅛ in. (0.32 cm) diameter. T1 requires 8 turns and T2 requires 5 turns. Two-tone third-order intermodulation distortion, as measured on the prototype unit, is 30 dB below the carrier level. (Use conventional VHF shielding and lead-length techniques to insure stable operation and proper performance of the completed unit.)

SECTION 6 Communications

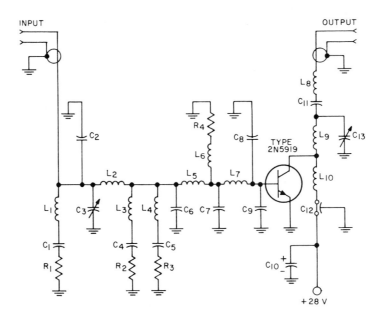

C_1 = Gimmick capacitor, 2.2 pF, Quality Components type 10% QC or equiv.
C_2 = 10 pF, silver mica
C_3 = Variable capacitor, 0.8 to 10 pF, Johanson No. 3957 or equiv.
C_4 = Gimmick capacitor, 1.0 pF, Quality Components type 10% QC or equiv.
C_5 = Gimmick capacitor, 1.5 pF, Quality Components type 10% QC or equiv.
C_6 = 36 pF, ATC-100 type or equiv.
C_7 = 51 pF, ATC-100 type or equiv.
C_8 = 68 pF, ATC-100 type or equiv.
C_9 = 47 pF, ATC-100 type or equiv.
C_{10} = 1 μF, electrolytic, 50 V
C_{11} = 12 pF, silver mica
C_{12} = Feedthrough capacitor, 1000 pF, Allen-Bradley No. FA5C or equiv.
C_{13} = Variable capacitor, 0.8 to 20 pF, Johanson No. 4802 or equiv.
L_1, L_3, L_4 = RF choke, 0.18 μH, Nytronics type P. #DD-0.18 or equiv.
L_2 = 1.5 turns*
L_5 = Copper strip, 5/8 inch long, 5/32 inch wide
L_6 = RF choke, 0.1 μH, Nytronics type P. #DD-0.10
L_7 = Transistor base lead, 0.5 inch long
L_8, L_{10} = 3 turns*
L_9 = 2 turns*
R_1 = 100 ohms, 1 watt
R_2, R_3 = 100 ohms, 0.5 watt
R_4 = 5.1 ohms, carbon, 0.5 watt

* All coils are wound from No. 18 wire with an inner diameter of 5/32 inch and a pitch of 12 turns per inch.

6.4.9 Motorola broadband 225 to 400 MHz power amplifier provides a constant power output of 16W with a gain variation of less than 1 dB for an input driving power of 3 to 4W. Two of these amplifiers can be connected in parallel to provide a constant power output of 25W over this frequency range.

The RCA 2N5919 transistor is operated class C. If the amplifier is to be used in an AM system, the linearity requirements can be met by the use of envelope correction or a slight forward bias (or both). The amplifier operates from a dc supply of 28V.

The collector efficiency is at least 63% across the frequency band. The second harmonic of a 225 MHz signal is 12 dB down and that of a 400 MHz signal is 30 dB down from the fundamental. This harmonic rejection is excellent for an amplifier that is required to have a bandwidth that covers almost an octave.

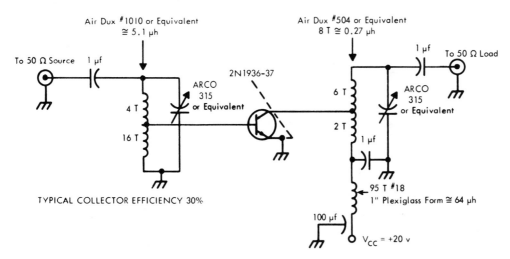

6.4.10 3MHz power amplifier. Use a Delbert-Blinn 113 heatsink for the T1 power transistor.

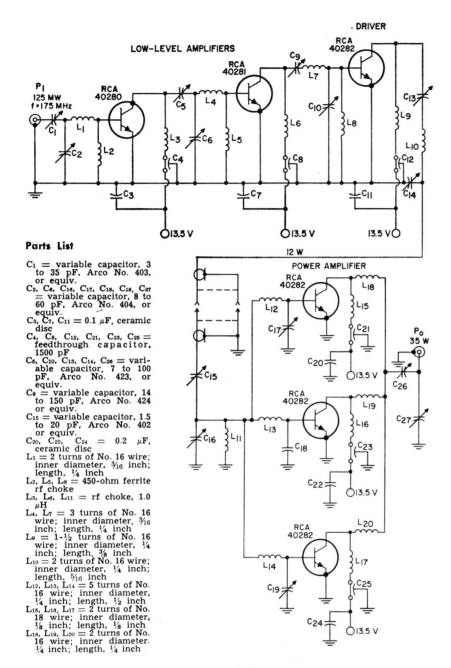

Parts List

C₁ = variable capacitor, 3 to 35 pF, Arco No. 403, or equiv.
C₂, C₆, C₁₆, C₁₇, C₁₈, C₁₉, C₂₇ = variable capacitor, 8 to 60 pF, Arco No. 404, or equiv.
C₃, C₇, C₁₁ = 0.1 μF, ceramic disc
C₄, C₈, C₁₂, C₂₁, C₂₃, C₂₅ = feedthrough capacitor, 1500 pF
C₅, C₁₀, C₁₃, C₁₄, C₂₆ = variable capacitor, 7 to 100 pF, Arco No. 423, or equiv.
C₉ = variable capacitor, 14 to 150 pF, Arco No. 424 or equiv.
C₁₅ = variable capacitor, 1.5 to 20 pF, Arco No. 402 or equiv.
C₂₀, C₂₂, C₂₄ = 0.2 μF, ceramic disc
L₁ = 2 turns of No. 16 wire; inner diameter, 3/16 inch; length, 1/4 inch
L₂, L₅, L₈ = 450-ohm ferrite rf choke
L₃, L₆, L₁₁ = rf choke, 1.0 μH
L₄, L₇ = 3 turns of No. 16 wire; inner diameter, 3/16 inch; length, 1/4 inch
L₉ = 1-1/2 turns of No. 16 wire; inner diameter, 1/4 inch; length, 3/8 inch
L₁₀ = 2 turns of No. 16 wire; inner diameter, 1/4 inch; length, 5/16 inch
L₁₂, L₁₃, L₁₄ = 5 turns of No. 16 wire; inner diameter, 1/4 inch; length, 1/2 inch
L₁₅, L₁₆, L₁₇ = 2 turns of No. 18 wire; inner diameter, 1/8 inch; length, 1/8 inch
L₁₈, L₁₉, L₂₀ = 2 turns of No. 16 wire; inner diameter, 1/4 inch; length, 1/4 inch

6.4.11 RCA four-stage RF power amplifier operates from a dc supply of 13.5V and delivers 35W of power output at 175 MHz for an input of 125 mW. The RCA silicon overlay transistors used in the amplifier supply, maximum output power at this level of dc voltage for use in mobile systems.

The low-level portion of the amplifier consists of three unneutralized class C amplifier stages interconnected by bandpass filters.

When the low-level stages and the output stage are mounted on separate chassis, the output from the driver stage is coupled to the output stage through a low-loss coaxial line terminated by C15, C16, and L11. The capacitors are adjusted to assure a good impedance match between the output of the driver and the input of the output stage at 175 MHz. The driving signal developed across inductor L11 is applied to the tuned input networks of three parallel transistors in the output stage. Adjust capacitors C26 and C27 to match the amplifier output to the load impedance at the operating frequency.

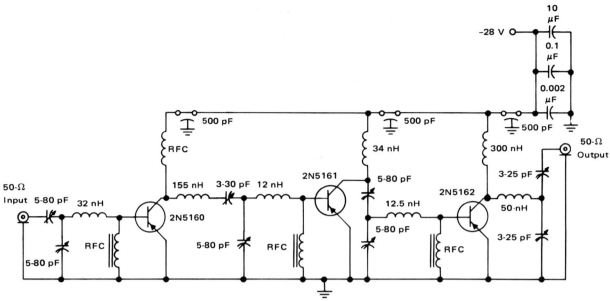

6.4.12 30W, 175 MHz class C amplifier using Motorola 2N5160, 2N5161, and 2N5162 VHF power transistors. The three-stage amplifier will deliver 30W of RF power to a 50-ohm load with an overall gain of 29.3 dB. The design has a high overall efficiency of 50.5% while maintaining all spurious outputs at least 40 dB below the 175 MHz level.

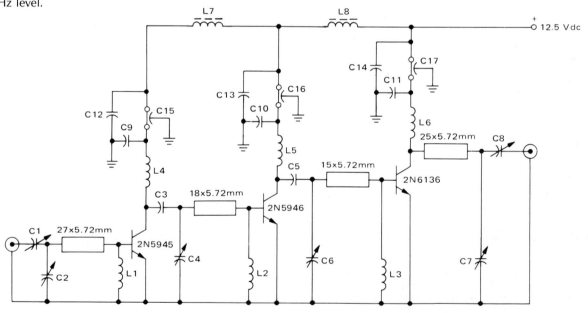

C1,2,4,6,7,8 — 1.5-20pf Arco 402 or equiv.
C3, 5 — 10pf dipped mica
C9,10,11 — 0.1 μf ceramic
C12,13,14 1μf Tantalum
C15,16,17 470pf feed thru
L1,2,3 — 3.9μhy molded choke w/ferrite core
L4,5,6 — 5 turns #20 awg closewound 3/16" I.D.
L7, 8 — Ferroxcube VK200 20/4B or equiv.
Board is 3/16" "G-10" epoxy fiberglass Dielectric with 1 oz copper on both sides.

6.4.13 Motorola 25W UHF RF amplifier operates from +12V supply. Photographs of prototype unit and board layout patterns are available from Motorola (AN-578). Parts placement can be critical.

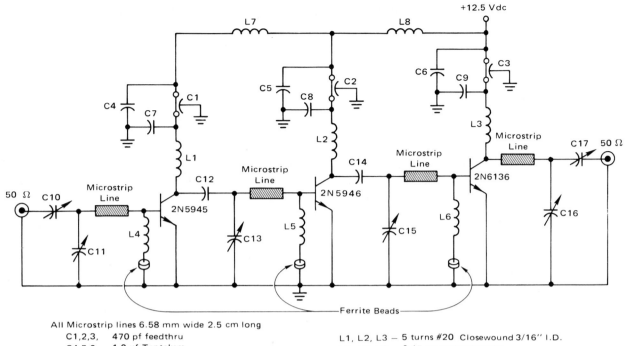

All Microstrip lines 6.58 mm wide 2.5 cm long
C1,2,3 470 pf feedthru
C4,5,6 1.0 µf Tantalum
C7,8,9 0.1 µf Ceramic
C10,11,13,15,16,17 1.5-20 pf Compression Trimmer ARCO 420
C12,14 10 pf Microwave capacitor ATC type 100-B-10-M-MS or equiv.

L1, L2, L3 — 5 turns #20 Closewound 3/16" I.D.
L4, L5, L6 — 0.15 µh molded choke
L7, L8 — Ferroxcube VK 200 20/4B or equiv.
Ferrite beads are Ferroxcube 56 590 65/3B or equiv.

6.4.14 Motorola UHF amplifier operates on +12V and produces 25W output in 450 to 512 MHz range. Design details and photographs of prototype are available from Motorola (application note AN-548).

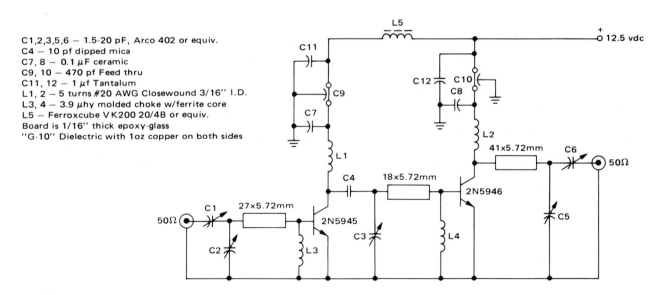

C1,2,3,5,6 — 1.5-20 pF, Arco 402 or equiv.
C4 — 10 pf dipped mica
C7, 8 — 0.1 µF ceramic
C9, 10 — 470 pf Feed thru
C11, 12 — 1 µf Tantalum
L1, 2 — 5 turns #20 AWG Closewound 3/16" I.D.
L3, 4 — 3.9 µhy molded choke w/ferrite core
L5 — Ferroxcube VK200 20/4B or equiv.
Board is 1/16" thick epoxy-glass
"G-10" Dielectric with 1oz copper on both sides

6.4.15 Motorola UHF amplifier uses a 2N5945 and a 2N5946 in cascade to obtain 10W of output power from 450 to 470 MHz. The power requirement is a single 12V source, making this ideal for land/mobile/amateur communications applications. In reproducing these designs, care should be taken in the placement of all parts, with particular attention given to the location and number of eyelets used. (Full-sized foil patterns appear in Motorola application note AN-578).

SECTION **6** Communications

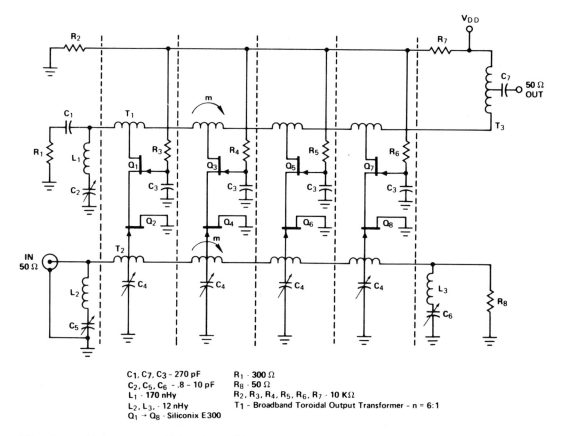

C₁, C₇, C₃ - 270 pF
C₂, C₅, C₆ - .8 - 10 pF
L₁ - 170 nHy
L₂, L₃, - 12 nHy
Q₁ → Q₈ - Siliconix E300

R₁ - 300 Ω
R₈ - 50 Ω
R₂, R₃, R₄, R₅, R₆, R₇ - 10 KΩ
T₁ - Broadband Toroidal Output Transformer - n = 6:1

6.4.16 Siliconix multiple-octave wideband amplifier employs image-parameter m-derived low-pass filter elements to achieve excellent group delay characteristics over the majority of the passband. The transmission line and its characteristic impedance need not be equal between the input and output, but it is important that the phase velocity of both lines be equal. For the image-parameter low-pass filter,

cutoff frequency $\quad f_c = \dfrac{1}{\pi\sqrt{LC}}$

characteristic impedance $\quad Z_o = \sqrt{\dfrac{L}{C}}$

phase constant $\quad \beta = \omega\sqrt{LC}$

The phase constant of each transmission line must match; it is also important to establish the characteristic line impedance for convenience in matching. The two transmission lines are designed by iteration. The following table provides typical values of inductance and capacitance for a cutoff frequency of 300 MHz. Any line selected from this group will have a matched phase constant. The amplifier is designed for a 50-ohm input impedance and 300-ohm drain impedance. A broadband 6:1 output transformer provides the match. FETs in the amplifier are arranged in cascode to insure low input conductance and good stability. The measured gain is 5.2 dB, and with two amplifiers cascaded, the overall gain of 10 dB is achieved across the 50 to 300 MHz passband. Flat group delay is nearly 90% of the total passband.

Z_o (ohms)	C (pF)	L (nH)
50	21.2	53
75	14.1	80
100	10.6	106
200	5.4	210
300	3.5	320
450	2.4	480

6.5 RF Preamplifiers and Converters

All circuits in this subsection are to be connected directly to an antenna. The output frequency varies; some converters are designed to work into a broadcast-band receiver and others are meant to be coupled to a communications receiver at a specified frequency. Be sure to check the output frequency for applicability to your own needs before starting on a converter project.

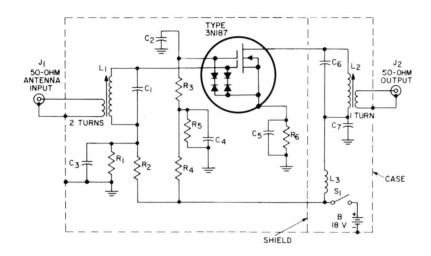

Parts List

B = Two RCA type VS323 batteries for transistor service; and one case, Bud-CU2103A or equivalent.
C_1 = 8 pF, mica or ceramic tubular
C_2, C_3, C_4, C_5, C_7 = 0.01 μF, ceramic
C_6 = 10 pF, mica or ceramic tubular
J_1, J_2 = Coaxial receptacle, Amphenol BNC type UG-1094 or equiv.
L_1, L_2 = 1.6 to 3.1 μH, adjustable, Miller 4404 or equiv.
L_3 = 22 μH, Miller 74F-225A1 or equiv.
R_1 = 27,000 ohms, 0.25 watt, 10%
R_2 = 150,000 ohms, 0.25 watt, 10%, carbon
R_3 = 1,800 ohms, 0.25 watt, 10%, carbon
R_4 = 100,000 ohms, 0.25 watt, 10%, carbon
R_5 = 33,000 ohms, 0.25 watt, 10%, carbon
R_6 = 270 ohms, 0.25 watt, 10%, carbon
S_1 = toggle switch, single-pole, single-throw

Tuned-Circuit Components for 21 and 50 MHz

Component	Value 21 MHz	50 MHz
C_1	22 pF	8 pF
C_2, C_3, C_4, C_5, C_7	No Change	1,000 pF, ceramic
C_6	22 pF	10 pF
L_1	No Change	8 turns, No. 30 E wire on 1/4-inch-diameter core (Miller 4500 or equiv.) Link: 2 turns, No. 30 E wire on ground end.
L_2	No Change	Same as L_1
L_3	No Change	6.8 μH (Miller 74F686AP or equiv.)

6.5.1 RCA preamplifier for 6, 10, or 15m amateur-band receiver uses a 3N187 dual-gate-protected MOS transistor to provide more than 26 dB of gain ahead of a receiver operated in the 6, 10, or 15m amateur band. This additional gain, together with the low noise figure of the preamplifier (less than 2.5 dB), substantially increases both the sensitivity and the signal-to-noise ratio of the receiver. The circuit as shown is intended for use for 10m and CB; the 3N187 MOS transistor, however, has excellent performance characteristics at frequencies well below the 10m band and up to 200 MHz. The preamplifier, therefore, can be readily adapted for use in other frequency bands with only a few changes in tuned-circuit components. A chart is provided to show the changes in components required for operation in the 15 and 6m bands. The dc operating voltage for the preamplifier may be obtained from a battery supply, as shown in the circuit diagram, or from any other reasonably well filtered dc supply voltage of 15 to 18V.

The tuning or the preamplifier is simplified because no special neutralization is required, even at frequencies as high as 155 MHz. Rough adjustments of coils L1 and L2 can be made by the use of a dip oscillator. The finishing adjustments are then made while listening to a weak station.

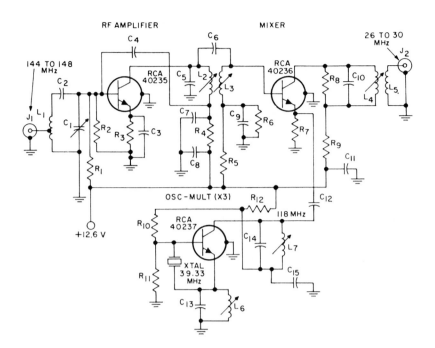

Parts List

C₁ = 0.5 to 5 pF, tubular trimmer, Erie 532-3R or equiv.
C₂ = 10 pF, ceramic tubular, Centralab TCZ-10 or equiv.
C₃, C₉, C₁₁ = 500 pF, silver button, Erie 662-003-501K or equiv.
C₄ = 4.7 pF, ceramic tubular, Centralab TCZ-4R7 or equiv.
C₅, C₁₂, C₁₄ = 3.3 pF, ceramic tubular Centralab TCZ-3R3 or equiv.
C₆ = 2.2 pF, ceramic tubular, Centralab TCZ-2R2 or equiv.
C₇ = 25 pF silver button, Erie 662-003-250 or equiv.
C₈, C₁₅ = 500 pF, ceramic disc, Centralab DD-501 or equiv.

C₁₀, C₁₃ = 30 pF, ceramic tubular, Centralab TCZ-30 or equiv.
J₁, J₂ = BNC-type coaxial jack
L₁ = 5 turns of No. 16 bare wire, ¼-inch diameter (spaced wire diameter), tap one turn up from bottom
L₂, L₃ = 4 turns of No. 26 enamelled wire, close wound on ¼-inch diameter ceramic slug-tuned form, Miller 4500 or equiv.
L₄ = 11 turns of No. 26 enamelled wire, close wound on ⅜-inch diameter phenolic slug-tuned form, Miller 21A000RBI or equiv.
L₅ = 3 turns of insulated wire, close-wound link

L₆ = 5 turns of No. 26 enamelled wire, close wound on ⅜-inch diameter phenolic slug-tuned form, Miller 21A000RBI or equiv.
L₇ = 7 turns of No. 26 enamelled wire, close wound on ¼-inch diameter ceramic slug-tuned form, Miller 4500 or equiv.
R₁ = 27,000 ohms, 0.5 watt
R₂ = 3,900 ohms, 0.5 watt
R₃, R₇ = 470 ohms, 0.5 watt
R₄, R₉ = 820 ohms, 0.5 watt
R₅ = 18,000 ohms, 0.5 watt
R₆ = 2,700 ohms, 0.5 watt
R₈ = 5,100 ohms, 0.5 watt
R₁₀ = 0.1 megohm, 0.5 watt
R₁₁ = 8,200 ohms, 0.5 watt
R₁₂ = 1,000 ohms, 0.5 watt
XTAL = 39.33 MHz, overtone crystal

6.5.2 RCA 2m converter can be used ahead of a 10m amateur-band radio receiver to provide amplification and the frequency conversion required to enable reception of signals in the 2m band. With minor circuit modification, the converter can also be used to adapt a 20m amateur-band receiver to receive 2m signals. The converter uses RCA 40235, 40236, and 40237 transistors in common-emitter circuit configuration to provide mobile as well as fixed-station operation.

Signals are coupled from the antenna through the coaxial connector J1 and the tuned input circuit (L1, C1, and C2) to the base of the 40235 transistor. Variable capacitor C1 is adjusted to tune the input circuit to select any desired signal in the 144 to 148 MHz frequency band. The selected signals are amplified by the 40235 transistor and coupled from the collector of this transistor by L2, L3, and C6 to the base of the mixer.

The RCA 40237 transistor is operated in an overtone-crystal oscillator–multiplier stage to develop the local-oscillator signal for the converter. The crystal used in the base-to-emitter circuit of the oscillator–multiplier has a fundamental frequency of 39.33 MHz; the collector load circuit is tuned to select the third harmonic of the crystal fundamental. The oscillator–multiplier stage develops a fixed-frequency 118 MHz local-oscillator signal that is coupled by C12 to the emitter 40236 mixer.

In the mixer stage, the signals from the antenna and local oscillator are heterodyned to derive the difference frequency used as the input to the receiver.

The 118 MHz local-oscillator frequency was selected so that the heterodyning action in the mixer provides a converter output of 26 to 30 MHz, depending upon the frequency of the selected RF input signal from the antenna.

This circuit uses coils that are not standard commercial items; such coils must be wound by the circuit builder.

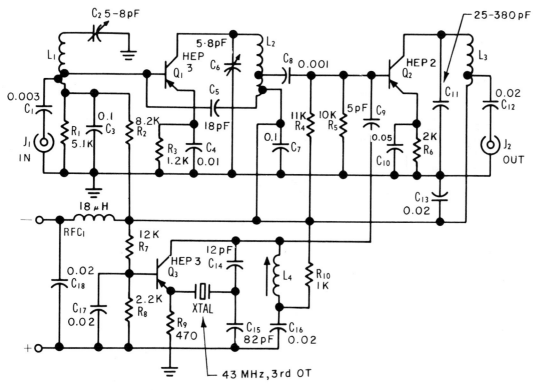

- L1 5 turns No. 20 enameled wire, ¼ in. diameter, closely wound; tapped at 1 and 2 turns from cold end (0.15H)
- L2 8 turns No. 20 enameled wire, ¼ in. diameter, closely wound; tapped at 2 and 4½ turns from cold end (0.19H)
- L3 26 turns No. 28 enameled wire, ¼ in. diameter, closely wound; centertapped (2.3H)
- L4 10 turns No. 26 enameled wire, ¼ in. diameter, closely wound; slug-tuned (0.55 to 0.85H)

6.5.3 Motorola 50 MHz receiver VHF-to-HF converter uses inexpensive transistors and is easy to construct. The overall gain exceeds 30 dB. Sensitivity is about 1 μV for a 10 dB signal-to-noise ratio at the receiver audio output with a 30% modulated signal, and well-modulated signals of less than 0.5 μV can be copied with ease. Receiver tuning frequency is 7 to 11 MHz (40m).

Tune the RF input and output circuits and the mixer output circuit to approximate resonance using a dip meter. Couple the meter to L1 and tune C2 for resonance at 50 MHz. In the same manner, couple L2 and tune C6 for resonance at 7 MHz. Definite dips should be obtained if the circuits are operating properly.

Adjust oscillator slug L4 to midrange in the coil. Connect an RF signal generator to the converter input jack and connect the converter output to the antenna terminals of any 40m receiver. Connect the power source to the converter; then apply a 50 MHz modulated signal to the converter and locate the signal with the receiver tuned to 7 MHz. If the oscillator is detuned too far, it may not oscillate; therefore, if the signal cannot be located at first, continue to search for it while slowly moving the oscillator slug. Once the signal is located, adjust the slug for maximum audio output in the receiver.

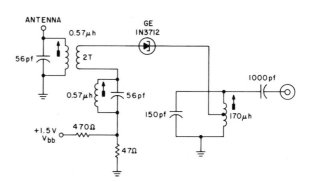

6.5.4 GE simple 27 MHz CB converter offers good stability with no crystal. The circuit is designed for connection to a car antenna at the input and to the antenna terminal of a car radio at the output. Tune the radio to about 1 MHz.

SECTION 6 Communications 299

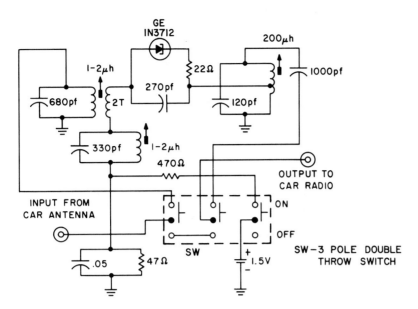

6.5.5 CAP converter employs a GE tunnel diode. The antenna is coupled into a tank circuit of 680 pF and an inductor of 1 to 2 µH. The 680 pF swamps the antenna capacitance and tunes with it and the inductor to 4.46 MHz. A two-turn secondary is loosely coupled to the primary of L1. By changing the number of loosely coupled turns in the secondary until the circuit oscillates, an optimum match can be established. It is necessary to back off somewhat from this optimum match condition to stabilize the circuit. Tune the car radio to about 1 MHz for CAP reception.

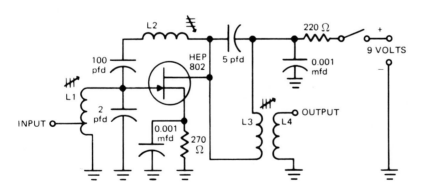

6.5.6 Motorola 2m amateur radio preamplifier offers 14 dB of signal amplification when used at the RF input of a VHF receiver. Construction on a printed circuit board is recommended. All coils should be wound on brass-slug ceramic forms. L1—5.25 turns, tapped at 1.25 turns, 26-gage wire L2—9.5 turns, 34-gage L3—5 turns, 26 gage L4—1.25 turns, 26-gage; wind around low end of L3

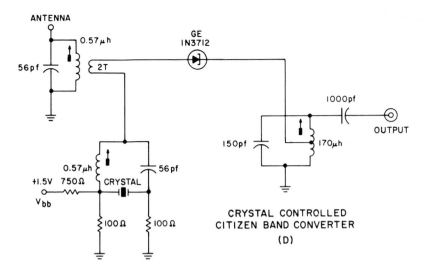

CRYSTAL CONTROLLED CITIZEN BAND CONVERTER (D)

6.5.7 27 MHz crystal-controlled converter using GE tunnel diode offers a high degree of stability with a minimum of components. Sensitivity at 28.05 MHz is typically 75 µV with 135 mV operating bias to the tunnel diode. The output is about 1 MHz, and a car radio should provide across-the-band tuning.

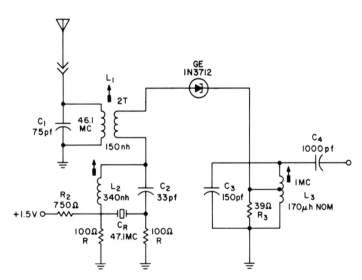

6.5.8 GE 46.1 MHz tunnel-diode crystal-controlled converter uses an automobile antenna and produces a 1000 kHz output. If additional frequency coverage is needed above 47 MHz, remove turns from L1.

6.6 Front Ends, RF/IF Amplifiers, and Mixers

These are modular circuit blocks for receivers covering a variety of operating frequencies. Unless otherwise specified, the input and output impedances are 50 ohms.

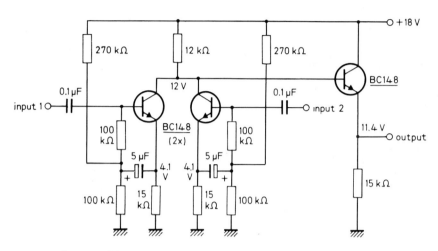

6.6.1 Mixer amplifier in which two inputs are fed to separate Amperex transistors with a common collector load resistor. An emitter follower stage insures a 70-ohm output impedance; input impedance is 2.5M. Voltage gain is unity.

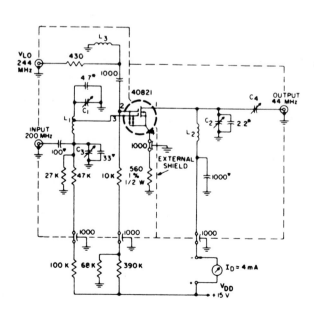

▼ Disc. ceramic.
• Tubular ceramic.
All resistors in ohms
All capacitors in pF
C1, C2 = 1.5-5 pF variable air capacitor: E.F. Johnson Type 160-102 or equivalent.
C3 = 1-10 pF piston-type variable air capacitor: JFD Type VAM-010, Johanson Type 4335, or equivalent.
C4 = 0.9-7 pF compression-type capacitor: ARCO 400 or equivalent
L1 = 5 turns silver-plated 0.02" thick, 0.07"-0.08" wide copper ribbon. Internal diameter of winding = 0.25"; winding length approx. 0.65". Tapped at 1-1/2 turns from C1 end of winding.
L2 = Ohmite Z-235 RF choke or equivalent.
L3 = J.W. Miller Co. #4580 0.1 μH RF choke or equivalent.
NOTE: If 50Ω meter is used in place of sweep detector, a low pass filter must be provided to eliminate local oscillator voltage from load.

6.6.2 Mixer circuit for 100 MHz to 44 MHz conversion using the RCA 40821 MOSFET IC.

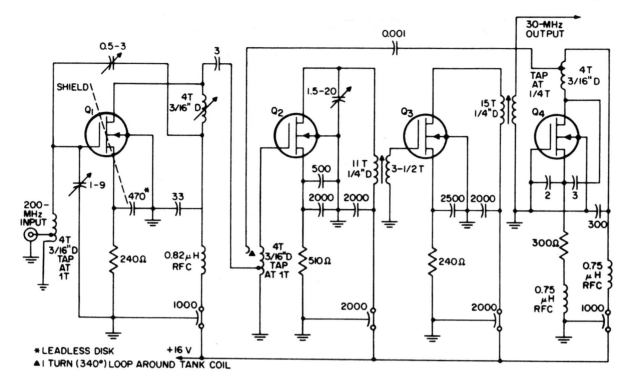

6.6.3 VHF receiver front end using the RCA 3N128 in all stages. The input stage is a straight-through 200 MHz amplifier employing a source resistor for gate bias. This configuration permits the gate to be at a dc ground and greatly reduces the possibility of damage to the MOS gate from input transients. The 240-ohm resistor allows a current of approximately 5 mA to flow through the device so that maximum VHF power gain is obtained. A variable inductor resonates with the output capacitance of the 3N128 to provide a bandwidth of approximately 12 MHz for the RF stage alone. (A narrower bandwidth could be obtained by the use of more capacitance in the tuned circuits and different loading on the output circuit.)

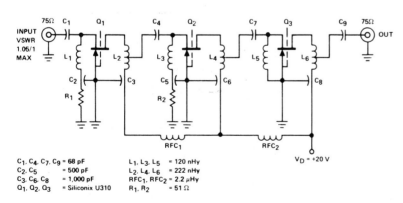

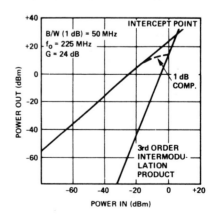

6.6.4 Siliconix VHF wideband amplifier has 75-ohm input and output impedances. The values given are for operation at 200 to 250 MHz. Broken-line separations show shielding to minimize the Miller effect. The three-stage amplifier exhibits 1 dB of ripple and performance as shown in the chart.

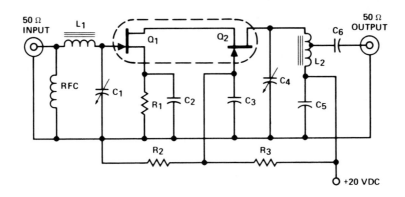

L1	1.7 μH, 22 turns No. 24 AWG enamel on Micrometals T50-10 toroidal core
L2	1.5 μH, 20 turns No. 24 AWG enamel on Micrometals T50-10 toroidal core
C1, C4	2–20 pF
C2, C3, C5, C6	0.01 μF
R1	390 ohms
R2	10K
R3	30K
Q1, Q2	Siliconix E420
RFC	30 μH Delvan choke

6.6.5 Siliconix cascode low-noise intermediate-frequency amplifier operates at 30 MHz in a 50-ohm system. The noise performance of the amplifier is less than 1.5 dB and the gain is in excess of 20 dB. The circuit has no critical adjustments. For maximum performance, short leads are required (point-to-point wiring is preferred). The two toroidal coil transformers should be located at right angles to one another to avoid the possibility of infringing on the magnetic fields and to eliminate the need for shielding to obtain stable operation. The protype circuit is unshielded and employs an input "el" match to transform the 50 ohms to the optimum value of source impedance necessary to achieve the best noise figure. The Siliconix E420 dual FET used in the amplifier displayed the best performance with a source-impedance value of 1.8K. The FETs were selected at random, providing a noise figure "window" between 0.9 and 1.2 dB; the gain was reasonably stable at 20 dB. A gain as high as 28 dB can be obtained if the tap on the output tank is removed, but shielding will probably be required for stable operation.

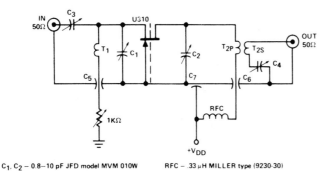

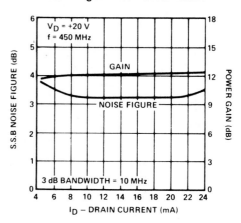

6.6.6 UHF common-gate amplifier for 420 to 450 MHz. The input and output impedances are 50 ohms. The chart shows the noise figure as a function of FET drain current; as shown, the power gain is on the order of 12 dB. The JFET is Siliconix U310.

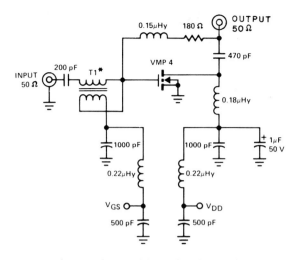

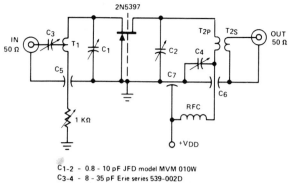

6.6.7 Siliconix-designed broadband amplifier offers 14 to 16 dB gain from 40 to 180 MHz. The drain current is 600 mA with V_{DD} of 24V. Transformer T1 consists of four turns of 22-gage twisted pair on an Indiana General form (F625-9Q2).

6.6.9 Siliconix UHF amplifier is designed for a 450 MHz center frequency, with a 3 dB bandwidth of 6 MHz. The amplifier stage is compatible with a 50-ohm system. It may be used as a receiver front end with excellent reverse isolation characteristics that reduce local-oscillator radiation from the antenna. Applications include fixed or mobile service, aeronautical navigation, UHF TV, police and fire communications, paging systems, and amateur radio. The amplifier uses a Siliconix 2N5397 N-channel junction FET.

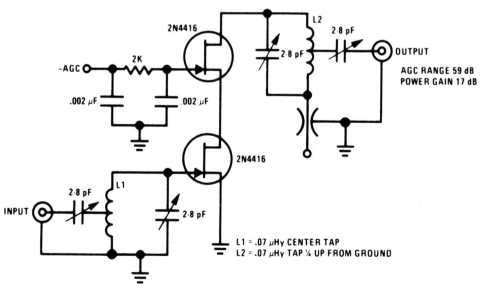

6.6.8 This Siliconix 200 MHz JFET cascode circuit features low cross modulation, large signal handling ability, no neutralization, and AGCing by biasing the upper cascode JFET. The only special requirement of this circuit is that I_{DSS} of the upper unit must be greater than that of the lower unit.

SECTION **6** Communications

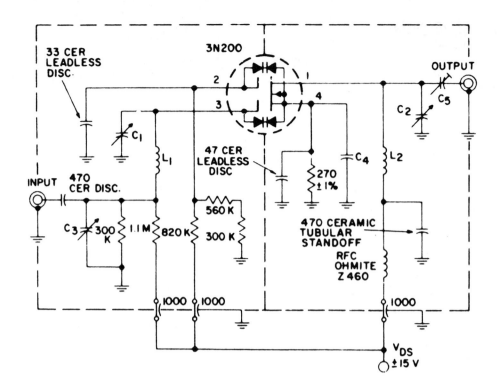

C1, C2 = 1.3-5.4 pF variable air capacitor: Hammerlund Mac 5 type or equivalent

C3 = 1.9-13.8 pF variable air capacitor: Hammerlund Mac 15 type or equivalent

C4 = Approx. 300 pF - capacitance formed between socket cover & chassis

C5 = 0.8-4.5 pF piston type variable air capacitor: Erie 560-013 or equivalent

L1, L2 = inductance to tune circuit

6.6.10 400 MHz amplifier using RCA 3N200 MOSFET IC provides a typical RF power gain of 12.5 dB with 4.5 dB noise factor at 400 MHz in a common-source configuration without the need for neutralization.

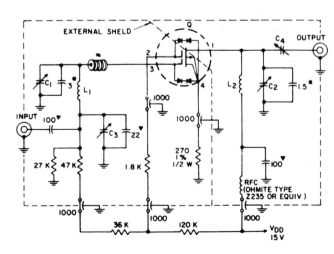

#Ferrite bead (4); Pyroterric Co. "Carbonl J"   Q = 3N187
  0.09 in. OD; 0.03 in. ID; 0.063 in. thickness.   ▼ Disc ceramic
All resistors in ohms   * Tubular ceramic
All capacitors in pF

C1 = 1.8-8.7 pF variable air capacitor: E.F. Johnson Type 160-104, or equivalent.

C2 = 1.5-5 pF variable air capacitor: E.F. Johnson Type 160-102, or equivalent.

C3 = 1-10 pF piston-type variable air capacitor: JFD Type VAM-010; Johanson Type 4335, or equivalent

C4 = 0.8-4.5 pF piston type variable air capacitor: Erie 560-013 or equivalent.

L1 = 4 turns silver-plated 0.02-in. thick, 0.075-0.085-in. wide, copper ribbon, internal diameter of winding = 0.25 in., winding length approx. 0.08 in.

L2 = 4½ turns silver-plated 0.02-in. thick, 0.085-0.095-in. wide, 5/16 in. ID. Coil = .90 in. long.

6.6.11 200 MHz amplifier using the RCA 3N187 MOSFET IC.

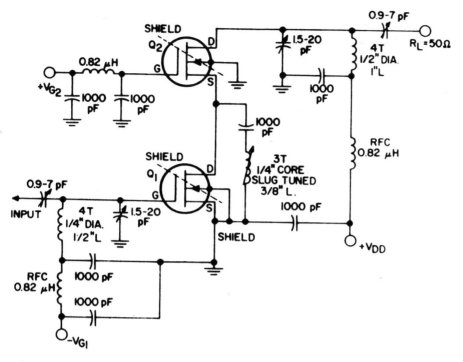

6.6.12 RCA 200 MHz cascode amplifier. This circuit normally requires a negative voltage on the gate of Q1, a positive voltage on the gate of Q2, and approximately equal drain-to-source voltages for each transistor. Although the gate of Q2 may require a positive voltage of 5 to 10V, the net gate-to-source voltage for this transistor should be approximately 0 to −1V. The transistors are RCA 3N128.

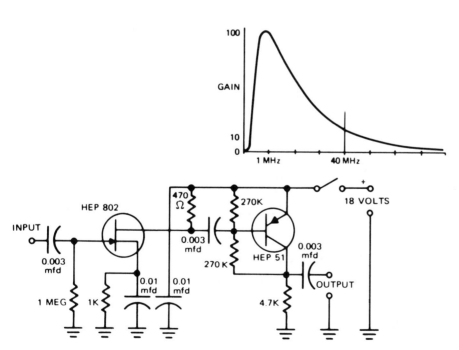

6.6.13 Motorola 1 to 40 MHz broadband RF amplifier features high input impedance, high signal-to-noise ratio. The resistors should have a tolerance of at least 10%; the capacitors should have a working voltage of 25V and should have a dielectric of ceramic.

SECTION 6 Communications

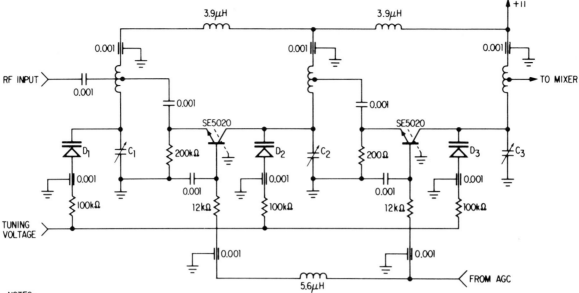

6.6.14 Two-stage RF amplifier provides both selectivity and AGC capability plus enough gain to maintain an acceptable noise figure. Its two Fairchild SE5020 transistors are biased at 3 mA, giving a noise figure of around 3 dB and a voltage gain of +10 dB with a change of 1 dB across the band. At 10 mA and 100 MHz, forward AGC-ing can furnish up to 40 dB of AGC control per stage.

The three varactor-tuned interstage networks operate from 105 to 120 MHz with a 1 MHz bandwidth. Varactor tuning offers two major advantages over conventional mechanical tuning: (1) tuning diode and frequency-selector controls can be physically separated (with only one dc lead required between the diode and controls), and (2) tuning is very rapid (order of microseconds), which makes it possible to use several frequency selector controls and to time-share among these controls.

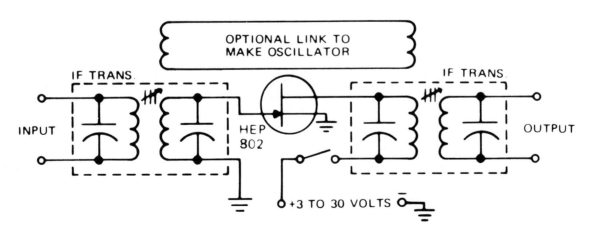

6.6.15 Motorola IF amplifier–oscillator can be used to amplify signals of 50 kHz, 455 kHz, 10.7 MHz, or any frequency within the limits of the FET. This high-impedance, low-current-drain circuit can be changed into a stable oscillator (tuned-drain-tuned-gate) by link-coupling the input to the output; coupling inputs and outputs of IF transformers can also be used. If the circuit doesn't oscillate, reverse the link on one end. (The link is 6 turns of wire at each end, approximately the same diameter as the coils.)

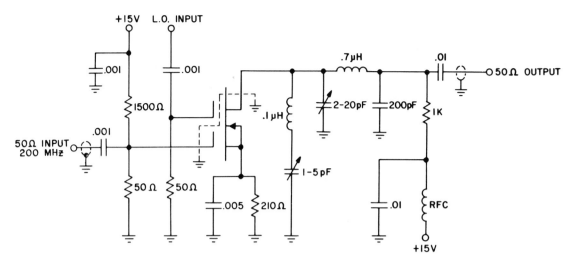

6.6.16 VHF mixer employs the General Instrument dual-gate MOSFET (MEM 564C) to achieve 30 dB of gate-to-gate isolation at 244 MHz. The mixer operating point is chosen so that the gate 2 Q point is close to the point of minimum-signal gate transconductance. As the local oscillator swings positive, the signal gate transconductance increases linearly. A minimum local-oscillator drive of approximately 2.5V is required at gate 2 for the gain to be approximately independent of the local oscillator drive level. The measurement of the interfering signal level for 1% cross modulation distortion is between 82 and 130 mV over the complete range of local oscillator injection. Thus the mixer is highly linear and relatively free of third-order distortion products.

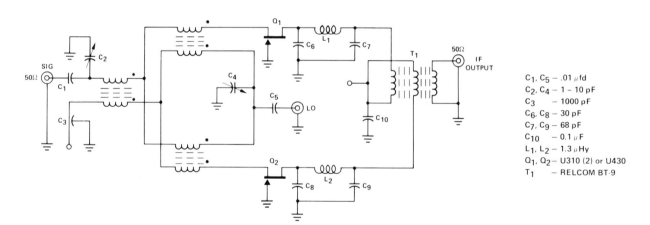

6.6.17 Active balanced mixer. A basic aim in mixer design is to avoid the effects of intermodulation product distortion and cross modulation. Part of the problem may be resolved by using a balanced mixer circuit.

The active transfer function of the Siliconix FET is represented by a voltage-controlled current source. For both cross modulation and intermodulation, the amount of distortion is proportional to the amplitude of the gate–source voltage. Since the input power is proportional to the input voltage and inversely proportional to the input impedance, the best FET performance is obtained in the common-gate configuration where the impedance is lowest.

When JFETs are used as active mixer elements, it is important that the devices be operated in their square-law region. Such operation emphasizes the importance of establishing proper drive levels for both quiescent bias and the local oscillator.

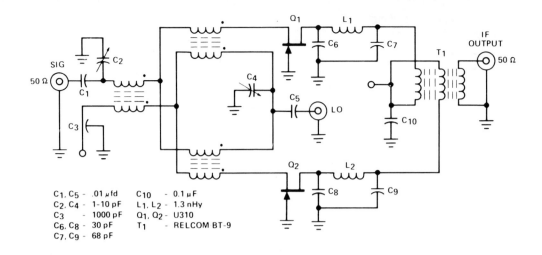

```
C1, C5 -  .01 µfd      C10  - 0.1 µF
C2, C4 -  1-10 pF      L1, L2 - 1.3 nHy
C3     -  1000 pF      Q1, Q2 - U310
C6, C8 -  30 pF        T1   - RELCOM BT-9
C7, C9 -  68 pF
```

50 to 250 MHz MIXER PERFORMANCE COMPARISON.

Characteristic	JFET	Schottky	Bipolar
Intermodulation intercept point	+32 dBm	+28 dBm	+12 dBm*
Dynamic range	100 dB	100 dB	80 dB*
Desensitization level (for an unwanted signal when the desired signal first experiences compression)	+8.5 dBm	+3 dBm	+1 dBm*
Conversion gain	+3 dB†	−6 dB	+18 dB
Single-sideband noise figure	6.5 dB	6.5 dB	6 dB

*Estimated.
†Conservative minimum.

6.6.18 Siliconix active balanced mixer is an improvement over hot-carrier diodes and bipolars. The inherent square-law transfer characteristics of the FET insure a high intermodulation intercept and signal desensitization. The grounded-gate connection is very stable, while source injection of both the signal and local oscillator make for easy impedance matching into the FETs. Balanced configuration reduces local-oscillator radiation from the signal port and suppresses the generation of even harmonics (which help to reduce intermodulation). The FET used in the mixer circuit is the Siliconix U310, which typically has a C_{gs} of about 1.9 pF. Such low gate capacitance is required for a wide mixer bandwidth. The U310 also has a typical g_{fs} of 14,000 microsiemens for useful conversion gain. Dynamic range is bracketed by the lowest drain current for an acceptable noise figure and the maximum drain current, typically $I_{DSS} = 40$ mA.

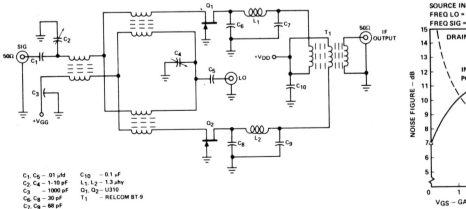
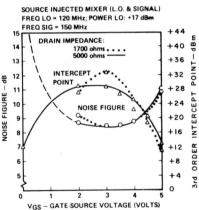

```
C1, C5 -  .01 µfd      C10  - 0.1 µF
C2, C4 -  1-10 pF      L1, L2 - 1.3 µhy
C3     -  1000 pF      Q1, Q2 - U310
C6, C8 -  30 pF        T1   - RELCOM BT-9
C7, C9 -  68 pF
```

6.6.19 Active balanced mixer employs Siliconix U310 JFETs. The chart shows a comparison of mixer intermodulation characteristics under specific input–output and frequency conditions.

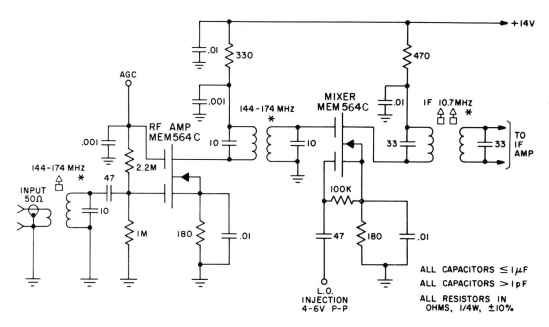

144-174 MHz Receiver Characteristics

Power Gain ≃ 28 dB

Noise Figure = 3.0 dB (typical) @ 174MHz

DC Power = 14V, 15mA

AGC = +4V Zero Sig.
   −4V Full Sig. (−50dB reduction)

Local Oscillator = 4-6V p-p into 3k ohms

Cross Modulation = 50,000 μV for 1% modulation of desired carrier by undesired signal. 30% modulation on undesired signal.

*Component values depend on frequency and required system characteristics, typical values shown.

6.6.20 General Instrument 144 to 174 MHz receiver front end, built with General Instrument dual-gate MEM 564Cs, offers improved cross modulation and AGC performance over that obtainable with bipolars. The power gain and noise figure are competitive with bipolar or Nuvistor circuitry.

The RF front end consists of a high-Q antenna circuit, double-tuned interstage coupling network, and mixer stage. The local-oscillator circuit is not shown, since injection is generally supplied externally.

SECTION 7
AUDIO

In many of the preceding sections audio circuits have appeared in a variety of forms—as portions of radio and TV subsystems, as modulators in transmitters, and as preamplifier units for microphones. This section contains a variety of circuits that can be similarly applied, but the general emphasis here is on the audio itself rather than on the application for it.

The circuits selected for this section are those I consider "choice" for one reason or another—because of an unusual degree of simplicity, for example, or economy. Some offer particularly appealing performance-versus-cost tradeoffs, and a few are compelling from the standpoint of sheer fidelity.

7.1 Preamplifiers, Followers, and Mixers

A true preamplifier accepts a low-level signal and delivers an amplified in-phase version of it at an impedance that matches that of the succeeding power amplifier. Since a conventional single-device amplifier stage results in phase inversion, it takes two such stages to deliver an in-phase signal replica of the input.

A follower, on the other hand, offers no gain and no phase inversion. The follower is valued for several reasons: It serves as an isolating buffer, for example, between audio processing stages whose characteristics might vary, depending on manufacturer and circuit class; and it serves as a virtually distortionless impedance transformer capable of driving a number of parallel stages without loading.

A distinction has been made between preamplifiers and followers—though both are often classed as preamps—to make it easier to assess the applicability of a low-level audio circuit for a given application.

A mixer, at least within the framework of this section, is an audio stage to which several inputs may be connected in parallel and controlled individually without mutual degradation and which has a single output that is the composite of the selected inputs. A mixer may be a follower, a true amplifier, or both.

A follower stage is easy to identify; the electron-emitting portion of the active device—the *cathode* of a tube, the *emitter* of a bipolar transistor, the *source* of a field-effect transistor—serves as the output electrode, which will contain a resistor but no bypass capacitor. (A capacitor is like a length of wire with ac signals, and placing one from the emitter to the ground would be the same as grounding the output. The resistor, without the bypass, keeps the emitter at some potential above the ground.)

The follower, by the way, gets its name from the no-phase-inversion characteristic: the output follows the input signal precisely; when the input voltage rises the output rises, and when it falls the output falls.

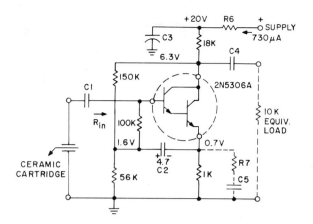

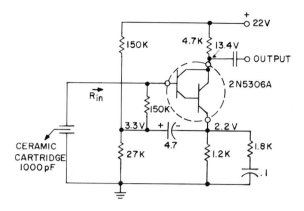

7.1.1 GE monolithic Darlington preamplifier offers higher input impedance and lower noise than a single bipolar transistor when used with a source impedance above 100K. An input impedance of 500K to 2M can be achieved with a single stage for improved low-frequency response.

The collector bias current should be between 0.4 and 2 mA for a low-noise circuit design. The noise voltage (e_n) cannot be appreciably reduced by going beyond 2 mA and the $e_n i_n$ product for 100 Hz increases below 0.4 mA.

The GE circuit shown is designed for a collector current of 0.7 mA. The base bias network is bootstrapped so that the input resistance (R_{in}) is about 500K. The bias network has dc feedback from the collector that will improve the bias stabilization for wide variations in supply voltage.

The circuit has a voltage gain of 6 when driving a 10K load. Thus 100 mV input gives a 600 mV output at 0.16% total harmonic distortion (THD) at 1 kHz. The THD is approximately 0.5% at 10 dB below the maximum signal output level of 3V. The maximum noise output is 84 dB below the 3V level. When this circuit is used with a ceramic phonograph cartridge, the value of C1 can be selected for the desired low-frequency response; also C5 and R7 can be chosen for equalization, since the low-level direct current will not harm the ceramic cartridge. Also the desired low-frequency response can usually be obtained by decreasing the value of C2 in place of using C1 for this purpose. A high-output cartridge (greater than 200 mV) can be loaded with a shunt capacitor to fix the reference output level at least 10 dB below the 3V clipping level.

7.1.3 GE Darlington preamplifier has an input resistance of 1.5M. A 100 mV input signal gives 490 mV output at approximately 0.1% total harmonic distortion (THD). The THD is about 0.3% at 10 dB below the maximum 6V signal output level. The maximum noise output is more than 88 dB below the maximum signal output level.

The amplifier will drive a load of greater than 50K. An emitter follower can be used to transform a lower impedance. This Darlington stage is useful with high-impedance transducers and for amplification following high-impedance filter networks.

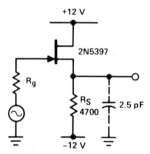

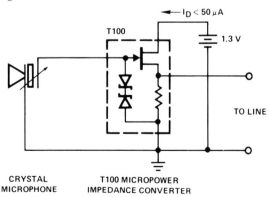

7.1.2 Micropower microphone preamplifier as a hearing-aid input element. The Siliconix device (T100) is an N-channel depletion-mode junction FET with diffused source resistor and a pair of back-to-back Schottky diodes to form the high-impedance gate bias resistance.

7.1.4 A Siliconix 2N5397 is an ideal FET source follower because of its low input capacitance and high g_{fs}, which remains high at the frequency range of interest. A source follower exhibits a high input impedance and low output impedance. The real part of the output impedance is the reciprocal of g_{fs}, which is independent of frequency up to about 600 MHz. The input capacitance is $C_{gd} + C_{gs}(1 - A_v)$ which, in this case, is approximately 1.5 pF maximum. The input capacitance is also independent of frequency and independent of the load when the load is larger than the output resistance.

The frequency response is dependent mainly on the generator internal impedance. For example, when R_g is increased to 1K the bandwidth falls to 80 MHz. In this circuit, the low-frequency voltage gain is 0.94.

The input resistance is proportional to $1/f^2$ and at some high frequency will go negative if the source resistor is large. For example, the input resistance is high at 10 MHz, but in the negative resistance region at 100 MHz. However, when R_s is 1000 ohms, the input resistance is real at this frequency.

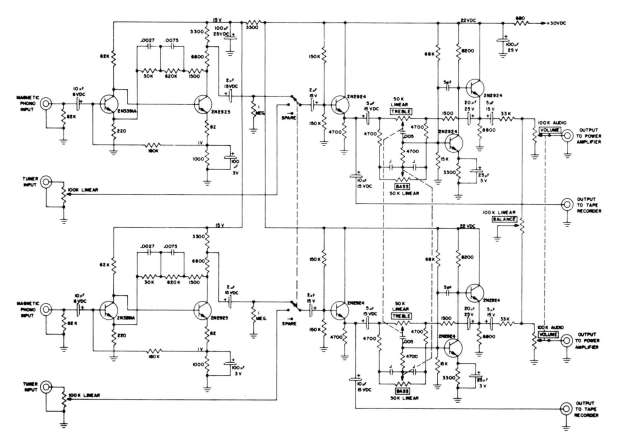

7.1.5 RCA magnetic phono preamplifier and tone control system. To minimize hum, insulate the phono input jacks from the chassis. Use shielded cables and ground the shields to the input jacks and to a ground point near the preamplifier circuit.

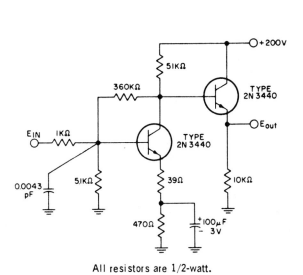

All resistors are 1/2-watt.

7.1.6 RCA final stage of an operational amplifier. An operational amplifier normally utilizes a chopper amplifier and other stabilizing circuits. This portion of the amplifier can be designed to operate at low supply voltages. The final stage, however, requires a high supply voltage because it must provide a large voltage swing to drive the high input impedance of the next operational amplifier.

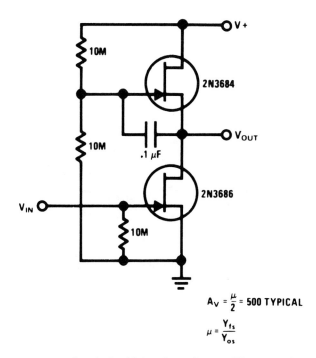

$$A_V = \frac{\mu}{2} = 500 \text{ TYPICAL}$$

$$\mu = \frac{Y_{fs}}{Y_{os}}$$

7.1.7 National ultrahigh-gain audio amplifier provides a very low-power, high-gain amplifying function. Since the μ of a JFET increases as the drain current decreases, the lower drain the current, the more gain you get. You do sacrifice input dynamic range with increasing gain, however.

SECTION 7 Audio 315

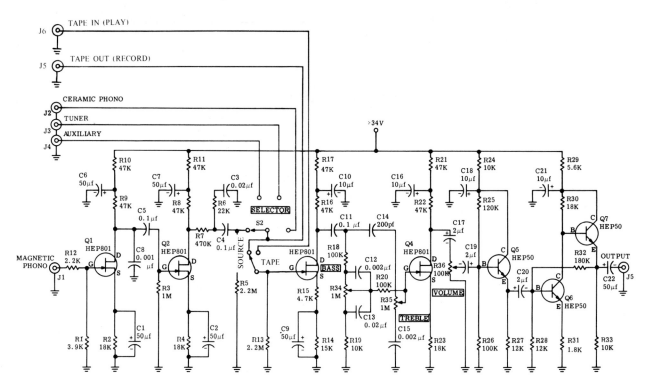

7.1.8 Motorola audio hi-fi preamplifier designed for low-impedance termination so that long cable runs to the power amplifier can be made without excessive noise pickup or high-frequency attenuation. Tape in/out jacks allow taping of any source while you listen to any program applied to J6.

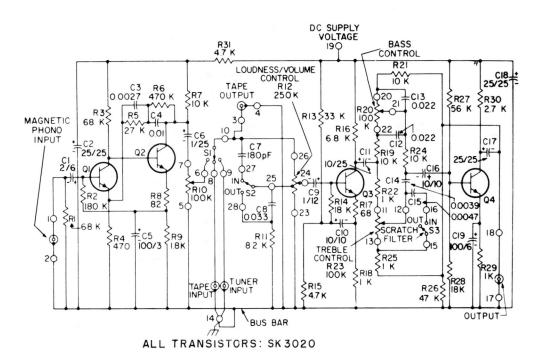

7.1.9 RCA phonograph preamplifier for use with a 5 mV magnetic pickup has provisions for both tape and tuner input. At 5 mV input signal level the preamplifier delivers an output of at least 1V.

316   DISCRETE / TRANSISTOR CIRCUIT SOURCEMASTER

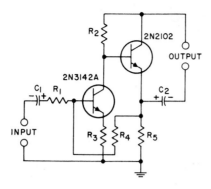

Parts List

C₁ = 10 microfarads, 6 volts, electrolytic
C₂ = 50 microfarads, 25 volts, electrolytic
R₁ = 1000 ohms, 0.5 watt
R₂ = 1200 ohms, 0.5 watt
R₃ = See chart, 0.5 watt
R₄ = See chart, 0.5 watt
R₅ = 270 ohms, 0.5 watt

Resistance Data for Different Voltage Gains and Input Impedances*

Voltage Gain	Input Impedance (ohms)	R₃ (ohms)	R₄ (kilohms)
166	2700	0	680
22	7300	39	470
17	9000	68	430
10	15000	100	390
3	55000	390	360
1	100000	1200	330

* Data obtained for an output of 1 volt rms into a 250-ohm line.

7.1.10 RCA simple two-stage amplifier is useful as a line driver for audio systems in which the power amplifier is located at a considerable distance from the signal source, as a driver for the line inputs of tape recorders, as an output stage for inexpensive radio receivers, etc. The amplifier has a frequency response that is flat from 20 to 20,000 Hz and can be used to drive any line that has an impedance of 250 ohms or greater. It operates from +12V and can supply a maximum undistorted output of 3V into a 250-ohm line.

The voltage gain and input impedance of the amplifier are determined by the values chosen for the emitter resistor (R3) and feedback resistor (R4) for the input stage. The chart shows values of these resistors for various voltage gains from unity to 166 and for input impedances from 2700 to 55,000 ohms.

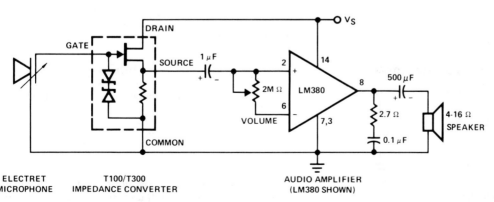

7.1.11 Microphone amplifier circuit built from a Siliconix JFET impedance converter (T300), which drives a National IC amplifier.

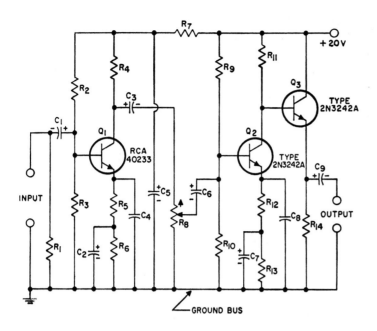

Parts List

C_1, C_6 = 15 μF, electrolytic, 6 V
C_2, C_7 = 300 μF, electrolytic, 6 V
C_3 = 10 μF, 15 V
C_4, C_8 = 0.05 μF, paper
C_5 = 250 μF, electrolytic, 25 V
C_9 = 50 μF, electrolytic, 15 V

R_1 = value required to match microphone line impedance (up to 10000 ohms), 10%, 0.5 watt
R_2, R_9 = 0.1 megohm, 10%, 0.5 watt
R_3, R_{10} = 6200 ohms, 10%, 0.5 watt
R_4, R_{11} = 10000 ohms, 10%, 0.5 watt

R_5, R_{12} = 68 ohms, 10%, 0.5 watt
R_6, R_{13} = 470 ohms, 10%, 0.5 watt
R_7 = 820 ohms, 10%, 0.5 watt
R_8 = potentiometer, 10000 ohms, 0.5 watt, audio taper
R_{14} = 1000 ohms, 0.5 watt

7.1.12 RCA three-stage superdynamic-range preamplifier is designed for use with high-level microphones. It has an overall voltage gain of 1500 to 2000 and can provide a maximum undistorted output voltage of 5V to a load impedance of 500 ohms or greater for a maximum undistorted input of 0.4V. The frequency response of the preamplifier is flat from 20 Hz to 30 kHz. The dc power requirements of the circuit are 20V at 30 mA.

The preamplifier uses a low-noise 40233 in a class A input stage and two 2N3242A transistors in direct-coupled class A driver and emitter-follower output stages. The circuit operates equally well with either low-impedance or high-impedance microphones, provided that the value of the input resistor R1 is selected to match the microphone line impedance (up to a maximum of 10K).

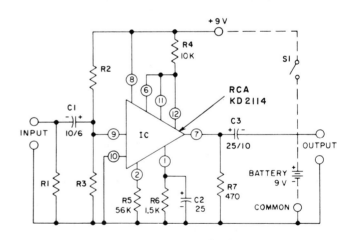

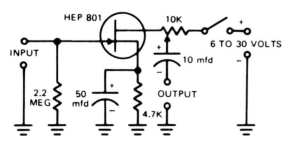

7.1.13 Motorola preamplifier for ceramic or crystal microphone, or phono cartridge, features excellent frequency response and operation from a wide range of supply voltages.

| | Microphone Type | |
Resistors	Low Impedance	High Impedance
R1	270 ohms	Not used
R2	220K	1000K
R3	56K	270K

7.1.14 RCA high-gain, low-noise high-fidelity preamplifier accommodates both low- and high-impedance dynamic microphones. It can be used with tape recorders and audio systems or with radio transmitters. The maximum output of the circuit is 1.4V.

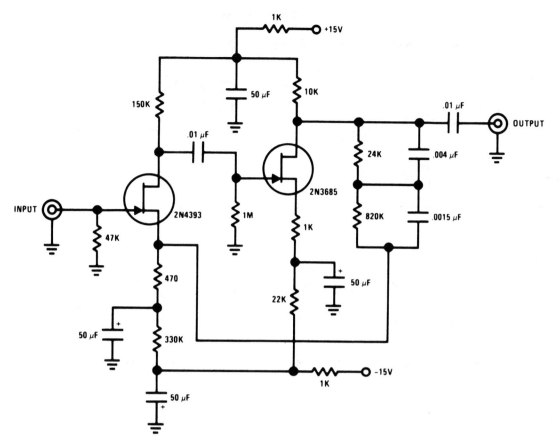

7.1.15 National magnetic-pickup phono preamplifier provides proper loading to a variable-reluctance phono cartridge. It provides approximately 35 dB of gain at 1 kHz (2.2 mV input for 100 mV output), features $S + N/N$ ratio of better than -70 dB (referenced to 10 mV input at 1 kHz), and has a dynamic range of 84 dB (referenced to 1 kHz). The feedback provides RIAA equalization.

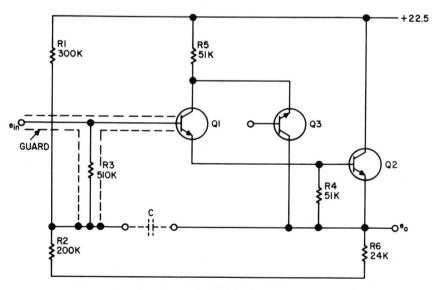

7.1.16 GE high-input-impedance unity-gain "follower" amplifier. It is basically a Darlington amplifier with collector and base-bias bootstrapping. Q1 and Q2 can be either 2N3391 or 2N3900 transistors; Q3 is a GE 2N2926.

SECTION 7 Audio **319**

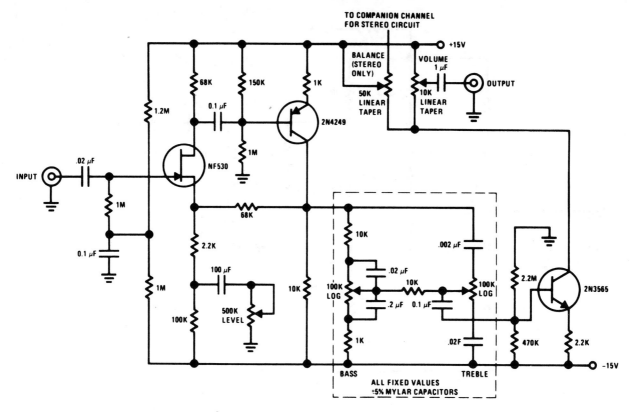

7.1.17 National low-cost high-level preamp and tone control circuit uses the JFET to its best advantage as a low-noise, high-input-impedance device. All device parameters are noncritical, yet the circuit achieves harmonic distortion levels of less than 0.05% with a signal-to-noise ratio of over 85 dB. The tone controls allow 18 dB of cut and boost; the amplifier has a 1V output for 100 mV input at maximum level.

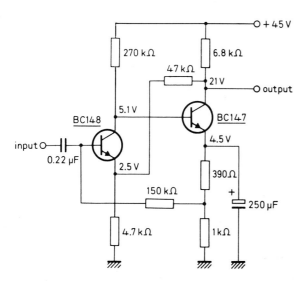

7.1.18 Amperex low-distortion, high output voltage amplifier has a voltage gain of 20 dB and a maximum output voltage of 10V. To achieve this high output voltage with low distortion, the supply voltage is set at 45V. The frequency response (−3 dB points) is from less than 20 Hz to greater than 20,000 Hz, and the input and output impedances are 140K and 200 ohms.

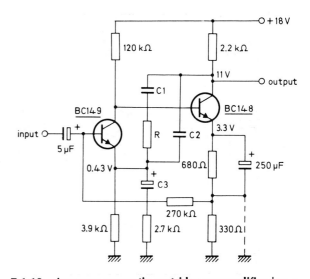

7.1.19 Amperex magnetic-cartridge preamplifier has a high input impedance and permits a magnetic pickup of any inductance to be used without change in the upper audio response. It is equalized for RIAA. R = 47K, C1 = 6.8 nF, C2 = 2.2 nF, and C3 = 5 μF.

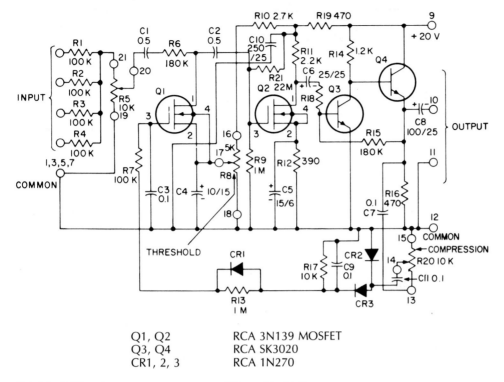

Q1, Q2	RCA 3N139 MOSFET
Q3, Q4	RCA SK3020
CR1, 2, 3	RCA 1N270

7.1.20 RCA mixer, compressor, amplifier with hi-fi characteristics. The circuit consists of a four-channel resistive mixer, a MOS transistor amplifier, and a two-stage bipolar line driver.

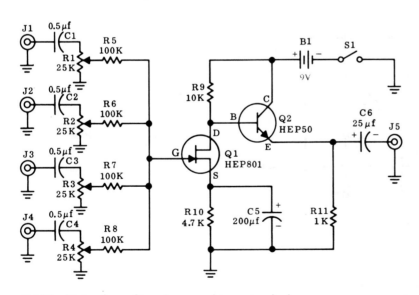

7.1.21 Motorola audio mixer combines signals from four separate high-impedance devices. The frequency response of the mixer is 20 Hz to 100 kHz (±2 dB). The voltage gain is 5 to 10 dB, depending upon the individual FET characteristics; 1V is the maximum output level to be expected at J5.

SECTION 7 Audio 321

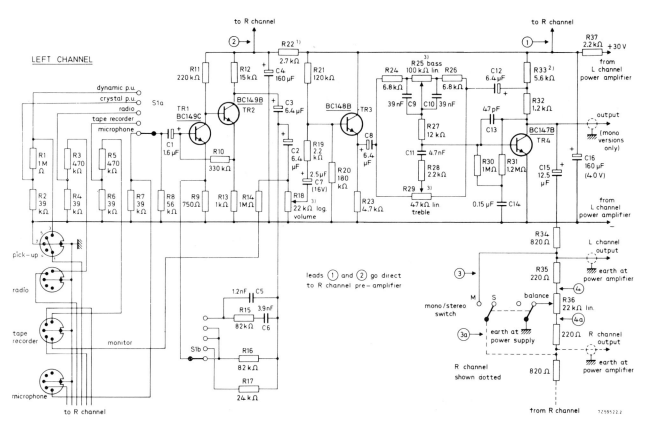

7.1.22 Amperex universal preamplifier circuit. R22 is omitted in the R channel together with C4, R37, and C16.

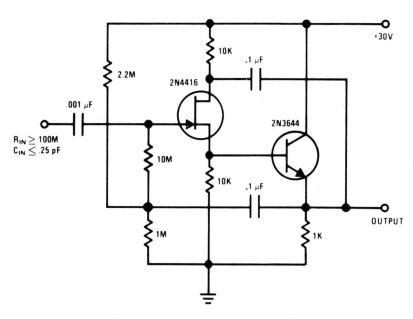

7.1.23 RCA superhigh-impedance ac unity-gain amplifier. Nothing is left to chance in reducing the input capacitance. The 2N4416, which has a low capacitance in the first place, is operated as a source follower with a bootstrapped gate bias resistor and drain. Any input capacitance you get with this circuit is due to poor layout techniques.

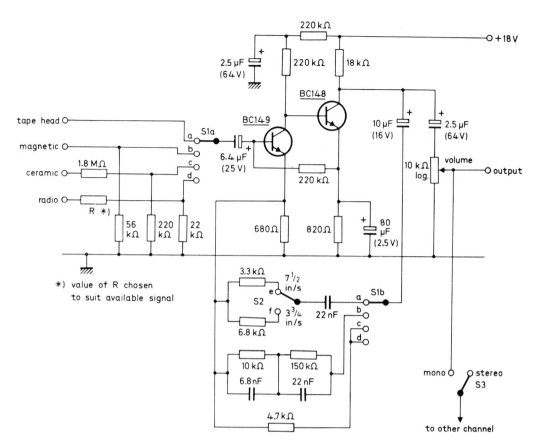

7.1.24 Amperex equalizing preamplifier for magnetic and ceramic pickup cartridges, radio, and tape playback heads.

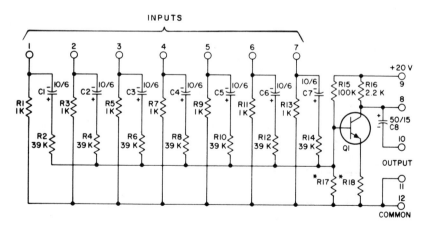

7.1.25 GE multi-input unity-gain mixer combines inputs from up to seven sources, for input to an amplifier, recorder, or other audio equipment. If more than seven inputs are required, as many as three mixers can be wired in parallel.

The amplifier portion of the circuit, shown at the right in the schematic diagram, is current-stabilized by the emitter resistor. This resistor is not bypassed, thereby providing a greater degree of degeneration and a reduction in the overall gain of the mixer to unity. This technique results in cancellation of stage distortion.

Number of Inputs	R17 (ohms)	R18 (ohms)
2	8.2K	120
3	7.5K	110
4	6.8K	91
5	6.8K	82
6	6.2K	75
7	6.2K	68

7.2 Amplifiers

The amplifier circuits in this subsection are arranged more or less according to output power. The first few are low-level types intended for general-purpose applications such as crystal-cartridge phono amplification or as headphone drivers. Those circuits that are power-rated are high-fidelity types suitable for use in quality stereo systems.

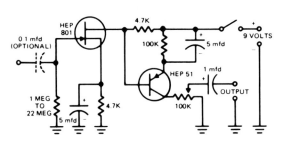

PARTS LIST:
1 HEP 801
1 HEP 51
1 Resistor, 1 meg to 22 meg, 1/2 Watt, ± 10%
1 Resistor, 100K, 1/2 Watt, ± 10%
2 Resistors, 4.7K, 1/2 Watt, ± 10%
1 Potentiometer, 100K, Audio Taper
2 Electrolytic Capacitors, 5 mfd, 15V
1 Electrolytic Capacitor, 1 mfd, 15V
1 SPST Switch
1 9 Volt Battery
1 Capacitor, 0.1 mfd, 25V (optional)

7.2.1 Motorola amplifier for low-level audio signals features high-impedance input, low current drain (200 to 400 μA), 10 Hz to 30 kHz frequency range, low-impedance output, gain of 200 to 400.

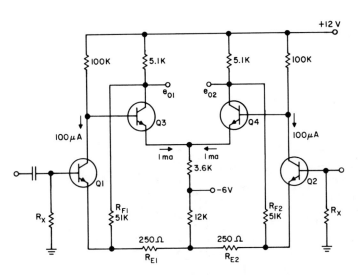

Q1 = Q2 = MATCHED 2N3390 OR 2N3391 OR 2N3900 OR 2N3901
Q3 = Q4 = 2N3392

7.2.2 GE building-block amplifier uses differential input configuration. Output e_{o1} is out of phase with the input, and e_{o2} is in phase with it.

The input resistance at the transistor base is greater than 0.5M because of the feedback to the emitters of the input transistors. If R_x is 100K, then the input impedance is greater than 80K. For a single-ended output, the output impedance is less than 2K. With a high input impedance and low output impedance, the amplifier becomes a stable and well-defined voltage amplifier that can be used as a building block.

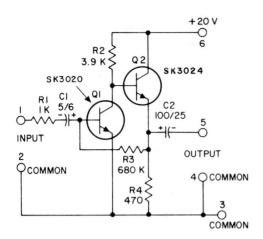

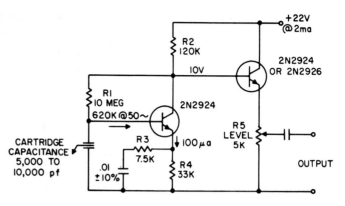

7.2.3 RCA headphone amplifier is very useful in audio work when the power amplifier is located at some distance from the headphones. It has a voltage gain of 100 and is capable of driving any line impedance of 250 ohms or greater. It has a maximum undistorted output of 3V rms into a 500-ohm line and has a flat frequency response from 20 to more than 25,000 Hz; the input impedance is 1800 ohms.

7.2.5 GE phono amplifier delivers good performance without sacrificing circuit simplicity. The GE 2N2924 transistor is operated at 100 μA with a simple biasing arrangement. This amplifier is designed specifically for high-output cartridges; for low-output cartridges or where passive filters are used between cartridge and transistor, a higher-gain device should be employed to minimize noise.

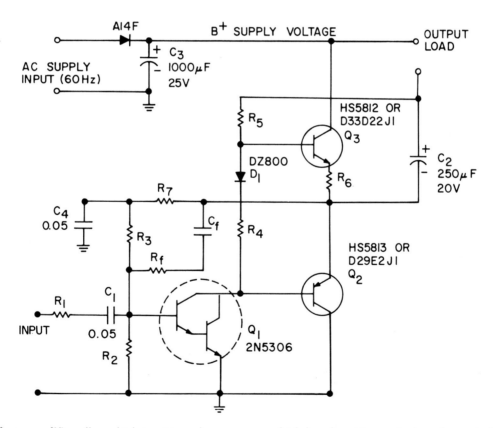

7.2.4 GE phono amplifier offers a high input impedance to obtain a good low frequency response and a bass boost network. The ac gain is controlled by the ratio of Z_f (the combined impedance of R_f and C_f) to R1. The separation of ac and dc feedback gives the circuit the advantage of a bass boost, which is often desired. The frequency at which bass boost is acquired can be controlled by properly choosing the RC time constant of R_f and C_f. The frequency-sensitive feedback networks permit about 1 dB of bass boost at 150 Hz when used with a transducer capacitance of approximately 1000 pF, and the boost is about 6 dB with a fixed voltage signal source.

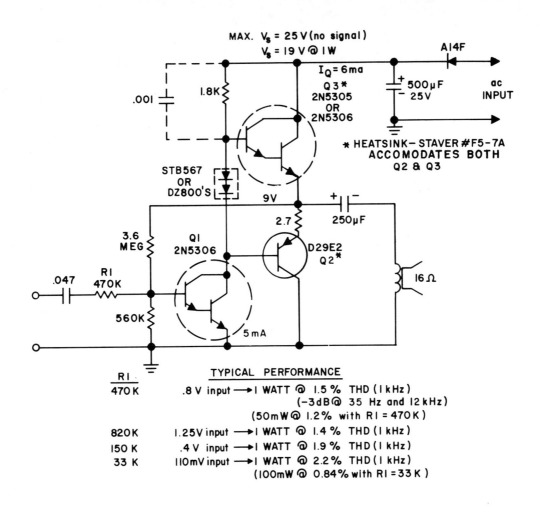

P_o and Load	1W, 8Ω	1W, 16Ω	2W, 16Ω
B+	12 V	19 V	22 V
R_1			
.5V Sensitivity	270K	330K	220K
1.0V	510K	680K	470K
1.5V	750K	1M	680K
R_2	330K	560K	330K
R_3	820K	1.8M	1.8M
R_4	27Ω	82Ω	39Ω
R_5	560Ω	1.8K	1K
R_6	.68Ω	2.7Ω	2.7Ω
R_7	470K	1.8M	1M
R_F	1.5M	2.7M	2.7M
C_F	270pF	150 pF	150 pF
Typical Distortion at rated P_o	2.0%	2.0%	2.0%
Frequency Response ±3 dB with transducer capacitance- 800 to 1000 pF	130 Hz–15 kHz	90 Hz–10 kHz	90 Hz–10 kHz

7.2.6 GE 1W monolithic Darlington amplifier eliminates the need to bootstrap the load resistor (1.8K) of Q1. This permits the amplifier load to be returned to ground without a minimum of parts.

The bias stabilization resistor (2.7 ohms) is used in the emitter of the PNP output, since there is less signal impedance from the load back to the collector of Q1 in this path compared to the path via the output Darlington.

The 0.001 µF capacitor shunting Q1 load resistor may not be required for high-frequency stabilization, depending on the layout and application.

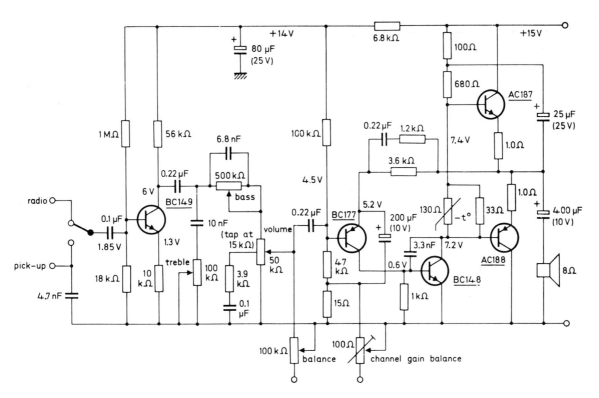

7.2.7 Amperex 2W stereo pickup/radio amplifier circuit. Most types of ceramic pickup and radio input levels can be accommodated. Tone controls are of the "cut" type, but the bass boost derived from feedback gives the effect that both the bass boost and the cut are available.

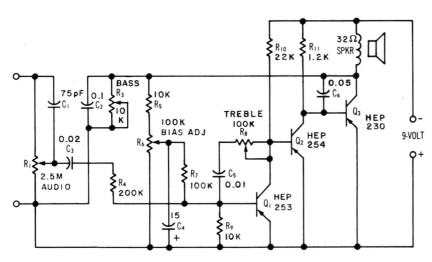

7.2.8 Motorola "mini-fi" transistor amplifier delivers 500 mW of clean audio from 40 Hz to 17 kHz. It requires 1.5V of drive. For stereo, use two and employ dual controls for volume, bass, and treble. Use four 8-ohm speakers in series for the load, which must be 32 ohms for full circuit efficiency.

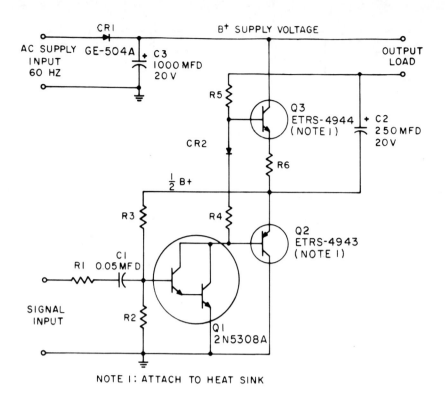

NOTE 1: ATTACH TO HEAT SINK

Parts List

Components	One Watt Output 8-ohms Load	One Watt Output 16-ohms Load	Two Watts Output 16-ohms Load
Battery (B+) Voltage and current OR	12 Vdc @ 160 MA	19 Vdc @ 110 MA	22 Vdc @ 160 MA
AC Supply Input (approximate)	9 — 12.6 Vac	14 — 18 Vac	16 — 20 Vac
C1	0.05 Mfd	0.05 Mfd	0.05 Mfd
C2	250 Mfd, 20V	250 Mfd, 20V	250 Mfd, 20V
C3	1000 Mfd, 20V	1000 Mfd, 20V	1000 Mfd, 20V
CR1	GE-504A	GE-504A	GE-504A
CR2	ETRS-4946*	ETRS-4946*	ETRS-4946*
Q1	2N5308A	2N5308A	2N5308A
Q2	ETRS-4943*	ETRS-4943*	ETRS-4943*
Q3	ETRS-4944*	ETRS-4944*	ETRS-4944*
R1 (sensitivity 0.5V)	220K ohms	470K ohms	270K ohms
R1 (sensitivity 1.0V)	470K ohms	820K ohms	510K ohms
R1 (sensitivity 1.5V)	680K ohms	1.2 megohms	750K ohms
R2	390K ohms	560K ohms	330K ohms
R3	1.5 megohms	3.6 megohms	2.7 megohms
R4	22 ohms	82 ohms	22 ohms
R5	560 ohms	1.8K ohms	1000 ohms
R6	0.68 ohm	2.7 ohms	2.7 ohms

* Available from General Electric Co., Dept B, 3800 North Milwaukee Avenue, Chicago, Ill. 60641

7.2.9 GE audio amplifier delivers outputs of 1 to 2W, depending on +V, the devices used, and the load.

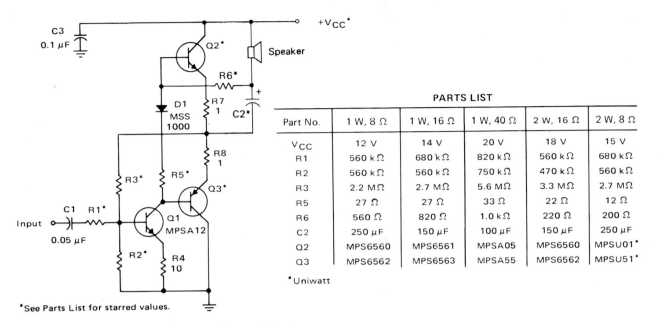

7.2.10 Motorola three-transistor audio amplifier for portable and toy phonographs that use ceramic cartridges of 2 to 3V output. The circuit provides a 2W output with suitable component and power supply values.

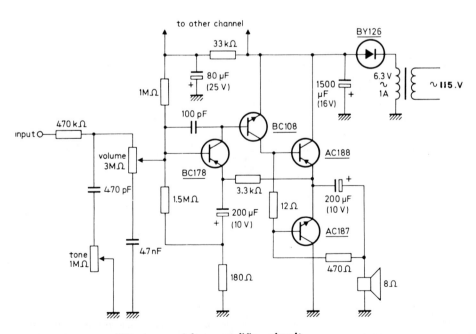

7.2.11 Amperex 1W stereo pickup amplifier circuit. The nominal supply voltage to transistors must not exceed 9V. Note that the supply voltage is negative.

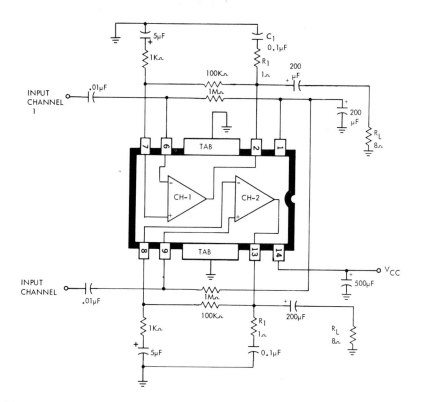

7.2.12 Simple IC hi-fi stereo amplifier employs a Sprague ULN-2277P device. When an unregulated supply voltage is used, the actual voltage present at pin 14 during full-signal conditions should not drop below the nominal supply voltage level if full power output is to be maintained. The collector voltage V_{cc} should be between 10 and 20V; at 16.8V, full-signal output power is not less than 2W. The closed-loop gain should be limited to 30 to 60 dB to maintain stable circuit operation.

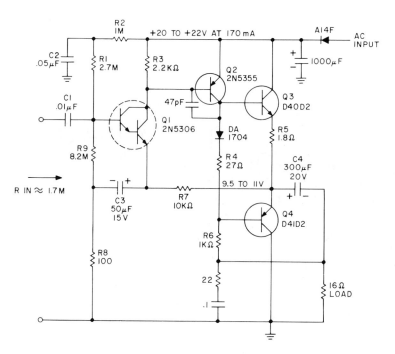

7.2.13 GE 2W audio amplifier is line-operated and has greater than 1.5 megohms input impedance, necessary for a good low-frequency response from a high-impedance transducer such as a crystal microphone or phonograph cartridge. The closed-loop gain is determined by the ratio R7/R8, and is 34 dB for R8 = 100 ohms. The total harmonic distortion is typically 1%. The optimum load is a 16-ohm speaker.

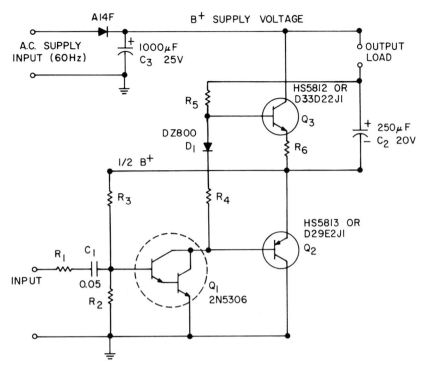

P_o & Load	1W, 8Ω	1W, 16Ω	2W, 16Ω
Supply B+	12V	19V	22V
Current, full load	160 mA	110mA	160 mA
Current, quiescent	11 mA	7.5 mA	11mA
R_1			
.5V	220K	470K	270K
1.0V	470K	820K	510K
1.5V Sensitivity	680K	1.2M	750K
R_3	1.5M	3.6M	2.7M
R_2	390K	560K	330K
R_4	22Ω	82Ω	22Ω
R_5	560Ω	1.8K	1000Ω
R_6	.68Ω	2.7Ω	2.7Ω
Typical Distortion			
at rated P_o (0.5V sensitivity)	2.0%	<2.0%	2.0%
at 50 mW (0.5V sensitivity)	2.4%	1.8%	1.3%
Supply input ac	9–12.6V	14–18V	16–20V
Open-loop voltage gain A_v	42dB	51 dB	54 dB

7.2.14 GE low-cost 1 or 2W audio power amplifier uses a direct-coupled complementary-symmetry output circuit that is capable of driving 8- or 16-ohm loads. The first consideration in biasing is the direct current through R5 in the driver stage. This must be large enough to supply the required peak base current of the NPN output transistor. This is determined by dividing the peak output current by the h_{FE} (measured by peak output current) of the output transistor.

The quiescent current in the complementary output stage is established by the voltage drop across diode D1 and resistor R3. The amount of idling current controls the amount of crossover distortion, but the stability of the amplifier is improved by using the minimum idling current that will meet the distortion requirements at low output level. Resistor R6 also is used to help stabilize the idling current in the output stage. Diode D1 offers a low dynamic impedance compared to an equivalent resistor giving the same dc voltage drop. This low ac impedance is desirable for equal signal drives at the two bases of the complementary outputs. The diode also helps compensate the decreasing V_{BE} of the output transistors with increasing ambient temperature.

The center point should be biased at, or slightly below, half the supply voltage. This will allow the maximum output voltage swing before clipping (beginning of high distortion). The center-point voltage is controlled by the ratio of resistor R3 to R2 and the two base–emitter drops of the Darlington driver stage. The value of R2 and R3 are chosen such that the bleeder current is large (greater than ×5) compared to the required base current in the driver. Then the ratio (R2 + R3)/R2 multiplied by the two base–emitter drops provides the center-point voltage. Resistor R3 provides both ac and dc feedback from the output to the input.

The amplifier voltage gain can be controlled by the ratio of R3 to R1, the upper limit being that available with no ac feedback current via R3 to the base of Q1. This open-loop voltage gain is approximately 46 dB with R5 = 1.8K.

The output load is returned to the power supply, which is an ac ground, so that the load coupling capacitor (C2) can also serve to bootstrap resistor R5. This eliminates the need for a separate capacitor to apply positive feedback to R5. The positive feedback increases the effective resistance of R5 and the open-loop gain. The half-power frequency response is 50 Hz to 20 kHz. GE's application note 90.89 contains board art and component layout pattern.

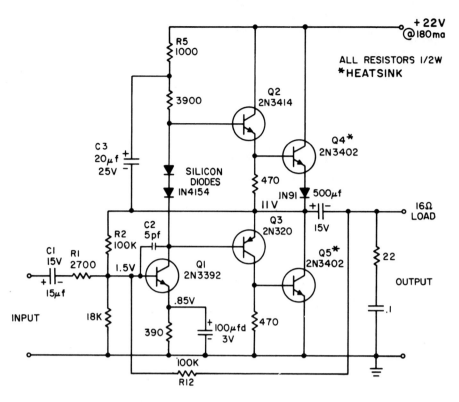

7.2.15 GE 2W amplifier for driving 16-ohm loads has harmonic distortion of less than 1%. Sensitivity is adequate to assure full volume with ceramic phono-cartridge input.

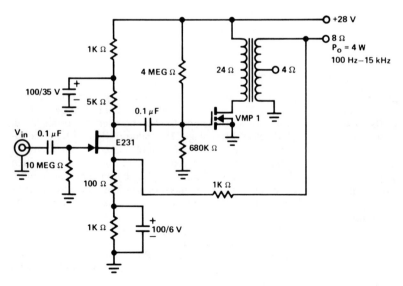

7.2.16 Siliconix simple power-FET audio amplifier is equivalent to the audio output stage of many inexpensive radios, televisions, and phonographs. Power output is about 4W from 100 Hz to 15 kHz. The design is greatly simplified by the use of an output transformer, and distortion is kept relatively low (2% at 3W) by 10 dB of negative feedback. No thermal stabilization components are needed, since the negative temperature coefficient of the Siliconix VMP1 drain current makes thermal runaway impossible.

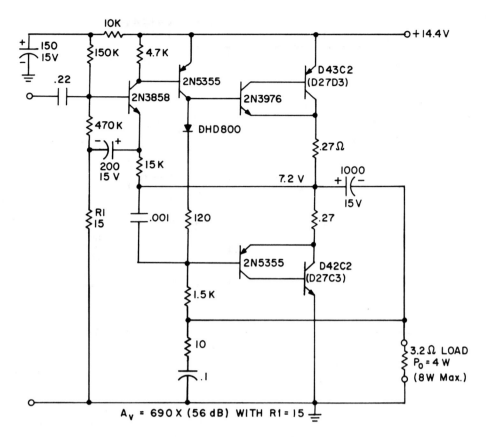

$A_v = 690 \times (56\ dB)$ WITH R1 = 15

7.2.17 GE 4W high-gain amplifier has a complementary push-pull drive stage added to a basic circuit configuration. This permits the class A drive stage to be operated at lower current and power dissipation, resulting in a higher voltage gain and lower cost transistors in the class A drive and output stages of a linear 4W amplifier. The additional sensitivity permits lower distortion where high gain is required, like an AM radio receiver.

The amplifier has a quiescent current of 30 mA and operates with an efficiency of 56% at 4W output, which is the onset of clipping. The amplifier can be driven to a maximum of 8W of clipped sine-wave output. An input signal of 5.4 mV gives 4W output at 1.8% THD with R1 = 15 ohms. There are 14 dB of negative feedback (with R1 = 15 ohms) and the power frequency response is −3 dB at 66 Hz and 75 kHz.

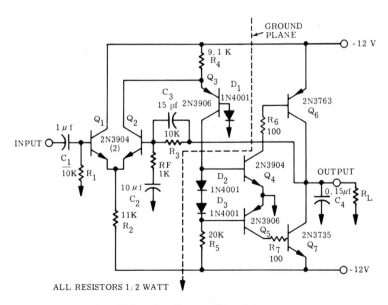

7.2.18 Motorola wideband audio amplifier delivers 4W into 8 ohms and has a frequency response of 35 to 100,000 Hz with less than 1% THD.

SECTION 7 *Audio* 333

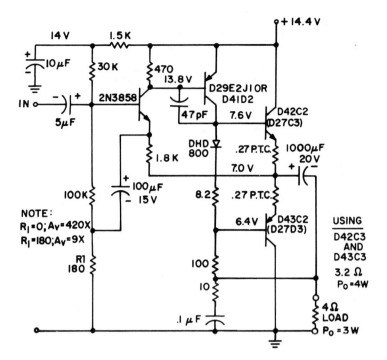

7.2.19 4W hi-fi amplifier utilizes complementary GE output transistors (D42C2, D43C2). Positive-temperature-coefficient resistors are used in series with the output emitters to protect the output transistor against overdrive and shorted load conditions. By using a PNP driver, positive feedback can be applied to the 100-ohm driver load resistor via the load coupling capacitor and still have the load returned to the ground.

The sensitivity and negative feedback can be changed with the value of R1. With R1 of 180 ohms the input impedance at 1 kHz is 25K; the THD at 3 watts output is 0.6% with 390 mV input signal. The power response is down 3 dB at 26 Hz and 100 kHz. With R1 = 0, the input impedance at 1 kHz is 3.3K; the THD at 3W output is 3.5% with 5.3 mV input signal. The power response is now down 3 dB at 60 kHz. The quiescent supply current drain is 75 to 80 mA.

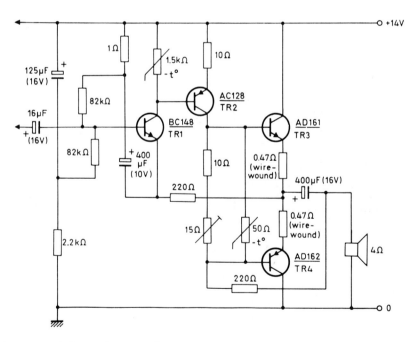

7.2.20 Amperex direct-coupled 4W class B audio amplifier for use in a car radio. The amplifier uses an AD161/AD162 complementary output pair, an AC128 driver transistor, and a BC148 or BC108 transistor as a first-stage amplifier (Amperex numbers). The amplifier is a conventional four-transistor circuit with one exception: the decoupling capacitor in the input stage is returned to the emitter of the BC148 instead of to the chassis. This neutralizes the effect of ripple at the supply line and increases the input impedance.

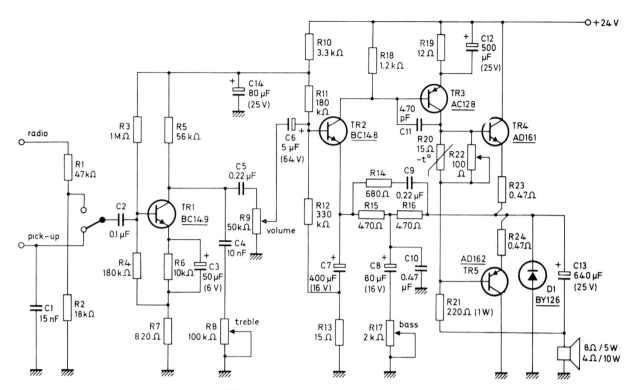

7.2.21 Amperex low-cost 5 to 10W amplifier.

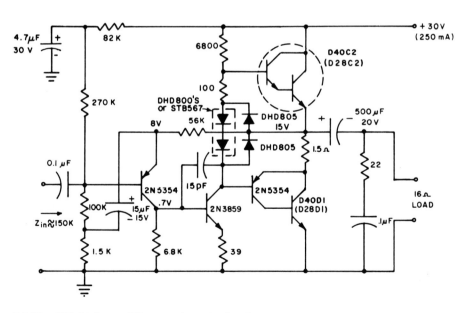

7.2.22 5W hi-fi amplifier employs diodes for current limiting. A very high-beta GE Darlington (D40C2) used in the output eliminates the need for bootstrapping. This circuit delivers 5W at 1.5% THD at 1 kHz with a 280 mV input signal. Lower distortion can be achieved with a 32V power supply. The THD at 1 kHz is less than 1% from 50 Hz to 15 kHz for output levels of 0.05 to 4W. The 2W power response is down 3 dB at 20 Hz and 170 kHz.

SECTION 7 Audio

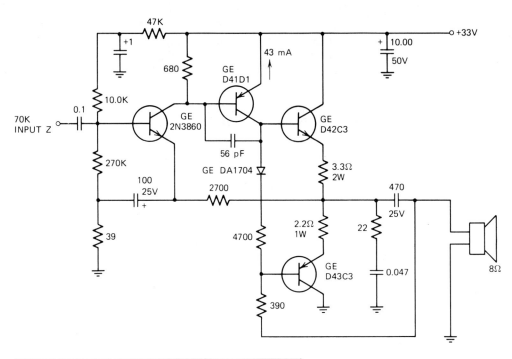

R1	Input	Output	Typical THD at 1 kHz
39 ohms	110 mV	5W	1.25%
82 ohms	210 mV	5W	0.92%

7.2.23 GE hi-fi amplifier delivers 5W into an 8-ohm load. The quiescent current of the amplifier is 45 to 50 mA and the full-load current is 400 mA. The duration that a load short can be sustained depends on the power supply regulation and the thermal resistance of the heat exchangers used for the GE output transistors. The amplifier sensitivity may be adjusted with the value of R1.

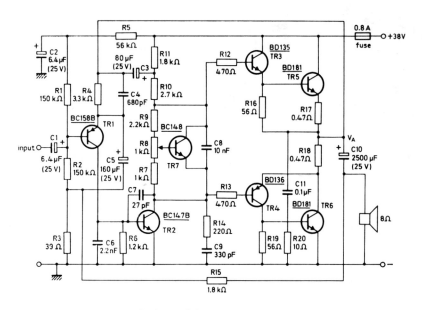

7.2.24 Amperex 15W hi-fi amplifier.

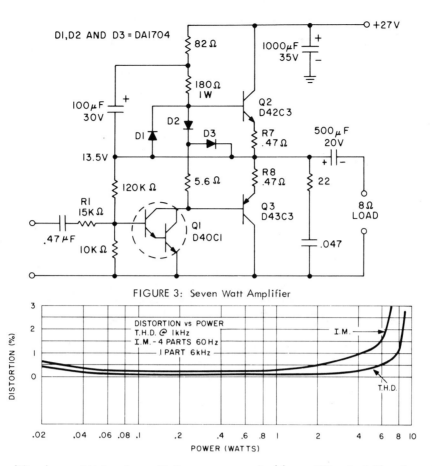

FIGURE 3: Seven Watt Amplifier

7.2.25 GE 7W amplifier has a 1V input sensitivity. Diode D3 in conjunction with D2 provides current limiting for Q2 by shunting the drive signal. D3 is nonconductive unless the current in R7 exceeds the peak value required for an 8W output. Thus the values of R7 and R8 determine the peak current level for limiting in Q2 and Q3. Diode D1 in conjunction with D2 provides current limiting for Q3 by shunting the drive signal.

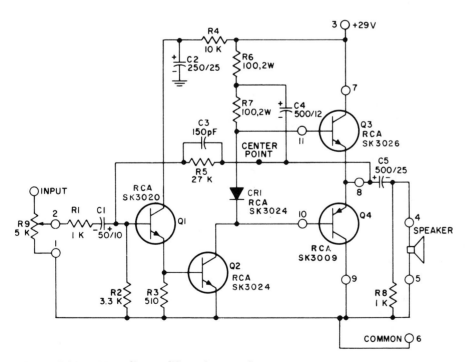

7.2.26 RCA 7.5W audio amplifier. The complementary-symmetry output circuit allows the speaker to be driven directly without the need of a transformer.

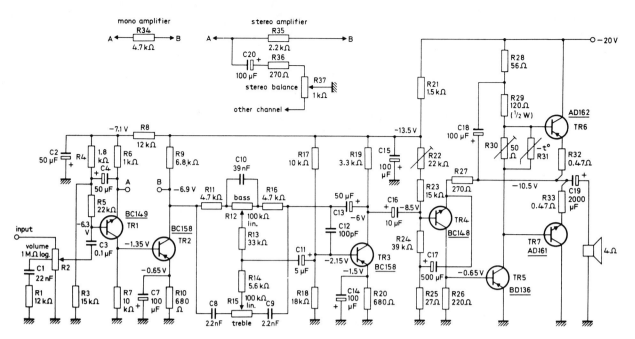

7.2.27 Amperex hi-fi eight-watter (use two for stereo).

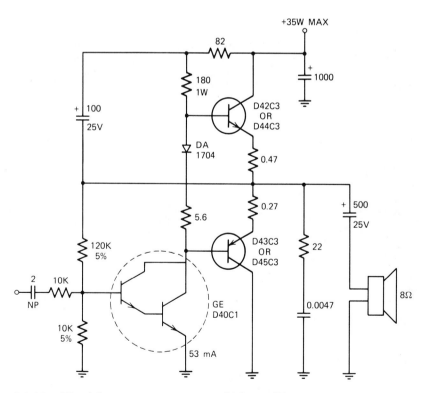

7.2.28 GE minimum-component 10W hi-fi amplifier can be driven to full output with 1V input signal swing. If the D44C3/D45C3 are used, smaller heat exchangers can be used to maintain the junction temperature within rating at both the maximum supply voltage and the operating ambient temperature. A small heat exchanger on the D40C driver will reduce the time for the bias to stabilize after applying dc power.

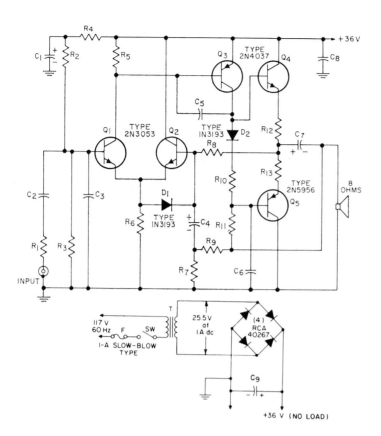

$C_1 = 10$ μF, electrolytic, 50 V
$C_2, C_4 = 5$ μF, electrolytic, 25 V
$C_3 = 150$ pF, ceramic
$C_5 = 75$ pF, ceramic
$C_6 = 470$ pF, ceramic
$C_7 = 1000$ μF, electrolytic, 25 V
$C_8 = 0.01$ μF, ceramic
$C_9 = 1000$ μF, electrolytic, 50 V

$F_1 =$ Fuse, 1 ampere, slow-blow type
$R_1 = 1800$ ohms, 0.5 watt
$R_2 = 39000$ ohms, 0.5 watt
$R_3 = 47000$ ohms, 0.5 watt
$R_4 = 10000$ ohms, 0.5 watt
$R_5 = 330$ ohms, 0.5 watt
$R_6 = 2700$ ohms, 0.5 watt
$R_7 = 47$ ohms, 0.5 watt
$R_8 = 18000$ ohms, 0.5 watt
$R_9 = 1000$ ohms, 0.5 watt

$R_{10} = 7.2$ ohms, 0.5 watt
$R_{11} = 330$ ohms, 2 watts
$R_{12}, R_{13} = 0.47$ ohms, 1 watt
$S_1 =$ On-off switch; single-pole, single-throw
$T_1 =$ Power transformer; primary, 117 volts; secondary 25.5 volts at 1 ampere; Thordarson No. 23V118, Stancor No. TP4, Triad No. F-93X, or equiv.

7.2.29 RCA 12W complementary-symmetry audio amplifier has very low distortion over a 20 to 20,000 Hz range. The 470 pF capacitor (C6) connected from the collector to the base of the 2N5956 transistor reduces the high-frequency response to approximately that of the 2N5497 plastic-package transistor. Both sections of the output stage, therefore, have essentially the same frequency-response characteristics—a feature that simplifies the addition of negative feedback.

The voltage divider (R7 and R9) connected across the speaker provides the proper amount of voltage for the loop feedback to transistor Q2 in the differential-amplifier input stage.

The dc supply voltage required for the amplifier is supplied by a full-wave transformer-coupled bridge. A single supply can provide the dc operating power for both amplifiers in a dual-channel system. The line voltage is stepped down to 25.5V by power transformer T1.

Output transistors Q4 and Q5 and diode D1 should be mounted on a common heatsink (Wakefield NC-403K or equivalent). Diode D1 should be attached to the underside of the heatsink by the use of small metal clamps. Transistors Q1 and Q2 should be matched for base-to-emitter voltage within 0.04V and should be selected for a beta between 100 and 300 at 1 mA and 5V.

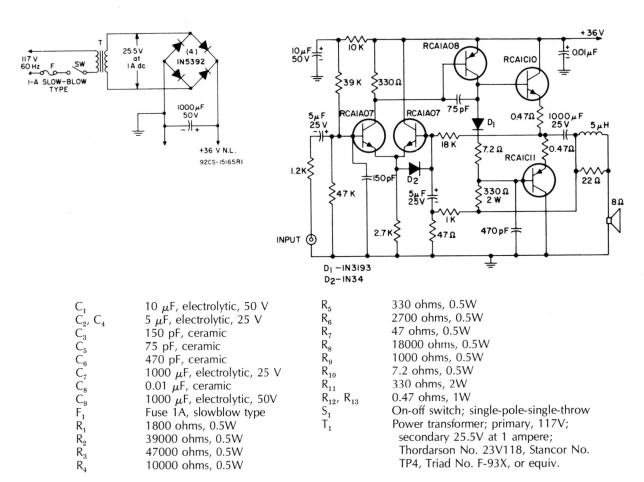

C_1	10 µF, electrolytic, 50 V	R_5	330 ohms, 0.5W
C_2, C_4	5 µF, electrolytic, 25 V	R_6	2700 ohms, 0.5W
C_3	150 pF, ceramic	R_7	47 ohms, 0.5W
C_5	75 pF, ceramic	R_8	18000 ohms, 0.5W
C_6	470 pF, ceramic	R_9	1000 ohms, 0.5W
C_7	1000 µF, electrolytic, 25 V	R_{10}	7.2 ohms, 0.5W
C_8	0.01 µF, ceramic	R_{11}	330 ohms, 2W
C_9	1000 µF, electrolytic, 50V	R_{12}, R_{13}	0.47 ohms, 1W
F_1	Fuse 1A, slowblow type	S_1	On-off switch; single-pole-single-throw
R_1	1800 ohms, 0.5W	T_1	Power transformer; primary, 117V; secondary 25.5V at 1 ampere; Thordarson No. 23V118, Stancor No. TP4, Triad No. F-93X, or equiv.
R_2	39000 ohms, 0.5W		
R_3	47000 ohms, 0.5W		
R_4	10000 ohms, 0.5W		

7.2.30 RCA practical 12W amplifier features excellent performance, simple proof-tested circuitry, and sensible power requirements (+36V at 1A). Power transformer T (inset) of power supply can be Thordarson 23V118, Stancor TP4, Triad F-93X, or any equivalent. With 600 mV input signal, the total harmonic distortion is 1%. With a shorted input, the hum and noise are 90 dB below the rated output level.

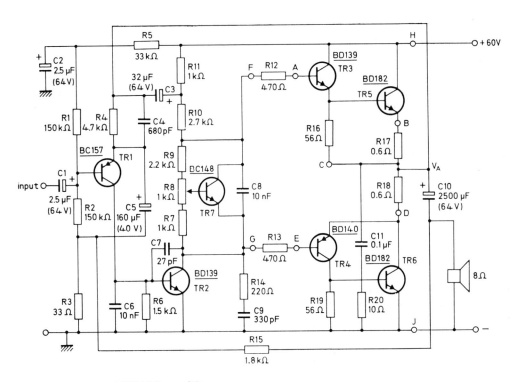

7.2.31 Amperex 40W hi-fi amplifier.

340 DISCRETE / TRANSISTOR CIRCUIT SOURCEMASTER

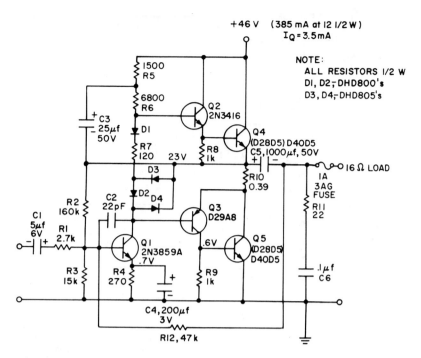

7.2.32 GE quasi-complementary push-pull amplifier requires 1.1V input signal for 12.5W output to a 16-ohm load. The total harmonic distortion at 10W power output is less than 0.5% from 25 Hz to 20 kHz. The 6V power output frequency response is −1.25 dB at 20 Hz and 90 kHz. The open-loop voltage gain is approximately 62 dB, and the ac feedback from R2 and R12 is 60 dB. The amplifier output impedance is 0.6 ohm and the input impedance about 3K.

When the output current in Q4 exceeds 1.5A, diode D4 conducts and limits the peak current. The current to the base of Q2 is shunted through D1, R7, D2, and D4. When the output current in Q5 exceeds 1.5A, diode D3 conducts and limits the peak current. The current to the base of Q3 is shunted through D3 and D2. The value of resistor R10 was selected to limit at 1.5A.

The electrical efficiency of the overall circuit is about 75% at 12.5W. The maximum transistor dissipation occurs at 50% efficiency. The tab on the output transistors should be attached to a heat radiating fin such as the Staver vertical thermovane No. V4-3-192 (Staver Company, Bay Shore, N.Y.) or equivalent heat radiator. This type of heat radiator (vertical fin) occupies a minimum of board area and will permit sine-wave operation of the amplifier in ambient temperatures up to 40°C. This size of heatsink would also be adequate for amplifier operation with program material in ambient temperatures up to 60°C. Silicone grease should be applied between the transistor tab and the heat fin when the application requires operation near the maximum dissipation as specified for the transistor and heat radiator.

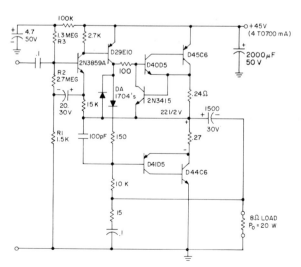

7.2.33 GE 20W hi-fi amplifier has an 8-ohm output and full short-circuit protection. The peak load current is limited to about 2.5A. With R1 of 1500 ohms the input signal for 20W output is 1.25V. Both the total harmonic and intermodulation distortion are well under 1%. With R1 = 560 ohms the sensitivity increases to about 500 mV for full output. The dc bias voltage (22.5V) is controlled by resistors R2 and R3, which should have a ±5% tolerance to insure symmetrical clipping at maximum power output.

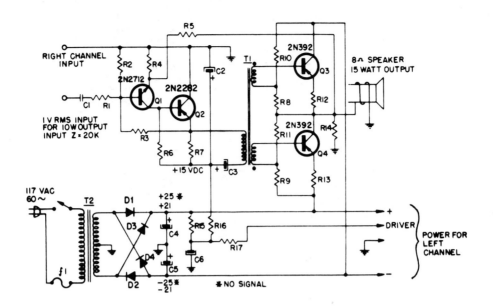

PARTS LIST

R_1	10K, ½W	R_{14}	100Ω, 1W	
R_2	33K, ½W	R_{15}	2Ω, 2W	
R_3	Select for Q_2 V_{CE}= 8VDC (100K is typical)	$R_{16,17}$	2Ω, 1W	
		C_1	.47µF, 12V	
R_4	75Ω, ½W	$C_{2,3}$	500µF, 15VDC	
R_5	1.5K, ½W	$C_{4,5,6}$	1000µF, 25VDC	
R_6	2.7K, ½W	$D_{1,2,3,4}$	Delco 1N2070 or 1N4004	
R_7	82Ω, 2W	f_1	1A, 3AG Fuse	
$R_{8,9}$	3.3Ω, ½W	Q_1	2N2712	
$R_{10,11}$	750Ω, 1W	Q_2	2N2282	
$R_{12,13}$	.47 or .5Ω, ½W	$Q_{3,4}$	Delco 2N392 Matched pair (Use Delco DTG-110 in high volume applications)	

T_1 Chicago Stancor TA-7 Core: EI-21, Radio Grade VI, ½" Stack, EI Laminations.
Coil Form: Use nylon bobbin for EI-21.
Winding Data: Simultaneously wind 2 #32 wires and 2 #28 wires to fill the nylon bobbin (approximately 250 turns). Interleave the EI Laminations in the bobbin. Connect the 2 #32 wires in series aiding to form the primary.

T_2 Triad F-92A Core: EI-12, M-19 Grade, 1" Stack, EI Laminations.
Winding Data: Pri: 660 t #26 AWG
Sec: 2 windings 120 T each #21 AWG Bifilar wind the secondary windings.

7.2.34 Stereo amplifier (15W) that uses Delco's economical 2N392 audio power transistor. With a good speaker system, sound is very good throughout the audio spectrum. Resistor R3 provides both ac and dc feedback. Alternating-current feedback lowers distortion in the driver stage and dc feedback allows the use of transistors of widely varying dc gain ranges. R3 also provides temperature compensation. The quiescent V_{CE} voltage of Q2 should be about 8V. If the V_{CE} is not in this range, R3 should be adjusted. By eliminating dc in the input transformer, its physical size is kept small and the low-frequency response benefits. The input transformer is wound multifilar for two important reasons. The most important reason is that the phase shift at high frequencies is kept small. This in turn allows feedback to be used effectively without creating high-frequency oscillations when the output is loaded with a high resistance (more than about 30 ohms). It is not uncommon for the speaker impedance to increase as much as five times its nominal impedance at frequencies across the audio spectrum. The amplifier must be completely stable at the highest impedance presented by the speaker. Multifilar winding also gives better frequency response.

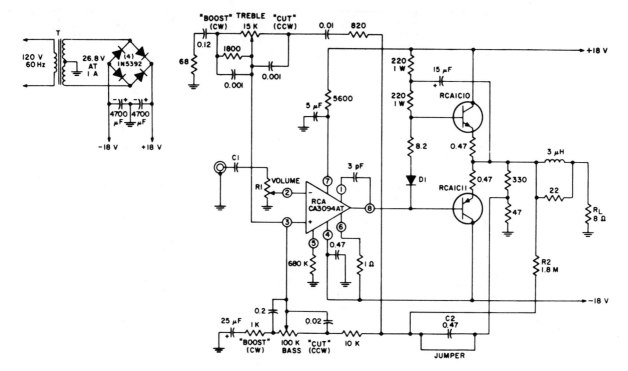

7.2.35 RCA-designed hi-fi amplifier delivers 15W music power into an 8-ohm load and uses a minimum of active components: one IC and a pair of complementary output transistors. Noninductive resistors are particularly important for the emitters of the output transistors, since this type has a positive temperature coefficient (resistance increases with increased power dissipation). A dual 18V supply is required for this circuit; the inset shows an adequate means of providing the required dc voltage. The power transformer secondary is 26.8V at 1A.

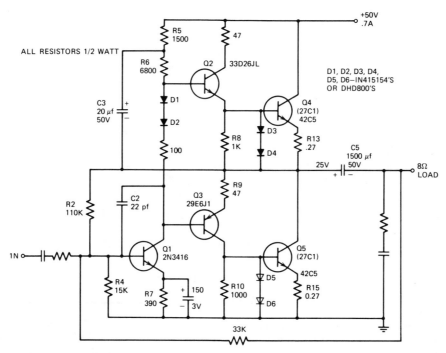

7.2.36 20W hi-fi amplifier from GE designers uses diodes D3, D4, D5, and D6 to protect Q4 and Q5 from overdissipation resulting from overdrive. Overdrive can result from switching, plugging, or extreme signal levels. The value of R13 is chosen so that when the peak current in Q4 reaches a certain value, the voltage drop across R13 plus the V_{BE} of Q4 will be equal to the conduction threshold of D3 and D4. As the peak current in Q4 tries to rise above this value, the drive current to Q4 is shunted by the increasing conductance of D3 and D4. This limiting value of output current should be selected to be just above the peak value required for maximum power output of the amplifier. D5 and D6 have the same function for protecting Q5. The current-limiting diodes start to limit at about 2.3A, which is right at 20W output into an 8-ohm load. The IM distortion is less than 1.5% at full output.

SECTION 7 Audio 343

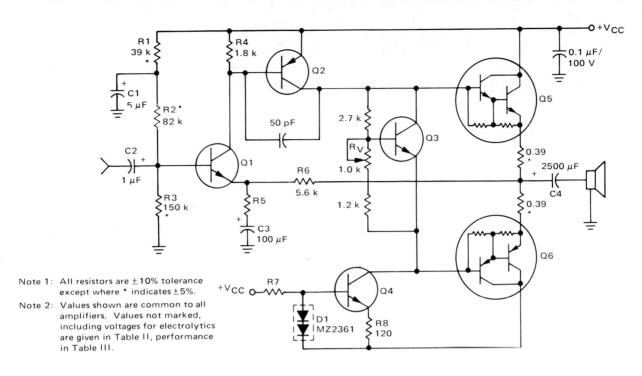

Parts List of 15 to 60 Watt Circuit

Power Watts (RMS)	15		20		25		35		50		60	
Load Impedance	4	8	4	8	4	8	4	8	4	8	4	8
V_{CC}	32 V	38 V	36 V	46 V	38 V	48 V	44 V	56 V	50 V	65 V	56 V	72 V
R5 (ohms)	620	510	560	470	560	390	470	330	390	270	330	220
R7 (ohms)	33 k	39 k	39 k	47 k	39 k	47 k	47 k	56 k	47 k	68 k	56 k	68 k
Q1	MPS A05	MPS A05	MPS A05	MPS A05	MPS A05	MPS A05	MPS A05	MPS A06	MPS A05	MPS A06	MPS A06	MPS A06
Q2	MPS A55	MPS A55	MPS A55	MPS A55	MPS A55	MPS A55	MPS A55	MPS A56	MPS A55	MPS A56	MPS A56	MPS A56
Q3	MPS U01	MPS U01	MPS U01	MPS U01	MPS U01	MPS U01	MJE 520	MPS U01	MJE 520	MJE 520	MJE 520	MJE 520
Q4	MPS A05	MPS A05	MPS A05	MPS A05	MPS A05	MPS A05	MPS A05	MPS A06	MPS A05	MPS A06	MPS A06	MPS A06
Q5	MJE 1100	MJE 1100	MJE 1100	MJE 1100	MJE 1102	MJE 1100	MJ 3000	MJ 1001	MJ 3000	MJ 3001	MJ 3001	MJ 3001
Q6	MJE 1090	MJE 1090	MJE 1090	MJE 1090	MJE 1092	MJE 1090	MJ 2500	MJ 901	MJ 2500	MJ 2501	MJ 2501	MJ 2501
Voltage rating on C1	35 V	40 V	40 V	50 V	40 V	50 V	45 V	60 V	50 V	65 V	60 V	75 V
Voltage rating on C2, C3	20 V	25 V	25 V	30 V	25 V	30 V	25 V	35 V	30 V	35 V	35 V	40 V
Voltage rating on C4	40 V	45 V	45 V	55 V	45 V	55 V	50 V	65 V	60 V	75 V	65 V	80 V
Min. heat sink for outputs @ 55°C ambient temperature and 10% high line voltage	9.5°C/W		7.0°C/W		5.0°C/W		6.0°C/W	5.5°C/W	4.0°C/W		3.0°C/W	

7.2.37 Motorola 15 to 60W ac-coupled hi-fi amplifier.
Center voltage at emitter resistor junction must be half V_{CC} for maximum output swing. Resistors R1, R2, and R3 form a voltage divider that sets the dc voltage at the base of Q1 1.5V above ½V_{CC}. This will maintain the center voltage at ½V_{CC}, since there is a constant 1.5V drop from the base of Q1 to the output center point. The input impedance is set by the parallel equivalent resistance of R2 and R3.

Transistor Q2 has a 60 dB voltage gain and determines the dominant pole in the amplifier. A 50 pF capacitor compensates the amplifier to prevent high-frequency oscillations.

Transistor Q3 is used to forward-bias the output devices to prevent crossover distortion.

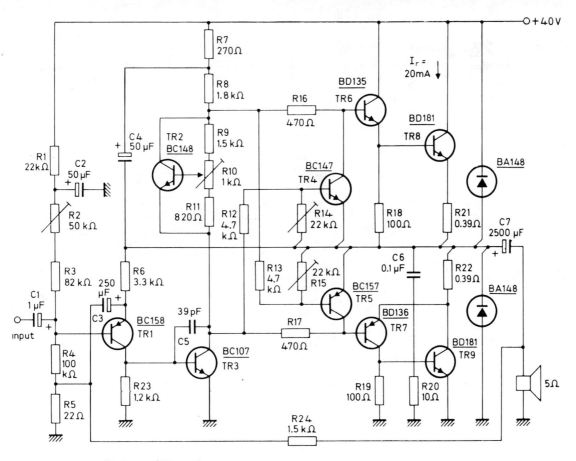

7.2.38 20-watt hi-fi amplifier. The Amperex output transistors are mounted on a 90 × 90 × 2 mm aluminum heatsink.

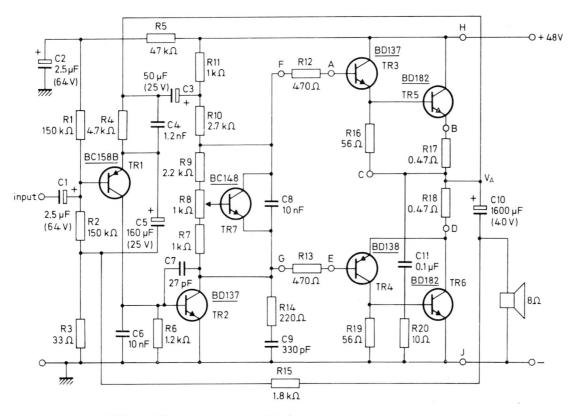

7.2.39 25-watt hi-fi amplifier uses Amperex bipolar transistors.

SECTION 7 Audio

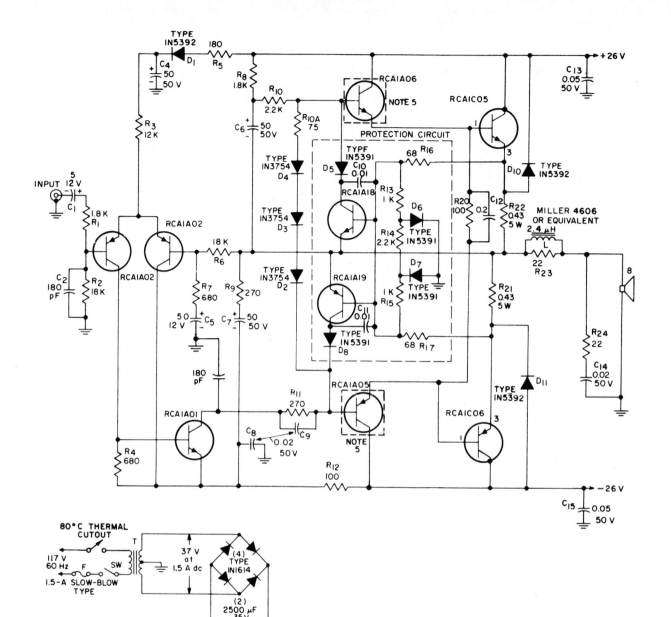

7.2.40 RCA 25W amplifier circuit featuring a complementary output with load-line limiting. With a dual 26V supply, a 600 mV input signal will produce full rated output across an 8-ohm load. The power into 4 ohms is 45W. Numbered schematic notes refer to the following:

1. T is Signal 36-2 (Signal Transformer Co., 1 Junius St., Brooklyn, N.Y. 11212).
2. Resistors are ½W unless otherwise specified; values are in ohms.
3. Capacitances are in microfarads unless otherwise specified.
4. Noninductive resistors.
5. Mount driver transistors on heatsink, Wakefield No. 209-AB or equivalent (this type may be obtained with a factory-attached integral heatsink).
6. Provide approximately 2°C/W heatsinking per output device.

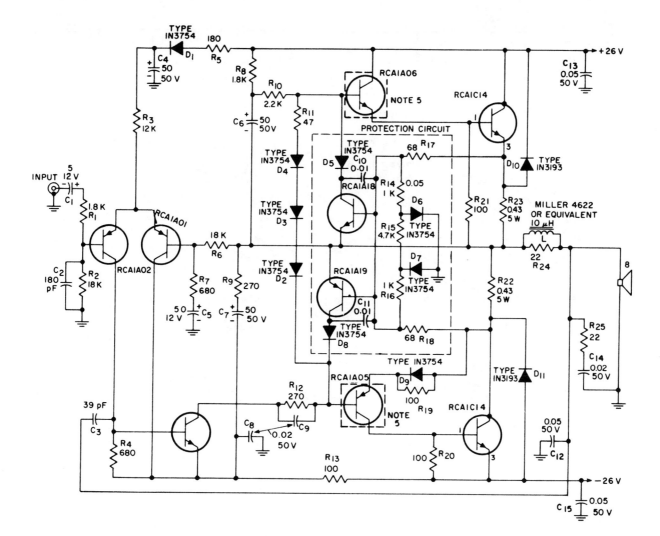
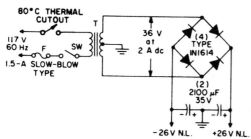

7.2.41 RCA 25W amplifier circuit featuring quasi-complementary-symmetry output delivers a full-rated output across an 8-ohm load (45W into 4 ohms) with a 600 mV input signal using a dual 26V supply. Numbered schematic notes refer to the following:

1. T is Signal 36-2 (Signal Transformer Co., 1 Junius St., Brooklyn, N.Y. 11212) or equivalent.
2. Resistors are ½W unless otherwise specified; values are in ohms.
3. Capacitances are in μF unless otherwise specified.
4. Denotes noninductive resistors.
5. Mount driver transistors on heatsink (Wakefield No. 209-AB or equivalent).
6. Provide approximately 2°C/W heatsinking per output device.

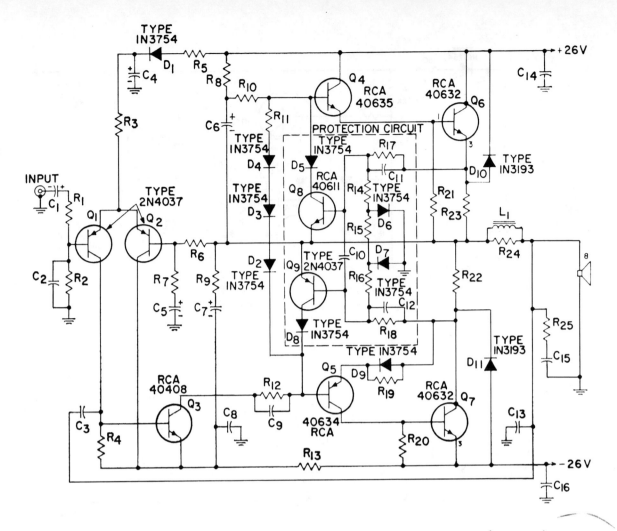

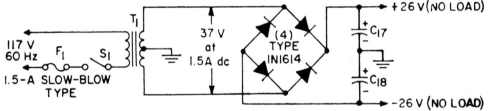

$C_1 = 5$ μF, electrolytic, 12 V
$C_2 = 180$ pF, ceramic, 50 V
$C_3 = 39$ pF, ceramic, 50 V
$C_4, C_6, C_7 = 50$ μF, electrolytic, 50 V
$C_5 = 50$ μF, electrolytic, 12 V
$C_8, C_9, C_{15} = 0.02$ μF, ceramic, 50 V
$C_{10}, C_{11}, C_{12}, C_{13}, C_{14}, C_{16} = 0.05$ μF, ceramic, 50 V
$C_{17}, C_{18} = 2100$ μF, electrolytic, 35 V
$F_1 =$ fuse, 1.5-ampere, slow-blow

$L_1 = 10$ μH, Miller 4622 or equiv.
$R_1, R_8 = 1800$ ohms, 0.5 watt
$R_2, R_6 = 18000$ ohms, 0.5 watt
$R_3 = 12000$ ohms, 0.5 watt
$R_4, R_7 = 680$ ohms, 0.5 watt
$R_5 = 180$ ohms, 0.5 watt
$R_9, R_{12} = 270$ ohms, 0.5 watt
$R_{10} = 2200$ ohms, 0.5 watt
$R_{11} = 47$ ohms, 0.5 watt
$R_{13}, R_{19}, R_{20}, R_{21} = 100$ ohms, 0.5 watt
$R_{14}, R_{16} = 1000$ ohms, 0.5 watt

$R_{15} = 4700$ ohms, 0.5 watt
$R_{17}, R_{18} = 68$ ohms, 0.5 watt
$R_{22}, R_{23} = 0.43$ ohms, 5 watts
$R_{24}, R_{25} = 22$ ohms, 0.5 watt
$S_1 =$ ON-OFF switch, single-pole, single-throw
$T_1 =$ power transformer; primary 117 volts; secondary, center-tapped, 37 volts at 1.5 amperes; Triwec Transformer Co. No. RCA-120 or equiv.

7.2.42 RCA 25W quasi-complementary audio power amplifier has a frequency response flat (within 1 dB) from 10 to 50,000 Hz. The amplifier requires no driver or output transformer and has built-in safe-area limiting protection that prevents damage to the driver and output stages from high currents and excessive power dissipation.

The differential-amplifier input transistors are matched for V_{BE} characteristics to give a minimum offset voltage between the input and output.

Output transistors Q6 and Q7 and diodes D2 through D4 should be mounted on a common heatsink (Wakefield NC-403K or equivalent). Diodes should be attached to the underside of the heatsink by small metal clamps. Transistors Q1 and Q2 should be matched for base-to-emitter voltage within 0.04V and should be selected for a beta between 100 and 300 at 1 mA and 5V.

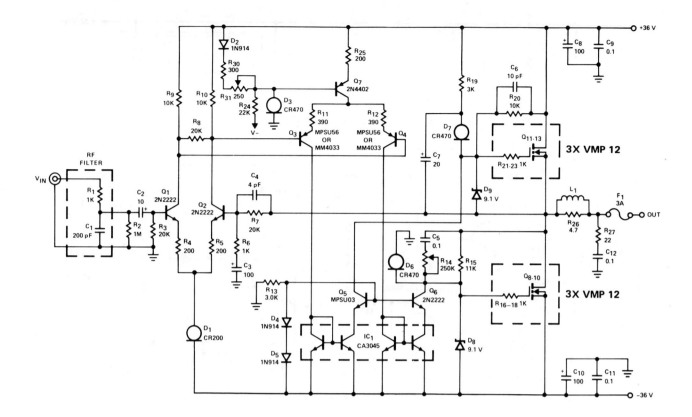

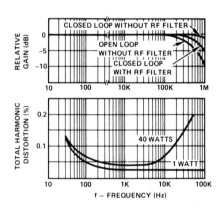

7.2.43 Siliconix superfidelity 40W audio amplifier suitable for high-quality stereo or quadraphonic systems. This amplifier uses Siliconix VMP 12s, which are similar to VMP 1s but selected for 90V breakdown. The operation is class AB, with about 300 mA of idling current used to bias the VMP 12s on. Three devices are paralleled for greater capacity.

The amplifier has low open-loop distortion, relatively small amounts of negative feedback (22 dB), and good open-loop frequency response to minimize the transient intermodulation distortion. Closed-loop frequency response (exclusive of the input filter) is flat to 1 MHz, and the slew rate is over 100 V/μs.

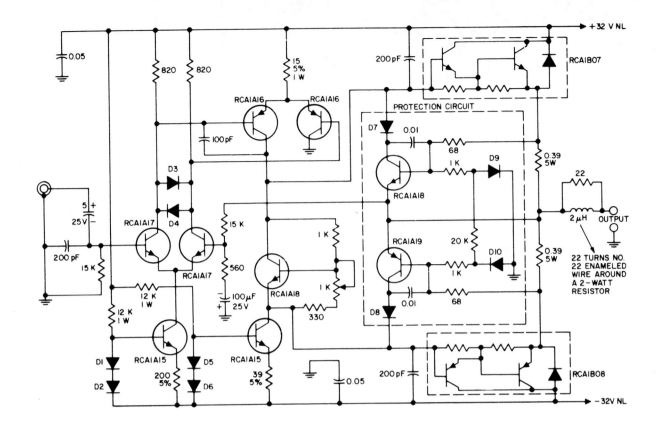

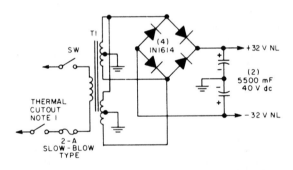

7.2.44 RCA 40W hi-fi amplifier features failsafe circuit protection and excellent performance with distortion at only 0.5% at full rated power output. The input resistance is 18K; the output impedance is 8 ohms. Schematic notes:

1. Provide 1.3°C/W heatsinking for each output device and use a mica washer with zinc oxide thermal compound (Dow-Corning 340).
2. 90°C thermal cutout attached to the heatsink for output transistors.
3. The power transformer is Signal 88-2 with parallel secondary (available from Signal Transformer Co., 1 Junius St., Brooklyn, N.Y. 11212).
4. All resistors are ½W types.
5. Capacitances, unless otherwise noted, are in microfarads.
6. Use noninductive resistors such as carbon composition or film. 7. Diodes D1 through D10 are 1N5391.

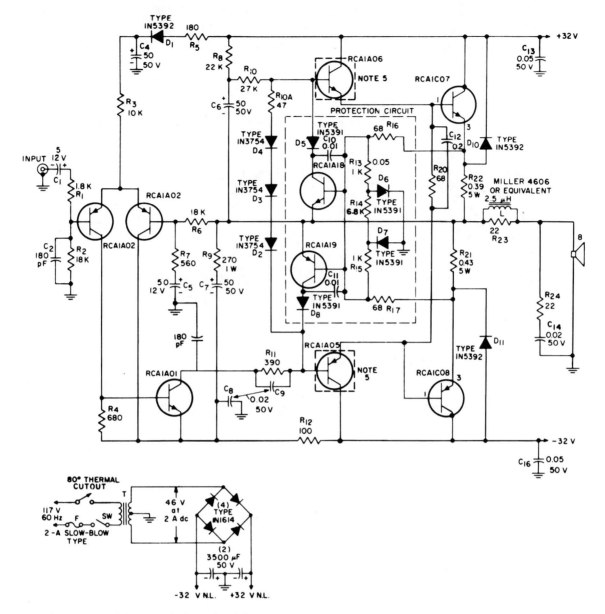

7.2.45 RCA 40W amplifier circuit featuring full complementary-symmetry output using load-line limiting. With a dual 32V supply, a 600 mV input signal will produce full rated output across an 8-ohm load; power into 4 ohms is 75W. Numbered schematic notes refer to the following:

1. Power supply transformer is Signal 88-2 (parallel secondary), available from Signal Transformer Co., 1 Junius St., Brooklyn, N.Y. 11212.
2. Resistors are ½W unless otherwise specified; values are in ohms.
3. Capacitances are in μF unless otherwise specified.
4. Denotes noninductive resistors.
5. Mount the driver transistors on the heatsink (Wakefield No. 209-AB or equivalent); alternately, these types may be obtained with a factory-attached integral heatsink.
6. Provide approximately 1.3°C/W heatsinking per output device.

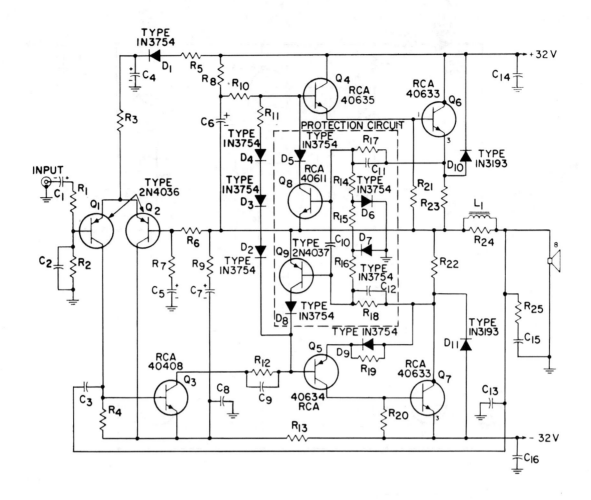

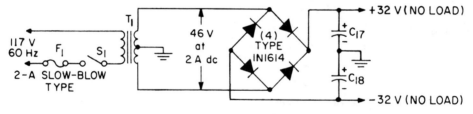

$C_1 = 5\ \mu F$, electrolytic, 12 V
$C_2 = 180$ pF, ceramic, 50 V
$C_3 = 39$ pF, ceramic, 50 V
$C_4, C_6, C_7 = 50\ \mu F$, electrolytic, 50 V
$C_5 = 50\ \mu F$, electrolytic, 12 V
$C_8, C_9, C_{15} = 0.02\ \mu F$, ceramic, 50 V
$C_{10}, C_{11}, C_{12}, C_{13}, C_{14}, C_{16} = 0.05\ \mu F$, ceramic, 50 V
$C_{17}, C_{18} = 3500\ \mu F$, electrolytic, 50 V
F_1 = fuse, 2-ampere, slow-blow

$L_1 = 10\ \mu H$, Miller 4622 or equiv.
$R_1 = 1800$ ohms, 0.5 watt
$R_2, R_6 = 18000$ ohms, 0.5 watt
$R_3 = 15000$ ohms, 0.5 watt
$R_4 = 680$ ohms, 0.5 watt
$R_5 = 180$ ohms, 0.5 watt
$R_7 = 560$ ohms, 0.5 watt
$R_8 = 2200$ ohms, 0.5 watt
$R_9 = 270$ ohms, 0.5 watt
$R_{10} = 2700$ ohms, 0.5 watt
$R_{11} = 47$ ohms, 0.5 watt
$R_{12} = 390$ ohms, 0.5 watt
$R_{13}, R_{19}, R_{20}, R_{21} = 100$ ohms, 0.5 watt

$R_{14}, R_{16} = 1000$ ohms, 0.5 watt
$R_{15} = 4700$ ohms, 0.5 watt
$R_{17}, R_{18} = 68$ ohms, 0.5 watt
$R_{22}, R_{23} = 0.39$ ohm, 5 watts
$R_{24}, R_{25} = 22$ ohms, 0.5 watt
S_1 = ON-OFF switch, single-pole, single-throw
T_1 = power transformer; primary 117 volts; secondary, center-tapped, 46 volts at 2 amperes; Triwec Transformer Co. No. RCA-119 or equiv.

7.2.46 RCA 40W hi-fi audio power amplifier can deliver 40W for an input of 0.6V. The frequency response of the amplifier is flat within 1 dB from 10 to 50,000 Hz. Total harmonic distortion at full rated output is less than 1% at 1000 Hz. The 40W amplifier operates from symmetrical positive and negative dc voltages of 32V.

Output transistors Q6 and Q7 and diodes D2 through D4 should be mounted on a common heatsink (Wakefield NC403K or equivalent). Diodes should be attached to the underside of the heatsink by small metal clamps. Transistors Q1 and Q2 should be matched for base-to-emitter voltage within 0.04V and should be selected for a beta between 100 and 300 at 1 mA and 5V.

Performance characteristics, measured at a line voltage of 120V, an ambient temperature of 25°C, and a frequency of 1 kHz:

Power (continuous, 8-ohm load)	40W
Hum and noise (below continuous power output)	80 dB
Input resistance	20K
Intermodulation distortion	0.1%

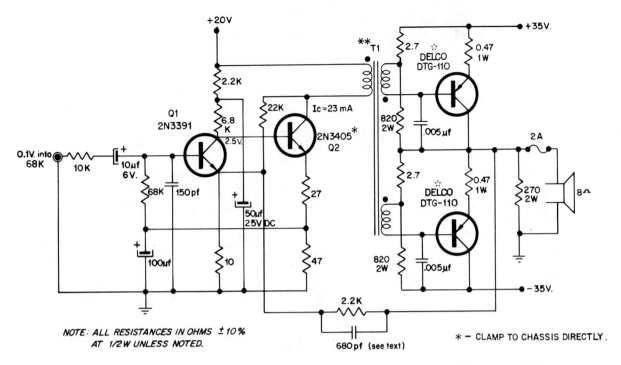

**T-1 DRIVER TRANSFORMER
 Triad Transformer P/N assigned: No. TY-160X.
 Orient T-1 relative to T-2 experimentally for minimum hum pickup and coupling.
 RATINGS
  Ipri = 50 mA DC
  Lpri = 4 h at 50 mA DC, 3V-60 Hz test signal
 SPECIFICATIONS
  Core: EI-75, fabricated with laminations of M-19 grade silicon steel. Use 1 mil air gap. Build up as square stack.
  Winding: Design is split primary sandwiching bifilar secondaries.

Details: On nylon bobbin, wind 600 turns of #30 enameled wire. Next, simultaneously wind 2 wires of #27 enameled wire for 200 turns (each). Finally, resume winding the primary for another 600 turns. Bring ends of wire out as desired, observing starts for proper connections as shown in circuit schematic.
Test measurements:
 Rpri 45 ohms
 Rsec 3.3 ohms
 N = 6:1:1

7.2.47 Delco high-performance audio amplifier delivers 50W of continuous power to an 8-ohm speaker load. The driver section is feedback-biased from the emitter of Q2 to the base of Q1 for optimum bias stabilization. About 16 dB of negative feedback is taken around the driver from the collector of Q2 to the emitter of Q1 to lower the output driving impedance and shape the local gain–bandwidth characteristic. The driver transformer uses a high turns ratio for maximum circuit efficiency and gain, and is fabricated with a split primary. The high coupling coefficient provided by this feature and the low output impedance of the driver combine to provide a voltage drive to the output transistors across a bandwidth exceeding the audio spectrum. This voltage drive has been proved to result in maximum gain linearity for the Delco DTG-110 because it exploits the inherent transconductance characteristic of the device and avoids the nonlinearities of the current gain. It also serves to reduce the power stage source impedance, optimizing the output admittance characteristics of the power transistors.

SECTION 7 Audio

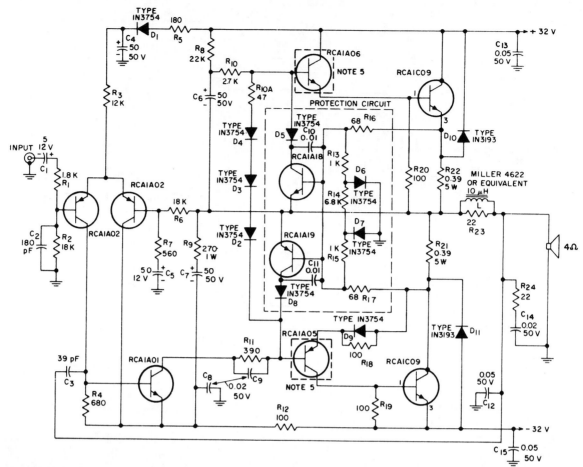

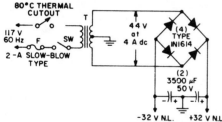

7.2.48 RCA 55W amplifier circuit, featuring quasi-complementary-symmetry output, delivers full rated output across a 4-ohm load with a 600 mV input signal using a dual 32V supply. Notes on the schematic refer to these numbered items:

1. T is Signal 88-2 (parallel secondary), Signal Transformer Co., 1 Junius St., Brooklyn, N.Y. 11212 (or equivalent).
2. Resistors are ½W unless otherwise specified; values are in ohms.
3. Capacitances are in μF unless otherwise specified.
4. Denotes noninductive resistors.
5. Mount the driver transistors on the heatsink (Wakefield No. 209-AB or equivalent); these types may be obtained with a factory-attached integral heatsink.
6. Provide approximately 1.3°C/W heatsinking per output device.

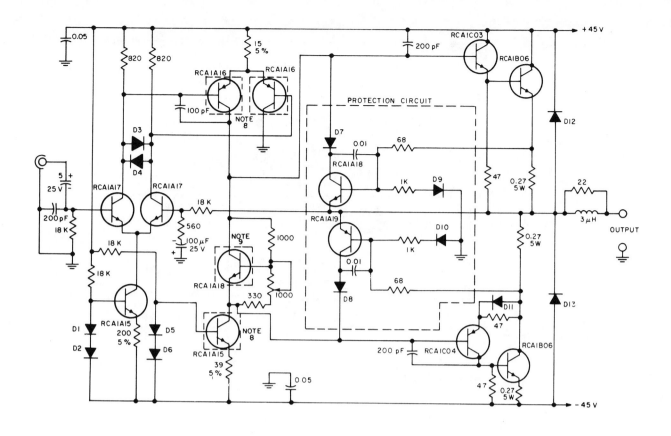

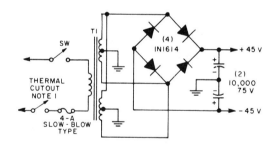

7.2.49 70W hi-fi audio amplifier, with full circuit protection, features a quasi-complementary output employing RCA's *pi-nu* output transistors. Power requirement is a dual 45V source. Notes on the schematic refer to these numbered items:

1. 90° thermal cutout attached to the heatsink used for the output transistors; this cutout will be in series with the power transformer primary, as shown in the inset.
2. The power transformer is 120-2 with a parallel secondary, from Signal Transformer Co., 1 Junius St., Brooklyn, N.Y. 11212.
3. All resistors are ½W types, and the resistances are in ohms.
4. Capacitances are in microfarads unless otherwise specified.
5. Use noninductive resistors.
6. Diodes D1 through D8 and D11 are 1N5391; diodes D9, D10, D12, and D13 are 1N5393.
7. Provide about 1°C/W heatsinking per output device and mount with a mica washer using Dow-Corning 340 thermal compound.
8. Mount on a heatsink such as Wakefield 209-AB.
9. Attach the heatsink cap (Wakefield 260-6SH5E) on the device and mount on the same heatsink with an output transistor.

SECTION 7 Audio

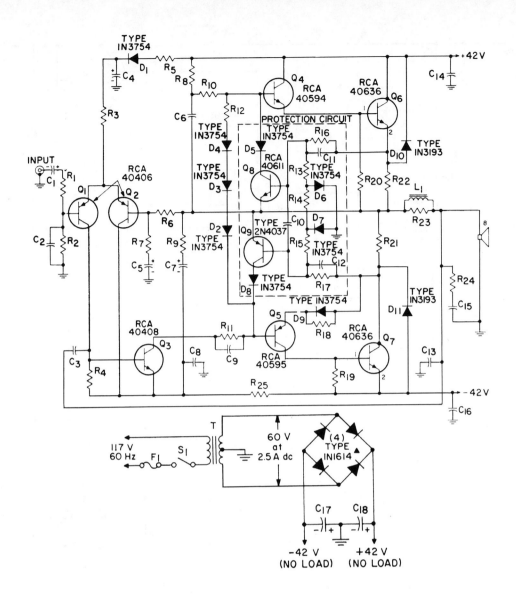

$C_1 = 5$ μF, electrolytic 12 V
$C_2 = 180$ pF, ceramic, 50 V
$C_3 = 39$ pF, ceramic, 50 V
$C_4, C_6, C_7 = 50$ μF, electrolytic, 50 V
$C_5 = 50$ μF, electrolytic, 12 V
$C_8, C_9, C_{15} = 0.02$ μF, ceramic, 50 V
$C_{10}, C_{11}, C_{12}, C_{13}, C_{14}, C_{16} = 0.05$ μF, ceramic, 50 V
$C_{17}, C_{18} = 3500$ μF, electrolytic, 50 V

$F_1 =$ fuse, 3 ampere, slow-blow type
$L_1 = 10$ μH, Miller 4622 or equiv.
$R_1 = 1800$ ohms, 0.5 watt
$R_2, R_3, R_6 = 18000$ ohms, 0.5 watt
$R_4 = 680$ ohms, 0.5 watt
$R_5 = 180$ ohms, 0.5 watt
$R_7, R_{11} = 470$ ohms, 0.5 watt
$R_8 = 2700$ ohms, 0.5 watt
$R_9 = 270$ ohms, 0.5 watt
$R_{10} = 3300$ ohms, 0.5 watt
$R_{12} = 47$ ohms, 0.5 watt
$R_{18}, R_{19}, R_{20}, R_{25} = 100$ ohms, 0.5 watt

$R_{13}, R_{15} = 1000$ ohms, 0.5 watt
$R_{14} = 4700$ ohms, 0.5 watt
$R_{16}, R_{17} = 68$ ohms, 0.5 watt
$R_{21}, R_{22} = 0.33$ ohm, 5 watts
$R_{23}, R_{24} = 22$ ohms, 0.5 watt
$S_1 =$ ON-OFF switch, single-pole, single-throw
$T_1 =$ power transformer; primary 117 volts; secondary, center-tapped, 60 volts at 2.5 amperes; Triwec Transformer Co. No. RCA 113 or equiv.

7.2.50 RCA 70W hi-fi quasi-complementary audio power amplifier provides 100 watts of IHFM music power output for an input of 1V. The frequency response of the amplifier is flat within 1 dB from 5 to 25,000 Hz. The total harmonic distortion at full rated power is less than 0.25% at 1000 Hz. Amplifier operates from symmetrical positive and negative dc supply voltages of 42V.

Output transistors Q6 and Q7 and diodes D2 through D4 should be mounted on a common heatsink (Wakefield NC403K or equivalent). Diodes should be attached to the underside of the heatsink by small metal clamps. Transistors Q1 and Q2 should be matched for base-to-emitter voltage within 0.04V and should be selected for a beta between 100 and 300 at 1 mA and 5V.

Perfomance characteristics, measured at a line voltage of 120V, ambient temperature of 25°C, and a frequency of 1 kHz:

Power (continuous)	70W
Hum and noise (below continuous power output)	85 dB
Input resistance	20K
Intermodulation distortion	0.1%

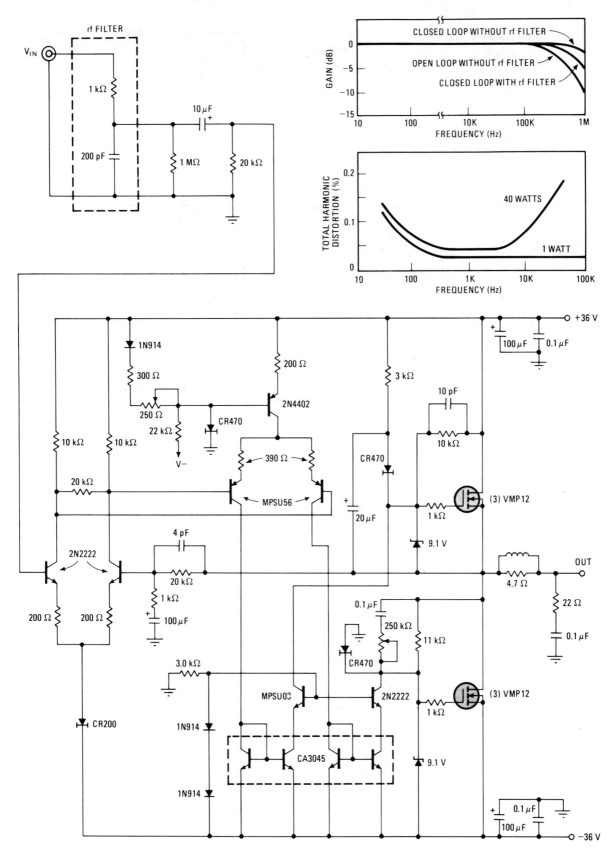

7.2.51 Siliconix 80W superfidelity amplifier uses six power FETs in a push-pull arrangement. Harmonic distortion is as low as 0.04% using very little negative feedback for a frequency response of 1 Hz to 800 kHz. A feedback of only 22 dB is needed with the FETs, whereas 40 dB is the value typically used with bipolar transistor stages.

The distortion, which depends on the output power, is shown for various combinations of open- and closed-loop operation, with and without RF filtering. An extra bonus when using the FETs is that the output of the amplifier is inherently short-circuit-protected and free from secondary breakdown and thermal runaway.

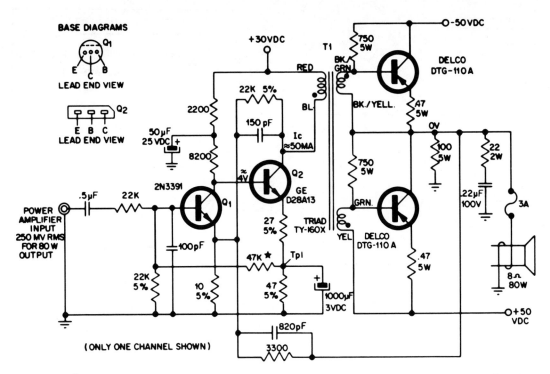

T-1 DRIVER TRANSFORMER*

RATINGS
$I_{pri} = 50$ mA DC
$L_{pri} = 4$H at 10 mA DC

SPECIFICATIONS

Core: EI-75, fabricated with laminations of M-19 grade silicon steel. Use 1 mil air gap.

Winding: Design is split primary sandwiching bifilar secondaries.

Details: On nylon bobbin, wind 600 turns of #30 enameled wire. Next, simultaneously wind 2 wires of #27 enameled wire for 200 turns (each). Finally, resume winding the primary for another 600 turns. Bring ends of wire out as desired, observing starts for proper connections as shown in circuit schematic.

Test measurements:
$R_{pri} \simeq 45$ ohms
$R_{sec} \simeq 3.3$ ohms
$N = 6:1:1$

*Triad Transformer Corp. P/N assigned: No. TY-160X.

7.2.52 Delco 160W stereo amplifier delivers 80W per channel to 8-ohm loads. The output circuit is a standard half-bridge that drives the load directly from a balanced ±50V supply. Because of this balanced supply, there is only a negligible dc component across the load caused by bias tolerances. The bias network is a stable voltage divider that presents a fixed voltage to the base of each power transistor to allow a collector bias current determined by the inherent transconductance characteristics of the devices. The 0.47-ohm emitter degeneration enhances this stiff bias voltage to provide a thermally stable circuit and offers a small amount of local ac feedback. The output circuit uses the transformer secondary resistance as part of the bias network, which minimizes the total drive voltage required from the transformer secondary and prevents the emitter diode of the power transistor from avalanching.

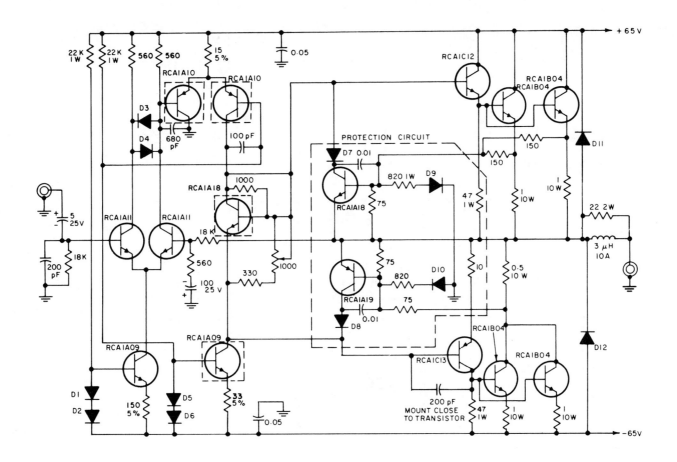

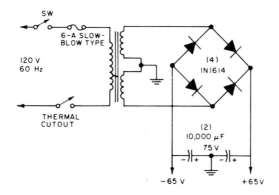

7.2.53 RCA 120W amplifier features quasi-complementary output transistors in parallel. Diodes D1 through D8 are 1N5391; D9 and D10 are 1N914; and D11 and D12 are 1N5393. Provide approximately 1°C/W heatsinking per output device based on mounting with mica washer and thermal compound such as Dow Corning 340 or equivalent. Mount broken-line-enclosed transistors on the heatsink (Wakefield 209-AB). For the 1A18, attach the heatsink cap (Wakefield 260-6SH5E or equivalent) on the device and mount on the same heatsink with an output transistor. The inset shows an adequate power supply. The power transformer is a Signal 88-6 (Signal Transformer Co., 1 Junius St., Brooklyn, N.Y. 11212).

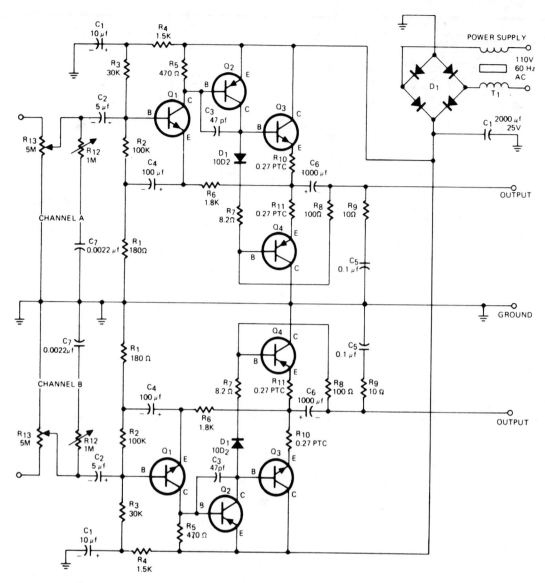

7.2.23 Workman complete stereo amplifier provides 5W per channel into a 4-ohm load. The circuit uses complementary output transistors, IR's IRTR55 (red) and IRTR56 (green), plus an IRTR56 as a PNP driver transistor.

7.3 Special-Purpose Circuits

This subsection is the audio "catchall"—those circuits which do not quite fit into either of the two prior categories.

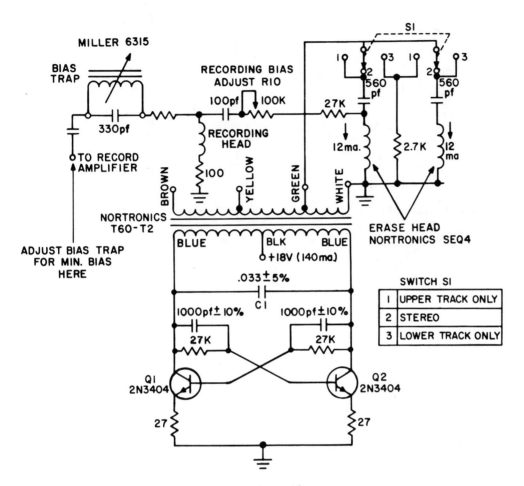

7.3.1 GE tape erase and bias-oscillator circuit provides ample power to give a minimum of 60 dB erasure of saturated tape (at 400 Hz) with a stereo erase head. This is accomplished with at least 10 mA of 70 to 80 kHz signal in the Nortronics SEQ4 erase head. The total power output of this circuit is 1.5W with an efficiency of about 60%.

The frequency of the oscillator circuit is determined by the value of C1 in the resonant tank circuit. The 27-ohm resistors provide negative feedback, which helps to compensate for component variations in the circuit. The emitter–base junction of Q1 and Q2 will operate in the avalanche mode for a short interval of the operating period. The circuit limits the peak power dissipation to 50 mW at the emitter–base junction in the avalanche interval.

This circuit is a cross-coupled multivibrator with a tuned load. The erase head winding is coupled to the transformer tap with 560 pF, which series-resonates with the erase-head winding. The load appears largely resistive on the transformer secondary winding. This permits switching a 2.7K resistor in place of the series-tuned erase head load without changing the loading or the frequency of the oscillator. With the proper switching this would permit erasing and recording on only one channel of the tape. Variations in either the L or the C will not alter appreciably the value of the erase current in the head.

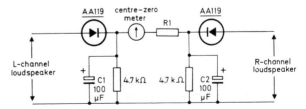

7.3.2 Amperex balance control meter. Lack of sensitivity of the ear in detecting small changes in volume levels can make the operation of balancing somewhat tedious. If precision is required, adding a visual indication that balance has been achieved will ease the task considerably.

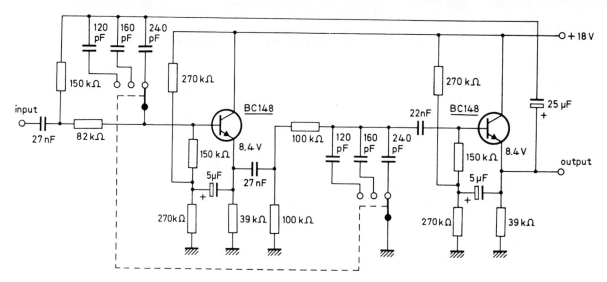

7.3.3 Amperex noise and rumble filter. Bass and treble cut are produced by an RC network connected between two emitter followers and a feedback loop from the output to the input through a second RC network. A high slope of around 13 dB/octave is achieved. The frequency limit of the rumble filter is fixed at 45 Hz, but the noise filter can be switched to limits of 16, 12, and 7 kHz.

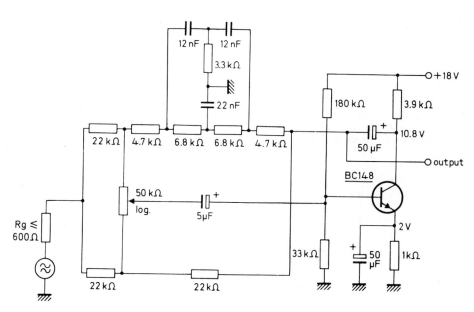

7.3.4 Amperex active presence control circuit uses frequency-selective negative feedback with amplitude control to achieve up to 13 dB boost at 2000 Hz. The feedback network is a bridged-tee filter. The nominal gain at flat respose is 0.95, and the input and output impedances are 12K and 100 ohms, respectively. The total distortion remains below 0.1% provided that the output voltage does not exceed 250 mV.

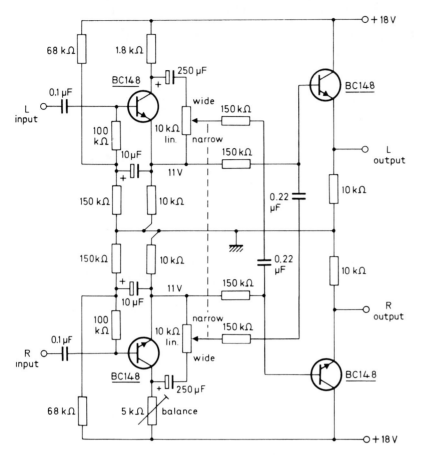

7.3.5 Amperex sound-source "width" control. The apparent width of the "sound stage" can be varied in stereo systems by deliberately introducing crosstalk between the channels, part of the signal voltage of one channel being added to the second channel. The circuit shown provides continuous control between the in-phase crosstalk of 100%, corresponding to mono operation, and the antiphase crosstalk of 24%. Greater antiphase crosstalk is not required, since the sound impression will fall apart for greater values.

The voltage gain of the circuit is 0.5. Input and output impedances are 750K and 45 ohms, respectively, and the frequency response (−3 dB points) is from less than 20 Hz to greater than 20 kHz.

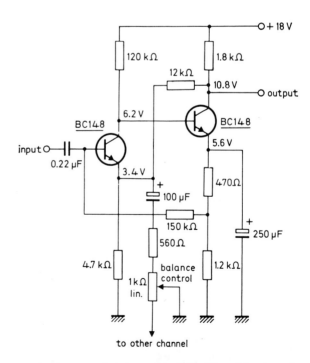

7.3.6 Amperex balance control circuit enables the voltage gain of both stereo channels to be varied by 6 dB in opposite directions. The controlling variable resistor forms part of the feedback circuit. The average gain is 23.4 dB.

SECTION 8
TEST AND MEASUREMENT

The semiconductor has been responsible for more growth in test, measurement, and calibration than any prior development. A single PN junction might well be termed the simplest of all precision "standards," since the voltage drop across it is a predictable and constant 700 mV (for silicon). Its ability to pass current in one direction only makes the simple diode an ideal go/no-go tester for continuity and polarity.

The transistor is even more versatile, since it permits circuits of considerable complexity to be crammed into tiny boxes that are self-powered and portable. The very structure of the device has permitted simplification of complex circuits with minimal compromises, and it has made practical and economical a wide variety of circuits that once were found only in well equipped labs. This section is a cross section of the total field.

8.1 Go/No-Go Testers

A go/no-go tester gives a straight yes or no answer to such questions as "Is this device good?" or "Is this wire broken?"—it does not give a quantitative indication of value. Despite the basic yes/no aspect of such testers, they're not always of simple circuitry; indeed, even the most sophisticated computers are sometimes programmed for such a function.

In selecting the circuits for this subsection, I have tried to avoid the more complicated candidates in favor of those I considered universally practical, using as my guide the fact that complex go/no-go testers are more economically constructed with integrated circuits than with discrete elements.

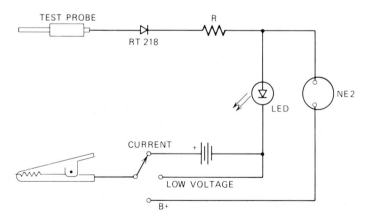

8.1.1 Selectable current/voltage continuity tester uses LED as the indicating element. Resistor R should be determined after the LED is selected; the value of this resistor should limit the current to LED's I_{max}. With an I_{max} of 250 mA and a maximum low voltage to be measured of 24V, let R = 100 ohms. A 9V battery would serve nicely; with 9V and continuity checks in nonvoltage carrying lines, the maximum LED current would be less than 90 mA—more than sufficient for illumination even in daylight.

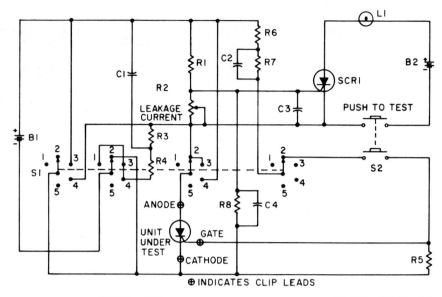

PARTS LIST

R1 — 22 K, 1W
R2 — 1000 OHM POTENTIOMETER
R3, R4 — 4.7 K, 1W
R5 — 1K, 1W
R6 — 680 OHMS, 1W
R7, R8 — 100 K, 1W
C1, C4 — .01 MFD
C2, C3 — .1 MFD
S1 — 4 POLE, 5 POSTION, NON-SHORTING ROTARY SWITCH
S2 — 2 POLE, N.O. PUSHBUTTON, BOTTOM CONTACT ADJUSTED AS INDICATED IN TEXT.
B1 — 67-½ VOLT DRY CELL
B2 — 3 VOLT DRY CELL
L1 — G.E. #49 LIGHT BULB
SCR1 — G.E. C5U CONTROLLED RECTIFIER

ROTARY SWITCH POSITIONS

1 BATTERY CHECK
2 FORWARD VOLTAGE DROP AND GATE TRIGGER (FOR SCR'S)
3 REVERSE LEAKAGE CURRENT
4 FORWARD LEAKAGE CURRENT (FOR SCR'S ONLY)
5 OFF

8.1.2 GE portable SCR and silicon rectifier go/no-go test set detects opens and shorts on diode rectifiers and SCRs, and checks forward blocking ability and triggering on SCRs. Convenient battery operation is particularly useful when working with automotive, marine, and aircraft equipments, as well as in other general industrial, military, and commercial applications of the semiconductor rectifier and the SCR. This tester is more objective than the ohmmeter because it applies at least 10V to the semiconductor under test.

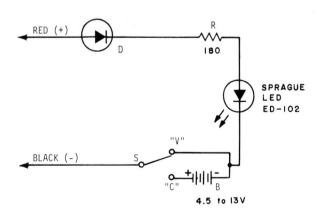

8.1.3 Sprague simple voltage and continuity tester is particularly suitable for automotive applications because the supply voltage range is 4.5 to 13.2V. To check for the presence of voltage, place the switch in **V** position and connect the black lead to a known ground; probe with red lead. LED lights green. To check for continuity, place the switch in **C** position.

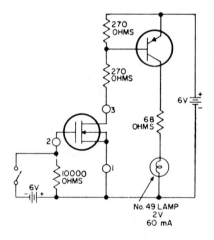

8.1.4 RCA go/no-go test circuit for MOS transistors can be used to test MOSFETs for opens or shorts. The substrate and source of the device being tested should be connected to terminal 1, the gate should be connected to 2, and the drain should be connected to 3. If the MOS transistor is a dual-gate type, the gates are tested separately. For N-channel depletion types, if the lamp lights when the switch is open and does not light when the switch is closed, the transistor is good. If the lamp lights with the switch in either position, the transistor is shorted. If the lamp remains off with the switch in either position, the transistor is open. For P-channel enhancement types, the reverse indications are obtained. (Circuit from RCA publication SC-15.)

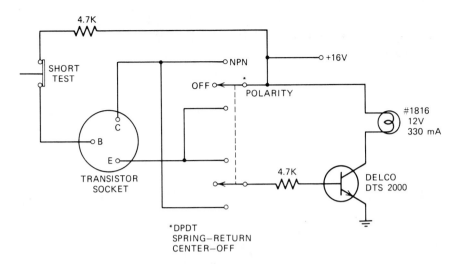

8.1.5 Small-signal transistor and diode tester uses the transistor under test to key the power transistor in the lamp circuit. The *short* test switch should be a normally closed momentary-contact type and the *polarity* switch should be a center-off spring-return DPDT. Use any available transistor socket or three separate pin jacks for the test receptacle.

To check the quality of a transistor, insert it into the socket so that its electrodes match the marked socket terminals and place the *polarity* switch in the appropriate position (NPN or PNP). If the transistor is good (not open), the light will come on. Press the *short* button to remove base drive from the device under test. If the light stays on, the device has an internal collector-to-emitter short; if the light goes out, the transistor is good.

This tester can also be used for checking diodes. Just insert the diode across the collector and emitter terminals of the test socket and place the *polarity* switch in first one position and then the other. A good diode will cause the lamp to come on in one position only. If the lamp does not come on in either position, the diode is open; if it comes on in both test positions, the diode is shorted.

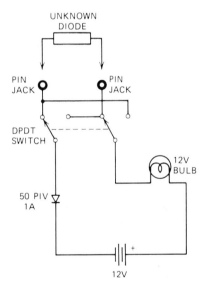

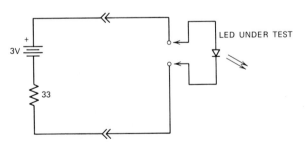

8.1.6 Sprague light-emitting-diode tester. LEDs operate at about 2V, but most will work satisfactorily with any supply voltage between 1.6 and 2.5V. The application of too great a voltage can result in permanent dimming or complete burnout. By substituting a 39-ohm resistor for the LED in the circuit shown and measuring the voltage drop across the resistor, the proper value for the current limiting resistor can be selected for any value of supply voltage and any circuit in which the LED is intended to be used. In most cases 15 to 20 mA will produce sufficient illumination for most purposes.

8.1.7 Diode and continuity tester tells the condition as well as the polarity of unknown devices. If your layout is as shown, the switch points to the cathode of the device under test when the light comes on.

For a *continuity* test, connect questionable leads to the pin jacks. If continuity in the leads exists, the light will come on regardless of switch position.

For a *polarity* test, place the diode across the pin jacks; if the light doesn't come on, flick the switch. When the light comes on, the switch pointer will be aimed at the cathode.

For a *quality* test, move the switch between its two positions; a good diode will permit the light to come on in one switch position only. If the lamp doesn't light, the diode is open; if it lights in both positions, it's shorted.

SECTION **8** *Test and Measurement*

8.2 Measurement Devices

Since an instrument is only as accurate as the components comprising it, it is important to use precision components in critical circuit elements. The range resistors in the voltmeters of this subsection should have a tolerance of not less than ±1% or better.

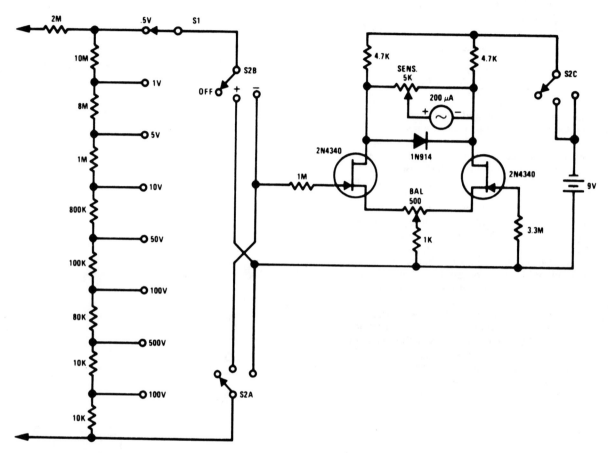

8.2.1 National FET voltmeter replaces the function of the VTVM while at the same time ridding the instrument of the usual line cord. In addition, drift rates are far superior to vacuum-tube circuits, allowing a 500 mV full-scale range (which is impractical with most vacuum tubes).

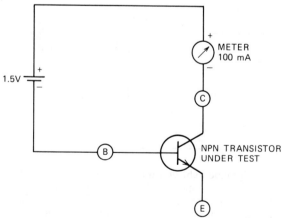

8.2.2 RCA I_{CBO} measuring circuit. I_{CBO} is the current flow, or leakage, from collector to base, with the emitter open. Here, a potential of 1.5V is applied to the collector and base of the transistor, and the meter is connected in the collector circuit. The collector-to-base leakage is indicated in microamperes.

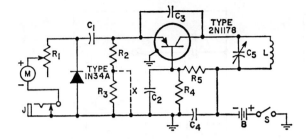

Parts List

B = 13.5 volts, RCA VS304
C₁ = 33 pF, mica, 50 V
C₂ = 0.01 μF, paper, 50 V
C₃ = 5 pF, mica, 50 V
C₄ = 0.01 μF, paper, 50 V
C₅ = variable capacitor, 50 pF, Hammarlund type HF-50 or equivalent

J = phone jack, normally closed
L = plug-in coil
M = microammeter, 0 to 50 μA, Simpson model 1227 or equivalent
R₁ = variable resistor, 0-0.25 megohm, 0.5 watt

R₂ = 220 ohms, 0.5 watt
R₃ = 3,000 ohms, 0.5 watt
R₄ = 3,900 ohms, 0.5 watt
R₅ = 39,000 ohms, 0.5 watt
X = jumper, omit for measurements below 45 MHz

Coil-Winding Data

Coil	Freq. Range	Wire Size	No. of Turns
1	3.4-6.9 MHz	#28, enamel	48¼, close wound
2	6.7-13.5 MHz	#24, enamel	22, close wound
3	13-27 MHz	#24, enamel	9⅛, close wound
4	25-47 MHz	#24, enamel	4⅛, close wound
5	46-78 MHz	#24, enamel	1½, close wound
6	74-97 MHz	#16, tinned	hairpin formed, 1⅞ inches long including pins, and ¼ inch wide

Coil forms are Amphenol type 24-5H or equivalent.

8.2.3 RCA transistor dip meter measures resonant frequencies from 3.5 to 100 MHz. This circuit, essentially a transistor version of the electron-tube grip-dip meter, consists of a 2N1178 common-base oscillator stage that can be tuned over a wide frequency range. A 1N34A diode and a dc microammeter are used to show when RF is being absorbed from the oscillator. The dc power for the oscillator is obtained from a 13.5V battery such as the RCA VS304.

Inductor L and capacitor C5 form the oscillator resonant circuit. Feedback to sustain oscillations in the resonant circuit is coupled by C3 from the collector to the emitter of the 2N1178. RF voltage in the emitter-to-base circuit is coupled by C1 to the 1N34A diode, and the rectified output appears on the dc microammeter. When power is absorbed from the oscillator, RF feedback is reduced, and the reading on the microammeter decreases.

The coil used for L is selected for the operating frequency desired. A frequency-tuning dial mounted on the same shaft with C5 indicates the operating frequency of the meter. For measurement of the frequency of a resonant circuit, a coil having a suitable frequency range is inserted in the dip meter, and the meter control knob is adjusted for a reading of about half-scale. The meter is then tightly coupled to the unknown tuned circuit and the tuning dial is rotated until a dip in the reading occurs. When the transmitter tank circuits are measured, the transmitter plate supply must be turned off to eliminate the danger of shock.

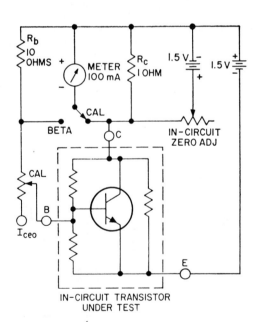

8.2.4 RCA in-circuit beta tester. The test circuit used to measure in-circuit current gain is similar to that used for out-of-circuit beta measurement. The in-circuit zero adjust control applies a voltage of reverse polarity to the collector metering circuit. This voltage compensates for the collector-to-emitter leakage through the components in the circuit under test and permits the meter to be set to zero.

The CAL adjustment and the metering circuit are the same as for out-of-circuit measurement.

The resistance of the measuring circuit is low in value so that no significant loading effect occurs from the circuit being tested.

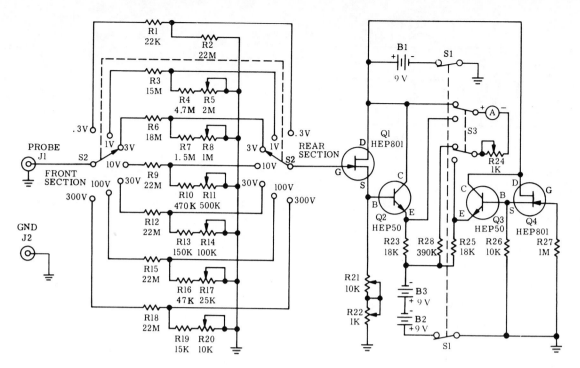

8.2.5 Motorola battery-operated dc voltmeter features an input impedance of 22 megohms on each of seven available ranges. The input ranges are 0.3, 1, 3, 10, 30, 100, and 300V for full-scale deflection.

Because of the lightly loaded source follower and emitter follower (100% voltage feedback), the circuit is capable of excellent linearity. Each input range is adjustable to provide correct calibration for that range. The linearity is determined primarily by the quality of the meter movement; accuracy is determined by the standard used for calibration.

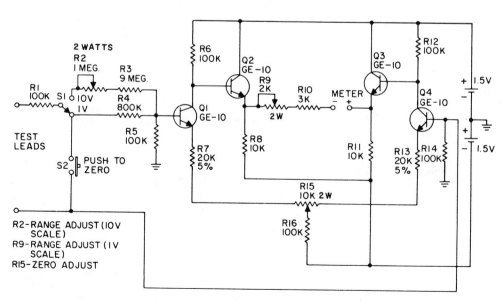

8.2.6 GE VTVM adapter for multitester offers an inexpensive method of obtaining low-voltage dc measurements at high impedance by using the meter movement only from the multimeter. Since this adapter draws only 1 μA from the measured circuit for full-scale deflection, it has a sensitivity of about 1 megohm per volt.

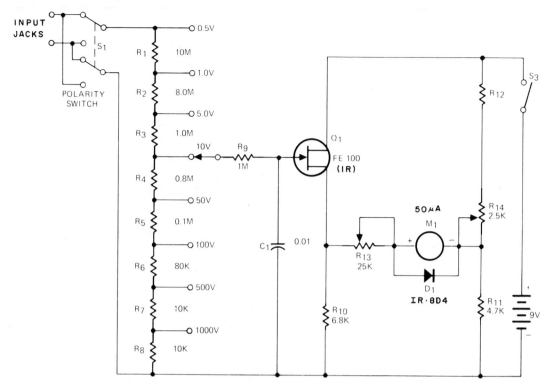

8.2.7 Workman quality dc voltmeter. Direct current voltage ranges are from the full-scale reading of 0.5 to 1000V. The low ratings of full-scale readings (0.5 and 1.0V) make this instrument especially useful for transistor work. Other features include easy portability, low battery drain, 11-megohm input impedance for all scales, 8 voltage ranges (established through voltage dropping resistors), meter protection through a shunting diode, and good accuracy over its entire range.

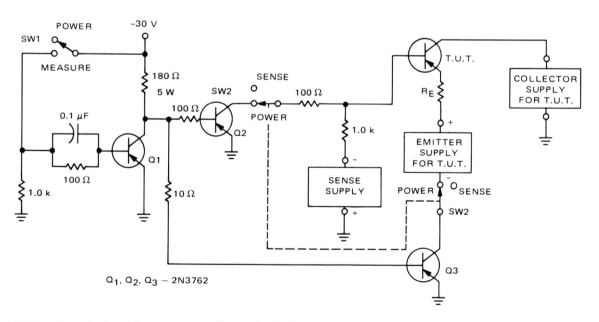

8.2.8 Motorola thermal response test fixture for PNP transistors dissipating up to 1A of collector current. A fixture for NPN transistors is built using NPN complements of the transistors shown and reversing all power supply polarities. The circuit can be scaled up to handle higher currents. (T.U.T. = transistor under test.)

SECTION **8** *Test and Measurement*

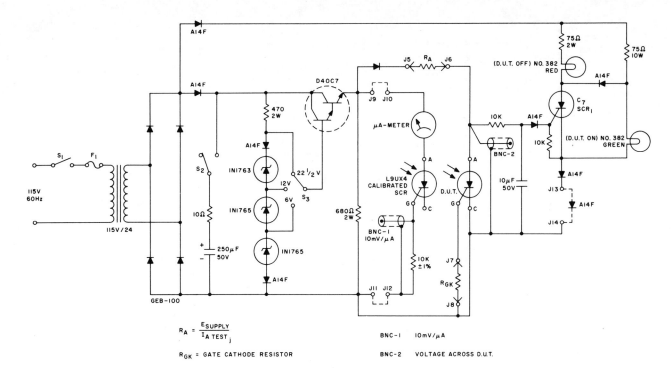

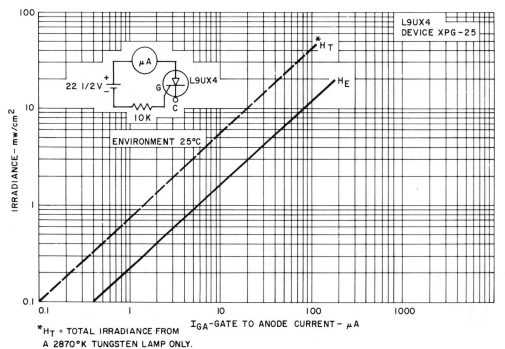

*H_T = TOTAL IRRADIANCE FROM A 2870°K TUNGSTEN LAMP ONLY.

8.2.9 GE test circuit measures the effective irradiance H_E and the "effective irradiance to trigger H_{ET}" on an LASCR. The heart of the circuit is a calibrated LASCR, the L9UX4, with calibration curve as shown.

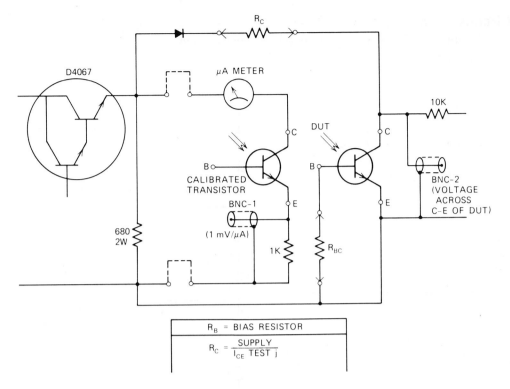

8.2.10 Circuit measures H_E and H_{ES} on a phototransistor. A calibrated transistor like the L14EX, available from General Electric, should be used. The detection circuit shown can be used, but a scope or voltmeter is recommended to insure that the transistor is really in saturation.

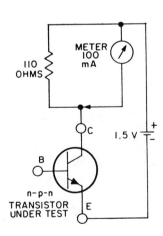

8.2.11 RCA I_{CEO}-measuring circuit. I_{CEO} represents the leakage from collector to emitter, with the base open. 1.5V is applied to the transistor, and the meter is connected in the collector circuit. The resistor shunting the meter reduces the meter sensitivity to 10 mA.

The measurement of I_{CEO} is normally made on the CAL position (circuit 8.2.9) of the 1 mA range. If I_{CEO} exceeds 1 mA, however, the range switch can be set to the 10 mA or 100 mA range. The collector-to-emitter leakage is indicated in milliamperes, depending on the current range that is used.

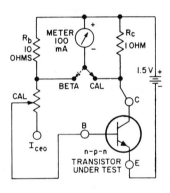

8.1.12 RCA beta-measuring circuit. Resistors R_b and R_c serve to establish collector current and to shunt the meter to the required sensitivity. Values for R_b and R_c are as follows:

Range	R_b	R_c
1 mA	1K	110 ohms
10 mA	110 ohms	10 ohms
100 mA	10 ohms	1 ohm
1A	1 ohm	0.1 ohm

When the range switch is set to the CAL function, the meter is in the collector circuit. The collector current is determined by the value of the collector resistor for the particular range and by the setting of the CAL control.

In the BETA function, the meter is switched to the base circuit. Direct-current beta is defined as the ratio of the steady-state collector current to the base current. Because the collector current is established at a known value by the CAL adjustment, the base-current meter reading can be interpreted in terms of dc beta for the transistor.

8.3 Miscellaneous Circuits

These circuits are those which cannot readily be classed as measuring devices; they include calibrators, references, and precision oscillators.

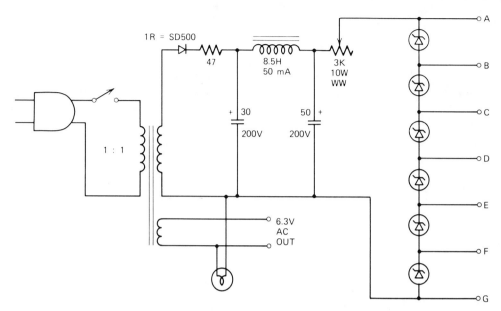

8.3.1 Workman zener voltage divider offers a wide range of potentials to be used for calibration and testing of voltmeters, as bias supplies, or as dc voltage output. It can supply adequate power for small-signal circuits. It is extremely useful to the determined experimenter and with its wide potential should find many uses in circuit work.

Almost any assortment of zener diodes may be used, provided that the sum total of their voltage ratings is in the range of 95 to 100V. If high-current output is desired, 10W zeners should be used; otherwise, 1W units would be satisfactory.

In operation, the desired output voltage is obtained by connecting to a selected pair of output terminals. For example, for 3.9V, connect across terminal F and G; for 6V, 6.8 or 7V across E and F. For 10V, actually 10.7, connect across E and G.

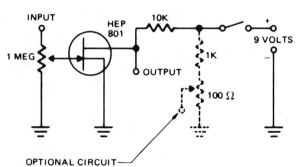

OPTIONAL CIRCUIT—
To obtain output down to zero, use this terminal as the "GROUND" point for the output.
(Adjust potentiometer to give zero output with input grounded)

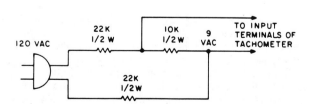

8.3.2 GE voltage-reducing network for calibrating electronic tachometers to 60 Hz line.

8.3.3 Motorola dc amplifier can be used with oscilloscope, VTVM, or other circuits requiring dc amplification; it can also be used as a relay amplifier. The input is 0 to −1.5V; the output, 0.4 to 9V. The output can go to zero if an offset circuit is supplied. Use a linear-taper potentiometer.

374   DISCRETE / TRANSISTOR CIRCUIT SOURCEMASTER

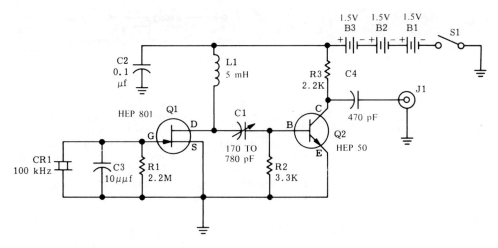

8.3.4 Motorola 100 kHz crystal oscillator offers excellent frequency stability and very low current drain. Oscillators of this type find their widest usage in the calibration of receivers and other electronic equipment. But it can also be used as a marker generator for TV repair in conjunction with a sweep generator and an oscilloscope. This unit will provide usable harmonics every 100 kHz up to 100 MHz.

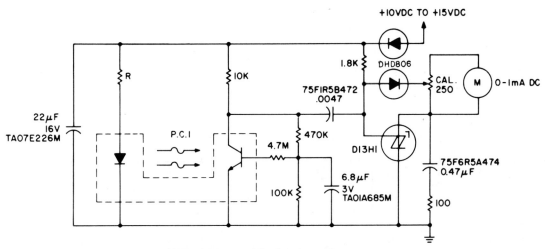

8.3.5 GE optical pickup tachometer offers remote non-contact measurement of the speed of rotating objects. The linearity and accuracy are extremely good and normally limited by the milliammeter used and the initial calibration. This circuit counts the leading edge of light pulses and ignores normal ambient light. Full scale at maximum sensitivity of the calibration resistance is read at about 300 light pulses per second. Longer range reflective operation may be obtained by using a focused incandescent lamp, operating straight from the supply voltage (filament time constant replaces filtering), to replace the IRED. A digital voltmeter may be used in place of the milliammeter by shunting its input with a 100-ohm resistor in parallel with a 100 μF capacitor (TA07A107M).

SECTION **8** Test and Measurement

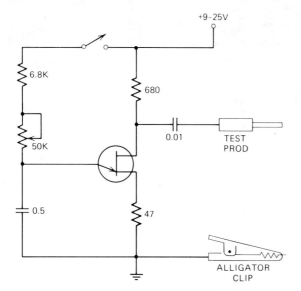

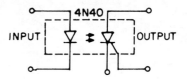

8.3.8 GE 400V symmetrical transistor coupler employs the high-voltage PNP portion of the 4N40 to provide a 400V transistor capable of conducting positive and negative signals with current transfer ratios of over 1%. This function is useful in remote instrumentation, high-voltage power supplies, and test equipment. Care should be taken not to exceed the 400 mW power dissipation rating when used at high voltages.

8.3.6 Motorola audio injector and signal generator. The frequency is varied by adjustment of the 50K potentiometer; for increased range change the value of the capacitor: add capacitance for increased frequency, and vice versa. For signal-generator applications exclusively, replace the test prod with an RCA jack and use the selector switch to choose between several capacitance values.

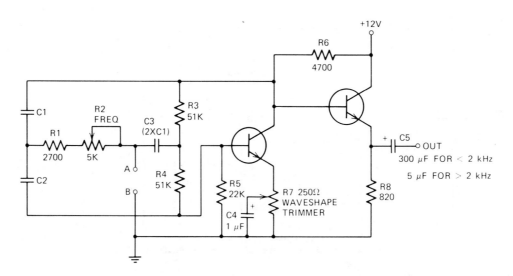

8.3.7 RCA audio oscillator provides a single-tone sine wave to well above 100 kHz. The circuit is excellently suited for use in the testing of high-fidelity audio equipment and amateur radio transmitters; it can also be adapted for use as a code-practice oscillator. (A key can be inserted between points **A** and **B**.) The oscillator operates from 12V and supplies a relatively distortion-free output waveform to any circuit that has an input impedance of 3000 ohms or more.

Potentiometer R2 provides an adjustment of ±10% in the oscillator frequency. Potentiometer R7 in the emitter circuit of transistor Q is adjusted to obtain the desired output waveform.

FREQ (H2)	C1, C2
≈100K	50 pF
50K	100 pF
10K	500 pF
5K	1000 pF
1K	0.005 µF
500	0.01 µF
100	0.05 µF
50	0.1 µF
10	0.5 µF
5	1 µF (nonpolar)

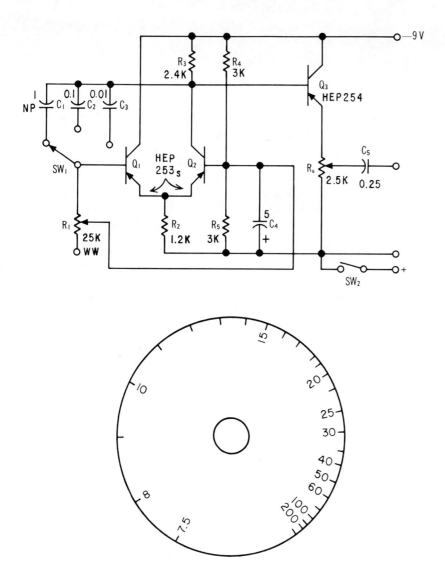

8.3.9 Motorola audio signal generator spans an audio frequency range from about 7.5 Hz to 20 kHz in three ranges, which covers the great majority of the common uses for audio signal generators. The output waveform is essentially a square wave with fast rise and fall times; thus the output signal has a high harmonic content, which is useful for checking audio amplifier response. The inset shows dial markings.

SECTION 9
AUTOMOTIVE

The circuits in this section are designed with the motorist in mind, but they draw on many of the ideas expressed in preceding sections. You can use the circuits presented here to stimulate your own creative thought processes and then refer to other appropriate sections with the idea of making slight modifications to tailor other circuits to these applications.

9.1 In-Car Circuits

The single common feature of all the circuits in this subsection is the +12V power source; for the most part, variation of these circuits can be studied by examining those in several of the foregoing sections. Burglar alarm variations can be created by modification of many of the control circuits of Sections 2.1, 2.5, and 2.6. Light-sensitive circuits can be those shown here or altered versions of those in Sections 2.1 and 2.2. Section 3 contains schematic diagrams for flashers and ring counters that can be adapted to achieve the same purpose as the safety flasher and the sequential turn signal in this subsection. Section 5 contains additional radio circuits.

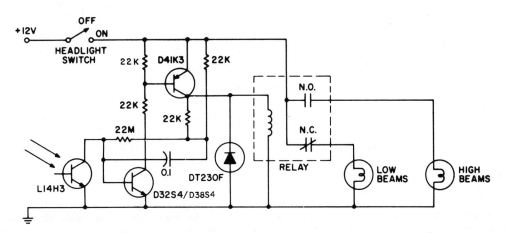

9.1.1 GE automatic headlight dimmer switches your car's headlights to low beam when the lights of an oncoming car are sensed. A relatively large amount of hysteresis is built into the circuit to prevent flashing. Sensitivity is set by the 22M resistor to about 0.5 candela at the transistor, while hysteresis is determined by the two 22K resistors across the D41K3, which drives the 22M resistor. The maximum switching rate is limited by the 0.1 μF capacitor to 15 per minute. The relay uses a 12V, 0.3A coil; the contacts should handle 20A. For the phototransistor, make the lens a minimum of 1 in. (2.54 cm) diameter, positioned for about a 10° view angle.

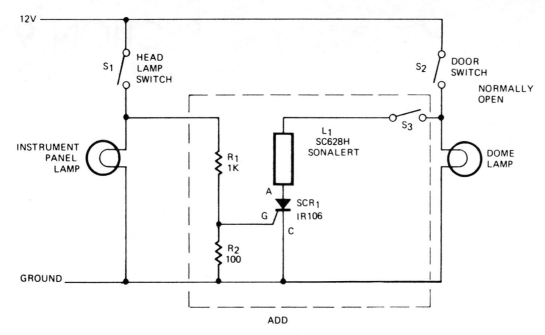

9.1.2 Workman simple "headlight-on reminder" alarm system uses a small IR silicon controlled rectifier and connections from the door switch and the headlamp switch to actuate a modular buzzer. Closing the door switches off the alarm. The alarm will activate when both the headlamp switch (which supplies a trigger current) and the door switch (which supplies current to the cathode of the SCR) are on at the same time. Closing the door turns off the buzzer.

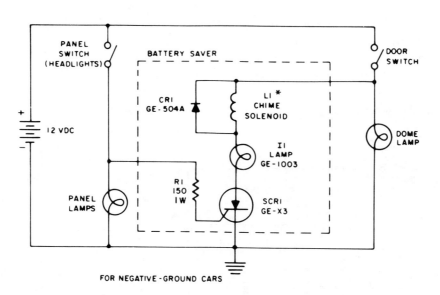

9.1.3 GE audible "circuit on" reminder saves the car battery. When the driver's door is opened, a courtesy switch on the door energizes the dome light and chime circuit. This action does not ring the chime unless the gate of the SCR is also energized. Since the SCR gate current comes from the panel lights, which are only on with the headlights or parking lights, the chime will only ring when both light circuits are energized.

In addition to the audible chime, a visual signal can be obtained from the lamp if the unit is mounted in a conspicuous location inside the car. This lamp is primarily a protective device: the chime is designed for momentary 6 to 8V operation only; if it is energized from 12V direct current continuously, the chime would soon overheat. Since the lamp has a low resistance when cold and a higher resistance when hot, it produces a high current in the chime solenoid when first energized. This current quickly reduces to a low level to protect the solenoid coil when the lamp comes on.

*L1 solenoid of Snapit Model 600R chime or equivalent.

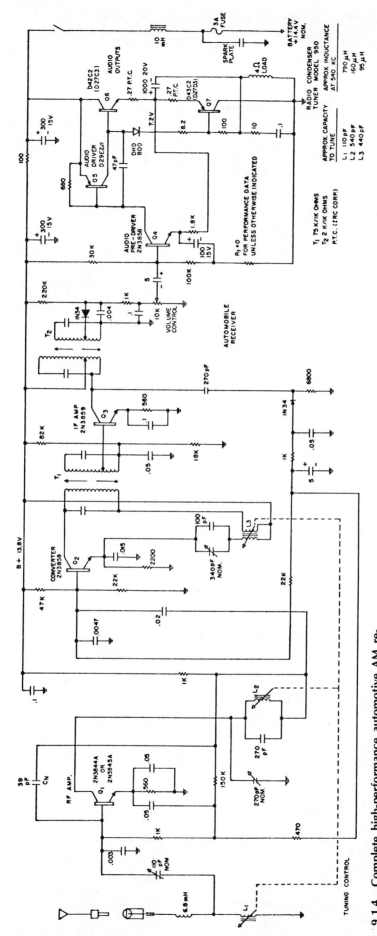

9.1.4 Complete high-performance automotive AM receiver with transformerless audio output. Semiconductor number are GE's.

SECTION 9 Automotive

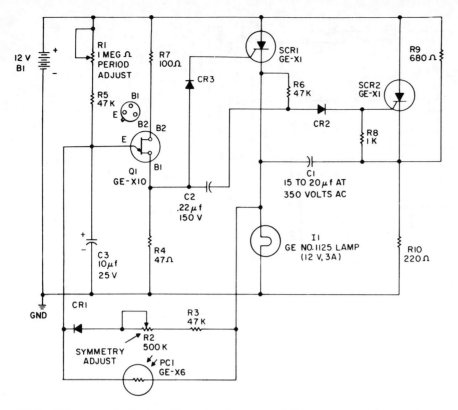

9.1.5 GE automatic flasher. Operating from your 12V car or boat battery (or two 6V dry packs in series), the flasher offers:

1. 36 to 40W output.
2. Variable flash rate up to 60 flashes per minute.
3. Independent control of both on and off cycles.
4. Photoelectric night and day control.

Use 2W potentiometers, ½W resistors.

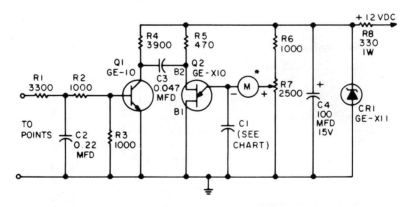

9.1.6 GE precision tachometer operates from either conventional ignition or electronic systems. The semiconductor timing circuit, connected directly to the distributor points, counts the times the distributor points close each minute. Since the number of current pulses is directly proportional to engine speed, the number of revolutions per minute is indicated by the meter. Each time a negative-going pulse from Q1 is coupled to base 2 of Q2, the unijunction transistor fires, discharging C1. During recharge, meter M indicates the brief period of charge current. Since the recharge pulses are all of equal duration, the average meter current reading is directly proportional to the number of breaker point closures per minute (rpm).

C1 CAPACITOR VALUES, μF:

Engine cycle	Cylinders			
	3	4	6	8
Two-stroke	0.47	0.33	—	—
Four-stroke	—	0.68	0.47	0.33

*M Meter (GE type D092, cat. No. 50 171 111EMEM) rated 0–500 μA; 220 ohms terminal resistance. New faceplate calibrated 0–6000 rpm.

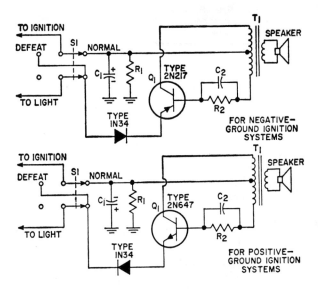

C1	30 µF, 25V	Speaker	1½-inch permanent magnet type; coil impedance, 3.2 ohms
C2	0.22 µF, 25V		
R1	680 ohms, ½W		
R2	1.5K, 1W	T1	Audio-output transformer; 400-ohm primary, 3.2-ohm secondary; Stancor No. TA-42 or equivalent
S1	DPDT switch		

9.1.7 RCA automotive light-minder circuit sounds an alarm if the lights of a car are left on when the ignition is turned off. The alarm stops when the lights are turned off. When the lights are intentionally left on, the alarm can be defeated so that no warning sounds. The alarm then sounds when the ignition switch is turned on as a reminder that the system has been defeated and that the switch should be returned to its normal position.

The circuit is essentially an oscillator that obtains its supply voltage from the ignition or the light system of the car. In the normal mode of operation, the ignition system is connected to the collector circuit and the light system is connected through the 1N34 diode to the emitter. When the ignition switch is on, the collector is at the supply voltage. If, at the same time, the lights are on, the emitter is also at the supply voltage. Because both the emitter and the collector are at the same voltage, the circuit does not oscillate and no alarm sounds. When the ignition is turned off, the collector is returned to ground through R1 and C1, but the emitter remains at the supply voltage and provides the necessary bias for the circuit to oscillate. Turning the lights out removes the supply voltage and stops the oscillation.

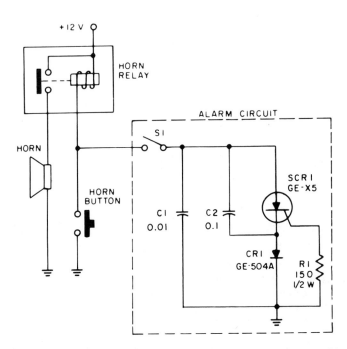

9.1.8 GE auto theft alarm operates whenever any part of the car's electrical system is energized, such as a light or starter motor. When a light or the engine starter is energized, the battery voltage suddenly drops. The charge on capacitor C2 causes the cathode of SCR1 (GE-X5) to go negative with respect to the gate. The current then flows in the gate and out the cathode and the SCR conducts to short out the horn button. This actuates the car's horn and it continues to blow. To silence the car horn, it is necessary to turn S1 off or momentarily depress the horn button. What burglar in his right mind would think of pushing the horn button to silence the horn? Since the alarm triggers when the electrical system is energized, you should open the car door before setting switch S1 to the on position.

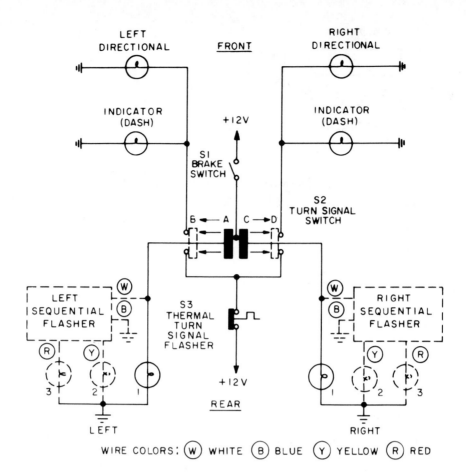

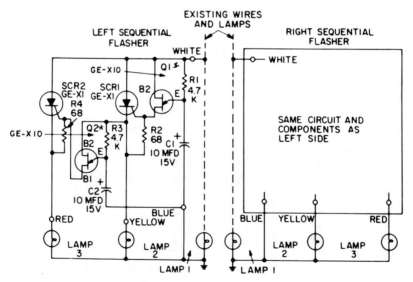

9.1.9 GE sequential turn-signal add-on. A standard turn-signal system found on conventional cars with negative ground is shown at top; the dotted lines show the added sequential flashers and lamps. When the brake is activated, S1 closes and applies +12V to lamp 1 at both the left and right rear through switch S2, in the off position. These are the stoplights which stay on continuously while the brake is depressed. With the additional sequential lamps connected, all six lamps serve as stoplights when the brake is actuated.

When switch S2 is operated for a right or left turn (A to B, or C to D), +12V is applied to the directional lamps through thermal flasher S3 and the turn-signal switch. The on-off action of the thermal flasher S3 controls the operation of the sequential flasher circuits. Should the brakes be applied during the time a sequential flasher is operating, all three lamps on the opposite side are turned on. The turn-signal lamps continue to operate sequentially until the turn is canceled at the steering wheel.

*Q1, Q2: GE-X10 unijunction transistor.

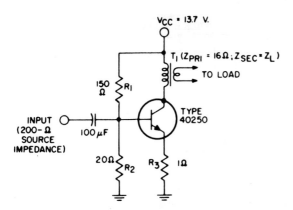

9.1.10 RCA simple audio amplifier operates from +12 to 13.5V, has a source impedance of 200 ohms, and delivers 400 mW of power at 1000 Hz to a 16-ohm load. The output increases to 4W if 6.5% distortion can be tolerated. The transistor is RCA type 40250.

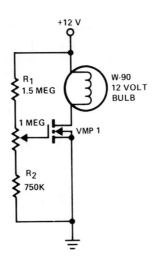

9.1.12 Siliconix ultrasimple automotive lamp dimmer for motor homes, vans, boats, and other vehicles with a 12V dc power source delivers power at a rate that is directly proportional to the potentiometer shaft position. If the Siliconix power FET is used in this circuit, it must be heatsinked extremely well; because it is operating in its linear region, the dissipation will be high as the light is dimmed.

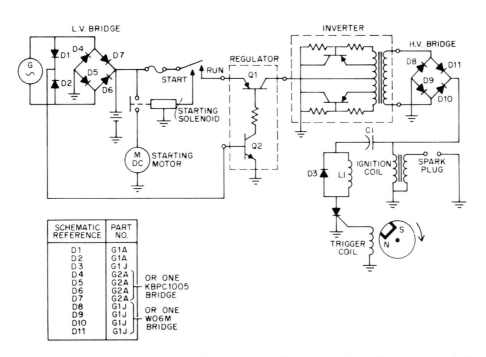

SCHEMATIC REFERENCE	PART NO.	
D1	G1A	
D2	G1A	
D3	G1J	
D4	G2A	OR ONE KBPC1005 BRIDGE
D5	G2A	
D6	G2A	
D7	G2A	
D8	G1J	OR ONE W06M BRIDGE
D9	G1J	
D10	G1J	
D11	G1J	

9.1.11 Capacitive-discharge ignition system. Switch S is the regular auto starting switch. It connects the starting solenoid to the battery, thus energizing the starting motor, after which it is in the normal running position.

The ac generator, driven from the engine shaft, is connected to the General Instrument low-voltage rectifier bridge, which supplies an inverter through a regulator. The inverter converts the direct current to 300V ac, which is again rectified by the high-voltage bridge to charge capacitor C1. The capacitor discharge, activated by the thyristor, passes through the primary of the ignition coil and is raised to 25 kV to fire the spark plug. The thyristor is turned on by a pulse generated in the trigger coil when a magnet on the flywheel passes the trigger coil. A de-spiking network (rectifier D3 and choke L1) limits the current when the capacitor discharges.

Because the ac generator voltage is proportional to the engine speed, the centertap rectifier (D1 and D2) supplies a corresponding voltage to the regulator as a bias to transistor Q2, which in turn controls pass transistor Q1. This provides a relatively constant output to the inverter circuit.

Advantages of this system are its low current drain and its ability to produce a relatively constant spark-plug firing voltage over widely varying engine speeds. The improved firing provides a better chance of starting with fouled spark plugs or a flooded engine. (The rectifiers are General Instrument.)

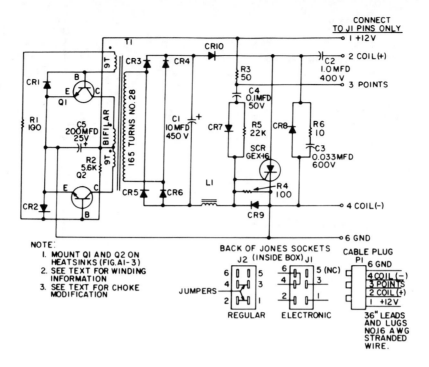

Parts List

C1	10 μF 450V electrolytic
C2	1.0 μF 400V nonpolarized
C3	0.033 μF 600V
C4	0.1 μF 50V
C5	200 μF, 25V electrolytic
CR1–10	GE-504A diodes
L1	250 mH choke; modified 0.5 mH choke with 0.030-inch shim (Triad C36X, or equivalent).
Q1, Q2	GE-X18 transistor
R1	100Ω 2W
R2	5600 Ω 1W
R3	50 Ω 5Watt wirewound
R4	100 Ω
SCR	Ge-X16
*T1	Pulse transformer
J1, J2	6-pin sockets
P1	Making plug
	Heatsinks (2)

*Available from General Electric Co., 3800 N. Milwaukee Avenue, Chicago, Ill. 60641. Specify ETRS-5450 for the two E cores and ETRS-5451 for the bobbin.

Note: Printed-wiring boards with completely wound and mounted T1 transformers may be purchased from Felmoe Electronics, N. Division Street Road, Auburn, N.Y. 13021.

9.1.13 GE capacitive discharge SCR ignition system overcomes many limitations of the conventional Kettering system and is efficient at all engine speeds. In the capacitive-discharge ignition, energy is stored at a high voltage in a capacitor and then dumped as a short-duration, high-amplitude current pulse into the ignition coil primary winding by triggering an SCR. Because the only current that flows in the coil primary during this process is the capacitor charge and discharge currents (no direct current), the system draws less than 1A from the car battery at maximum rpm, less than 500 mA at idle. Because the output voltage pulse developed across the coil secondary rises in a few microseconds to a high peak value that is relatively independent of engine speed, the system is able to fire plugs that would misfire and have to be discarded with conventional ignition systems. Engine timing errors are minimized, since the only current the points must handle is a low-power SCR trigger signal that is insufficient to cause contact arcing and erosion.

The original automobile coil, breaker points, and condenser are retained for use with the electronic system. Conventional ignition operation can be restored at any time.

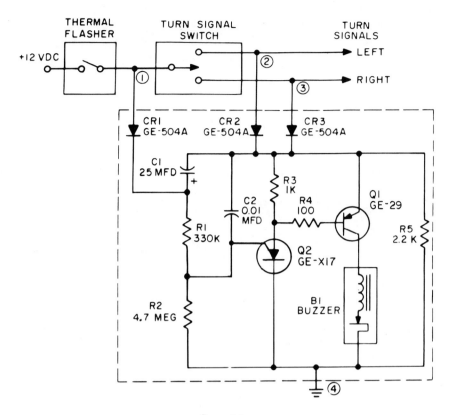

Parts List

B1 — 12-volt d-c buzzer
C1 — 25-mfd, 15-volt electrolytic capacitor
C2 — 0.01-mfd, 50-volt capacitor
CR1, CR2, CR3 — GE-504A rectifier diode
Q1 — GE-29 transistor
Q2 — GE-X17 programmable unijunction transistor

R1 — 330K-ohm, 1/2-watt resistor
R2 — 4.7-megohm, 1/2-watt resistor
R3 — 1K-ohm, 1/2-watt resistor
R4 — 100-ohm, 1/2-watt resistor
R5 — 2.2K-ohm, 1/2-watt resistor

9.1.14 GE turn-signal reminder. With the circuit connected in the car, capacitor C1 charges to the full battery voltage through rectifier CR1 and resistor R5. When the turn signal is activated, capacitor C1 starts to charge in the opposite direction through rectifiers CR2 or CR3 and the two resistors R1 and R2. This finally drops the gate of Q2 to a low value and the programmable unijunction transistor turns on, thus providing base current for transistor Q2 through resistor R4. Transistor Q1 conducts and activates the buzzer. As the flasher opens and closes, the buzzer beeps.

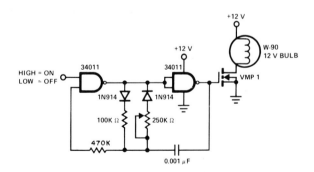

9.1.15 Siliconix high-efficiency/automotive lamp dimmer employs an IC oscillator and is controlled by low-level logic signals, with a high at the input for *on* and a low at the input for *off*. Dimming, achieved with a 250K potentiometer, is a function of the ratio of R1 to R2. This circuit is ideal for vans and mobile homes that use 12V lamps. Since the VMP1 power FET (Siliconix) is operated either at pinchoff or saturation, very little power must be dissipated by the device itself.

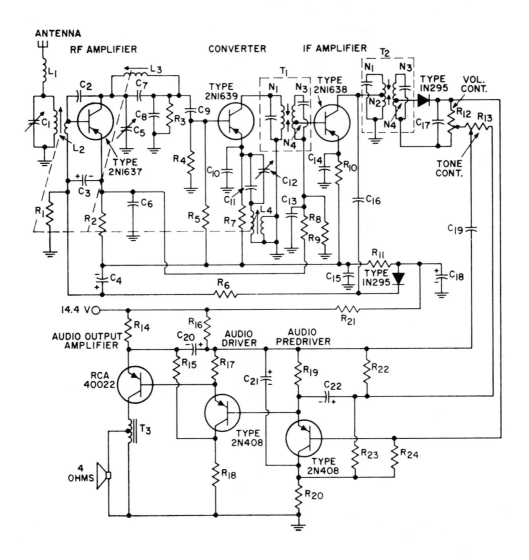

9.1.16 RCA superheterodyne auto radio receiver. The RF amplifier uses a high-gain 2N1637 transistor to provide the increased sensitivity and high signal-to-noise ratio required in automobile radio receivers. The tuned RF amplifier selects and amplifies the amplitude-modulated signals from the desired broadcast station picked up by the automobile whip antenna. In the 2N1639 converter stage, the signal from the RF amplifier is mixed with a local-oscillator signal developed by the tuned circuit of L4, C11, and C12 to provide a signal at the receiver intermediate frequency of 262.5 kHz (this value, rather than 455 kHz, is used because the IF amplifier provides greater gain and selectivity at the lower frequency).

The antenna circuit, RF amplifier, and converter are

Parts List

C1	5 to 80 pF trimmer capacitor (Arco 462 or equivalent)
C2	2 pF, silver mica
C3	2.2 µF, 3V electrolytic
C4	25 µF, 6V electrolytic
C5, 12	110 to 580 pF trimmer capacitor (Arco 467 or equivalent)
C6, 9, 13, 14, 15, 19	0.05 µF ceramic disc
C7	200 pF, silver mica
C8	0.005 µF, ceramic disc
C10	0.0075 µF, ceramic disc
C11	330 pF, Silver mica
C16	180 pF, silver mica
C17	0.02 µF, ceramic disc
C18	100 µF, 15V electrolytic
C20	500 µF, 3V electrolytic
C21	50 µF, 6V electrolytic
C22	100 µF, 3V electrolytic
L1	5 µH RF choke
L2, 3, 4	Ganged tuning coil assembly (F. W. Sickles Co. and Radio Condenser Corp.)
L2	Antenna coil; primary: variable inductor (tunes with 110 pF capacitance from 535 to 1610 kHz); secondary: 3½ turns
L3	RF coil; variable inductor, tunes with 600 pF capacitance from 535 to 1610 kHz.
L4	Oscillator coil; primary: variable inductor, tunes with 470 pF capacitance from 797.5 to 1872.5 KHz; secondary: 30 turns
R1	82K, ½W
R2	560 ohms, ½W
R3	180 ohms, ½W
R4	56K, ½W
R5	5.7K, ½W
R6	8.2K, ½W
R7	1.5K, ½W
R8	5.6K, ½W
R9	100K, ½W
R10	470 ohms, ½W
R11	100 ohms, ½W
R12	2.5K, ½W audiotaper pot
R13	1K, ½W audio taper pot
R14	3.3 ohms, 1W
R15	82 ohms, ½W
R16	68 ohms, ½W
R17	120 ohms, ½W
R18	220 ohms, ½W
R19	1.2K, ½W
R20	4.7K, ½W
R21	680 ohms, ½W
R22, 24	3.3K, ½W
R23	3.3K, ½W
T1	First IF transformer (includes 220 pF capacitor across each winding); turns ratio of tapped secondary: N3/N4 = 18.25
T2	Second IF transformer (includes 110 pF capacitor across each winding); turns ratio of tapped primary: N1/N2 = 4.28; turns ratio of tapped secondary: N3/N4 = 10.2
T3	Output transformer; transforms 22 ohms at 425 mA dc to 3.5 ohms (Thordarson-Meissner No. TR-168 or equiv.)

tuned by means of mechanically ganged variable inductors L2, L3, and L4 so that the local-oscillator frequency is always 262.5 kHz above the frequency to which the other circuits are tuned. Trimmer capacitors C1, C5, and C12 are adjusted to provide the proper tracking relationship.

This circuit uses coils and transformers that are not available as stock items from any manufacturer. Home construction of this circuit should not be attempted unless the builder has had considerable experience in the winding of inductive components and has access to the special equipment required.

9.2 In-Garage Circuits

These circuits consist of 6 and 12V battery chargers, a garage light control, and a garage-door controller. For additional chargers, see Section 1.2. The more sensitive of the photo switches of Section 2.2 will serve nicely as garage-light control systems. Other applicable door-control circuits appear in Section 2.8. (See also Sections 4 and 8.)

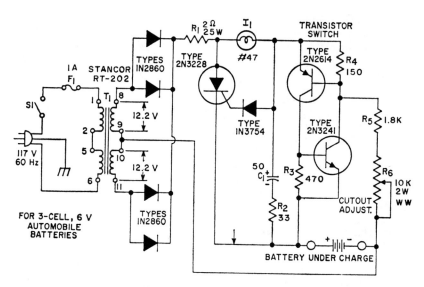

9.2.1 RCA 6V battery charger can be used to charge 3-cell 6-volt lead–acid storage batteries at a maximum charging rate of 3.2A. The four 1N2860 diodes are connected in a full-wave centertapped rectifier circuit that provides charging current of 3.2A to the 6V battery. Heatsinks are required for 1N2860 diodes; fuse clips will do the job.

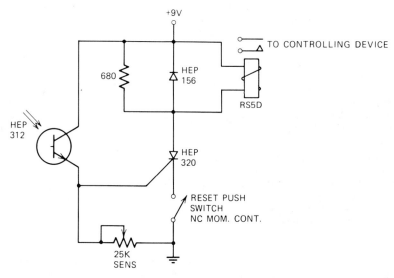

9.2.2 Motorola headlight-operated garage light control can be used to control the door opener or any other switching function with lights at night. Sensitivity adjustment prevents false triggering. The power consumption is low enough to make operation from a 9V "transistor" battery practical. The relay used for this device should have a coil resistance of between 300 and 400 ohms; the prototype used a Potter and Brumfield RS5D 6V relay, which has a coil resistance of about 335 ohms.

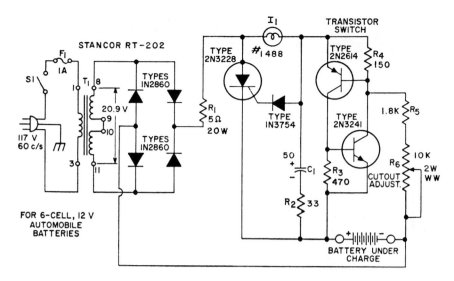

9.2.3 RCA 12V battery charger can be used to charge lead–acid storage batteries at a maximum charging rate of 2A. When switch S1 is closed, the rectified current produced by the four diodes in the bridge charges C1 through R1 and R2 and indicator lamp I1. As C1 charges, the anode of the 1N3754 diode is rapidly raised to a positive voltage high enough so that the diode is allowed to conduct. Gate current is then supplied to the SCR to trigger it into conduction. The SCR and the battery under charge then form essentially the full load on the bridge, and a charging current flows through the battery that is proportional to the difference in potential between the battery and the rectifier. Resistor R1 limits the current to a safe value to protect the diodes in the event that the load is an exhausted battery. The energy stored in C1 insures that the SCR conducts and that the charging current flows or the full 180° of each successive half-cycle of input until the battery is fully charged. (The SCR is actually cut off near the end of each half-cycle but is retriggered shortly after the beginning of each succeeding half-cycle by the gate current applied through the 1N3754 diode.)

When the battery is fully charged, the two-transistor regenerative switch is triggered into conduction (the triggering point is preset by R6). As a result of the regenerative action, the transistors are rapidly driven to saturation and thus provide a low-impedance discharge path for C1. The capacitor discharges to about 1V, too low to sustain conduction of the 1N3754 diode, and the SCR is not triggered on the succeeding half-cycle of the input. The saturated transistor switch also provides a low-resistance path for the current to the No. 1488 indicator lamp, which glows to signal the fully charged condition of the battery. The current in the lamp circuit provides a trickle charge of approximately 150 mA to the battery.

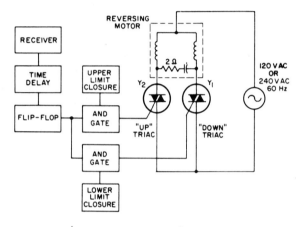

	120VAC, 60Hz	240VAC, 60Hz
Y_1	T2800B	T2800D
Y_2	T2800B	T2800D

9.2.4 RCA electronic garage-door system uses the principle of motor reversing for garage-door direction control. The system contains a transmitter, a receiver, and an operator to provide remote control for door opening and closing. The block diagram shows the functions required for a complete solid-state system. When the garage door is closed, the gate drive to the *down* triac is disabled by the lower-limit closure and the gate drive to the *up* triac is inactive because of the state of the flip-flop. If the transmitter is momentarily keyed, the receiver activates the time-delay monostable multivibrator so that it then changes the flip-flop state and provides continuous gate drive to the *up* triac. The door then continues to travel upward until the upper-limit switch closure disables gate drive to the *up* triac. A second keying of the transmitter provides the *down* triac with gate drive and causes the door to travel downward until gate drive is removed by the lower-limit closure. The time in which the monostable multivibrator is active should override normal transmitter keying for the purpose of eliminating erroneous firing. A feature of this system is that, during travel, transmitter keying provides motor-reversing independent of the upper- or lower-limit closures.

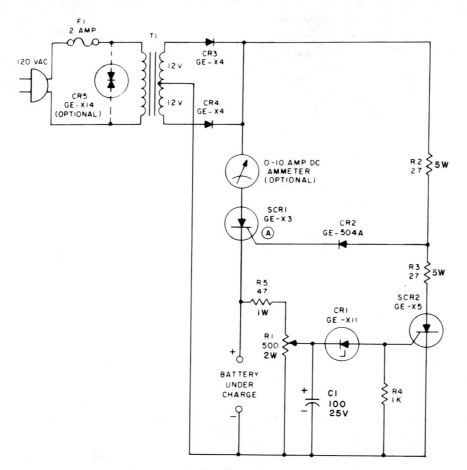

9.2.5 GE 12V auto-battery charger. This inexpensive unit will rapidly charge a 12V lead–acid battery at maximum design current until the battery is fully charged; then it will automatically shut itself off. Should the battery become discharged while the charger is connected, the charger will automatically switch itself back on again. This particular feature makes this charger excellent for maintaining emergency standby battery power supplies. For car or boat battery charging, it permits a rapid charge without possible battery damage due to overcharge.

INDEX

Ac adapter for cars, 42
Ac chaser, 187, 190
Ac chaser, 100 W lamps, 188
Ac chaser, half-wave, 187, 189
Ac chaser, high-current, 191
Ac-controlled triac switch, 75
Ac control relay, solid-state, 60-63
Ac-coupled FET integrator, 224
Ac-coupled hi-fi amplifier, 60 W, 346
Ac/dc AM radio, 242
Ac/dc crowbar protection ckt, 166
Ac/dc high/low voltage timer, 206
Ac/dc load controller, 115 V, 97
Ac/dc motor speed control, 108
Ac/dc servo amplifier, 170
Ac/dc variable-power control, 129
Ac flasher, 180, 182
Ac flasher, sequential, 191
Ac lamp dimmer, 92, 95, 100
Ac lamp dimmer, full-range, 12
Ac lamp dimmer, low-cost, 99
Ac lamp dimmer, low-hysteresis, 100
Ac lamp dimmer, no hysteresis, 92
Ac lamp dimmer, photocell-controlled, 103
Ac lamp dimmer, RFI-proof, 84, 103
Ac lamp dimmer, soft-start, 101
Ac lamp dimmer, UJT, 110
Ac lamp dimmer, with feedback, 110
Ac large-angle compensator, 109
Ac latch relay, solid-state, 65, 72
Ac light-controlled switch, 78, 79-83
Ac line-operated 1-shot timer, 206
Ac line-voltage, 140
Ac line-voltage generator, 46
Ac load control, 1000 W, 85
Ac motor speed control, 98, 109, 114
Ac motor speed control, 3.5 A, 106
Ac motor speed control, 500 W, 98
Ac motor speed control, induction motor, 87
Ac motor speed control, shunt motor, 94
Ac motor speed control, with dc signal, 119
Ac-operated AM radio, 242, 246
Ac-operated flasher, 120 V, 181
Ac-operated relay control, 61
Ac phase control, 101
Ac phase control, 1-5 A, 111
Ac phase control, 400 Hz, 93
Ac phase control, 800 W, 130
Ac phase control, for dc shunt motor, 131
Ac phase control, with feedback, 105
Ac phase control, hysteresis-free, 106
Ac phase control, low-gain, 100
Ac phase control, low-hysteresis, 126
Ac phase control, no-hysteresis, 92, 129
Ac phase control, plug-in, 115
Ac phase control, RFI-proof, 125
Ac phase control, for series motors, 89
Ac phase control, symmetrical, 114
Ac phase control, three-phase (4-wire), 112
Ac phase control, with timer, 119
Ac phase control, wide-range, 124
Ac power control, compensated, 100, 122
Ac power control, full wave, 118
Ac power control, pulsating dc, 97
Ac power control, RFI filtered, 84
Ac power switch, automatic, 77
Ac power switch, Darlington-coupled, 68
Ac power switch, light-controlled, 78, 79, 81
Ac power switch, with time delay, 203, 210
Ac power switching, with light, 83
Ac ramp-and-pedestal control, 116
Ac regulator, 90 V, 4, 9
Ac regulator, 115 V, 107
Ac regulator, line-voltage, 6, 15, 16
Ac relay, low-power dc controlled, 61
Ac relay, optoelectronic, 72
Ac sequential flasher, 115 V, 187
Ac soft-start control, 120
Ac solid-state power "relay," 103
Ac source from 32 V dc, 44
Ac static contactor, 74

Ac switch, multiposition, 74
Ac switch, SCR controlled, 73, 75
Ac switch, 1-triac, 75, 71
Ac switching, with dc power, 77
Ac switching photo timer, 219
Ac switching timer, 204
Ac switching timer, 0.2 to 10 s, 208
Ac time delay, 204, 210
Ac time-delay relay, 120 V, 209, 495
Ac timer, blender-motor, 211
Ac triac power switch, 103
Ac unity-gain amplifier, 324
Ac voltage, from batteries, 41
Ac voltage control, 89
Ac voltage control circuits, 84
Ac voltage control l kW, 128
Ac voltage speed control, 91
Ac ZVS, SCR-controlled, 73, 75
Accurate-tracking 3-phase load control, 98
Active balanced mixer, 311, 312
Adapter, 115 V automotive, 45, 46
Adapter, 12 V-to-117 V ac, 42, 46
Adapter, flash slave, 218
Adapter, sequential turn-signal, 386
Adapter, variable-sweep, 31
Adapter, VOM-to-VTVM, 372
Adjustable ac timer, 208
Adjustable battery charger, 32
Adjustable current regulator, 5
Adjustable-delay power switch, 203, 210
Adjustable-feedback inverter, 45
Adjustable kitchen timer, 211, 503
Adjustable low-voltage regulator, 12
Adjustable mike preamp, 223
Adjustable-period flasher, 6 V, 185
Adjustable-period 5-sec timer, 204
Adjustable-period timer, 219
Adjustable-period timer (up to 1 min), 204
Adjustable-period timer (20 min), 204
Adjustable-period timer (3 min), 214
Adjustable-period timer, universal, 209
Adjustable power source, 22
Adjustable-rate ac chaser, 191
Adjustable-rate flasher, automotive, 384
Adjustable-rate flasher, 1000 W, 188
Adjustable regulated supply, 10-34 V, 17, 29
Adjustable regulated supply, 9 V, 27
Adjustable regulated supply, 10-28 V, 27
Adjustable regulator, 5, 6, 10-28
Adjustable dc regulator, 8-16 V, 5
Adjustable dc regulator, 13 V, 15
Adjustable regulator, 60 V, 1 A, 14
Adjustable-speed keyer, 222
Adjustable dc supply, 4-12 V, 20
Adjustable temperature alarm, 134
Adjustable timer, 2 s to 4 min, 183
Adjustable voltage alarm, 166
Adjustable wide-range timer, 203
Advertising-light control, 83
AGC, FET-and-bipolar, 257
AGC amplifier, color TV, 267
AGC and buffer, TV, 258
AGC keyer, 258
AGC keyer, amp, and phase inverter, 267
Aimed-light sensor, 152
Aircraft power amplifier, 40 W, 291
Aircraft power supply, 200 V, 36
Aircraft B+ supply, 36
Aircraft transmitter, 284
Aircraft transmitter, 4 W, 289
Aircraft transmitter, 13 W, 289
Aircraft transmitter, 130 MHz, 283
Aircraft transmitter, broadband, 285
Air-temperature regulator, 135
Alarm, audio (strobed), 166
Alarm, automotive circuit-on, 382
Alarm, auto theft, 385
Alarm, body-proximity, 163, 164
Alarm, headlamps-on, 382
Alarm, headlight-operated, 83
Alarm interrupted-beam, 161

Alarm, intrusion, 161
Alarm, intrusion, 12 V dc, 162, 164
Alarm, intrusion/hazard, 160
Alarm, light-controlled, 78, 80
Alarm, light-controlled SCR, 78
Alarm, liquid-level, 147
Alarm, moisture, 147
Alarm, multi-entry burglar, 160
Alarm, overtemperature, 134
Alarm, overvoltage, 166
Alarm, phone-ring, 167
Alarm, phototransistor, 161
Alarm, power-failure, 154
Alarm, power supply, 21
Alarm, smoke, 162, 163, 165
Alarm, smoke and gas, 165
Alarm, universal (12 V), 166
Alarm clock, "electronic," 232
Alarm control, with light, 79, 80, 81, 83
Alarm control, N.C. photo, 82
Alarm control, photocell, 81
Alarm control, phototransistor, 79
All-frequency oscillator/amplifier, 310
Alternating switcher, 44
Amateur radio circuits, 219
Amateur radio UHF amplifier, 306
Amperex 4 W audio amplifier, 336
Amperex 40 W hi-fi amplifier, 342
Amperex mike amplifier, 223
Amperex mixer/amplifier, 304
Amplified crystal set, 6 V, 244
Amplified crystal set, solar, 244, 245
Amplifier, AGC and keyer, 267
Amplifier, audio, 326
Amplifier, audio (1 W), 328, 329
Amplifier, audio (1-2 W), 330
Amplifier, audio (2 W), 329, 331, 334
Amplifier, audio (4 W, for radio), 335
Amplifier, audio (5-10 W), 337
Amplifier, audio, half-watt, 387
Amplifier, audio (portable-TV), 259
Amplifier, audio voltage, 322
Amplifier, broadband, 1-40 MHz, 309
Amplifier, building-block, 326
Amplifier, car-radio-sound, 4 W, 336
Amplifier, cascode (30 MHz), 306
Amplifier, cascode (200 MHz), 307, 309
Amplifier, cascode buffer, 220
Amplifier, cascode video, 250
Amplifier, color difference, 266
Amplifier, color TV booster, 267
Amplifier, complementary 2 W, 333
Amplifier-compressor, 323
Amplifier, data transmission, 171
Amplifier, dc, 376
Amplifier, driven, color-organ, 235
Amplifier, FET (4 W), 334
Amplifier FET 80 W, 359
Amplifier, FET, high-gain, 317
Amplifier, headphone, 327
Amplifier, hi-fi, 4 W, 336
Amplifier, hi-fi, 5 W, 337
Amplifier, hi-fi, 5 W (8 ohms), 338
Amplifier, hi-fi, 7 W, 339
Amplifier, hi-fi, 7.5 W, 339
Amplifier, hi-fi, 8 W, 340
Amplifier, hi-fi, 10 W, 340
Amplifier, hi-fi, 12 W, 341, 342
Amplifier, hi-fi, 12.5 W, 343
Amplifier, hi-fi, 15 W, 338, 345
Amplifier, hi-fi, 15 W (complete), 344
Amplifier, hi-fi, 15-60 W, 346
Amplifier, hi-fi, 20 W, 343, 347
Amplifier, hi-fi, 28 W, 347, 348, 349, 350
Amplifier, hi-fi, 40 W, 352, 354
Amplifier, hi-fi, 40 W (8 ohms), 342
Amplifier, hi-fi, 50 W, 355
Amplifier, hi-fi, 55 W, 356
Amplifier, hi-fi, 70 W, 357
Amplifier, hi-fi, 120 W, 361
Amplifier, hi-fi, low-power, 326

393

Amplifier, 455 kHz IF, 243
Amplifier, 40-180 MHz, 307
Amplifier, 40-180 MHz (12 W), 292
Amplifier, 100-120 MHz RF, 310
Amplifier, 150 MHz RF, 313
Amplifier, CW (30 W at 175 MHz), 296
Amplifier, 175 MHz, 35 W, 295
Amplifier, 200 MHz RF, 308
Amplifier, 200-250 MHz, 305
Amplifier, 200-400 MHz, 294
Amplifier, 400 MHz RF, 308
Amplifier, 420-450 MHz, 306
Amplifier, 450 MHz RF, 307
Amplifier, UHF, 10 W, 297
Amplifier, UHF, 25 W, 296, 297
Amplifier, linear (5 W, 2 m), 293
Amplifier, line-operated, 2 W, 332
Amplifier, low-capacitance, 250
Amplifier, luminance, 265
Amplifier, magnetic, 39
Amplifier, 6-meter, 50 W, 293
Amplifier, microphone, 223
Amplifier/mixer, 18 V, 304
Amplifier-mixer-compressor, 323
Amplifier, op-amp final, 317
Amplifier, operational, 317
Amplifier/oscillator, all frequency, 310
Amplifier, phono, 327
Amplifier, power (136 MHz, 40 W), 291
Amplifier, power (200 W @ 50 KHz), 292
Amplifier power supply, 60 V, 28
Amplifier power supply, 160 W, 26
Amplifier, push-pull servo, 171
Amplifier, quadrature, 269
Amplifier, RGB, 268
Amplifier, rf, 263
Amplifier, RF (80 meters), 294
Amplifier, servo (6 W), 175
Amplifier, servo (100 W), 169
Amplifier, servo (28 V), 172
Amplifier, servo (ac/dc), 170
Amplifier, servo (transformer-coupled), 174
Amplifier, 2-stage audio, 319
Amplifier, stereo (2 W), 332
Amplifier, stereo (5 W/channel), 337
Amplifier, stereo, 160 W, 360
Amplified solar relays, 76
Amplifier, TV audio (2 W), 260
Amplifier, TV IF, 264
Amplifier, TV video, 264
Amplifier, unity-gain, 324
Amplifier, unity-gain FET, 316
Amplifier, vertical deflection, 249
Amplifier, video, 252
Amplifier, video IF, 258
Amplifier, video IF, 44 MHz, 256
Amplifier, voltage-controlled, 233
Amplifier, wideband audio, 4 W, 335
Amplifier, wide-dynamic-range, 320
Amplifier w/tone controls, 329
Amplitude modulation, *see entries under AM*
Amplitude modulator, 290
AM circuits, 241
AM converter, CAP, 302
AM IF amplifier, 455 kHz, 243
AM radio, automobile, 383
AM radio, dual-gate MOSFET, 248
AM radio, for cars, 390
AM radio, line-operated, 242
AM radio, 2-stage 6 V, 244
AM radio, sun-powered, 244, 245
AM regenerative receiver, 243
AM series modulator, 290
AM transmitter, aircraft, 286
AM transmitter, 175 MHz, 287
AM transmitter, 2-meter, 289
AM transmitter, tunnel diode, 226
AM transmitter, wireless mike, 228
AM wireless mike, 226, 228
Analog-controlled ac switch, 65, 72
Anderson 3rd overtone crystal, 274
Anode-triggered ring counter, 193
Answer-device ring sensor, 167, 168
Antenna preamp, 2 m, 302
Anticoincidence circuit, 152
Anticoincident load selector, 152
Aperture-disc motor control, 87

Appliance control with light, 77
Appliance-motor control, 108
Appliance-motor speed control, 91
Appliance speed control, 115
Appliance speed control, timer, 119
Arc preventer for power relays, 70
Arc-proof motor startup, 86
Arc-proof power switch, 115 V, 73
Arc-proof static contactor, 74
Assembly line counter, 82
Assembly line jam sensor, 157, 158
Astable multivibrator, 44
Asymmetrical amplifier, wide-range, 129
Attenuator, tee, 233
Audible circuit-on reminder, 382
Audio alarm, strobed, 166
Audio amplifiers, 326
Audio amplifier, hi-fi, 2 W, 332
Audio amplifier, hi-fi, 4 W, 336
Audio amplifier, hi-fi, 5 W, 337, 338
Audio amplifier, hi-fi, 7 W, 339
Audio amplifier, hi-fi, 7.5 W, 339
Audio amplifier, hi-fi, 8 W, 340
Audio amplifier, hi-fi, 10 W, 340
Audio amplifier, hi-fi, 12 W, 341, 342
Audio amplifier, hi-fi, 12.5 W, 343
Audio amplifier, hi-fi, 15 W, 338, 344, 345
Audio amplifier, hi-fi, 15-60 W, 346
Audio amplifier, hi-fi, 20 W, 343, 347
Audio amplifier, hi-fi, 25 W, 347, 350
Audio amplifier, hi-fi, 40 W, 342, 352, 353, 354
Audio amplifier, hi-fi, 50 W, 355
Audio amplifier, hi-fi, 55 W, 356
Audio amplifier, hi-fi, 70 W, 357
Audio amplifier, hi-fi, 120 W, 361
Audio amplifier, hi-fi (lo-pwr), 326
Audio amplifier, building-block, 326
Audio amplifier, for car radio, 336
Audio amplifier, FET, 80 W, 359
Audio amplifier, for headphones, 327
Audio amplifier, for radio, 335
Audio amplifier, 1 W, 328, 329, 331
Audio amplifier, 1-2 W, 330
Audio amplifier, 2 W, 260, 329, 331
Audio amplifier, compl., 2 W, 331, 333, 334
Audio amplifier half-watt, 387
Audio amplifier, high-gain, 317
Audio amplifier, line-operated, 2 W, 332
Audio amplifier, low-cost, 10 W, 337
Audio amplifier/oscillator, 221
Audio amplifier, phono, 334
Audio amplifier, portable-TV, 259
Audio-amplifier power supply, 28, 30
Audio amplifier/servo amplifier, 170
Audio amplifier, 20 stage, 319
Audio amplifier, stereo, 5 W, 337
Audio amplifier, stereo, 160 W, 360
Audio amplifier, wideband, 4 W, 335
Audio amplifier, with tone controls, 329
Audio-balance meter, stereo, 363
Audio circuits, 315
Audio click generator, 236, 237
Audio emitter follower, 321
Audio generator, 10 Hz to 175 kHz, 221
Audio generator, sawtooth, 229
Audio generator, single-tone, 224
Audio generator, sun-powered, 232
Audio generator, tone, 220
Audio IF, portable TV, 261
Audio mixer, 315
Audio mixer-compressor, 323
Audio mixer, 4-input, 323
Audio mixer, 7-input, 325
Audio noise and rumble filter, 364
Audio oscillator, sine-wave, 378
Audio oscillator, twin-tee, 235
Audio preamp, 84 dB, 321
Audio preamp, complete, 318
Audio preamp, crystal/ceramic, 320
Audio preamp, Darlington, 316
Audio preamp, hearing-aid, 316
Audio preamp, hi-fi, 320
Audio preamp, magnetic, 317
Audio preamp, microphone, 319
Audio preamp, phono, 325
Audio preamp, and tone control, 322, 324
Audio preamp, wide-range, 320

Audio presence control, 364
Audio signal generator, 379
Audio signal injector, 378
Audio tester, hi-fi, signal source, 378
Audiovisual slide fader, 232, 236
Audio voltage amplifier, 322
Auditorium-light dimmer, 123
Auto-answer phone-ring sensor, 167, 168
Auto battery charger, 6 V, 392, 394
Auto battery converter, 300 V, 37
Auto battery simulator, 29
Auto battery simulator regulator, 16
Autodropout repeater timer, 214
Automatic CATV switching source, 19
Automatic flashing light, 186
Automatic headlight dimmer, 381
Automatic keyer, 222
Automatic lamp-brilliance control, 151
Automatic lamp flasher, 185
Automatic liquid-level control, 148
Automatic night light, 77-83
Automatic porch light, 77, 81
Automatic power shutdown, 166
Automatic-shutoff speed control, 119
Automatic swim-pool refill, 150
Automatic turnoff motor control, 127
Automatic turn-on with darkness, 80, 81
Automatic turnoff with light, 77, 78, 79, 81, 82
Automobile lights-on reminder, 382
Automobile radio audio ampl, 336
Auto theft alarm, 164, 385
Automotive adapter, 115 V, 45, 46
Automotive alarm, 166
Automotive AM radio, 383, 390
Automotive burglar alarm, 162, 385
Automotive CB transmitter, 5 W, 281
Automotive CDI system, 387
Automotive circuits, 381
Automotive circuits, in-garage, 392
Automotive continuity tester, 368
Automotive flasher, 180, 182, 384
Automotive headlight dimmer, 381
Automotive interior-lamp dimmer, 387
Automotive intrusion alarm, 160
Automotive inverter, 110 W, 52
Automotive lamp flasher, 185
Automotive lights-on reminder, 385
Automotive load regulator, 16, 34
Automotive SCR ignition system, 388
Automotive sequential turn signal, 386
Automotive smoke alarm, 165
Automotive tachometer, 384
Automotive tach-strobe source, 53
Automotive tape-player speed control, 121
Automotive transmitter, 175 MHz, 280, 287
Automotive turn-signal reminder, 389
Automotive UHF power amplifier, 296, 297
Auxiliary flash unit, 217
Average-voltage regulator, 120 V, 6

Backlighting flash unit, 217
Back-to-back SCRs, 65, 72
Balance control, 363, 365
Balance control meter, stereo, 363
Balanced-emitter UHF tripler, 277
Balanced mixer, FET, 311, 312
Balanced modulator, 290
Ballast, electronic, 48
Base-collector leakage tester, 370
Battery-activated triac switch for ac, 71
Battery charger, automotive, 6 V, 392
Battery charger, auto, 12 V, 394
Battery charger, 12 V, 29
Battery charger, 12 V (2 A), 393
Battery charger, half-wave, 32
Battery charger, photoflash, 2.4, 35
Battery-charging power source, 21
Battery condition sensor, 159
Battery converter, 6-12 V, 40
Battery converter, 150/300 V, 37
Battery-driven-motor speed control, 96
Battery eliminator, 9 V, 30
Battery-operated ac source, 117 V, 42
Battery-operated AM radio, 244
Battery-operated burglar alarm, 160, 162
Battery-operated code oscillator, 220, 221, 223, 224

Battery-operated dc voltmeter, 372
Battery-operated flasher, 24 V, 186
Battery-operated flash slave, 215, 217
Battery-operated fluorescent lamp, 44
Battery-operated heat alarm, 134
Battery-operated intercom, 226
Battery-operated smoke sensor, 162
Battery-operated tachometer, 156
Battery-saver circuit-on alarm, 382
Battery-saver lights-on reminder, 385
Battery-simulating regulator, 10, 16
Battery tester, 159
Battery-to-ac (117 V) inverter, 46, 47
Battery-voltage flash compensator, 216
Battery-voltage monitor, 159
BCD ring counter, 199
Beam-interruption alarm, 161
Beam-interrupt sensor, 148
Beam modulator, light, 227
Bench car-battery simulator, 29
Beta tester, in-circuit, 371
Beta tester, transistor, 375
Bias oscillator, tape-head, 363
Billboard flasher, 120 V, 181
Billboard light control, 77, 78, 80
Billboard-light control circuit, 83
Billboard lighting, automatic, 77
Billboard sequential flasher, 187
Billboard turn-on at night, 80
Binary-coded decimal counter, 199
Bipolar beta tester, 371, 375
Bipolar clock sync circuit, 25
Bipolar/FET audio amplifier, 326
Bipolar/FET audio mixer, 323
Bipolar-FET broadband amplifier, 309
Bipolar-FET cascode amplifier, 254
Bipolar-FET dc voltmeter, 372
Bipolar-FET Schmitt trigger, 70
Bipolar-FET video amplifier, 254
Bipolar follower, audio, 321
Bipolar "grid-dipper," 371
Bistable multivibrator, 199
Bistable multivibrator (PUTs), 189
Bistable switch with memory, 184
Black and white TV circuits, 256
Blender motor control, timing, 127
Blender-motor speed control, 91, 96, 106, 202
Blender-motor timer/switch, 24
Blender speed control, 115
Blender speed control plug-in, 111
Blender speed control with timer, 119
Blinker, electronic, 185
Blinking-eyes LEDs, 185
Blocking oscillator, 1 us pulses, 220
Blower control, thermostat-gated, 143
Blower motor control, RFI-proof, 122, 125
Blower-motor speed control, 99
Blower speed control, 115
Boat/aircraft B+ supply, 36
Boat/aircraft power supply, 200 V, 36
Boat flasher, 384
Boat siren, 237, 238
Body-capacitance sensor, 163, 164
Body-capacitance switch, 232, 238
Booster, color TV signal, 267
Bootstrapped audio follower, 321
B-plus converter, 28 to 300 V, 38
B-plus supply, 150 V and 300 V, 20, 37
B-plus supply, 300 and 600 V (vehicular), 37
B-plus supply, for 6 V cars, 36
B-plus supply, mobile, 36, 39
Break-in alarm, 160
Bridge converter, 31
Bridge inverter, 180 V, 54
Bridge, speed-control, 103
Brightness compensation:
  for color organs, 235
Brightness control, lamp, 151
Brightness control, LED, 152
Brightness and luminance, 265
Broadband aircraft transmitter, 283, 284, 285, 289
Broadband amplifier, 200-400 MHz, 294
Broadband AM transmitter, 2 m, 289
Broadband RF amplifier, 1-40 MHz, 309
Broadband VHF amplifier, 307
Broadcast-band AM transmitter, 226
Broadcast radio, 6 V, 244

Braodcast radio, FM, 247, 248
Broadcast radio, for cars, 390
Broadcast radio, line-operated, 242
Broadcast radio, mobile, 383
Broadcast radio, regenerative, 248
Broadcast radio, 1-transistor, 245
Broadcast radio, 2-transistor, 244
Broadcast radio, solar, 244, 245
Braodcast receiver, 244
Broadcast receiver, FM, 245
Broadcast wireless mike AM, 226, 228
Broadcast wireless mike FM, 225
Buffer, 2-FET cascode, 220
Buffer, hi-Z and lo-C, 224
Buffer, video, 258
Building-block amplifier, audio, 326
Buoy flasher, 186
Buoy-light turn-on, 80
Burglar alarm, 160
Burglar alarm, 12 V dc, 164
Burglar alarm, automotive, 385
Burglar alarm, body-sensing, 163, 164
Burglar alarm control, 79
Burglar alarm control, N.C., 82
Burglar alarm dc light control, 80
Burglar alarm light control, 80
Burglar alarm, multi-entry, 160
Burglar alarm photocell control, 81
Burglar alarm, photoelectric, 80
Burglar alarm photoelectric device, 78
Burglar alarm power supply, 21
Burglar alarm siren, 234
Burglar foiling with light, 77
Burst gate, color TV, 266
Butterworth filter, 256
Buzzer, overtemperature, 134

Cable-TV switching, 319
Cadmium-sulfide relay, 76
Calibrator, crystal, 100 kHz, 377
Calibrator, tachometer, 377
Call monitor, telephone, 167, 168
Camera-flash slave, 215, 217
Camera-viewfinder amplifier, 252
Candle, electronic, 239
Capacitance diode see Varactor
Capacitive-load 12 V supply, 22
Capacitance multiplier, 224
Capacitance switch, latching, 238
Capacitance touch switch, 232
Capacitor-charge compensator, 216
Capacitor charging unit, photo, 35
Capacitor-discharge ignition, 388
Capacitor-discharge ignition system, 387
CAP converter for car radio, 302
Car battery inverter, 400 Hz, 43
Car/boat battery charger, 12 V, 393
Car battery generator, 117 V, 42
Car battery simulating regulator, 162
Car-headlamp control of switch, 392
Car radio, AM, 383, 390
Car-radio audio amplifier, 336
Car radio CAP converter, 302
Car radio CB converter, 303
Carrier-dropout timer, 214
Carrier mobility, 71
Car-theft alarm, 385
Car-voltage tester, 386
Cascode amplifier, 220
Cascode amplifier, 200 MHz, 307, 309
Cascode IF amplifier, 30 MHz, 306
Cascaded-stage LED counter, 190
Cascode video amplifier, 250
Cathode-ray tube, see entries under CRT
CATV switching supply, 19
CB converter, tunnel diode, 301
CB crystal oscillator, 274
CB regulator for ac supply, 16
CB transmitter, 5 W, 281
CB transmitter, solar, 288
CDI system, SCR, 387, 388
CDI system, solid-state, 387
Cell, color organ, 235
Cell-voltage monitor, 159
Centrifugal-switch eliminator, 86
Ceramic-cartridge amplifier, 327, 329, 331, 334
Ceramic-cartridge amplifier, 2 W, 332

Ceramic-cartridge amplifier, FET (4 W), 334
Ceramic cartridge preamp, 320, 325
Ceramic-magnetic preamp, 325
Chamber, double-ionization, 165
Chamber, ionization, 162, 163
Chamber, single ionization, 165
Chamber, smoke-sensing, 162, 163
Channel-expand stereo control, 365
Charge converter, photoflash, 216
Charger, battery, 6 V, 392
Charger, battery, 12 V, 29, 393, 394
Charger, battery, half-wave, 32
Charger, photoflash, 2.4 V, 35
Charging standby power source, 21
Chart, lamp-dimmer performance, 110
Chaser, 187
Chaser, ac, 187, 190, 191
Chaser, ac, 100 W lamps, 188
Chaser, ac, half-wave, 187, 189
Chaser, dc, 192, 202
Checker, diode-transistor, 369
Checker, LED continuity, 367
Checker, SCR & diode, 368
Checker, voltage, automotive, 368
Checking device, LED, 369
Chemical-bath temperature control, 139
Chroma, see entries under Color
Chroma processing circuitry, 266
Chrominance, see entries under Color
Circuits for phase control, 113
Circuit-on reminder, 389
Cistern level control, 150
Citizen Band, see entries under CB
Civil Defense siren, 237, 238
Class A, see entries under Audio
Class B audio amplifier, 336
Blass B servo amplifier, 170
Class C amplifier, 175 MHz (35 W), 296
Class C amplifier, 200-400 MHz, 294
Click generator, 236
Clock sync circuit, 25
Closed-loop 90 V rms regulator, 4
CMOS oscillator, 2 MHz, 274
CMOS power converter, 33
CMOS supply, 33
CMOS-triac switching combo, 194
Code practice oscillator, 220, 221, 223, 224, 237
Coffeemaker switch, light-controlled, 77
Coffeepot temperature control, 133
Collector-current checker, 375
Collector-emitter leakage tester, 375
Collector-to-base leakage tester, 370
Color decoding, color TV, 266
Color difference amplifiers, 266
Color-lamp music modulator, 231
Color organ, 231
Color organ cell, 235
Color TV booster, 267
Color TV circuits, 265
Color TV crystal filter, 266
Color TV luminance amplifier, 265
Color TV phase inverter, 267
Color TV RGB amplifier, 268
Commercial killer, TV, 260
Common-anode ring counter, 192
Common-base mobile B+ supply, 39
Common-collector amplifier, 321
Common-collector mobile B+ supply, 36
Common-emitter converter, 36, 39
Common-emitter mobile B+ supply, 39
Common-gate FET amplifier, UHF, 306
Common-source 400 MHz amplifier, 308
Communications circuits, 271
Communications receiver, 244
Commutating ring counter, 193
Commutation, 33
Compact electronic keyer, 222
Comparator, long-time, 207
Comparator, for motor speed control, 88, 117
Comparator, for Schmitt trigger, 70
Compensated lamp dimmer, 100, 123
Compensating regulator, 4
Compensation, temperature, 85
Compensator, large-angle, 109
Complementary audio ampl, 331, 334, 335
Complementary, audio amplifier, 2 W, 260, 333
Complementary audio ampl, 4 W, 335, 336

INDEX **395**

Complementary audio ampl, 4 W hi-fi, 336
Complementary audio ampl, 5-10 W, 337
Complementary audio ampl, 5 W hi-fi, 338
Complementary audio ampl, 7 W hi-fi, 339
Complementary audio ampl, 7.5 W hi-fi, 339
Complementary audio ampl, 8 W hi-fi, 340
Complementary audio ampl, 10 W hi-fi, 340
Complementary audio ampl, 12 W hi-fi, 341, 342
Complementary audio ampl, 15 W hi-fi, 338
Complementary audio ampl, 15 W hi-fi, 345
Complementary audio ampl, 20 W hi-fi, 343, 347
Complementary audio ampl, 25 W hi-fi, 348
Complementary audio ampl, 40 W hi-fi, 353
Complementary light control, 81
Complementary MOS, see entries under CMOS
Complementary SCR flasher, 185
Complementary-SCR ring counter, 202, 192
Complementary-SCR Schmitt trigger, 72
Complementary-SCR supply, 29
Complementary servo amplifier, 169, 170
Complementary servo preamp, 173
Complementary stereo ampl, 5 W, 337
Complementary symmetry, 331, 334-343, 345, 347, 348, 352, 353
Complementary UJT, see entries under CUJT
Complementary vert deflection, 249
Complete audio amplifier, 15 W, 345
Compound series-feedback ampl, 250
Compound series-feedback buffer, 224
Compressor, audio, 323
Computer-power failure alarm, 154
Computer power supply, 41
Contact-arc preventer, 70
Contactor, static, 74
Constant-brightness control, 151
Constant-current source, 18
Constant-output smoke chamber, 162
Constant voltage source, 12 V, 10
Construction-hazard flasher, 182
Continuity, voltage, current tester, 367
Continuity tester, 369
Continuity tester, automotive, 368
Continuous-tone on/off switching, 178
Control, 3-phase dc voltage, 124
Control, 3-phase 4-wire phase, 112
Control, ac/dc power, 129
Control, ac dimmer, 400 W, 97
Control, for ac/dc loads, 115 V, 97
Control, blender, with timer, 119
Control, blender motor, 126
Control, dc motor speed, 88
Control, dc power, 103
Control, dispensed-drink, 150
Control, fader, 232, 236
Control, fan motor speed, 99
Control, feedback motor speed, 105
Control, full-range ac power, 92
Control, full-wave ac, 900 W, 118
Control, full-wave ac power, 118
Control, full-wave phase, 92
Control, full-wave series-motor, 89
Control, garage-door, 393
Control, garage-light, 392
Control, heating-element, 146
Control, high-gain phase, 95
Control, hysteresis-free phase, 106
Control, hysteresis-free power, 129
Control, induction-motor, 99
Control, induction-motor speed, 91, 127
Control, isolated, 100%, 110
Control, kitchen-appliance speed, 106
Control, lamp, with feedback, 111
Control, lamp brightness, 151
Control, lamp, remote, 174, 178
Control, lamp, 2-time-constant, 92
Control, LED brightness, 152
Control, light, 76-83
Control-line remote switching, 175
Control, liquid-level, 148, 150
Control, low-cost phase, 99
Control, low-hysteresis ac voltage, 86
Control, mini phase, 25 W, 117
Control, motor, 5 A, 104
Control, motor RFI-proof, 122
Control, motor-speed, 91, 84, 100, 114
Control, motor speed, with feedback, 95
Control, motor speed, half-wave, 96

Control, motor speed, reversing, 90
Control, motor speed, wide-range, 105
Control, optical feedback, 90
Control, oven temperature, 135
Control, phase, 89, 110
Control, phase, 120/240 V, 103
Control, phase, 800 W, 130
Control, phase, with feedback, 110
Control, phase, low-gain, 100
Control, phase, low-hysteresis, 100, 126
Control, phase, PUT, 101
Control, phase, SUS-triggered, 110
Control, phase, with RFI filter, 84
Control, plug-in speed, 115
Control, PM motor, 96
Control, projector, 126
Control, proportional heating, 7.5 kW, 144
Control, proportional lighting, 114
Control, PWM dc motor-speed,
Control, ramp-and-pedestal, 116
Control, reversing, for motors, 116
Control, reversing motor-speed, 90
Control, RFI-proof phase, 103, 125
Control, SCR phase, 94
Control, self-timing blender, 127
Control, and sensing, 59
Control, shaded-pole, 99, 121
Control, shop-motor, 126
Control, soft-start lamp dimmer, 125
Control, speed, tachometer, 159
Control, shunt-motor speed, 94, 120
Control, for shunt and PM motors, 121
Control, shunt-wound motor, 131
Control, stereo balance, 363, 365
Control, stereo presence, 364
Control stereo width, 365
Control, switching, 61
Control, tachometer speed, 118
Control, temp.-sensitive, 141
Control, temperature (ZVS), 138, 142
Control, time-dependent phase, 120
Control, tool-speed, plug-in, 111
Control, triac lamp-dimming, 86
Control, universal-motor-speed, 108
Control, voice or sound, 232
Control, voltage, 400 Hz, 93
Control, voltage-compensated, 100
Control, wide-range, 1 kW, 128
Control, wide-range phase, 92
Control, wide-range phase, 1 kW, 129
Control circuit, light-operated, 81
Control circuit, for phase controls, 113
Controller, 84
Controller, extended range motor, 124
Controller, 3-phase stepless dc, 98
Control-voltage-switched power, 61
Control waveforms, synchronous, 102
Control system, tone, 318
Control timer, ac/dc, 206
Conversion, power, 33
Converter, 33
Converter, 2-meter, 300
Converter, 6-meter, 301, 303
Converter, 11-meter, 301, 303
Converter, CAP, 302
Converter, dc, 1.5 V to 12 V, 33
Converter, dc, 6 V-to-12 V, 40
Converter, dc, 6 V-to-250 V, 36
Converter, dc, 14 V-to-150/300 V, 37
Converter, dc, 25 V at 20 A, 37
Converter, dc, 28 V, 56
Converter, dc, 28/150 V, 34
Converter, dc, 28 V-to-150/300 V, 37
Converter, dc, 180 V, 57
Converter, dc, 28 to 200 V, 36
Converter, dc, 28 V-to-300 V, 37
Converter, dc, 28/400 V, 34
Converter, dc, 100 W, 39
Converter, dc, 150 V to 500 V, 38
Converter, dc, 275 V-to-50 V, 39
Converter/mixer, 100-to-44 MHz, 304
Converter, multitester, to VTVM, 372
Converter, photoflash, 35, 216
Converter, pulse-to-dc, 31
Conveyor-belt pileup sensor, 157, 158
Conveyor pileup monitor, 157
Copy-machine power supply, 24

Counter, decade, 199, 202
Counter, ring 5 mW, 188
Counter, ring, 4-stage, 190
Counter, 30-stage ring, 188
Counter, ring commutating, 193
Counter, ring, CSCR, 202, 192
Counter, ring, hi-speed, 196
Counter ring, LED, 190
Counter, ring, wide tolerance, 193
Counter, ring, variable timing, 188
Counter, SCR ring, 199
Counter, SCS ring, 192, 200, 201, 202
Counter, and shift register, 198
Counter-driven sequential timer, 208
Coupler, diode-diode, 153
Coupler, transistor, 400 V, 378
Coupling, optical, 61
Coupling, photon, 171
Coupling transister, 73
CPO, see Code practice oscillator
Cross-connected switch, 44
Cross-coupled erase oscillator, 363
Cross fader for slide, 232, 236
Crossover inhibit circuit, 41
Crosstalk introducer, sound, 365
Crosstalk, telephone, 167
Crowbar, electronic, 166
CRT cathode driver, 254
CRT hi-voltage supply, 262
CRT magnetic deflection, 250
Crystal calibrator, 100 kHz, 377
Crystal-cartridge amplifier, 2 W, 332
Crystal-cartridge phono ampl, 327, 331, 334
Crystal-cartridge preamp, 320
Crystal converter, for CB, 303
Crystal converter, 47 MHz, 303
Crystal filter, color TV, 266
Crystal oscillator, 100 kHz, 377
Crystal oscillator, 2 MHz, 274
Crystal oscillator, 47 MHz, 274
Crystal oscillator, CB, 274
Crystal oscillator, Pierce, 274
Crystal oscillator/transmitter, 73.5 MHz, 288
Crystal set, amplified, 6 V, 244
Crystal set, sun-powered, 244, 245
Crystal sun-powered CB transmitter, 288
CSCR, 72
CSCR anti-coincidence circuit, 152
CSCR lamp flasher, 185
CSCR ring counter, 192, 193, 200
CSCR time delay, 204, 210
CSCR time-delay relay, 1-30 sec, 213
CSCR timer, adjustable-period, 208
CSCR timer, 1 min delayed turn-on, 209
CSCR timer, 20 minutes, 204
CTCSS power switching, 128
CTCSS tone generator, 221
CUJT-inhibited sequential timer, 208
CUJT organ tone generator, 233
Current detector for phone line, 167, 168
Current-limited supply, 10-34 V, 29
Current-limited, supply, dc 28 V, 173
Current-limiting circuit, 18
Current-limiting LED supply, 26
Current regulator, 50-200 mA, 5
Current source, 9
Current source, constant, 18
Current, and voltage tester, 367
CW amplifier, 30 W at 175 MHz, 196
CW amplifier, 200-400 MHz, 16 W, 294
CW transmitter keyer, 222
CW transmitter, 136 MHz, 289
CW transmitter, 40 W 6 m, 282

Dampness sensor, 147
Darkness sensor and latch, 80
Darkroom temperature control, 139
Darkroom timer, 147, 211, 219
Darkroom timer, delayed dropout, 209
Darkroom timer, simple, 208
Darkroom timer, 2 seconds to 4 minutes, 183
Darkroom timer, up to 3 minutes, 214
Darlington audio amplifier, 330
Darlington audio amplifier, 1 W, 328
Darlington audio amplifier, 5 W hi-fi, 337
Darlington audio amplifier, 60 W hi-fi, 346

Darlington audio preamp, 316
Darlington-controlled 24 V supply, 28
Darlington-coupled triac switch, 60, 69
Darlington dc converter, 37
Darlington dc regulator, 100 W, 8
Darlington-driven ampl, 260
Darlington-driven amplifier, 1-2 W, 333
Darlington-driven ampl, 7 W hi-fi, 339
Darlington-driven ampl, 10 W, 340
Darlington-driven ampl, hi-fi, 332
Darlington-driven regulator, 100 V, 13
Darlington follower, audio, 321
Darlington phono amplifier, 327
Darlington power switcher, 69, 70
Darlington servo amplifier, 170, 171
Darlington switch, 5 kW, 68
Darlington vertical stage, 253
Dash-mountable tachometer, 384
Data transmission amplifier, 171
Data xmssn isolator, 171
Dc/ac industrial control timer, 206
Dc alarm, 12 V, 166
Dc amplifier, 376
Dc and ac load control, 115 V, 97
Dc anticoincidence circuit, 12 V, 152
Dc B+ converter, 6 V input, 36
Dc B+ supply, dual voltage, 20
Dc burglar alarm, low-voltage, 164
Dc chaser, 192, 193, 202
Dc control of ac motor, 119
Dc-controlled attenuating amplifier, 233
Dc converter, 33
Dc converter, 10 kHz, 34
Dc converter, 1.5 to 12 V, 33
Dc converter, 6 V-to-12 V, 40
Dc converter, 25 V at 20 A, 37
Dc converter, 28 to 150 V, 34
Dc converter, 28 V-to-200 V, 36, 37
Dc converter, 70 V at 50 W, 54
Dc converter, 180 V, 57
Dc converter, 40 W, 38
Dc converter, 100 W, 39
Dc converter, 225 W, 37
Dc converter, 1 kW, 39
Dc converter, com. collector, 36
Dc converter, com. emitter, 36, 39
Dc converter, low-power, 56
Dc converter, photoflash, 35
Dc crowbar protector, 166
Dc current regulator, 50-200 mA, 5
Dc flasher, 6 V, 186
Dc flasher, 24 V, 186
Dc flasher, low-voltage, 182
Dc flasher, SCR/CSCR, 185
Dc flash slave, light-triggered, 215, 217
Dc gas-and-smoke alarm, 165
Dc intrusion alarm, 162
Dc lamp flasher, 6 V, 185
Dc lamp flasher, 12 V, 184
Dc latch relay, 65
Dc light control, low-voltage, 80
Dc motor speed control, 88, 96, 108, 117
Dc motor speed control, shunt and PM, 121
Dc motor speed control, shunt-wound, 131
Dc multi-B+ power supply, 21
Dc-output 3-phase voltage control, 124
Dc overheat alarm, 134
Dc phase shifter, variable, 178
Dc power control, 103, 129
Dc power source, for emergency systems, 21
Dc power supply, 12 V, 29, 32
Dc power supply, 20 V, 30
Dc power supply, 24 V regulated, 28
Dc power supply, 26 V dual, 348
Dc power supply, 28 V regulated, 27
Dc power supply, 36 V, 342
Dc power supply, 800 V, 21
Dc power supply, half-wave, 32
Dc power supply, multivoltage, 376
Dc-pulsed ac switch, 65
Dc pulsed audio alarm, 166
Dc radiation detector, 155
Dc reference, 8
Dc regulated supply, 4-12 V, 20
Dc regulated supply, 9 V, 27, 30
Dc regulated supply, 10-34 V, 29
Dc regulated supply, 24 V, 17, 39

Dc regulated supply, 28 V (var), 17
Dc regulated supply, 30 V, 26
Dc regulated supply, 34-45 V, 18
Dc regulated supply, 80 V, 25
Dc regulator, 4
Dc regulator, 8 V at 100 mA, 5
Dc regulator, 8 and 16 V, 6
Dc regulator, 8-16 V at 100 mA, 5
Dc regulator, 12 V at 2 A, 11
Dc regulator, 13 V at 6 A, 15
Dc regulator, 28 V at 500 mA, 12
Dc regulator, 60 V at 1 A, 14
Dc regulator, 100 V at 400 mA, 13
Dc regulator, 1 kX, 3
Dc relay, for ac control with light, 79
Dc relay, for control with light, 81
Dc relay, controlled by 0.5 mA, 61
Dc smoke alarm, low-voltage, 165
Dc source, 9
Dc stepless, 3-phase control, 98
Dc supplies, 17
Dc supply, dual 32 V, 354
Dc supply, 60 V regulated, 28
Dc switching regulator, 7
Dc switching regulator, 28 V at 100 W, 8
Dc switching regulator, 150-250 V, 10
Dc switching regulator, o'load protected, 11
Dc synchronous power control, 121
Dc tachometer, light-triggered, 156
Dc timer, delayed-turn-on, 12 V, 209
Dc timer, load-switching, 210
Dc timer, 1-minute, 28 V, 211
Dc timer, 100-second, 12 V, 207
Dc-to-ac devices, *see entries under Inverters*
Dc-to-ac inverters, 41
Dc-to-dc converters, *see entries under Dc converters*
Dc-tuned RF amplifier, 100 MHz, 310
Dc voltage monitor, 159
Dc voltage regulator, 12 V at 2 A, 16
Dc voltage regulator, 28 V at 10 A, 12
Dc voltmeter, battery-operated, 372
Dc voltmeter, single-FET, 373
Deadband adjustment for servo, 119
Decade counter, 199, 202
Decimal counter (BCD), 199
Decoder, chroma, 266
Decoder-operated power switch, 178
Deenergizing relays with light, 79, 81, 82
Deflection, vertical, 249, 255
Deflection, magnetic, 250
Deflection oscillator, vertical, 253
Delay, 20 V universal time, 209
Delay, CSCR time, 204, 210
Delay circuit, 1-minute, 211
Delayed-dropout relay, 208, 209, 214
Delayed-dropout timer, 203
Delayed-off timer, 219
Delayed one-shot, 205
Delayed-on latch circuit, 210
Delayed-on power switch, 1200 W, 210
Delayed-on timer, 219
Delayed-on timer, 1 min, 209
Delayed-on timer, universal, 206
Delayed power switch, 203 210
Delayed-pull-in RC timer, 206
Delayed-pull-in relay, 203
Delay link, 102
Delay relay, 10 ms to 1 min, 205
Delay relay, 20 minutes, 204
Delco ripple filter, 7
Demodulator, FM, 225
Design, power supply, 17
Despiking technique, inverter, 49
Detector, AM broadcast, 248
Detector-amplifier, 6 V, 244
Detector-amplifier, solar, 244, 245, 254
Detector, assembly-line jam, 158
Detector, dialed-number, 168
Detector, fire, 134, 232
Detector, flame brightness, 151
Detector, fluid drop, 149
Detector, fluid-level, 147
Detector, gas-and-smoke, 165
Detector, infrared, 155
Detector, intrusion, 160, 161
Detector, intrusion/hazard, 160

Detector, light, 155
Detector, light interruption, 80, 82
Detector, liquid-level, 148
Detector, low-light-level "drop", 154
Detector, moisture, 147
Detector, overcurrent, 153
Detector, overvoltage, 153, 166
Detector, phone current, 167, 168
Detector, phone-ring, 167
Detector, power-failure, 154
Detector, production line jam, 157
Detector, proximity, 163, 164, 232
Detector, rate, 148
Detector, regenerative, 243
Detector, second, BW TV, 258
Detector, smoke, 162
Detector, smoke (FET), 165
Detector, smoke, line-operated, 163
Detector, sound, 232
Detector, spot-of-light, 152
Detector, telephone-ring, 168
Detector, temperature, 136
Detector, water-level, 149
Developing-chemical temp control, 139
Diac replacement, dimmer, 106
Diac-triggered motor control, 91
Dialed-number detector, 168
Dial pulse indicator, 168
Diffamp sensing regulator, 90 V ac, 4
Difference amplifier, color, 266
Differential input amplifier, 326
Differential 12 V dc regulator, 2 A, 10
Digital-clock sync circuit, 25
Digital clock sync/rectifier, 25
Digital pulse sensor, telephone, 168
Digital-readout seconds timer, 213
Dimmer, 84. *See also entries under Lamp dimmer*
Dimmer, 1 kW, 104
Dimmer, 1 to 5 A, 111
Dimmer, 400 W, 97
Dimmer, compensated, 100, 123
Dimmer, double-time-constant, 92
Dimmer, extended-range, 124
Dimmer, with feedback control, 110
Dimmer, fluorescent, 93
Dimmer, fluorescent-lamp, 130
Dimmer, full-wave, 122
Dimmer, headlamp, 381
Dimmer, high-gain, 95
Dimmer, hi-intensity lamp, 99
Dimmer, without hysteresis, 92
Dimmer, hysteresis-free, 92, 106, 129
Dimmer, interior-car-light, 387
Dimmer, lamp, 89, 92, 100, 109, 114
Dimmer, lamp, 800 W, 86
Dimmer, lamp, 1000 W, 85
Dimmer, lamp, with feedback, 111
Dimmer, lamp, full-range, 98
Dimmer, lamp, low-gain, 100
Dimmer, lamp, with RFI filter, 84
Dimmer, low-cost, 99
Dimmer, low-hysteresis, 86, 100, 126, 128
Dimmer, miniature, 25 W, 117
Dimmer, no-hysteresis, 1000 W, 129
Dimmer, performance chart, 100
Dimmer, photocell-controlled, 103
Dimmer, 2-position, 239
Dimmer, proportional-lighting, 114
Dimmer, PUT-driven, 101
Dimmer, RFI filter, 103
Dimmer, RFI-proof, 125
Dimmer, RFI-proof, 240 V, 122
Dimmer, soft-start, 101, 120, 125
Dimmer, SUS-triggered, 110
Dimmer, switch, automatic, 381
Dimmer, table-lamp, 122
Dimmer, time-dependent, 125
Dimmer, transformer-gated, 110
Dimmer, wide-range, 92
Dimmer, wide-range, 1 kW, 128, 129
Dimmer, wide-range, 800 W, 130
Diode and DCR checker, 368
Diode-diode coupler, 153
Diode-polarity checker, 369
Diode tester and continuity tester, 369
Diode-transistor tester, 369
Diode, tuning, *see entries under Varactor*

INDEX **397**

Dip meter, bipolar-transistor, 371
Dip meter, transistor, 371
Direct-coupled servo amplifier, 169, 170
Direct current power conversion, 33
Direct current regulator, 50-200 mA, 5
Direct current regulator, 100 V, 13
Direct current source, 9
Directional signals, sequential, 386
Direction-reversing motor control, 116
Direction and speed control, 90
Direct-voltage phase control, 113
Dispensed-drink level control, 150
Display-brightness controller, 152
Displayed-digits seconds timer, 213
Divider, zener voltage, 376
Door-activator control, 82
Doorbell, multibutton, 230
Door opener, garage, 393
Double-chamber smoke sensor, 165
Double-time-constant trigger, 126
Doubler, 50-to-100 MHz, 277
Double, 200 to 400 MHz, 277
Doubler, VHF frequency, 277
Double-phase-shift ac control, 118
Double-pole electronic switch, 71
Double-pole switching with light, 79
Double-throw switching with light, 79, 80
Double-time-constant dimmer, 90, 92, 124
Double-time-constant dimmer, 1 kW, 129
DPDT switching circuit, 71
DPDT switching with light, 79, 81
Drill-motor speed control, 115
Drill speed control, 111
Drink dispenser volume control, 150
Drink-trough level control, 148
Drink-vending liquid control, 150
Drip detector, fluid, 149
Drip detector, low-light-level, 154
Drip monitor, LED, 154
Dripping-fluid drop detector, 149
Drive, CRT, 254
Drive, positioning servo, 119
Drive, servo motor, 119
Driver, audio line, 319
Driver, fluorescent lamp, 47
Driver, lamp, 68
Driver, picture-tube, 253
Driver, power transistor, 68
Driver, and timer (1 minute), 211
Driver, TV horizontal, 251
Drop detector, fluid, 149
Drop detector, low-light-level, 154
Dropout, delayed, 209
Dual-amplifier power supply, 30
Dual-amplifier power supply, 60 V, 28
Dual-B+ power supply, 20
Dual-chamber smoke sensor, 165
Dual-device 100 V regulator, 13
Dual FET cascode amplifier, 220
Dual-gate MOSFET FM radio, 248
Dual-gate MOSFET FM tuner, 241
Dual-gate MOSFET front end, 247
Dual-gate MOSFET radio, 248
Dual input, dual-output converter, 37
Dual-level 8 and 16 V regulator, 6
Dual dc power supply, 18 V, 345
Dual power supply, 26 V, 348
Dual power supply, 32 V, 352
Dual power supply, 32 V dc, 354
Dual power supply, 45 V, 357
Dual power supply, 65 V, 361
Dual stereo supply, 80 W, 26
Dual-supply servo preamp, 172
Dual-time-constant dimmer, 92, 124
Dual-time-constant dimmer, 1 kW, 129
Dual-tone power switch, 178
Dual-voltage test supply, 31
Dual-voltage transmitter supply, 20, 21
dv/dt network trigger prevention, 85
dv/dt snubber network, 93
dv/dt soft-start dimmer, 101
dv/dt suppressed blower control, 143
Dynamic-mike preamp, 220, 223, 320

Earphone amplifier, 327
Easy-on, see entries under Soft-start
ECL tuned-collector oscillator, 275

Economy fire alarm, 137
Economy power inverter, 180 V, 57
Economy 700 V, 7 A switch, 70
Effective-irradiance tester, 374
Effective-voltage regulator, 90 V, 4
Efficiency versus power, inverter, 51, 52
Egg timer, 1-minute, 211
Electret-mike preamp, 319
Electric-heat control system, 103
Electric-train speed control, 103
Electrolytic-capacitor ring counter, 193
Electromagnetic interference, see entries under EMI
Electromechanical switching, 61
Electromechanical timer, 203, 209
Electromechanical timer, photo, 183
Electronic ballast, 48
Electronic crowbar, 166
Electronic doorbell, 230
Electronic-flash slave, 217, 218
Electronic flickering candle, 239
Electronic garage-door control, 393
Electronic keyer, compact, 222
Electronic metronome, 236, 237
Electronic organ oscillator, 233
Electronic power "relay," 65
Electrostatic precipitator supply, 24
Electronic-relay performance, 63
Electronic relay simulator, 60
Electronic ripple filter, 7
Electronic rooster, 232
Electronic siren, 234, 237, 238
Electronic switch, 25 V at 10 A, 64
Electronic-tach calibrator, 376
Electronic thermometer, 137
Electronic timer, 147
Electronic toy organ, 235
Electrons versus holes, FETs, 71
Element, plug-in, tool speed control, 111
Eliminator, battery, 12 V, 29
Eliminator, transistor-battery, 30
Emergency flasher for cars, 185, 384
EMI filter for dimmer, 109
EMI-proofing, dimmer, 122
EMI-proof lamp dimmer, 84, 125
EMI-proof SCR switch, 65, 73, 75
Emitter-coupled logic, see ECL
Emitter-coupled Schmitt trigger, 70
Emitter-follower, audio, 321
Emitter-follower amplifier, 324
Emitter-follower preamp, 316
Encoder, tone, 220, 221, 226
Encoder, tone, simple, 223
Energizing circuits with light, 78
Enhanced-switching inverter, 28 V/70 V, 54
Enhancer, pulse speed, 153
Enlarger timer, 208, 211, 219
Equalization of phono ampl, 325, 327
Equalized audio preamp, 325
Erase/bias oscillator, 363
Excessive voltage alarm, 166
Excessive voltage shutdown, 153
Expansion control, stereo sound, 365
Explosion-proof pulse switch, 73
Exposure-control timer, 219
Extended-range dimmer, 92, 124, 129
Extended-range dimmer, 800 W, 130
Extended-range dimmer, 1000 W, 128, 129

Fade-rate-controlled dimmer, 123
Fader control for slides, 232, 236
Failsafe power supply, 21
Fairchild video amplifier, 252
Fan-and-coil motor speed control, 145
Fan-motor control, 91, 99, 115, 124
Fan-motor control, half-wave, 96
Fan-motor control, low-hysteresis, 126
Fan-motor control, RFI-proof, 122, 125
Fast Schmitt trigger, 70
Fast-switching inverter, 25 kHz, 51
Fast-switching inverter, 70 V, 54
Fast-switching inverter, 15 W, 42
Fast-switching inverter, 120 W, 44
Feedback, diac, for inverter, 47
Feedback ac dimmer, 115 V, 110
Feedback control, optical, 90

Feedback lamp control, 100, 111, 123
Feedbackless motor-speed control, 96
Feedback lighting control, 114
Feedback motor-speed control, 87, 105, 110
Feedback motor-speed control, dc, 117, 121
Feedback phase control, 90, 95
Feedback plug-in motor control, 111
FET amplifier, hi-gain, audio, 317
FET amplifier, 4 W phono, 334
FET amplifier, 80 W super-fi, 359
FET amplifier, 40-180 MHz, 292
FET amplifier, 200 MHz RF, 307, 308
FET amplifier, 400 MHz RF, 308
FET amplifier, 450 MHz RF, 306
FET/bipolar AGC amplifier, 257
FET/bipolar amplifier, audio, 326
FET/bipolar ampl, broadband, 309
FET/bipolar amplifier, cascode, 254
FET/bipolar amplifier, video, 254
FET/bipolar mixer, audio, 323
FET/bipolar dc voltmeter, 372
FET/bipolar follower, audio, 324
FET/bipolar Schmitt trigger, 70
FET cascode amplifier, 200 MHz, 309
FET cascode amplifier, video, 250
FET checker, 368
FET crystal oscillator, 274
FET current source, 9
FET dc voltage reference, 8
FET dc voltmeter, 373
FET DPDT power switch, 71
FET frequency doubler, 277
FET hi-fi preamplifier, 320
FET integrator, 224
FET mixer, 244 MHz, 311
FET mixer, active balanced, 311
FET mixer-compressor, 323
FET oscillator, 20 MHz, 273
FET oscillator, 800 MHz, 275
FET oscillator/amplifier, 310
FET phase shifter, 360°, 78
FET quadrature amplifier, 269
FET-source CSCR timer, 204
FET source follower, 316
FET UHF RF amplifier, 307
FET VFO, 272
FET video amplifier, 250
FET voltmeter, 370
FET "VTVM," 370
Fiber-optics FM receiver, 225
Fiber optics for speed control, 88
Fiber optics for speed regulation, 117
Fiber optics transmitter, 227
Field-effect transistor switch, 71
Filament-life extender, 152
Filament-transformer lamp control, 174
Filter, Butterworth, 256
Filter, color TV crystal, 266
Filter, m-derived 4.5 MHz, 254
Filter, m-derived low-pass, 298
Filter, noise and rumble, 364
Filter, RFI, for dimmers, 103
Filter, ripple, 7
Filtered power supply, 800 V, 21
Filtered lamp dimmer, 84
Filtering, 33
Final amplifier, 35 W 175 MHz, 295
Final amplifier, TV video, 264
Final op-amp, 317
Fire alarm, 134, 137
Fire alarm, smoke portion, 163
Fire detector, 232
Firing windows, SSR, 63
Fire/police RF amplifier, 307
Fixed dc regulator, 8 V, 5
Fixed-period timer, 210
Fixed-rate flasher, 181
Fixed-rate dc flasher, 12 V, 185
Fixed-rate lamp flasher, 183
Fixture, thermal-response test, 373
Flame-controlled 120 V load, 151
Flame monitor, 151
Flame-sensing control, 151
Flame simulator, 239
Flashers, 180
Flasher, 3 V lamp, 183
Flasher, 6 V portable dc, 185

Flasher, 12 V, 1-lamp, 182
Flasher, 12 V dc lamp, 184
Flasher, 115 V ac, 181
Flasher, 115 V, 60 Hz, 184
Flasher, 120 V, 1000 W load, 182
Flasher, 120 V flip-flop, 181
Flasher, ac (2-triac), 180
Flasher, automotive, 384
Flasher, dc (single-lamp), 185
Flasher, economy, 186
Flasher, high-brightness, 186
Flasher, high-power ac, 181
Flasher, 1-lamp low-power, 186
Flasher, LED, 185
Flasher, multiple-lamp, 187
Flasher, phototransistor, 186
Flasher, 2-SCR, 186
Flasher, sequential, 192, 202
Flashing detector, spotlight, 152
Flashing indicator, spot-of-light, 152
Flashlight-controlled model, 173
Flashlight-gated commercial killer, 260
Flashlight transceiver, 227
Flashlight-triggered relay, 80
Flash slave unit, 215, 217
Flash trigger, xenomo-gas, 218
Flash unit, charger, 2.4 V, 35
Flash unit, self-firing, 217
Flickering candle, 239
Flip-flop, free-running, 44
Flip-flop, SUS, 184
Flip-flop counter, 199
Flip-flop flasher, 183
Flip-flop flasher, 12 V dc, 182
Flip-flop flasher, 120 V, 181
Flip-flop inverter, 56
Flip-flop lamp flasher, 185
Flip-flop LED blinker, 185
Flip-flop trigger, 189
Flip-flop, with variable delay, 205
Flow-stop detector, 117
Fluid drop detector, 154
Fluid level control, 150
Fluid level sensor, 147, 149
Fluid-temperature regulator, 135
Fluorescent-lamp dimmer, 93, 130
Fluorescent-lamp driver, 47
Fluorescent-lamp inverter, 48
Fluorescent light, battery-operated, 44
Flyback switching regulator, 8
FM circuits, 241
FM optical transmitter, 227
FM optical receiver, 225
FM phase modulator, 291
FM radio front end, 241
FM transmitter, wireless mike, 225
FM tuner, broadcast, 248
FM tuner, 2-MOSFET, 245
FM wireless mike, 225
Focused-light intrusion alarm, 161
Foldback current limiting, 29
Follower, bipolar, 321
Follower, FET source, 316
Food mixer speed control, 115, 126
Free-running multivibrator, 44
Frequency doubler, 277
Frequency doubler, 50-to-100 MHz, 277
Frequency doubler, 200-to-400 MHz, 277
Frequency multiplier, 500-to-400 MHz, 276
Frequency-regulated inverter, 45
Frequency-selective break-in alarm, 160
Frequency tripler, 0.25 to 0.75 GHz, 278
Frequency tripler, 200 to 600 MHz, 279
Frequency tripler, to 450 MHz, 277, 278
Frequency-variable flip-flop, 44
Frequency versus power, inverter, 52
Fringe-area TV preamplifier, 267
Front end, 304
Front end, FM broadcast, 247
Front end, 2-meter, 313
Front end, VHF, 305
Full-isolation phase control, 110
Full-isolation ring detector, 167, 168
Full-power low-speed motor control, 99
Full-range adjustable supply, 22
Full-range lamp dimmer, 86, 98
Full-range power control, 92, 129

Full-range power controller, 97
Full servomotor control, 119
Full-wave bridge converter, 31
Full-wave phase control, 85, 92
Full-wave phase control, 900 W, 118
Full-wave phase control, motor speed, 131
Full-wave power control, 118
Full-wave 115 V pulse switch, 73
Full-wave series-motor control, 89
Full-wave soft-start lamp control, 85
Full-wave synchronous supply, 23, 24
Full-wave triac dimmer, 86
Full-wave triac lamp dimmer, 122

Gain-of-3 peak detector, 164
Gain-of-100 servo amplifier, 172
Garage-door opener, electronic, 393
Garage-light control, 83, 392
Gas-and-smoke sensor, 165
Gas detector, 165
Gas-lamp dimmer, 93
Gas-temperature regulator, 135
Gate, color-burst, 266
Gated-anode ring counter, 193
Gated-load ring counter, 193
Gate-driven-load counter, 193
Gated-thyristor charger, 32
Gated-thyristor 80 V supply, 25
Gated-triac 90 V rms regulator, 4
Gated triac switch, 94
Gate-load ring counter, 193
GE irradiance tester, photodevice, 374
GE portable fluorescent lamp, 44
General-purpose alarm, 164
General-purpose regulated supply, 9 V, 24
General-purpose video amplifier, 250
Generator, audio, 223
Generator, audio signal, 378
Generator, audio, with dial, 379
Generator, click, 236, 237
Generator, current, 9
Generator, organ tone, 233
Generator, pulse, 220
Generator, rectangualr-pulse, 71
Generator, sawtooth, tone, 229
Generator, sine-wave, 378
Generator, square-wave, 44
Generator, tone, 220, 221, 224, 230, 235
Generator, tone, sun-powered, 237
Generator, trigger pulse, 48
Germanium-diode rectifier, 22, 23
Germanium-transistor supply, 23
Glass-fiber-optic transmitter, 227
Go no-go rate sensor, 148
Go no-go testers, 367
Goof-proof car-battery charger, 29
"Grid-dipper," transistor, 371
Grounded-base current regulator, 5
Ground-sensing relay, 166
Guide, solar-panel sun sensing, 81

h_{FE} tester, transistor, 371, 375
Half-amp regulator, 28 V, 12
Half-kilovolt dc regulator, 7
Half-kilowatt ac power "relay," 78
Half-kilowatt ac regulator, 6, 9
Half-kilowatt line inverter, 48
Half-minute timer, 210
Half-percent line voltage regulator, 16
Half-watt audio amplifier, 329
Half-wave ac chaser, 187, 189, 387
Half-wave battery charger, 12 V, 32
Half-wave motor speed control, 95, 96
Half-wave power switching, 76
Half-wave SCR phaser, 111
Half-wave shunt-motor control, 94
Half-wave synchronous rectifier, 25
Hand drill speed control, 111
Happy-Face blinker, 185
Harmonic-rich oscillator, 223
Harmonic-suppressing doubler, 277
Harmonic tripler, to 600 MHz, 279
Hash-free voltage source, 8
Hazard detectors, 160
Hazard-indicating flasher, 182
Hazardous-environment switch, 73
HCD, in 15 A line-op. inverter, 50

HCD-detector video amplifier, 264
Headlamp dimmer, automatic, 381
Headlamp-operated garage light, 83, 392
Headlights-on reminder, 382
Headphone amplifier, 327
H_E and H_{ES} photodevice tester, 375
Hearing-aid preamp, 316
Heat alarm, 134
Heat control, for 115 V ac, 103
Heat control, proportional, 144
Heater control, room, 132
Heater control, space, 136
Heater control, temp. sensitive, 141
Heater control, ZVS, 138
Heater-fan control, 143
Heater-power control, 135
Heater temperature regulator, 133, 135
Heating-element control, 103, 132
Heating-element control, 240 V, 136
Heating-element regulator, 133
Heating-element switch, 146
Heating system phase control, 112
Heat regulator, RFI-proof, 146
Heat sensor, 142
Heat-triggered sensor, 147
HEP panic button, 237, 238
Hi-fi amplifier, 1 W or 2 W, 333
Hi-fi amplifier, 2 W, 332
Hi-fi amplifier, 4 W, 335, 336
Hi-fi amplifier, 5 W, 335
Hi-fi amplifier, 5 W, 337
Hi-fi amplifier, 5 W (8-ohm), 338
Hi-fi amplifier, 7 W, 339
Hi-fi amplifier, 7.5 W, 339
Hi-fi amplifier, 8 W, 340
Hi-fi amplifier, 10 W, 340
Hi-fi amplifier, 12 W, 341, 342
Hi-fi amplifier, 12 W, (16-ohm), 343
Hi-fi amplifier, 15 W, 338, 344, 345
Hi-fi amplifier, 15 W (8-ohms out), 345
Hi-fi amplifier, 15-60 W, 346
Hi-fi amplifier, 20 W 343, 347
Hi-fi amplifier, 25 W, 347, 348, 349
Hi-fi amplifier, 25 W (8 ohms), 350
Hi-fi amplifier, 40 W, 352, 353, 354
Hi-fi amplifier, 40 W Amperex, 342
Hi-fi amplifier, 40 W, 351
Hi-fi amplifier, 50 W, 355
Hi-fi amplifier, 55 W (4 ohms), 356
Hi-fi amplifier, 60 W, 346
Hi-fi amplifier, 70 W, 357
Hi-fi amplifier, 70 W quasi-comp, 357
Hi-fi amplifier, 80 W FET, 359
Hi-fi amplifier, 120 W, 361
Hi-fi amplifier, 160 W, stereo, 360
Hi-fi audio follower, 321
Hi-fi audio mixer, 323
Hi-fi audio preamp, 322
Hi-fi low-power amplifier, 326
Hi-fi mike preamp, 220, 223
Hi-fi mixer-compressor-ampl, 323
Hi-fi preamp, crystal/ceramic, 320
Hi-fi preamp, IC, 320
Hi-fi preamp, magnetic cartridge, 317
Hi-fi preamp, wide-range, 320
Hi-fi presence control, 364
Hi-fi signal injector, 378
Hi-fi stereo preamp, 318, 324
High-brightness lamp flasher, 186
High-B+ power supply, 24
High-B+ transistor radio, 242
High-B+ video amplifier, 252
High-capacitance varactor tuner, 623
High-current, low-voltage inverter, 45
High-current 25 V switch, 64
High-current 28 V dc regulator, 12
High-current 700 V switcher, 53
High-current dc motor control, 117
High-current ring relay, 167
High-current switching with 5 A relay, 70
High-di/dt SCR trigger, 71
High-efficiency converter, 1 kW, 39
High-efficiency dc converter, 39
High-efficiency regulator, 12 V, 11
High-efficiency switching regulator, 10
High-frequency B+ supply, 23
High-frequency converter, 28/150 V, 34

INDEX 399

High-frequency converter, 200 V, 36
High-frequency inverter, 12 V, 44
High-frequency inverter, 60 W, 54
High-frequency pulse generator, 71
High-frequency supply, 7.5 kV, 24
High-feedback inverter, 254
High-gain audio amplifier, 317
High-gain audio ampl, 4 W, 335
High-gain phase control, 95
High-harmonic signal generator, 378
High-impedance buffer, 224
High-impedance, low-C ampl, 250
High-impedance vibrato, 229
High-input-impedance buffer, 224
High-input-voltage dc converter, 38
High-input-voltage inverter, 57
High-input-voltage inverter, 80 W, 54
High-intensity-lamp dimmer, 99
High low power light-controlled relay, 77
High-low power switch, 239
High-output audio voltage ampl, 322
High-performance car radio, 383
High-performance lamp control, 111
High-performance TV mixer, 257
High-power flasher, 120 V, 181
High-power ac flasher, 120 V, 182
High-power dc converter, 20 A, 37
High-power dc motor control, 117
High-power heat control, 144
High-power inverter, 49
High-power regulator, 1 kV, 3
High-power switching, 5 kW, 68
High-power lamp dimmer, 83, 112, 128
High-power line inverter, 50
High-power regulator, 7
High-power speed control, 111
High-power stereo, 30 V supply, 26
High-power switch, 700 V at 7 A, 69, 70
High-power-switching ring counter, 188
High-power temp control, 136
High-power temperature controller, 132
High-power theater lamp dimmer, 123
High-speed converter, 38
High-speed data transmission, 171
High-speed inverter, 25 kHz, 51
High-speed inverter, 28 V, 42
High-speed paper-tape reader, 159
High-speed ring counter, 196
High-speed power switch, 700 V, 53
High-speed switch, 700 V at 7 A, 69
High-symmetry lighting control, 114
High-temperature alarm, 134
High-threshold-logic supply, 41
High-threshold logic triac switch, 194
High-torque-motor speed control, 95, 105, 117
High-voltage ac light control, 79
High-voltage dc converter, 36
High-voltage dc regulator, 2-3 A, 10
High-voltage dual B+ supply, 20
High-voltage electronic switch, 64
High-voltage flash slave, 217
High-voltage 250 mA "relay," 64
High-voltage phase control, 113
High-voltage servo system, 70
High-voltage solar radio, 245
High-voltage 7 A switcher, 53
High-voltage 800 V supply, 21
High-wattage-lamp control, 83
High-Z VOM-to-VTVM adapter, 372
Hi-level audio preamp, 322
Holes versus electrons in FETs, 71
Home fire alarm, 137
Horizontal driver, 251
Horizontal output, 251
Horizontal output circuit, 251
Horizontal sweep circuit, 251
Horizontal-sync B+ supply, 23
Horn-honk theft alarm for cars, 385
Horn-operating car alarm, 164
Hot-plate temperature controller, 135
Hot-pot temperature regulation, 133
Hot-carrier diode, *see entries under HCD*
Hot-carrier-diode inverter, 50
Hot-shoe slave attachment, 48
HTL, *see entries under High Threshold Logic*
HTL-compatible inverter, 41
HTL switching of triac, 194

Human sensor, 163, 164
Humidity controlled 120 V load, 157
Hybrid CATV switching supply, 19
Hybrid current generator, 9
Hybrid 12 V dc flasher, 184
Hybrid-feedback inverter, 47
Hybrid flasher, ac, 181
Hybrid line-operated servo ampl, 170
Hybrid liquid-level control, 150
Hybrid proximity detector, 164
Hybrid Schmitt trigger, 70
Hybrid servo preamplifier, 172
Hybrid timer, 10 ms to 1 min, 204
Hybrid-TV horizontal output, 251
Hygrometer control of 120 V load, 157
Hysteresis control of switch, 65
Hysteresis explained, 126
Hysteresis-free dimmer, 92, 106, 111, 129
Hysteresis-free dimmer, 800 W, 130
Hysteresis-free dimmer, 1000 W, 129
Hysteresis-free dimmer, soft-start, 125
Hysteresis, lamp dimmer, 84

I_{CBO}-measuring device, 370
IC audio amplifier and tone osc, 221
IC code practice oscillator, 221
IC-controlled current source, 9
IC-controlled FET alarm, 166
IC-controlled proximity alarm, 164
IC-controlled regulated source, 9 V, 27
IC-controlled regulated supply, 24 V, 28
IC-controlled regulated supply, CATV, 19
IC-controlled regulator, 6 A, 13 V, 15
IC-controlled temperature control, 140
IC-controlled temperature regulator, 132
IC-controlled triac dimmer, 111
IC-controlled triac flasher, 181
IC-driven 15 W hi-fi amplifier, 345
I_{CEO}-measuring device, 375
IC-oscillator car-lamp dimmer, 387
IC power supply, 5 V, 26
IC preamp, hi-fi audio, 320
IC stereo amplifier, 2 W, 332
IC timer with LED readout, 213
IC wireless FM mike, 225
IF, portable-TV sound, 261
IF amplifier, 455 kHz, 243
IF amplifier, 44 MHz video, 256
IF amplifier, B and W video, 258
IF amplifier, cascode, 30 MHz, 306
IF amplifier/oscillator, 310
IF amplifier, TV, 264
Ignition system, capacitor-discharge, 387
Ignition system, SCR, 388
Ignition tachometer, 384
Image-parameter filter, 298
Impulse-actuated switch, 134
Incandescent-lamp dimmer, 87
Incandescent-lamp ring counter, 200, 201
In-car automotive circuits, 381
In-circuit beta tester, 371
Indicator, overvoltage, 166
Indoor-outdoor thermometer, 137
Induction motor speed control, 87, 91, 99, 127
Inductive-ballast dimmer, 130
Inductive-load control, 93, 103, 116, 118, 154
Inductive-load dc controller, 124
Inductive load coupler, 154
Inductive-load snubber, 87, 93
Industrial-control servo amplifier, 169, 170
Industrial-control timer, 206
Industrial-power flasher, 182
Industrial power switch, 5 kW, 68
Industrial process control, 81
Industrial regulating converter, 37
Industrial siren, 234
Infrared detector, 155
Infrared light transceiver, 227
In-garage car circuits, 392
Inhibited sequential timer, 208
Injector, audio-signal, 378
Injector, hi-fi tone, 378
Injector, signal, 223
Input-voltage-dependent current source, 9
Inrush-current elimination, 85
Inrush-free lamp switch, 152

Integral CATV 80 W switching supply, 19
Integral-cycle heat control, 136
Integrated voltage regulated, 27
Integrator, ac-coupled, 224
Intercom, truck/camper, 226
Intercommunications system, 226
Interference, *see entries under RFI or EMI*
Interference-free dimmer, 103
Interference-proof phase control, 125
Interference, from switching, 33
Interior-lamp dimmer for cars, 387
Interrupted-beam alarm, 161
Interrupted-beam drop detector, 149
Interrupted-beam rate sensor, 148
Interruption sensor, light, 80, 81, 82
Interval timer, 147
Interval timer, 0.5 s to 3 min, 203
Interval timer, 0.6 to 6 seconds, 204
Interval timer, 2 s to 3 min, 214
Interval timer, 15-second, 203
Interval timer, adjustable, 219
Interval timer, darkroom, 183
Interval timer, and phase control, 119
Interval timer, simple, 208
Interval timer/switch, 211
Intruder-detector light, 160
Intrusion alarm, 160
Intrusion alarm, 12 V dc, 162, 164
Intrusion alarm, body-sensing, 163, 164
Intrusion alarm, control, 79, 82
Intrusion alarm, light control, 80
Intrusion alarm, phototransistor, 160, 161
Intrusion alarm, SCR, 78
Inverters, 41
Inverter, 2500 Hz, 44
Inverter, 28 kHz, 180 V, 54
Inverter, 1 kW line-operated, 41
Inverter, 12 V-input, 56
Inverter, 12 V input, 110 W, 52
Inverter, 12 V-to-115 V at 400 Hz, 45
Inverter, 12 V-to-110 V, 400 Hz, 43
Inverter, 12 V to 117 V 60 Hz, 42
Inverter, 12 V to 117 V at 110 W, 46
Inverter, 12 V to fluorescent, 44
Inverter, 12 V square-wave, 44
Inverter, 28 V at 25 kHz, 42
Inverter, 28 V-to-70 V, 54
Inverter, 28 V-to-240 V, 52
Inverter, 115 V in and 15 V/15 A out, 50
Inverter, 150 V input, 200 W, 51
Inverter despiking, 49
Inverter driver, motor control, 91
Inverter efficiency versus power, 51, 52
Inverter, fluorscent lamp, 48
Inverter frequency versus power, 52
Inverter gating circuit, 48
Inverter, high-input-voltage, 57
Inverter, high-power (line-op.), 50
Inverter, hybrid-feedback, 47
Inverter, 3-phase, 55
Inverter, phase, color TV, 267
Inverter, power, multirange, 49
Inverter, power, versus frequency, 52
Inverter, pulse shaper, 55
Inverter, rectified for 20 A out, 39
Inverter, regulated 400 Hz, 45
Inverter, strobe lamp, 53
Inverter, tailoring efficiency of, 51
Inverter, ultrasonic, hi-power, 49
Inverter, universal, 54
Inverting 7.5 kV supply, 24
Ionization chamber, double, 165
Ionization chamber, single, 165
Ionization chamber smoke sensor, 162
IR photosensitive transceiver, 227
Irradiance tester, photodevice, 374
Isolated phase control, 110
Isolated phone-ring detector, 167, 168
Isolated power switching, 61, 62
Isolated 115 V pulse switch, 73
Isolated-tab triac ring detector, 167
Isolator, data transmission, 171
Isolation, logic level switching, 69
Isolation, for switching, 61
Isolation, telephone-line, 167
IV-controlled temp. regulator, 142
IVR power supply, 27

JFET, 70

Keyed solar oscillator, 237
Keyed tone generator, 221
Keyed tone oscillator, 223, 224, 226
Keyer, AGC, 258
Keyer, AGC and phase inverter, 267
Keyer, electronic, compact, 222
Keyer, and video buffer, 258
Killer, commercial (TV), 260
Kilowatt ac flasher, 181, 182
Kilowatt converter, 275 V-50 V, 39
Kilowatt generating inverter, 49
Kilowatt inverter, 20 kHz, 41
Kilowatt lamp dimmer, 85, 104
Kilowatt lamp dimmer, ac, 128, 129
Kilowatt switching regulator, 7
Kilovolt regulator, 3
Kitchen-appliance control and timer, 127
Kitchen appliance motor control, 96
Kitchen-appliance speed control, 91, 106, 115
Kitchen-appliance timed speed control, 119
Kitchen-blender timer, 211
Kitchen timer, 147
Kitchen timer, 0-3 min, 214

Lamp, strobe, source, 53
Lamp brightness control, *see entries under Dimmer*
Lamp-bulb life extender, 152
Lamp control, 9 V dc, 178
Lamp control, by remote, 174
Lamp control circuit, 25 W, 134
Lamp dimmer, 89, 100
Lamp dimmer, 400 Hz, 87
Lamp dimmer, 1 kW, 85, 104, 128, 129
Lamp dimmer, 25 W miniature, 117
Lamp dimmer, 400 W, 97
Lamp dimmer, 500 W, 92
Lamp dimmer, 800 W, 86, 130
Lamp dimmer, ac, 95
Lamp dimmer, car interior lights, 387
Lamp dimmer, compensated, 100, 123
Lamp dimmer, with feedback, 110, 111, 124
Lamp dimmer, fluorescent, 93, 130
Lamp dimmer, full-wave, 85, 86, 98, 122
Lamp dimmer, hysteresis-free, 92, 106, 129
Lamp dimmer, low-cost, 99
Lamp dimmer, low-hysteresis, 92, 100, 122, 126
Lamp dimmer, low-voltage, 99
Lamp dimmer, no-hysteresis, 106
Lamp dimmer, photocell-controlled, 103
Lamp dimmer, PUT, 100
Lamp dimmer, RFI-proof, 84, 125
Lamp dimmer, soft-start, 101, 120, 125
Lamp dimmer, SUS-triggered, 110
Lamp dimmer, symmetrical, 114
Lamp dimmer, time-dependent, 125
Lamp dimmer, triac, 109, 114
Lamp dimmer, unijunction, 110
Lamp dimmer, 2-time-constant, 92
Lamp driver, 68
Lamp driver, fluorescent, 47
Lamp-filament stress reliever, 152
Lamp flasher, 3 V, 183
Lamp flasher, 12 V dc, 185
Lamp flasher, 120 V ac, 184
Lamp flasher, hi-brightness, 186
Lamp flasher, low-voltage, 183
Lamp flasher, multivibrator, 185
Lamp flasher, power (12 V), 184
Lamp flasher, sequential, 187
Lamp-indicating burglar alarm, 160
Lamp inverter, fluorescent, 98
Lamp, portable fluorescent, 44
Lamp-power measuring circuit, 16
Lamp preheating, 152
Lamp regulator, 15, 16
Lamp ring counter, 200, 201
Lamps-on reminder, automobile, 382
Lamp switch, inrush absorption, 152
Lamp target, photocell, 152
Lamp voltage regulator, 16
Large-angle compensator, ac, 109
LASCR-controlled lamp switch, 152
LASCR irradiance tester, 374
LASCR line switch, 76

LASCR motor speed control, 90
LASCR photoflash slave, 215, 217
LASCR switch, 73
LASCR voltmeter tachometer, 156
Latching ac "relay," 65, 72
Latching photo switch, 80
Latching relay, dc, 65
Latching time delay, 115 V, 210
Latching touch switch, 238
LC commutating ring counter, 193
Lead-acid-battery charger, 392, 393, 394
Leakage tester, collector-to-base, 370
Leakage tester, transistor, 375
LED blinker, 185
LED continuity tester, 367
LED continuity and voltage tester, 368
LED-controlled darkness sensor, 82
LED coupled Schmitt trigger, 156
LED-display brightness control, 152
LED drop detector, 154
LED-readout seconds timer, 213
LED ring counter, 190
LED-sensing relay, 82
LED sequential flasher, 190
LED supply, 26
LED tester, 369
Level control, liquid, 150
Level sensing for liquids, 149
Level sensor, liquid, 147
Level shifter for speed control, 117
Light, automatic night, 81, 83
Light, automatic off-on, 77
Light, automatic turn-on at night, 80
Light, night, automatic, 77
Light, phototransistor-controlled, 79, 81
Light-activated SCR, *see entries under LASCR*
Light-activated switch, 147
Light-beam communication, 226, 227
Light-beam drip detector, 149
Light-beam-interrupted alarm, 161
Light-beam interruption sensor, 82
Light-commutating motor speed control, 90
Light-controlled ac phaser, 110
Light-controlled appliances, 77
Light-controlled dimmer, 85
Light-controlled model motor, 173
Light-controlled motor speed, 88
Light-controlled relay, 79, 81, 82
Light-controlled relay, N. C., 78, 79, 80
Light-controlled solenoid, 79, 81
Light-controlled switch, 81. See also *Light-controlled relay*
Light-controlled switch, 76
Light-controlled triac switch, 600 W, 78
Light-controlled turn-on, 79, 81
Light-coupled transceiver, 227
Light-deenergized relay, 79, 80, 81, 82
Light detector, logarithmic, 155
Lighted-display brightness control, 152
Light-emitting-diode, *see entries under LED*
Light-energized relay, 81, 82
Light flasher, *see entries under Lamp flasher*
Lighting for billboards, automatic, 77
Lighting control, proportional, 114
Light-interrupting motor control, 87
Light-interruption sensor, 80, 81
Light-modulated transmitter, 22, 67
Light modulator, music-driven, 231
Lights-on reminder, 385, 389
Lights-on reminder, automobile, 382
Light-operated circuits, *see Optoelectronic*
Light-operated relay, 80, 82. See also *Light-controlled relay*
Light-operated SCR alarm, 78, 80
Light-operated switch, 83
Light-operated triac switch, 81
Light organ, 231
Light organ cell, 235
Light-output monitor, 155
Light pulse counter, 82
Light-sense alarm, 166
Light-sensitive flasher, 186
Light-spot detector, 152
Light target, 152
Light-triggered commercial killer, 260
Light-triggered flash slave, 215, 217
Light-triggered object counter, 148

Light-triggered switch, 134
Light-triggered tachometer, 156
Light-wave communicator, 225, 227
Limited-range low-cost dimmer, 99
Limiting, current, 18
Linear 244 MHz mixer, 311
Linear amplifier, 5 W on 2 meters, 293
Line driver, audio, 319
Line-operated AM radio, 242
Line-operated audio amplifier, 2 W, 332
Line-operated HCD inverter, 50
Line-operated inverter, 47
Line-operated inverter, 1 kW, 41
Line-operated servo amplifier, 170
Line-operated 1-shot timer, 206
Line-operated smoke sensor, 163, 165
Line-operated timer, 206
Line-powered alarm-system charger, 21
Line status monitor, 167, 168
Line switch, light-activated, 76, 79, 81
Line switch, light controlled, 79, 80, 81, 83
Line switch, light-operated, 79
Line voltage, from car battery, 42
Line-voltage-compensated dimmer, 100
Line-voltage compens, phase control, 123
Line-voltage control, 5 A, 111
Line-voltage control, 400 Hz, 93
Line-voltage control, 400 W, 97
Line-voltage control, 600 W, 120
Line-voltage control, 800 W, 130
Line-voltage control, 1000 W, 128, 129
Line-voltage control, hysteresis-free, 106
Line-voltage control, for motors, 109
Line-voltage control, RFI-proof, 125
Line-voltage control, shunt-motor, 120
Line-voltage control, for small motor, 114
Line-voltage control, symmetrical, 114
Line-voltage control, wide-range, 92, 129
Line voltage, from car battery, 42
Line-voltage generator, 46
Line voltage lamp flasher, 184
Line-voltage-operated light, 83
Line-voltage phase control, 88, 89, 99, 116
Line voltage regulator, 15, 16
Line-voltage regulator, 500 W, 6
Line-voltage sources, 41
Line-voltage-switching timer, 208
Liquid-drop detector, 154
Liquid-level alarm, 147
Liquid-level control, 148
Liquid-level control, hybrid, 150
Liquid-level sensor, 149
Liquid-temperature regulator, 139
Liquid-volume control, 150
Load control, flame-sensed, 151
Load controller, ac and dc, 97
Load dropout detector, 154
Load driver and time delay, 211
Load-gated ac switch, 74
Load-in-gate ring counter, 193
Load-line-limited 25 W hi-fi, 348
Load monitor, 120 V, 154
Load monitor, and alarm, 154
Load switcher with adj delay, 210
Load-switching delayed dropout, 209
Load-varying motor speed control, 95
Logarithmic light detector, 155
Logic alarm, low=off, high=on, 166
Logic-compatible audio alarm, 166
Logic-compatible bistable switch, 184
Logic-compatible coupler, 156
Logic-compatible data isolator, 171
Logic-compatible electronic reed relay, 69
Logic-compatible power-out detector, 154
Logic-compatible ring detector, 167
Logic-compatible sun tracker, 81
Logic-controlled motor control, 127
Logic-controlled switch, 62, 64
Logic-driven power switch, 69
Logic-driven ring counter, 199
Logic-driven transistor driver, 68
Logic-driven triac, 194
Logic-level tape reader, 159
Logic-output dial pulser, 168
Logic-output proximity alarm, 164
Logic signal switching isolation, 69
Logic switching DPDT, 71

Logic switching, HTL and triac, 194
Logic-voltage lamp flasher, 183
Log-jam control, 158
Log-output radiation sensor, 155
Log transducer, light-to-voltage, 155
Long-delay power switch, 203, 210
Long-distance-call alarm, 168
Long-duration timer, 20 min, 204
Long-life 6 V dc flasher, 186
Long-line headphone amplifier, 327
Long-term timer, up to 20 min, 204
Long-time comparator, 207
Long-time-constant integrator, 224
Loudspeaker alarm, logic-strobed, 166
Loudspeaker-driven color organ, 231
Low-capacitance amplifier, 250
Low-capacitance audio follower, 324
Low-capacitance buffer, 224
Low-capacitance varactor tuner, 263
Low-conduction-angle regulator, 107
Low-cost fire alarm, 137
Low-cost full-range dimmer, 86
Low-cost lamp dimmer, 99
Low-cost ring counter, 193
Low-cost shaded-pole-motor control, 99
Low-cost shunt motor control, 131
Low-cost 80 V supply, 1.5 A, 25
Low-current ac "relay," 65
Low-current dc flasher, 6 V, 186
Low-current power driver, 68
Low-current ring detector, 168
Low-distortion amplifier, 4 W audio, 335
Low-distortion amplifier, 5 W hi-fi, 337
Low-distortion amplifier, 40 W hi-fi, 352
Low-distortion amplifier, 50 W hi-fi, 355
Low-distortion audio preamp, 322
Low-distortion oscillator, 20 MHz, 273
Low-drain power-transistor driver, 68
Low-frequency power switch, 70
Low-frequency tone generator, 221
Low-gain phase control, 100
Low-hysteresis dimmer, 92, 122, 129
Low-hysteresis 800 W dimmer, 130
Low-hysteresis lamp dimmer, 86, 92, 100, 106 111, 126
Low-hysteresis motor control, 105
Low-hysteresis phase control, 92
Low-hysteresis table-lamp dimmer, 122
Low input-capacitance buffer, 224
Low-level signal switcher, 69
Low-light-level drop detector, 154
Low-loss synchronous supply, 23
Low-noise 30 MHz amplifier, 306
Low-pass m-derived filter, 298
Low-power automatic night light, 77
Low-power hi-fi amplifier, 326
Low-power hi-gain audio amplifier, 317
Low-power light-controlled relay, 77
Low-power inverter, 400 Hz, 45
Low-power inverter, 28 V, 42
Low-power regulator reference, 8
Low-power ring counter, 188
Low-power transistor driver, 68
Low-resistance temp. sensor, 140
Low-reverse-transfer buffer, 220
Low-RFI temperature controller, 146
Low-speed, high-torque control, 99
Low-speed-motor control, 105
Low-speed power switch, 700 V, 70
Low-voltage-controlled ac switch, 71
Low-voltage dc flasher, 183, 186
Low-voltage dc light control, 80
Low-voltage-input ac inverter, 45
Low-voltage lamp flasher, 183
Low-voltage phase control, 113
Low-voltage regulator, 5, 6, 7
Low-voltage regulator, 8 V and 16 V, 6
Low-voltage regulator, 10 A, 12
Low-voltage regulator, 100 mA-16 V, 5
Low-voltage regulator, 50-200 mA, dc, 5
Low-voltage ring counter, 188, 202
Low-voltage servo amplifier, 171
Low-voltage smoke alarm, 165
Low-voltage TV supply, 23
Luminance amplifier, 265

Mag-amp dc converter, 39

Magnetic amplifier, 39
Magnetic-cartridge preamp, 317, 318, 321, 322, 325
Magnetic deflection circuit, 250
Magnetic phone preamp, 317
Majority versus minority carriers, 71
Manual 3-phase ac control, 113
Marker-light activator, 80
Marking-buoy flasher, 1-lamp, 186
Master oscillator, organ, 233
Master tone generator, 233
Matrixing, chroma, 266
m-Derived 4.5 MHz filter, 254
m-Derived low-pass filter, 298
Measurement devices, 370
Metallic-pair function switching, 175
Meter, balance control, stereo, 363
Meter, dip, wide range, 371
Meter, FET millivolt, 373
Meter, temperature-reading, 137
Metronome, PNP, 236
Metronome, UJT, 237
Microamp integrator, 224
Microphone amplifier, 319
Microphone amplifier, 40 dB, 223
Microphone amplifier, hi-fi, 320
Microphone preamplifier, 220, 316
Microphone, wireless (AM), 226, 228
Microphone, wireless (FM), 225
Micropower dc converter, 33
Mike, see entries under Microphone
Mike amplifier, 319
Mike amplifier, hi-fi, 320
Mike preamp, 316
Miller integrator, 224
Miller integrator timer, 207
Miniature B+ supply, 23
Miniature inverter, 200 W, 51
Miniature lamp dimmer, 100 W, 100
Miniature low-hysteresis dimmer, 122
Miniature phase control, 117
Mini-fi transistor amplifier, 329
Minimum-hysteresis lamp dimmer, 92
Minority versus majority carriers, 71
Minute timer, 204, 211
Miscellaneous circuits, 376
Mixer, 100-to-44 MHz, 304
Mixer, active balanced, 311
Mixer amplifier, 304
Mixer, audio, 323
Mixer, audio, 4-input, 323
Mixer, audio, 7 inputs, 325
Mixer, balanced, 312
Mixer control, with timer, 127
Mixer motor speed control, 106, 115
Mixer speed control, plug-in, 111
Mixer speed control, timed, 119
Mixer, TV, 259
Mixer, TV tuner, 263
Mixer, VHF, 244 MHz, 311
Mixer, VHF TV, hi-performance, 257
Mobile ac power source, 42
Mobile AM broadcast radio, 383
Mobile aircraft transmitter, 289
Mobile audio amplifier, 4 W, 336
Mobile B+ supply, 150/200 V, 37
Mobile B+ supply, 500 V, 39
Mobile CB transmitter, 5 W, 281
Mobile inverter, 400 Hz, 43
Mobile inverter, 60 W, 53
Mobile 1-lamp flasher, 384
Mobile-light controlled relay, 83
Mobile line-voltage source, 42, 46
Mobile power inverter, 110 W, 52
Mobile transmitter, 118-150 MHz, 289
Mobile transmitter, 175 MHz, 280
Mobile transmitter, 175 MHz, 25 W, 287
Mobile UHF amplifier, 25 W, 297
Mobile UHF power amplifier, 10 W, 297
Moblie UHF power amplifier, 296, 297
Mobility of carriers, 71
Model-device remote controller, 173
Model-railroad scale speed control, 234
Model-railroad speed control, 103
Model speed control, scale, 234
Model train scale speed control, 234
Modulated oscillator, AM, 228

Modulated oscillator, FM, 225
Modulated oscillator, 73.5 MHz, 274
Modulated oscillator, sun-powered, 288
Modulated-triac ZVS, 138
Modulator, 290
Modulator, balanced, 290
Modulator, light-beam, 227
Modulator, phase, 291
Modulator, Schmitt, 117
Modulator, series, 290
Modulator, for transmitter, 4 W, 290
Modulator, for transmitter, 25 W, 290
Moisture detector, 147
Monitor, battery-voltage, 159
Monitor, conveyor pileup, 157, 158
Monitor, flame, 151
Monitor, ground-level, 166
Monitor, lights-on, for cars, 382
Monitor, light-output, 155
Monitor, phone-use, 168
Monitor, tach-controlled motor speed, 169
Monitor, voltage, 166
Monitor amplifier, video, 252
Monitor/control, lamp brightness, 151
Monolithic adjustable regulator, 8-16 V, 5
Monolithic Darlington preamp, 316
Monolithic regulator, 100 mA 8 V, 5
Monostable multivibrator, 205
MOS/bipolar clock sync, 25
MOSFET amplifier, 200 MHz RF, 308
MOSFET amplifier, 400 MHz, 308
MOSFET AM radio, 248
MOSFET FM front end, 247
MOSFET FM tuner, 248
MOSFET mixer, 100 MHz, 304
MOSFET mixer, 244 MHz, 311
MOSFET tester, go/no-go, 368
MOSFET TV IF amplifier, 264
MOSFET variable-frequency oscillator, 272
MOSFET VHF front end, 305
MOS-transistor checker, 368
Motor, solar-cell-controlled, 173
Motor control, design data, 105
Motor control, induction, 127
Motor control, inverter, 91
Motor control, reversing, 119
Motor control, shaded-pole, 99, 121
Motor control, soft-start, 120
Motor control, with timer, 127
Motor controller, 25 W, 134
Motor-home lamp dimmer, 387
Motorola CATV switching supply, 19
Motorola metronome, 237
Motorola panic button, 237, 238
Motor rate sensor, 88
Motor speed control, 84, 87, 91, 100, 106
Motor speed control, 5 A, 104
Motor speed control, 400 Hz, 93
Motor speed control, 115 V 60 Hz, 106
Motor speed control, 120 V, 100
Motor speed control, 500 W, 98
Motor speed control, ac, 92
Motor speed control, dc, 88, 96
Motor speed control, fan, 99
Motor speed control, feedback, dc, 117
Motor speed control, half-wave, 95, 96
Motor speed control, induction, 91, 99
Motor speed control, low-hysteresis, 126
Motor speed control, low-power, 109
Motor speed control, low-speed, 105
Motor speed control, optofeedback, 90
Motor speed control, oscillator, 91
Motor speed control, plug-in, 111, 115
Motor speed control, PM-motor, 121
Motor speed control, reversing, 90, 116, 119
Motor speed control, RFI-proof, 103, 122, 125
Motor speed control, series, 89
Motor speed control, series wound, 90
Motor speed control, shop tools, 106
Motor speed control, shunt, 121
Motor speed control, shunt-wound, 90, 94, 120, 131
Motor speed control, soft-on, 125
Motor speed control, synchronous, 102, 121
Motor speed control, triac, 114
Motor speed control, universal, 90, 108, 126
Motor speed control, wide-range, 124

Motor speed control w/feedback, 105, 110
Motor speed control, tachometer, 158, 159
Motor starter, split-C, 86
Movie projector control, brightness, 151
Movie projector control, speed, 115
Multibutton doorbell, 230
Multi frequency oscillator, 235
Multifunction switching over wire pair, 175
Multikilowatt inverter, 49
Multilamp ac chaser, 187, 189, 192
Multilamp ring counter, 188
Multilamp sequential flasher, 187
Multilevel universal shift register, 198
Multioctave wideband amplifier, 298
Multi-output dc power supply, 376
Multiphase inverter, 115 V, 55
Multiplier 450 MHz, 278
Multiplier, 500-to-400 MHz, 276
Multiplier, 750 MHz, 278
Multiplier, 276
Multiplier, capacitance, 224
Multiplier, FET (VHF), 277
Multiplier, VHF to UHF, 277
Multiple-B+ power supply, 23
Multiple-contact light control circuit, 81
Multiple-contact switching with light, 79
Multiple-entry intrusion alarm, 160
Multiple-LED counter, 190
Multiple-load ac chaser, 192
Multiple-load flasher, 190, 192, 193, 202
Multiple-load ring counter, 188
Multiple-load sequential flasher, 187, 189
Multiple-octave amplifier, 298
Multiple-sensor alarm, 147
Multiple-stage counter, 192, 202
Multiple-switch burglar alarm, 160
Multiple-time-constant dimmer, 129
Multiple-transducer alarm, 166
Multiple-voltage power supply, 376
Multiple-relay light control, 79, 81
Multiposition triac switch, 74
Multirange power inverter, 49, 54
Multirange timer, 204
Multitapped B+ supply, 23
Multitester converter, to VTVM, 372
Multitester-to-VTVM converter, 372
Multivibrator, astable, 44
Multivibrator, bistable, 184
Multivibrator, bistable, PUTs, 189
Multivibrator-controlled regulator, 11
Multivibrator, erase and bias, 363
Multivibrator flasher, 12 V dc, 182
Multivibrator lamp flasher, 185
Multivibrator LED flasher, 185
Multivibrator keyer, 222
Multivibrator pulse shaper, 55
Music modulator for lights, 231, 235

Narrowband servo preamp, 172, 173
Negative temperature coefficient, 85, 132
Negative-voltage current source, 9
Neon-gated-SCR lamp flasher, 184
Neon relaxation oscillator, 239
Neon-sign power supply, 24
Network, Butterworth filter, 256
Network, *m*-derived filter, 298
Network, snubber, 93
Network, timing, 179
Network, voltage-reducing, 376
Neutralizationless 200 MHz ampl, 307
Neutralizationless 450 MHz ampl, 308
Night light, automatic, 77-83
No-arc 50 A switcher, 70
No-arc pulse-activated switch, 73
No-arc split-capacitor motor control, 86
No-hysteresis dimmer, 92, 106, 129
No-hysteresis dimmer, 800 W, 130
No-hysteresis dimmer, 1 kW, 128, 129
No-inrush lamp switch, 152
Noise filter, audio, 364
Noise-proof phone monitor, 167, 168
No neutralization RF ampl, 200 MHz, 309
No neutralization ampl, 400 MHz, 308
Noninductive hi-power switch, 69, 70
Noninductive-load power switch, 70
Noninductive load switcher, 53
Noninductive-load voltage control, 89

Nonlatching touch switch, 232
Nonloading 0-360° phase shifter, 178
Nonloading phone-ring detector, 167, 168
Nonmagnetic trigger for SCRs, 118
Nonsine tone generator, 378
No-RFI lamp dimmer, 84
No-RFI thermostat for heater, 142
No-RFI triac power control, 125
Normally closed switch, 10 A, 60, 64
Normally closed "relay," 250 mA, 64
Normally closed electronic relay, 60
Normally closed light control relay, 80
Normally closed light relay, 77, 78, 79, 81, 82
Normally closed light switch, 80
Normally closed triac switch, 67
Normally closed ZVS, 154
Normally open 10 A switch, 64
Normally open "relay," 250 mA, 64
Normally open electronic relay, 60
Normally open light-controlled relay, 77
Normally open light relay, 79-82
Normally open "relay," 66, 71
Normally open SCR, light-controlled, 78
No-thermal-overshoot temp control, 136
Novelties and toys, 229
Novelty blinker circuit, 185
No-voltage-drop dimmer, 100
NPN audio mixer, 7 inputs, 325
NPN code oscillator, 221
NPN emitter follower, 321
NPN op-amp final amplifier, 317
NPN regulator, 100 V dc, 13
NPN regulator, 100 V at 400 mA, 13
NPN thermal response tester, 373
NPN transistor preamp, 220
NPN 175 MHz transmitter, 280
NPN UHF tripler, 278
NPN video amplifier, 252
NTC sensor, 132
NTC thermistor dimmer, 85

Object counter, light-triggered, 148
Object counter photodevice, 79, 82
Object counting, light-controlled, 81
Off-at-dark circuit, 81
Off-at-dawn power flasher, 182
Office-light control, automatic, 77
Off-on heater control, ZVS, 142
On-at-dark hazard flasher, 182
On-at-dawn 1-lamp flasher, 186
On-at-dawn oscillator, 232
One-amp regulator, 60 V dc, 14
One-chamber smoke sensor, 165
One-lamp flasher, 186
One-only load selection control, 152
One-shot multivibrator, 205
One-shot timer, line-operated, 206
One-transistor regulator, 13
One-transistor regulator, 9 V, 24
One-transistor regulator, 50-200 mA dc, 5
One-triac ac "relay," 71, 75
On-off dimmer waveforms, 106
On off switch, high-frequency, 254
On-off temperature switching, 135, 146
Op-amp and triac switch combo, 194
Op-amp-driven servo amplifier, 170
Op-amp-driven servo preamp, 172, 173
Opener, garage-door, 393
Open-loop ac compensator, 109
Open-loop compensator, small-angle,
Open-loop ac regulator, 90 V, 9
Operational amplifier, 317
Optical coupler, 153
Optical coupling, 61
Optically coupled switch, 10 A, 61
Optical drip detector, 149
Optical feedback control, 90
Optical-feedback dc motor control, 117
Optical pickoff controller, dc, 88
Optical-pickoff dc motor control, 117
Optical-pickup tachometer, 377
Optical-programmer control, 134
Optical pulse shaper, 88
Optical receiver, FM, 227
Optical tachometer, 156
Optical tape reader, 155, 159
Optical transmitter, FM, 227

Opto commutating phase control, 90
Opto-coupled Schmitt trigger, 156
Opto-coupler, *see entries under Optoelectronic*
Opto-coupler-driven SCR, 154
Optoelectronic ac relay, 72
Optoelectronic coupler, 153
Optoelectronic Darlington-coupled relay, 68
Optoelectronic data link, 171
Optoelectronic dc motor control, 117
Optoelectronic dc latch relay, 65
Optoelectronic drip detector, 149
Optoelectronic feedback control, 90
Optoelectronic garage-light control, 392
Optoelectronic lighting control, 114
Optoelectronic parts-flow monitor, 157
Optoelectronic phase control, 90
Optoelectronic phone ring detector, 168
Optoelectronic power switch, 61, 73
Optoelectronic reed relay, 69
Optoelectronic relay, 61
Optoelectronic ring detector, 167, 168
Optoelectronic Schmitt trigger, 156
Optoelectronic tachometer, 156, 371, 377
Optoelectronic tape reader, 155, 159
Optoelectronic TTL relay, 62
Optoelectronic ZVS, 73
Organ, color, 231
Organ, electronic toy, 235
Organ, light, 231
Organ tone generator, 232
Oscillator, 271
Oscillator, 100 kHz crystal, 377
Oscillator, 2 MHz CMOS, 274
Oscillator, 20 MHz, 273
Oscillator, 27.255 MHz, 274
Oscillator, 47.100 MHz, 274
Oscillator, 73.5 MHz TD, 288
Oscillator, 700-800 MHz, 275
Oscillator/amplifier, all frequencies, 310
Oscillator/amplifier inverter, 55
Oscillator, audio, nonsine, 378
Oscillator, audio-radio, 379
Oscillator, bias, tape-head, 363
Oscillator, blocking, 220
Oscillator, code practice, 220, 221, 224
Oscillator, crystal, Pierce, 274
Oscillator, metronome, 236, 237
Oscillator, for motor speed control, 102
Oscillator, multivibrator, 44
Oscillator, neon relaxation, 239
Oscillator/pulse shaper for inverter, 55
Oscillator, push-pull, 200 W, 51
Oscillator, relaxation, PUT, 189, 230
Oscillator, sawtooth tone, 229
Oscillator, sine-wave, 378
Oscillator, siren, 237, 238
Oscillator, sun-powered, 237
Oscillator, sun-powered CB, 288
Oscillator, tone, 230
Oscillator, tone-producing, 274
Oscillator, tunnel diode, 273
Oscillator, twin-tee, 229, 235
Oscillator, unijunction, 173
Oscillator, variable-amplitude, 273
Oscillator, variable-frequency, 272
Oscillator, vertical deflection, 249, 253
Oscillator, voltage-controlled, 230
Oscillator-mixer, TV, 263
Oscilloscope dc amplifier, 376
Outage detector, power, 154
Outdoor light on-off control, 83
Output, horizontal, 251
Output circuit, vertical, B and W, 262
Output-protected 40 W hi-fi amplifier, 352
Oven control circuit, 135
Overload-protected regulator, 11
Overload-protected supply, 34-45 V, 18
Overload protection circuit, 11
Overtemperature alarm, 134
Overvoltage alarm, 166

Panic button, 237, 238
Paper-tape reader, 155, 159
Peak-detector, gain-of-3, 164
Pedestal-and-ramp load control, 157
Pedestal-and-ramp phase control, 116
Performance chart, dimmer, 110

INDEX **403**

Permanent-magnet, *see entries under PM*
Permanent-magnet-motor control, 96
Permanent-split-capacitor-motor control, 127
Phase, and voltage control, 97
Phase control, 1 to 5 A, 111
Phase control, 400 Hz, 93
Phase control, 400 W, 91
Phase control, 800 W, 86, 130
Phase control, 900 W, 118
Phase control 1 kW, 128
Phase control, 115 V, 89
Phase control, 115 V ac, 101
Phase control, 115 V, full-wave, 118
Phase control, 115 V, 2-SCR, 110
Phase control, 115 V, with feedback, 110
Phase control, 120 V ac, 100
Phase control, 120/240 V, 103
Phase control, direction-reversing, 116
Phase control, for fans, 1.5 A, 99
Phase control, with feedback, 110
Phase control for foreign line voltages, 103
Phase control, of heater power, 135
Phase control, without hysteresis, 106
Phase control, for inductive load, 116
Phase control, for motor, 91, 95
Phase control, for motor, 500 W, 98
Phase control, with opto feedback, 90
Phase control, for SCRs, 90
Phase control, for shunt motor, 131
Phase control, full-wave, 92
Phase control, full-wave, 100%, 85
Phase control, half-wave, 96
Phase control, high-gain, 95
Phase control, induction motor, 91
Phase control, inductive-load, 103
Phase control, 100% isolated, 110
Phase-controlled lamp regulator, 16
Phase-controlled line-voltage regulator, 6
Phase-controlled SCR switch, 73, 75
Phase-controlled temperature regulator, 141
Phase control, low-cost, 99
Phase control, low-gain, 100
Phase control, low-hysteresis, 86, 92, 100, 126
Phase control, miniature, 25 W, 117
Phase control, no-hysteresis, 92, 106, 129
Phase control, 3-phase (4-wire), 112
Phase control, 3-phase SCR, 124
Phase control, plug-in, 115
Phase control, ramp-and-pedestal, 116
Phase control, with RFI filter, 84
Phase control, RFI-proof, 125
Phase control, with SCRs, 88
Phase control with 2 SCRs, 94
Phase control, self-timing, 127
Phase control, series-motor, 89, 90
Phase control, shaded-pole motor, 99
Phase control, shunt and PM motors, 121
Phase control, shunt-motor, 90, 94
Phase control, soft-start, 85, 101
Phase control, SUS-triggered, 110
Phase control, symmetrical, 114
Phase control, synchronous, 102
Phase control, 2-time-constant, 92
Phase control, time-dependent, 120
Phase control, with timer, 119
Phase control, triac dimmer, 100
Phase control, wide-range, 92, 124
Phase control, wide-range, 800 W, 130
Phase control, wide-range, 1 kW, 129
Phase control, universal motor, 90
Phase control, voltage-compensated, 123
Phase control circuits, 84
Phase inverter, color TV, 267
Phase-inverting Schmitt trigger, 72
Phase modulator, 291
Phaser, half-wave SCR, 111
Phase-shift ac power control, 118
Phase shifter, 0-to-360°, 178
Phone-answer ring detector, 168
Phone-answer ring sensor, 167
Phone-call monitor, 168
Phone-compatible pulser, 168
Phone-cartridge preamp, magnetic, 317
Phone dial pulse indicator, 168
Phone-line add-ons, 167
Phone-line ring detector, 167, 168
Phone-line sensors, 167

Phone status monitor, 167, 168
Phone-use monitor, 168
Phone amplifier, 327
Phono amplifier, 2 W, 331, 334
Phono amplifier, 4 W (FET), 334
Phone amplifier, 22 V, 327
Phono amplifier, ceramic, 332
Phono equalization data, 327
Phonograph speed control, 96, 121
Phono-motor speed control, 121
Phono preamp, crystal-cartridge, 320
Phono preamp, magnetic, 318
Phono preamp, magnetic cartridge, 321, 322
Phono preamp, mag and ceramic, 325
Photocell-controlled dimmer, 103
Photocell-controlled solar relay, 76
Photocell controlled switch, 81
Photocell-controlled triac, 78
Photocell-gated triac, 78, 79, 81
Photocell lamp target, 152
Photocell lighting control, 114
Photocell-operated triac/diac, 81
Photo-Darlington, 61
Photo-Darlington switch, 61, 82
Photo-Darlington switch (N. O.), 82
Photodevice irradiance tester, 374
Photodevices, 76
Photodevice tester, 375
Photo developing temp control, 139
Photodiode lamp control relay, 83
Photo electric auto flasher, 384
Photoelectric-control flasher, 182
Photoelectric load control, 157
Photoelectric switches, 76
Photo-enlarger timer, 208
Photoflash converter, 35, 216
Photoflash, self-firing, 217
Photoflash slave, 215, 217
Photoflash slave trigger, 218
Photographic interval timer, 183
Photographic temperature regulator, 139
Photon-coupled normally closed ZVS, 154
Photon-coupled ring detector, 167, 168
Photon-coupling *see entries under*
  Optoelectronic
Photon coupling, 171
Photon-fired switch, 73
Photo strobe adapter, 218
Photo timer, darkroom, 219
Photo timer, delayed dropout, 209
Phototransistor burglar alarm, 161
Phototransistor commercial killer, 260
Phototransistor-controlled SCR, 80
Phototransistor flasher, 186
Phototransistor flash slave, 218
Phototransistor intrusion alarm, 160, 161
Phototransistor IR detector, 155
Phototransistor motor control, 88
Phototransistor relay, 79, 80, 81
Phototransistor relay, N. C., 79
Phototransistor SCR alarm, 78
Phototransistor sun sensor, 81
Phototransistor switch, 79
Photovoltaic motor, 173
Pickoff, optical, dc motor, 88
Picture-monitor amplifier, 252
Picture-tube cathode drive, 254
Picture-tube driver, 253
Pierce crystal oscillator, 274
Pier-marker flasher, 120 V ac, 182
Pileup sensor, assembly-line, 157, 158
Pilot SCRs, 118
Pi-Nu 70 W hi-fi amplifier, 357
Plate supply, 12 V-to-250 V, 39
Plate supply, 300 V/600 V, 20
Playback-head preamp, 325
Plug-in motor speed control, 115
Plug-in tool-speed control, 111
PM-motor dc speed control, 96
PM motor speed control, 121
PNP code oscillator, 224
PNP metronome, 236
PNP regulator, 50-200 mA dc, 5
PNP thermal-response tester, 373
PNP tone oscillator, 220, 224
PNP transmitter, 175 MHz, 280
Portable flasher, 6 V dc, 186

Power ac regulator, 90V at 5.5 A, 9
Power amplifier, 3 MHz, 294
Power amplifier, 40-180 MHz (12 W), 292
Power amplifier, 50 MHz, 293
Power amplifier, 200-400 MHz, 294
Power amplifier, 10 W UHF, 297
Power amplifier, 25 W UHF, 296, 297
Power amplifier, 40 W, 136 MHz, 291
Power amplifier, 70 W audio, 357
Power amplifier, UHF (25 W), 297
Power amplifier, VLF (200 W), 92
Power amplifier, servo, 169, 170, 171, 172, 174
Power control, 3-phase, 112
Power control, 25 W at 115 V, 117
Power control, 400 Hz, 93
Power control, 900 W variable, 118
Power control, ac, 123
Power control, ac, 1 kW, 128
Power control, by lamp brightness, 151
Power control, compensated, 100
Power control, dc, 103
Power control, full-range ac, 92
Power control, full-wave ac, 118
Power controller, full-range, 97
Power control, photoelectric, 157
Power control, with filtering, 87
Power conversion, 33
Power converter, 1 kW, 39
Power converter, 6 V to 250 V, 36
Power converter, 28 to 200 V, 36
Power converter, 28 V-to-300 V, 37
Power converter, dc, 39
Power converter, for industrial, 37
Power-Darlington hi-fi ampl, 5 W, 337
Power-Darlington amplifier, 60 W, 346
Power dimmer for theaters, 123
Power-failure detector, 154
Power-FET 4 W, audio ampl, 334
Power flasher for cars, 384
Power inverter, 13.6 V input, 52
Power inverter, 115 V-to-15 V, 50
Power inverter, 28 V-to-70 V, 56
Power inverter, 28 V to 240 V, 52
Power inverter, 32 V-to-150 V, 44
Power inverter, 150 V input, 49, 51
Power inverter, 180 V-to-180 V, 54
Power inverter, 30 W, 57
Power inverter, 400 W, 42
Power inverter, multikilowatt, 49
Power inverter, multirange, 49, 54
Power inverter, 3-phase 28 V, 55
Power lamp flasher, 12 V dc, 184
Power regulator, 500 V, 7
Power regulator, 1 kV at 0.1 A, 3
Power-saver intercom, 226
Power sources, 17
Power suppplies and rectifiers, 17
Power supply, regulated, 4-12 V, 20
Power supply, 7.5 kV, 24
Power supply, 9 V, 30
Power supply, regulated, 9 V, 24, 27
Power supply, regulated, 10-34 V, 17, 29
Power supply, 12 V, 29
Power supply, 12 V at 1 A, 22
Power supply, 12 V (½-wave), 32
Power supply, 18 V, 345
Power supply, regulated, 24 V, 17, 28, 39
Power supply, 26 V, split, 348
Power supply, 28 V dc, 344, 352, 353
Power supply, regulated, 28 V, 27
Power supply, 32 V (dual), 342, 345
Power supply, regulated, 34-45 V, 18
Power supply, 36 V dc, 342
Power supply, 45 V (dual), 357
Power supply, regulated, 60 V, 28
Power supply, 65 V (dual), 361
Power supply, regulated, 80 V, 25
Power supply, 300/600 V dc, 20
Power supply, 800 V, 21
Power supply, 80 W/80 W, 26
Power supply, CATV, 19
Power supply, design, 17
Power supply, multiple-voltage, 376
Power supply, for security systems, 21
Power supply, for stereo amp, 30
Power supply, for TV, 262
Power supply, universal, 31

Power switch, 10 A, 60, 64
Power switch, 10 A at 25 V, 64
Power switch, 5 kW, 68
Power switch, Darlington-coupled, 68
Power switch, for small signals, 68
Power switch, high-low, 239
Power switch, with integral delay, 210
Power switch, light-controlled, 81
Power switch, light-controlled ac, 78, 79
Power switch, long-delay, 203, 210
Power switch, low speed, 70
Power switch, N. O. and N. C., 60, 69
Power switch, opto, 61, 62
Power switch, remotely activated, 178
Power switch, triac, 103
Power switch, TTL-compatible, 62
Power switcher, 700 V at 7 A, 53, 69, 70
Power switching, ac, 60-67
Power switching, from 10 V pulse, 61
Power switching, with 10 V, 61
Power switching, with light, 78
Power switching, minute timer, 205
Power switching, N. O. and N. C., 64
Power switching, ring detector, 167
Power-tool motor control, 126
Power transistor driver, 68
Power versus efficiency, inverter, 51, 52
Power versus frequency, inverter, 52
Polarity checker, diode, 369
Polarity tester, diode/transistor, 369
Police/fire UHF RF amplifier, 307
Portable-device battery monitor, 159
Portable GE fluorescent lamp, 44
Portable hi-fi amplifier, 326
Portable-motor speed control, 111
Portable strobe lamp for tach, 53
Portable-tool speed control, 106, 111
Portable-TV audio amplifier, 259
Portable-TV sound IF, 261
Positioning servo drive, 119
Position sensor, solar, 81
Preamplifier, crystal/ceramic, 320
Preamp, dynamic-mike, 223
Preamp, hi-dynamic-range, 320
Preamplifier, phono, mag. and ceramic, 325
Preamp, tone control, 317, 318
Preamp/amp servo system, 170
Preamplifier, 2-meter, 302
Preamplifier, audio, 316
Preamplifier, Darlington, audio, 316
Preamplifiers, followers, 315
Preamplifier, hi-fi audio, 320
Preamplifier, magnetic-cartridge, 321, 322
Preamplifier, microphone, 220, 319
Preamplifier, phono, 318
Preamplifier, RF (6, 10, 15 m), 299
Preamplifier, servo, 173
Preamplifier, for servo unit, 192
Preamplifier, stereo, 324
Preamplifier, and tone controls, 322, 324
Preamplifier, TV signal, 267
Precision current source, 9
Precision electronic thermometer, 137
Precision kilowatt regulator, 7
Precision motor speed control, 90
Precision regulator, 1 kV at 0.1 A, 3
Precision 30s timer, 210
Precision temperature control, 135
Precision temp regulator, 139
Precision variable timer, 208
Precipitation sensor, 147
Premises protection, with light, 83
Presence control circuit, 364
Presettable lamp dimmer, 123
Preventer, contact-arc, 70
Processor, chroma, 266
Process counter control element, 79, 81, 82
Production line jam sensor, 157, 158
Programmable UJT, *see entries under PUT*
Programmable-UJT dimmer, 101
Programmable UJT oscillator, 230
Programmed lamp controller, 152
Programmer, timed sequence, 198
Progressive-Darlington switch, 68
Projection-lamp regulator, 15, 16, 33
Projector cross fader, 232, 236
Projector-lamp brightness control, 151

Projector motor control, 126
Projector speed control, 115
Proportional heat control, 136, 144
Proportional heat control, element, 143
Proportional lighting control, 114
Proportional-speed motor control, 99
Proportional voltage regulator, 10
Protected supply, 10-34 V dc, 17
Protected 70 W hi-fi amplifier, 357
Protected CATV switching supply, 19
Protected switching regulator, 11
Protection circuit, amplifier short, 352
Protection circuit, audio, 357
Protection, overload, 11
Protector, overcurrent, 153
Protector, overvoltage, 153
Proximity detector, 163, 232
Proximity detector, self-biased, 164
PTC-resistor temperature control, 142
Public address amplifier, 169
Pulsating-dc flasher, 184
Pulsating-dc power controller, 99, 129
Pulsating-dc power switch, 76
Pulsating-dc shunt motor control, 94
Pulse-actuated switch, 73, 134
Pulsed-on-10-sec time delay, 204, 210
Pulsed ring counter, 50 mA, 193
Pulsed ring counter, 192, 202
Pulsed SCR oscillator, 173
Pulsed-SCS timer, 204, 210
Pulsed-transformer dimmer, 130
Pulse counter, light, 82
Pulse-counting voltmeter, 156
Pulse generator, PUT, 227
Pulse generator, rectangular, 71
Pulse generator, trigger, 48
Pulse-indicating dial monitor, 168
Pulsed-output line-operated inverter, 41
Pulse power switching, 61
Pulse shaper, 117, 153
Pulse shaper, integrator, 224
Pulse shaper, for inverters, 55
Pulse shaper, optical, 88
Pulse speed enhancer, 153
Pulse supply, 31
Pulse switch, 73
Pulse-to-dc converter, 31
Pulse-transformer phase control, 94
Pulse-triggered alarm, 166
Pulse-triggered full-wave control, 118
Pulse-width-modulated flyback regulator, 8, 11
Pump control, 150
Pump-motor control, 149
Push-pull ampl, 12.5 W hi-fi, 343
Push-pull ampl, FET 80 W hi-fi, 359
Push-pull amplifier, 50 kHz power, 292
Push-pull amplifier, servo power, 171, 175
Push-push doubler, 50-to-100 MHz, 277
PUT capacitance switch, 163
PUT-controlled battery charger, 29
PUT-driven dimmer, 101
PUT flasher, small-bulb, 183
PUT-gate phase control circuit, 100
PUT motor speed control, 87
PUT power supply, 80 V, 25
PUT pulse generator, 227
PUT timer, 213
PUT vertical deflection oscillator, 253
PWM dc regulator, 12 V at 2 A, 11
PWM flyback regulator, 8, 11
PWM inverter, rectified, 20 A, 39
PWM regulator, 150-250 V, 10

Quadrature amplifier, 267
Quadrupler, 50-to-400 MHz, 279
Quality checker for diodes, 369
Quarter-amp, regulated supply, 9 V, 24, 27
Quasi-complementary servo amplifier, 167
Quasi-complementary amplifier, 12.5 W, 343
Quasi-complementary hi-fi, 25 W, 349, 350
Quasi-complementary amplifier, 55 W, 356
Quasi-complementary amplifier, 70 W, 357
Quasi-complementary amplifier, 120 W, 361

Radiation alarm, 166
Radio, 6 V, 2-stage AM, 244
Radio, automotive AM, 390
Radio circuits, 241

Radio, dual-gate MOSFET, 248
Radio, FM, 247
Radio, FM broadcast, 245, 248
Radio, line-operated AM, 242
Radio, regenerative, 243
Radiation detector, infrared, 155
Radio preamp, 325
Radio receiver, AM, 383
Radio remote on/off switch, 178
Radio-set audio ampl, 4 W, 335
Radio supply, vehicular, 37
Radio switching, by remote control, 178
Radio/TV circuits, 570
Railroad wig-wag blinker, 185
Rain detector, 147
Ramp-and-pedestal load control, 157
Ramp-and-pedestal phase control, 116
Rate sensor, go/no-go, 148
RCA 70 W hi-fi amplifier, 357
RCA beta tester, 375
RCA dual-gate-MOSFET tuner, 245
RCA electronic keyer, 222
RC-model solar-cell control, 173
RC timer, 209
RC timer, 10-second, 204, 210
RC timer, universal, 206
Reader, paper-tape, 155, 159
Receiver, broadcast, 6 V, 244
Receiver, broadcast, AM, 390
Receiver, broadcast, line-operated, 242
Receiver, FM, 245
Receiver, FM broadcast, 248
Receiver, FM optical, 225
Receiver, light-coupled, 227
Receiver, line-operated radio, 242
Receiver, regenerative, 243
Receiver, regenerative AM, 248
Receiver, sun-powered, 244, 245
Receiver, sun-powered AM, 244, 245
Receiver converter, for CAP, 302
Receiver converter, for CB, 301, 303
Receiver crystal calibrator, 377
Receiver front end, 150 MHz, 313
Receiver front end, 200 MHz, 305
Receiver front end, FM, 247
Receiver supply, mobile, 37
Rectangular pulse generator, 71
Rectification circuits, 17
Rectified line-voltage shunt-motor control, 131
Rectified push-pull inverter, 40
Rectifier, half-wave, 12 V, 32
Rectifierless converter, 6-12 V, 40
Rectifiers, and power supplies, 17
Rectifier pulse, 31
Rectifier-regulator, 9 V, 30
Rectifier, speed controller, 103
Rectifier, synchronous, 23, 25
Rectifying switched dimmer, 239
Red-green-blue, *see entries under RGB*
Reduced-hysteresis lamp dimmer, 122, 126
Reed relay, solid-state, 69
Reed switch controlled lamp, 178
Reed switch control power relay, 79
Reference amplifier, chroma, 266
Reference source, 8
Referencing lamp regulator, 16
Regenerative detector, 243
Regenerative MOSFET radio, 248
Regenerative ring counter, 198
Regenerative shift register, 198
Regenerative tone oscillator, 220
Register-counter, 198
Register, shift, 198
Regulated inverter, 400 Hz, 45
Regulated CATV supply, 80 W, 19
Regulated dc motor control, 117
Regulated power supply, 9 V, 30
Regulated power supply, 10-34 V, 29
Regulated supply, dc, 4-12 V, 20
Regulated supply, 9 V, 24
Regulated supply, dc, 9 V, 27
Regulated supply, dc, 60 V, 28
Regulated supply, dc, 10-28 V, 27
Regulated supply, 12 V, 1 A, 22
Regulated supply, 24 V, 28
Regulated supply dc, 24 V, 17, 39
Regulated supply, dc, (var) 28 V, 17

INDEX **405**

Regulated supply, 34-45 V, 18
Regulated supply, 80 V, 25
Regulated B+ supply, 23
Regulated line-operated inverter, 41
Regulated motor speed control, 91
Regulated stereo-amp supply, 30 V, 26
Regulated stereo-amp supply, 30
Regulating ac voltage transformer, 107
Regulating converter, 20 A, 37
Regulating dc converter, 1 kW, 39
Regulator, 8 V and 16 V dc, 6
Regulator, 8 V at 100 mA, 5
Regulator, 8-16 V dc at 100 mA, 5
Regulator, 12 V at 2 A, 10
Regulator, 12 V at 2 A dc, 16
Regulator, 13 V at 6 A, 15
Regulator, 22-30 V, at 0-10 A, 12
Regulator, 28 V dc at 500 mA, 11
Regulator, 60 V at 1 A, 14
Regulator, 90 V rms, 4, 9
Regulator, 100 V dc at 400 mA, 13
Regulator, 115 V ac, 107
Regulator, 120 V ac, 6
Regulator, temperature, 120 V, 140
Regulator, 120-250 V, dc switching, 10
Regulator, 1 kV, 3
Regulator, 100 W flyback, 8
Regulator, 1 kW dc, 7
Regulator, current, 50-200 mA dc, 5
Regulator, heat, 7.5 kW, 144
Regulator, lamp brightness, 151
Regulator, lamp voltage, 46
Regulator, motor speed, 87
Regulator, oven-temperature, 135
Regulator, projection lamp, 16
Regulator, reference, 8
Regulator, room-temperature, 141
Regulator, series, 4
Regulator, temperature, 146
Regulator, temperature, solder-iron, 133
Regulator, voltage-protected, 11
Relaxation oscillator, 253
Relaxation-oscillator flasher, 183, 184
Relaxation oscillator, neon, 235
Relaxation oscillator, PUT, 189, 230
Relay, ac, optoelectronic, 72
Relay, break-in detector, 161
Relay, commercial killer, 260
Relay, delayed-dropout, 214
Relay, electronic, TTL-compat., 62
Relay, ground-sensor, 166
Relay, intrusion-detection, 161
Relay, light-activated, 79
Relay, light-controlled, 81
Relay, light-controlled (N.C.), 78, 79, 82
Relay, light-deenergized, 79, 81, 82
Relay, light-energized, 81
Relay, light-operated, 80, 82, 83
Relay, normally open power, 66, 71
Relay, photocell-controlled triac, 79, 81
Relay, phototransistor, 79, 80, 81
Relay, resonant-reed, 178
Relay, 15-second-timed, 203
Relay, sensitive, 147
Relay, simulated, 60
Relays, solar, 76
Relays, solid-state, 60
Relay, solid-state, ac power, 78, 103
Relay, solid-state, Darlington-coupled, 68
Relay, solid-state latch, 65, 72
Relay, solid-state reed, 69
Relay, solid-state, TTL compatible, 62
Relay, sound-activated, 232
Relay, thermostat-controlled, 135
Relay, time-delay, 208
Relay, time delay, 10 ms to 1 min, 205
Relay, time delay, 10-second, 204, 210
Relay, time delay, up to 3 min, 203
Relay, time delay, up to 4 min, 183
Relay, voice-controlled, 232
Relay, ZVS normally closed, 154
Relay amplifier, dc, 376
Relay-SCR light control, 80
Reminder, lights-on, 385
Reminder, turn-signal, 389
Remote control, 169
Remote control, for 500 W k amplifier, 78

Remote control, lamp, 174
Remote-control, lighting dimmer, 123
Remote-control, switching, 61
Remote power switching, 178
Remote status monitor, 167, 168
Remote strobe adapter, 218
Remote trigger, flash, 218
Repeater control circuitry, 61
Repeater control timer, 203
Repeater main power control, 61
Repeater-off timer, 214
Repeater phone pulser, 168
Repeater tone encoder, 220, 221, 224
Repeater tone switch, 178
Resistive-load control, ZVS, 135
Resistive photocell power control, 78, 79, 81
Resistive photocell switch, 78, 81
Resistive-sensor alarm, 166
Resistor, photoelectric, 78
Resonant-reed relay, 178
Reversible ring counter, 196
Reversing motor speed control, 90, 116, 119
RF amplifier, 1-40 MHz, 309
RF amplifier, 120 MHz, 310
RF amplifier, at 175 MHz, 35 W, 245
RF amplifier, 200 MHz, 308
RF amplifier, 200 MHz cascode, 307, 309
RF amplifier, 200-250 MHz, 305
RF amplifier, 400 MHz, 308
RF amplifier, 420-450 MHz, 306
RF amplifier, TV, 263
RF amplifier, 2-meter, 313
RF amplifier, and converters, 299
RF amplifier/mixer, 248
RF amplifier/mixer, FM, 245
RF converter, 27 MHz, 301, 303
RF converter, 46 MHz, 303
RF converter, 150 MHz, 300
RF converter, CAP, 302
RF converter, 6-meter, 301
RF frequency doubler, 400 MHz, 277
RF power amplifiers, 290
RF/IF amplifiers, 304
RF/IF amplifier, FM, 248
RF/IF amplifier, FM broadcast, 247
RFI filter for dimmers, 103
RFI problems, 33
RFI-proof heater-fan control, 143
RFI-proof lamp brightness control, 151
RFI-proof lamp dimmer, 84, 122, 125
RFI-proof SCR switch, 65, 73, 75
RFI-proof temperature regulator, 146
RF preamplifier, 2 m, 302
RF preamplifier, 6, 10, 15 m, 299
RF tripler, to 450 MHz, 278
RF tripler, to 750 MHz, 278
RGB amplifier, color TV, 268
Rhythm generator, 236, 237
Ring counter, 187, 199
Ring counter, 5 mW, 188
Ring counter, commutating, 193
Ring counter, CSCR, 192, 202
Ring counter, 3-stage, 188
Ring counter, 4-stage, 190
Ring counter, high-speed, 196
Ring counter, LED, 190
Ring counter, reversible, 176
Ring counter, SCR, 199
Ring counter, SCS, 192, 200, 201
Ring counter, SUS, 202
Ring counter, variable timing, 188
Ring counter, wide-tolerance, 193
Ring-counter-driven sequential timer, 208
Ring counter/shift register, 198
Ring detector, telephone, 167, 168
Ripple filter, 7
Rise-time improver, pulse, 153
RIZA mag/ceramic preamplifier, 325
Rms regulator, 80 V ac, 15
Rms regulator, 90 V, 4, 9
Rms regulator, 90 V at 5.5 A, 9
Rms regulator, 120 V, 6
Roadside flasher, 186
Room-heater fan control, 143
Room-temperature regulator, 141
Rooster, electronic, 232
RPM monitoring voltmeter, 156

Rumble filter, stereo, 364
Run-start motor control, 86

Saber-saw speed control, 115
Saddle-yolk vertical deflection, 255
Safety alarm, load monitoring, 154
Safety flasher, battery-operated, 185
Safety flasher for cars, 384
Safety voltage alarm, 166
Sampling regulator, 4
Sampling, temperature, 140
Sampling temperature control, 142
Sawtooth tone generator, 229
SBS-gated dimmer, 100
SBS gating of thyristor, 76
SBS-gated triac dimmer, 123
SBS-gated triac lamp dimmer, 86
SBS light-interrupt sensor, 82
SBS triac trigger, 61
SBS-zener "diac," 106
Scale speed control, 234
Schmitt modulator, 88
Schmitt trigger, 70, 72, 164
Schmitt trigger, for dc speed control, 117
Schmitt trigger, opto-coupled, 156
Schmitt trigger waveforms, 72
Schottky hearing-aid preamp, 316
SCR ac regulator, 120 V, 6
SCR alarm, light-controlled, 78
SCR alarm, light-operated, 80
SCR chaser, half-wave, 187, 189
SCR controlled supply, 80 V, 25
SCR controlled battery charger, 32
SCR-controlled ZVS, 73, 75
SCR dimmer for ac, 101
SCR and diode checker, 368
SCR flasher, 186
SCR ignition system, 388
SCR inductive-load control, 154
SCR intrusion alarm, 164
SCR-inverter pulse shaper, 55
SCR lamp dimmer, 95
SCR lamp flasher, 186
SCR lamp flasher, 12 V, 185
SCR lamp regulator, 16
SCR light dimmer, 400 W, 97
SCR motor speed control, 91, 105, 126
SCR multi-kW inverter, 49
SCR nonmagnetic trigger, 118
SCR, opto-coupler-driven, 154
SCR phase control, 94
SCR phase control circuit, 88
SCR phaser, half-wave, 111
SCR power switching, 66, 73
SCR relays, 60
SCR ring counter, 192, 199
SCR smoke alarm, 165
SCR solid-state relay, 72
SCR speed control, 5 A, 111
SCR speed control, plug-in, 115
SCR switch, 60-75
SCR time-delay relay, 209
SCR timer, and switch (1 min), 205
SCR timer, UJT-keyed, 203
SCR trigger, high di/dt, 71
SCS ring counter, 188, 200, 201
SCS ring counter, 20 kHz, 196
SCS ring counter and timer, 208
SCS timer, delayed-turn-on, 206
SCSR time delay, 1-minute, 211
Secode-compatible off/on switch, 178
Second detector, Black-and-White TV, 256
Seconds-pulsing lamp flasher, 184
Seconds timer with LED readout, 213
Security-system battery charger, 21
Selective TV signal booster, 267
Selective-voltage switching, 175
Self-amplifying tone oscillator, 220
Self-biased proximity alarm, 164
Self-compensating photoflash converter, 216
Self-firing photoflash, 217
Self-flashing lamp, 186
Self-modulated oscillator, 73.5 MHz, 224
Self-modulated transmitter, 73.5 MHz, 288
Self-starting photoflash converter, 35
Self-timing motor speed control, 119
Self-timing phase control, 127

Semiadjustable supply, 24 V, 28
Semiconductor switching circuits, 60
Sensing, and control circuits, 59
Sensing regulator, 90 V rms, 4
Sensitive-gate SCR alarm, 78
Sensitive-gate SCR switch, 62
Sensitive-gate triac switch, 194
Sensitive N.C. light control, 82
Sensitive N.O. light relay, 82
Sensitive opto burglar alarm, 161
Sensitive photo switch, 80
Sensitive relay, 147
Sensitive SCS alarm, 166
Sensitivity multiplier, dc, 376
Sensitivity multiplier, VOM, 372
Sensor, aimed-light, 152
Sensor, assembly-line jam-up, 157
Sensor, assembly-line pileup, 158
Sensor, battery-voltage, 159
Sensor, dampness, 147
Sensor, drop, fluid, 154
Sensor, flame, 151
Sensor, ground-level, 166
Sensor, hazard, 160
Sensor, infrared, 155
Sensor, intrusion, 160
Sensor, light-detection, 83
Sensor, light-interruption, 80, 82
Sensor, liquid drop, 149
Sensor, liquid-level, 148, 149
Sensor, overcurrent, 153
Sensor, overheat, 134
Sensor, overvoltage, 153, 166
Sensor, phone current, 167, 168
Sensor, phone-dial digital, 168
Sensor, phone-line ring, 167
Sensor, photoelectric break-in, 161
Sensor, power-failure, 154
Sensor, proximity, 163, 164
Sensor, rate, 148
Sensor, smoke, 165
Sensor, smoke-and-gas, 165
Sensor, sun-position, 81
Sensor, telephone-ring, 167, 168
Sensor, temperature, 136
Sensor, touch, 232
Sensor, water-drip, 149
Sequential ac flasher, 187, 189, 191
Sequential flasher, SCS, 192
Sequential multilamp flasher, 187
Sequential ring counter, 193
Sequential timer with inhibits, 208
Sequential turn-signal add-on, 386
Series-connected dc converter, 36
Series dc regulator, 150-250 V, 10
Series-feedback amplifier, 250
Series-feedback buffer, 224
Series modulator, 290
Series modulator, AM, 290
Series-motor reversing control, 116, 119
Series motor reversing speed control, 90
Series-motor synchronous control, 102
Series-pass regulator, *see entries under Series regulator*
Series-regulated inverter, 400 Hz, 45
Series regulator, 4
Series regulator, 100 V, 13
Series regulator, 1 kV, 3
Series regulator, o'load-protected, 11
Series regulator, voltage, 22-30 V, 12
Series-wound-motor control, 89
Servo amplifier, 169
Servo amplifier, 6 W, 175
Servo amplifier, ac/dc, 170
Servo amplifier, low-voltage, 171
Servo amplifier, transformer-coupled, 172, 174
Servo circuits, 169
Servo-compatible phase shifter, 178
Servo drive, positioning, 119
Servo preamplifier, 172, 173
Servo preamp, single-ended, 170
Servo system, complete, 170
Servo system, line-operated, 170
Sewing-machine speed control, 115
Shaded-pole motor control, 99, 121
Shaper, pulse, 88, 153
Shaper, pulse, for inverters, 55

Shaving adapter for cars, 43
Shielding techniques, 33
Shifter, phase (0-360 degrees), 128
Shift register/counter, 198
Shock-absorbing lamp switch, 152
Shop-tool speed control, 106, 115
Short-circuit amplifier protection, 352
Short-circuit tester, diode, 369
Short-circuit tester, transistor, 369
Short-detector, diode and SCR, 368
Short-detector, for SCRs, 368
Short-duty-cycle flasher, 186
Short-proof charger, 12 V, 29
Short-proof dc converter, 28 V-400 V, 34
Short-proof dc supply, 10-34 V, 17
Short-proof hi-fi amplifier, 20 W, 343
Short-proof hi-fi ampl, 70 W, 357
Short-term timer, simple, 208
Shortwave receiver, 244
Shunt, and PM motor speed control, 121
Shunt-motor speed control, 120, 121, 131
Shunt-motor speed control, 1/2-wave, 94
Shunt regulator, 28 V at 500 mA, 12
Shunt-wound motor speed control, 90
Shutdown, overvoltage, 166
Sideband balanced modulator, 290
Signal booster, color TV, 267
Signal-controlled triac switch, 73, 75
Signal conversion, 100 MHz, 304
Signal detector, telephone, 167, 168
Signal generator, with dial, 379
Signal generator/injector, 378
Signal injector, 223
Signal source, hi-fi audio, 378
Signal source, square wave, 44
Silicon bilateral switch, *see entries under SBS*
Silicon bilateral switch, 61, 106
Silicon controlled switch, *see entries under SCS*
Silicon unilateral switch, *see entries under SUS*
Simple battery-voltage monitor, 159
Simple CSCR Schmitt trigger, 72
Simple high-voltage supply, 21
Simple human-sensor alarm, 163
Simple hybrid Schmitt trigger, 70
Simple inverter, 110 W, 3.5 kHz, 52
Simple LED tester, 369
Simple light control circuit, 78
Simple opto-SCR power switch, 65
Simple phase control, 99
Simple photo relay, 79, 81
Simple photo switch, 78
Simple power control, 600 W dc, 103
Simple power switch, 10 A, 64
Simple radio, 100 uV/m, 242
Simple regulated supply, 9 V, 24
Simple regulated supply, 24 V, 18
Simple regulator, 8 V, 5
Simple regulator, 8-16 V 100 mA, 5
Simple regulator, 100 V at 400 mA, dc, 13
Simple SCR intrusion alarm, 164
Simple tachometer, 12 V, 156
Simple 1-transistor power switching, 61
Simple 2-position dimmer, 239
Simulated-movement ac chaser, 187
Simulator, electronic relay, 60
Simulator, flame, 239
Sine-wave audio oscillator, 378
Sine-wave generator, 235
Sine-wave oscillator, 221, 224, 273
Sine-wave phase control, 86
Sine-wave VCO, 230
Single-chamber smoke sensor, 163, 165
Single-Darlington audio preamp, 316
Single-device converter, 46 MHz, 303
Single-device CAP converter, 302
Single-device CB converter, 301, 303
Single-device regulator, 13
Single-ended op-amp servo preamp, 170
Single-FET dc voltmeter, 373
Single-lamp dc flasher, 185
Single-lamp flasher, 12 V dc, 384
Single-lamp SCR flasher, 186
Single-phase induction-motor control, 91
Single sideband, *see entries under SSB*
Single-sideband modulator, 290
Single-tone generator, 220, 221, 224
Single-tone oscillator, 378

Single-tone power switch, 178
Single-transistor amplifier, 4 W, 387
Single-transistor current regulator, 5
Sinusoidal tone generator, 378
Siren, automatic-wail, 234
Siren novelty, 237, 238
Size-saving multi-B+ supply, 23
Slave, photoflash, 215
Slave adapter, photo strobe, 218
Slide fader control, 232, 236
Slide-projector brightness control, 151
Slow-on, slow-off phase control, 120
Slow-speed motor control, 95
Slow turn-on lamp control, 120
Slow-turn-on lamp dimmer, 1 kW, 85
Small-angle phase control, 118
Small-appliance adapter, for cars, 42
Small-appliance motor control, 96
Small-appliance power source, 42, 46
Small-conduction angle ac regulator, 107
Small-lamp phase control, 115 V, 117
Small-load dimmer circuit, 122
Small-model remote controller, 173
Small-motor speed control, 115
Small motor speed control, with timer, 119
Small-signal power switch, 68
Small-volume table-lamp dimmer, 122
"Smile" blinker, 185
Smoke detector, FET, 165
Smoke detector ionization chamber, 162
Smoke detector, TGS-sensor, 165
Smoke sensor, line-operated, *see entries under Lamp dimmer*
Snubber, for dimmers, 87
Snubber, for dimmer circuit, 122
Snubber, inductive-load, 93
Snubber network, 109
Soft-start, 85
Soft-start dimmer, 101, 125
Soft-start dimmer, 1 kW, 85
Soft-start lamp dimmer, 92, 111, 120
Soft-start phase control, 120
Solar amplifier, audio, 244
Solar audio oscillator, 232
Solar CB transmitter, 288
Solar-cell-controlled motor, 173
Solar-cell-operated motor, 173
Solar-cell-powered transmitter, 226
Solar-device remote control, 173
Solar oscillator, 237
Solar-panel guiding sensor, 81
Solar radio, sun-powered, 245
Solar radio, 1-transistor, 245
Solar radio, 2-transistor, 244
Solar relays, amplified, 76
Solar tracker, 81
Soldering-iron phase control, 117
Solder iron regulation, 133
Solder-iron voltage control, ac, 103
Solenoid, light-controlled, 79, 81
Solid-state ac control relay, 66
Solid-state hi-power lamp control, 83
Solid-state motor starter, 86
Solid-state reed relay, 69
Solid-state relays, 60
Solid-state relays, 10 A, 25 V, 64
Solid-state relay, 10 A, 220 V, 64
Solid-state relay, Darlington-coupled, 68
Solid-state relay, TTL, 62
Solid-state triac relay, 62
Solid-state vibrator, 44
Sound-activated relay, 232
Sound IF, portable-TV, 261
Sound-source width control, 365
Source, constant-current, 18
Source, current, 9
Source, follower, 316
Source, low-voltage, 8
Source, square-wave signal, 44
Space-heater power control, 142
Space-heater temperature control, 133
Spark-free motor control, 86
SPDT switching with light, 79, 80
Speaker-driven color organ, 231
Special-effects flash unit, 217
Special-purpose circuits, 363
Speed control, 5 A, 104

INDEX **407**

Speed control, 400 Hz, 93
Speed control, dc motor, 88
Speed control, for fan, 99
Speed control, for induction motor, 87
Speed control, for inductive loads, 118
Speed control, and timer, 119
Speed control, for shunt motors, 120, 131
Speed control, for tools, plug-in, 111
Speed control, for wide speed ranges, 105
Speed control, half-wave, 95
Speed control, induction-motor, 91, 99, 127
Speed control, low-hysteresis, 126
Speed control, model railroad, 103
Speed control, model-train, 234
Speed control, motor, 84, 100, 109, 114
Speed control, motor, 115 V, 91, 106
Speed control, motor, 400 W, 106
Speed control, 500 W, motor, 98
Speed control, motor, with feedback, 105
Speed control, motor, half-wave, 96
Speed control, motor, reversing, 90
Speed control, with no hysteresis, 92
Speed control, with optical feedback, 90
Speed control, plug-in, 115
Speed control, PM-motor, 96
Speed control, reversing, 116
Speed control, RFI-proof, 103, 122, 125
Speed control, series motor, 89
Speed control, shaded-pole motor, 99
Speed control, shunt and PM motor, 121
Speed control, shunt-wound motor, 94
Speed control, small-motor, 106
Speed control, SUS-triggered, 110
Speed control, synchronous dc, 121
Speed control, synchronous motor, 102
Speed control, tachometer, 158, 159
Speed control, time-dependent, 125
Speed control, with timer, 127
Speed control, universal-motor, 108
Speed control, wide-range, 124
Speed regulation, of induction, 87
Spike-free ac switch, 73, 75
Split-capacitor motor control, 127
Split-capacitor motor starter, 86
Split power supply, 45 V, 357
Split-site repeater power switch, 178
Split-site repeater switching, 175
Split supply, 18 V, 345
Split supply, 65 V, 361
Sprague phase control circuit, 110
Square-wave inverter, 12 V-110 V, 43
Square-wave inverter, 117 V, 42, 46
Square-wave inverter, 120 V, 47
Square-wave oscillator, 44
Square-wave power inverter, 120 W, 44
Square-wave-pulsed ring counter, 192, 202
Squelch tone switching, 178
SSB balanced modulator, 290
Stable-voltage lamp dimmer, 100
Stable voltage source, 8
Stage-widening control, stereo, 365
Standby-battery charger, automatic, 394
Standby-battery charging device, 21
Starter for split-capacity motor, 86
Static contactor, 74
Static switch, 3-position, 74
Stepdown switching regulator, 14
Stepless 3-phase dc controller, 98
Stereo amplifier, 2 W IC, 332
Stereo amplifier, 5 W/channel, 337
Stereo amplifier, 15 W, 344, 345
Stereo amplifier, hi-fi, 160 W, 360
Stereo amplifier, complete, 337
Stereo-amplifier supply, 26
Stereo-amplifier supply, 60 V, 28
Stereo balance control, 363, 365
Stereo bias, and erase oscillator, 363
Stereo power supply, 2 x 50 W, 306
Stereo preamp, hi-fi, 324
Stereo preamp, and tone control, 318
Stereo presence control, 364
Stereo width control, 365
Stressless lamp switch, 152
Stripe-activated speed control, 88
Stripe-pulsed motor regulation, 117
Strobe adapter, remote, 218
Strobed audio alarm, 166

Strobe-flash slave, 218
Strobe lamp power source, 53
Strobe, self-firing photo, 217
Subcarrier sync, color TV, 266
Substitute, dimmer diac, 106
Sump-pump level controller, 148, 149, 150
Sun-operated relay, 76
Sun-position sensor, 81
Sun-powered AM radio, 244, 245
Sun-powered CB transmitter, 288
Sun-powered oscillator, 400 Hz, 237
Sun-powered radio, 285
Sun-powered wireless mike, 226
Sun tracker, 81
Superdynamic range preamp, 320
Superectifier, general instrument, 163
Superefficient inverter, 180 W, 49
Super-fidelity amplifier, 40 W, 351
Super-fidelity amplifier, 80 W, 359
Superheterodyne AM car radio, 390
Superpower switcher, 68
Supersonic oscillator, 223
Supplementary ripple filter, 7
Supple, power (TV), 262
Supply, 10-34 V at 1 A, 29
Supply, half-wave, 12 V, 32
Supply, 28 V dc, 344, 352, 353
Supply, 36 V dc, 342
Supply, 300/600 V at 300 mA, 20
Supply, 7500 V dc, 24
Supply, 800 V, 21
Supply, power (12 V), 29
Supply, regulated, 4-12 V dc, 20
Supply, regulated, 9 V, 24
Supply, regulated, 9 V at 250 mA, 27
Supply, regulated, 10-34 V dc, 17
Supply, regulated, 24 V, 28
Supply, regulated, 24 V dc, 18
Supply, regulated, 28 V, 27
Supply, regulated, 30 V, 26
Supply, regulated, 34-45 V, 18
Supply, regulated, 60 V, 28
Supply, regulated, 80 V, ac, 15
Supply, regulated, 80 V dc, 25
Supply, transistor-battery, 9 V, 30
Supply, adjustable, 12 V at 1 A, 22
Supply, aircraft, 200 V, 36
Supply, CATV, 19
Supply, dual, 32 V, 352, 354
Supply, dual, 45 V, 357
Supply, dual, 65 V, 361
Supply, dual stereo, 20 V, 30
Supply, switching, 33
Supply, universal, 31
Supply, zener-output, 376
Suppressor, arc, 70
Suppression of RFI, on dimmer, 122
Suppressed-RFI phase control, 115 V, 125
SUS flip-flop with memory, 184
SUS-gated ring counter, 190
SUS ring counter, 202
SUS sequential flasher, 190
SUS-triggered motor speed control, 110
Sweep, horizontal, 251
Sweep supply, variable, 31
Swimming pool level control, 148, 150
Switch, 3-position static, 74
Switch, 5 kW, 68
Switch, 300 V at 250 mA, 64
Switch, 700 V at 7 A, 69, 70
Switch, ac power, with triac, 103
Switch, bistable, with memory, 184
Switch, capacitance, 238
Switch, controlled power, 10 V, 61
Switch, Darlington-coupled triac, 69
Switch, delayed-on triac power, 210
Switch, high-current, 64
Switch, high-frequency, 254
Switch, high-low power, 239
Switch, high-power, 70
Switch, hi-I, hi-E, 53
Switch, hi power and low speed, 70
Switch, incandescent lamp, 68
Switch, light-activated, 147
Switch, light-controlled, 79, 81
Switch, light-controlled, 120 V, 81
Switch, light-controlled, 120/240 V, 79

Switch, light-controlled, (N.C.), 79
Switch, light-operated, 83
Switch, logic-and-triac, 194
Switch, low-drain power, 61
Switch, normally closed, 60, 61
Switch, normally closed, 10 A, 60
Switch, normally closed light, 82
Switch, normally open, 60, 61
Switch, photocell-controlled, 78
Switch, photoelectric, 76
Switch, phototransistor, 79, 80, 81
Switch, pulse-actuated, 73
Switch, regulator, flyback, 8
Switch, remote-controlled lamp, 174, 178
Switch, solid-state, 60
Switch, thermostat-controlled, 135
Switch, time-delayed, 203, 210
Switch/timer, 10 ms to 1 min, 205
Switch, transient-proof ac, 73, 75
Switch, transistor-coupled, 73
Switch, triac, 71, 75
Switch, TTL compatible, 62
Switch, zero-point, 135
Switch, zero-point (SCR), 73, 75
Switched-zener remote control, 175
Switching, DPDT, 71
Switching, zero-voltage, 66
Switching circuits, photoelectric, 76
Switching converter, 1.5 to 12 V, 33
Switching converter, 1 V to 12 V, 33
Switching converter, 6-12 V, 40
Switching converter, 25 V at 20 A, 37
Switching converter, 28-200 V, 36
Switching converter, 275 V-to-50 V, 39
Switching converter, 300 V dc, 37, 38
Switching converter, 225 W, 34
Switching interference, 33
Switching, opto controlled, 61, 62
Switching, photo timer, 219
Switching, power, 10 A, 64
Switching, power supply, 33
Switching, pulse shaper, 55
Switching regulator, 11
Switching regulator, 1 kW, 7
Switching regulator, 12 V at 2 A, 11
Switching regulator, 60 V, 1 A, 14
Switching regulator, 120 V ac, 6
Switching regulator, 150-250 V, 10
Switching supply, 400 V dc, 34
Switching supply, CATV, 19
Switching timer, 0.2 to 10 seconds, 208
SWR bridge, 282
Summetrical lighting control, 114
Symmetrical transistor coupler, 378
Synchronized flash slave, 215, 217
Synchronous biased supply, 25
Synchronous control, series-motor, 102
Synchronous dc power control, 121
Synchronous heating-element switch, 146
Synchronous rectifier, 23
Sync/rectifier circuit, digital clock, 25
Synchronous rectifier, half-wave, 25
System, line-operated servo, 170

Table-lamp dimmer, 122
Tach-controlled motor speed monitor, 159
Tach-less induction-motor control, 87
Tach strobe inverter, 53
Tachometer calibrator, 376
Tachometer circuit module, 88
Tachometer, ignition, 384
Tachometer module for speed control, 117
Tachometer, optical, 156
Tachometer, optical-pickup, 377
Tachometer rate sensor, 148
Tachometer speed control, 158, 159
Tachometer strobe-lamp driver, 53
Tachometer voltmeter, 156
Tank-level control, 148, 149
Tape-deck speed control, 12 V, 121
Tape erase and bias oscillator, 363
Tape-head erase/bias, 363
Tape-head preamp, 325
Tape-player motor speed control, 96
Tape-player regulator, 16
Tape-player speed control, 121
Tape reader, optical, 159

Tape reader, paper, optical, 155
Target, lamp, 152
Tee attenuator, 233
Telephone compatible alarms, 167
Telephone-dial pulse sensor, 168
Telephone ring detector, 167, 168
Telephone sensing devices, 167
Telephone unauthorized-use alarm, 168
Television, *see entries under TV*
Television circuits, 570
Temperature alarm, 134, 137
Temperature-compensated dimmer, 85
Temperature control, 7.5 kW, 144
Temperature control, VZS, 140
Temperature control, ZVS, 138, 142
Temperature control, darkroom, 139
Temperature-controlled heater fan, 143
Temperature controller, 120 V, 132, 136
Temperature controller, 240 V, 136
Temperature-regulated, outlet, 115 V, 133
Temperature regulator, by phase, 135
Temperature regulator, oven, 135
Temperature regulator, RFI-proof, 146
Temperature sampling circuit, 140
Temperature sampling heat control, 142
Temperature-sense alarm, 166
Temperature-sensitive control, 141
Test, and measurement equipment, 367
Tester, battery, 159
Tester, collector-base, 370
Tester, continuity, 369
Tester, diode, 369
Tester, diode-transistor, 369
Tester, hi-fi audio, 378
Tester, for LEDs, 369
Tester, MOS-transistor, 368
Tester, photodevice irradiance, 374
Tester, SCR and diode, 368
Tester, transistor beta, 371, 375
Tester, transistor leakage, 375
Tester, voltage and continuity, 368
Tester, voltage, current and continuity, 367
Test fixture, thermal response, 373
Test oscillator, nonsine, 223
Test power supply, transistor, 31
TGS gas-and-smoke sensor, 165
TGS sensor source, 165
Theatrical-lighting dimmer, 123
Theft alarm, 12 V dc, 164
Theft alarm, automotive, 385
Thermal-response test fixture, 373
Thermal stress relief for lamp, 152
Thermistor-controlled temp switch, 135
Thermistor heat control, ZVS, 138
Thermistor heat regulator, 146
Thermistor liquid-temperature control, 139
Thermistor thermometer, 137
Thermistor-triggered alarm, 147
Thermometer, electronic, 137
Thermometer, indoor/outdoor, 137
Thermostat, darkroom-chemical, 139
Thermostat, electronic, 136
Thermostat, electronic room, 132
Thermostat, heater (ZVS), 142
Thermostat, room-temperature, 141
Thermostat, ZVS, 138
Thermostat-controlled outlet, 133
Thermostat-operated blower control, 143
Thermostat, and power control, 144
Third-overtone crystal oscillator, 274
Three-terminal regulator, 8-16 V dc, 5
Thyristor ac power switch, 103
Thyristor ac switch, 65, 74
Thyristor regulator, ac 120 V, 6
Thyristor fluores, lamp driver, 48
Thyristor-inverter pulse shaper, 55
Thyristor kilowatt inverter, 49
Thyristor lamp dimmer, 92
Thyristor lamp regulator, 16
Thyristor light control, 78
Thyristor night lights, 77
Thyristor power switch, 67
Thyristor power switch, 600 W, 78
Thyristor power switching, 73
Thyristor relay, 60
Thyristor "relay" circuit, 72
Thyristor trigger, high-di/dt, 71

Thyristor ZVS, 65, 75
Time constant, double, 92
Timed-cycle blender speed control, 119
Time delay, CSCR, 204, 210
Time-delayed 1-shot, 205
Time-delay relay, 203, 208
Time delay relay, 1 to 30 seconds, 214
Time delay relay, 2 sec to 3 min, 214
Time delay relay, 40 to 60 sec, 209
Time delay relay, up to 1 min, 204
Time delay relay, 4 min, 183
Time-delay relay, ac 120 V, 209
Time-delay switch, 203, 210
Time delay triac, adjustable, 210
Time-dependent lamp dimmer, 125
Time-dependent motor control, 125
Time-dependent phase control, 120
Timed-off delay relay, 203
Timed-on delay relay, 203
Timer, 179, 203
Timer, 28 V dc or 240 V ac, 206
Timer, 10 ms to 1 min, 204, 205
Timer, 0.5 sec to 3 min, 203
Timer, 0.6 sec to 1 min, 204
Timer, 2 seconds to 4 minutes, 183
Timer, 10-second, 204, 210
Timer, 20-second, 214
Timer, 30-second, 210
Timer, 100-second, 207
Timer, 0 to 20 minutes, 204
Timer, 1-minute, 211
Timer, ac, adjustable, 208
Timer, for blender, 211
Timer, and-blender-control circuit, 119
Timer, and blender motor control, 127
Timer, delayed-on (1 min), 209
Timer, enlarger, 219
Timer/phase control, 119
Timer, RC, power switching, 206
Timer, second (LED readout), 213
Timer, sequential, 208
Timer, simple, 208
Timer, universal, precision, 209
Timer, universal (0.6-6 sec), 204
Timer, versatile, 147
Timing circuits, 179
Timing power switch, 204
Timing sequence programmer, 198
Tone burst generator, 220, 224
Tone-burst oscillator, 221, 223, 224
Tone control, preamp, 317, 322, 324
Tone control, preamp, audio, 318
Tone-decoder power switch, 178
Tone generator, 220, 223, 224, 230
Tone generator, 10 Hz to 175 kHz, 221
Tone generator, nonsine, 378
Tone generator, organ, 233
Tone generator, sawtooth, 229
Tone generator, sine-wave, 378
Tone generator, solar, 237
Tone oscillator, 0.01 to 175 kHz, 221
Tone oscillator, keyed, 220, 224
Tone oscillator, sun-powered, 232
Tone oscillator/amplifier, 221
Tone-producing VHF oscillator, 274
Tool-motor control, 126
Tool speed control, plug-in, 111
Toroidal-yoke vertical deflection, 255
Toroid winding data, 44
Touch switch, 232
Touch switch, isolated, 163
Touch switch, latching, 238
Touch-to-switch circuit, 232
Tower-light flasher, 120 V ac, 182
Toy organ, 235
Tracker, sun, 81
Train speed control, models, 103, 234
Transceiver, light-coupled, 227
Transducer, light-to-voltage, 155
Transformation, ac voltage, 107
Transformer-controlled triac switch, 75
Transformer-coupled servo amplifier, 172, 174
Transformer-gated dimmer, 130
Transformer-gated phase control, 118
Transformer-gated SCR control, 94
Transformer-gated triac dimmer, 101
Transformer-gated triac lamp switch, 174

Transformer-triggered phase control, 110
Transformer-type hi-fi ampl 50 W, 355
Tranformer vertical amplifier, 253
Transformerless amplifier, audio, 331, 334, 335
Transformerless ampl, audio, 2 W, 260, 332
Transformerless amplifier, hi-fi, 2 W, 332
Transformerless amplifier, hi-fi, 4 W, 336
Transformerless amplifier, hi-fi, 5 W, 337, 338
Transformerless amplifier, hi-fi, 7 W, 339
Transformerless amplifier, 7.5 W, 339
Transformerless amplifier, hi-fi, 8 W, 340
Transformerless amplifier, hi-fi, 10 W, 340
Transformerless amplifier, hi-fi, 12 W, 341, 342
Transformerless amplifier, hi-fi, 12.5 W, 343
Transformerless amplifier, hi-fi, 15 W, 338, 345
Transformerless amplifier, hi-fi, 20 W, 343, 347
Transformerless amplifier, hi-fi, 25 W, 347, 349, 350
Transformerless amplifier, hi-fi, 40 W, 342, 352, 354
Transformerless line-voltage control, 88
Transformerless servo amplifier, 169-171
Transformerless servo amplifier, 6 W, 175
Transformerless vertical deflection, 255
Transient-prone full-wave control, 88
Transient-proof SCR ac switch, 73, 75
Transistor amplifier, 500 mW, 329
Transistor-battery eliminator, 27, 30
Transistor beta tester, 375
Transistor-beta tester, in-circuit, 371
Transistor-coupled switch, 73
Transistor-diode tester, 369
Transistor dip meter, 371
Transistor "grid-dipper," 371
Transistor intercom, 226
Transistor leakage tester, 375
Transistor optocoupler, *see entries under Optoelectronic*
Transistor photo switch, 79, 80, 81
Transistor power switch, controlled, 10 V, 61
Transistor radio, 2-stage, 244
Transistor-radio power supply, 24
Transistor relay simulators, 60
Transistor switches, 60
Transistor-test adapter, 31
Transistor-test power supply, 31
Transistor thermal-response test, 373
Transmission isolator, data, 171
Transmitter, 280
Transmitter, 2-meter AM, 289
Transmitter, 73.5 MHz, 274, 288
Transmitter, 175 MHz, 280, 287
Transmitter, 175 MHz, 30 W, 287
Transmitter, AM broadcast, 228
Transmitter, 40 W 6 m, 282
Transmitter, aircraft, 284
Transmitter, aircraft, 118-150 MHz, 283
Transmitter, aircraft, 13 W, 289
Transmitter B+ supply, 20
Transmitter, CB, 5 W, 281
Transmitter CW keyer, 222
Transmitter, FM broadcast, 225
Transmitter, FM optical, 227
Transmitter, light-modulated, 227
Transmitter, low-power, 136 M, 285
Transmitter, power supply, 20, 21
Transmitter, sun-powered, 226
Transmitter, sun-powered, CB, 288
Transmitter, supply, mobile, 37
Transmitter, tunnel diode, 226
Trapless video amplifier, 252
Tremolo circuit, 229
Triac, photocell-controlled, 78
Triac, as relays, 60
Triac, sensitive-gate, 194
Triac ac power switch, 60-66, 71, 75, 103
Triac ac regulator, 120 V, 6
Triac ac static contactor, 74
Triac chaser, 191
Triac 2-circuit kilowatt flasher, 181
Triac color-organ cell, 235
Triac contact arc prevention, 70
Triac crowbar circuit, 166
Triac dimmer, 92
Triac flasher, IC-controlled, 181
Triac lamp dimmer, 85. *See also entries under Thyristor*

INDEX 409

Triac lamp dimmer, 89, 109, 114
Triac lamp dimmer, 100 W, 100
Triac lamp dimmer, 500 W, 98
Triac lamp dimmer, 600 W, 87
Triac lamp dimmer, 800 W, 86
Triac lamp dimmer, full-wave, 122
Triac lamp dimmer, low-hysteresis, 126
Triac lamp dimmer, no-hysteresis, 92, 106
Triac lamp dimmer, RFI-proof, 125
Triac 2-lamp flasher, 180
Triac lamp switch, gated, 9 V, 178
Triac light control circuit, 83
Triac light-controlled relay, 81
Triac light-controlled switching, 78
Triac logic buffer, 194
Triac mode II and III switching, 194
Triac modes I and IV switching, 194
Triac motor starter, 86
Triac night light, automatic, 77
Triac phase control, filtered, 84
Triac phase control, no-hysteresis, 129
Triac power switch, 10 A, 60
Triac power switch, with delay, 210
Triac power switch, remote-control, 178
Triac remote switching, 178
Triac reversing speed control, 116, 119
Triac-RFI filter, 103
Triac sequential flasher, 187
Triac solid-state relay, 62
Triac switch, Darlington-coupled, 69
Triac telephone-ring detector, 167
Triac ZVS temp control, 138
Trickle charger, auto battery, 32
Trickle-charging supply circuit, 21
Trigger, double-time-constant, 126
Trigger, nonmagnetic, for SCRs, 118
Triggers, for phase control, 113
Trigger, Schmitt, opto-coupled, 156
Trigger, xenon-flash, 218
Trigger circuit, for 900 W load, 118
Trigger-cord flash unit, 218
Triggered-anode ring counter, 193
Triggered-triac power switch, 103
Triggered UJT inductive-load control, 118
Trigger, high-di/dt, for SCR, 71
Trigger pulse generator, 48
Trigger transformer, for phase control, 110
Trigger, Schmitt, 72, 70
Trimmable 4-12 V regulated supply, 20
Trimmable RC timer, 209
Tripler, 150 to 450 MHz, 277, 278
Tripler, 200 to 600 MHz, 278
Tripler, to 750 MHz, 278
True-output power-failure sensor, 154
TTL-compatible power switch, 62
TTL-compatible "relay," 62
TTL-compatible solar tracker, 81
TTL switching of triac, 194
TTL-triac switch combo, 194
Tunable oscillator, 5.46 MHz, 273
Tuned-collector oscillator, 275
Tuner, FM (MOSFET), 245
Tuner, FM (2-MOSFET), 245
Tuner, FM (3-transistor), 248
Tuner, front end, FM, 247
Tuner, varactor (VHF), 263
Tuning diode, see entries under Varactor
Tunnel-diode CAP converter, 302
Tunnel-diode CB converter, 301, 303
Tunnel-diode CB oscillator, 274
Tunnel diode converter, 46 MHz, 303
Tunnel diode converter, 47 MHz, 303
Tunnel-diode oscillator, 273
Tunnel diode oscillator, 47 MHz, 274
Tunnel diode oscillator, 73.5 MHz, 274
Tunnel diode transmitter, 226
Tunnel diode transmitter, 73.5 MHz, 288
Turntable-motor speed control, 121
Turn signal adapter, sequential, 386
Turn-signal reminder, 389
TV audio amplifier, 259
TV audio amplifier, 2 W, 260
TV commercial killer, 260
TV horizontal sweep circuit, 251
TV IF amplifier, 264
TV low-voltage supply, 23
TV mixer, hi-performance, 257

TV mixer stage, 259
TV power supply, 262
TV/radio circuits, 570
TV rf amplifier, 263
TV signal booster, 267
TV sound IF strip, 261
TV vertical deflection circuits, 255
TV vertical deflection oscillator, 253
TV video-amplifier final, 264
Twin-tee electronic organ, 235
Twin-tee oscillator, 221, 224, 229
Twisted-pair sensing devices, 167
Two-amp switching regulator, 7
Two-level regulator, 8 V and 16 V, 6
Two-network dimmer, 124

UHF amplifier, 450 MHz, 307
UHF amplifier, 450 MHz RF, 307
UHF amplifier, 10 W, 297
UHF amplifier, 25 W, 296, 297
UHF common-gate amplifier, 306
UHF FET oscillator, 275
UHF frequency doubler, 277
UHF tripler, 0.75 GHz, 278
UHF tripler, to 600 MHz, 279
UHF-TV RF amplifier, 307
UJT code practice oscillator, 223
UJT-controlled lamp dimmer, 97
UJT firing circuit, 98
UJT-fired triac dimmer, 130
UJT-gated SCR flasher, 186
UJT lamp dimmer, 110
UJT metronome, 237
UJT motor speed control, dc, 96
UJT oscillator, 173
UJT oscillator flasher, 180
UJT pulse amplifier, 71
UJT sawtooth oscillator, 229
UJT SCR trigger, noninductive-load, 118
UJT-shifted SCS ring counter, 188
UJT signal injector, 378
UJT timer, 15-second, 203
Ultraminiature phase control, 25 W, 117
Ultrasensitive light control (N.C.), 82
Ultrasensitive light control (N.O.), 82
Ultrasensitive light-operated relay, 82
Ultrasimple CSCR Schmitt Trigger, 72
Ultrasonic inverter, high-power, 49
Ultrastable lamp dimmer, 100
Unijunction, see entires under UJT
Unijunction code practice oscillator, 223
Unijunction motor speed control (dc), 96
Unijunction pulse shaper, 55
Unijunction SCR 3-phase control, 98
Unijunction SCR trigger, 71
Unitized-gun picture tube driver, 253
Unity-gain amplifier, 324
Unity-gain amplifier/buffer, 224
Unity-gain audio amplifier, 316
Unity-gain audio mixer, 323, 325
Unity-gain audio mixer, 7-input, 325
Unity-gain mixer, 304
Universal alarm, for any sensor, 166
Universal CSCR timer, 208
Universal dc power converter, 39
Universal inverter, 54
Universal-motor control, 108
Universal-motor control, 117 V, 110
Universal motor control, and timer, 119
Universal motor reversal/speed control, 90
Universal motor speed control, 90, 91, 96, 103, 104, 105, 126, 127
Universal power supply, 31
Universal shift register, 198
Universal timer, 219
Universal timer, 0.6-6 sec, 204
Universal timer, RC, 206
Unwanted-load detector, 154
Using 5 A relay for 50 A switching, 70

Vacuum-tube plate supply, 20
Vacuum-tube plate supply, 12 V, 39
Van interior-light dimmer, 387
Varactor amplifier, RF, 120 MHz, 310
Varactor multiplier, 400 MHz, 279
Varactor multiplier, 1-step, 276
Varactor tripler, out, 600 MHz, 279

Varactor tuner, VHF, 263
Varactor UHF tripler, 278
Variable ac voltage control, 84, 91
Variable-amplitude oscillator, 273
Variable-amplitude pulse supply, 31
Variable-click generator, 237
Variable-click oscillator, 236
Variable current generator, 9
Variable-current 13 V regulator, 15
Variable-dc 3-phase control, 124
Variable-delay flip-flop (1-shot), 205
Variable-delay timer, 211
Variable-duty-cycle photoflash unit, 35
Variable-frequency inverter, 49
Variable frequency oscillator, 272
Variable-interval and -duration flasher, 185
Variable-load motor control, 95
Variable-on-time pulse switch, 73
Variable-on-time switch, 134
Variable output regulated supply, 28 V, 29
Variable-period timer, 208
Variable-period 5-sec timer, 208
Variable phase shifter, 178
Variable-pitch tone oscillator, 220
Variable-power inverter, 54
Variable-rate flasher, 6 V dc, 185
Variable-rate flasher, 1 kW, 181
Variable-rate sequential flasher, 191
Variable-rate auto flasher, 384
Variable-reluctance preamp, 321
Variable shunt-wound-motor control, 90
Variable signal generator, 379
Variable-speed control, induction, 127
Variable-speed control, shunt motor, 90
Variable-sweep supply, 31
Variable-time ring counter, 188
Variable-tone audio generator, 223
Variable-torque motor speed control, 95
Variable-voltage ac control, see entries under Dimmer
Variable-voltage control, 115 V, 118
Variable voltage control, for ac, 100
Variable-voltage control, dc, 600 W, 103
Variable-voltage power supply, 22
Variable-voltage regulated supply, 17
Variable-voltage regulator, 5
Variable-voltage regulator (8-16 V) 100 mA, 5
Variable-voltage source, ac, 115 V, 106
Variable-voltage triac switch, 74
Varicap, see entries under Varactor
VCA, 233
VCO, 230
VCO, tuned-collector, 275
VCO sine-wave generator, 230
Vehicular B+ supplies, 36, 39
Vehicular generator, ac, 117 V, 46, 47
Vehicle-interior-lamp dimmer, 387
Vehicular inverter, 400 Hz, 43
Vehicular power inverter, 60 W, 53
Vehicular power supply, 150/300 V, 37
Vending-machine drink control, 150
Ventilating blower control, 143
Vertical Black-and-White output circuit, 262
Vertical deflection circuit, 249, 255
Vertical deflection oscillator, 253
VFO, 80 through 2 meters, 272
VHF active balanced mixer, 311
VHF amplifier, 40-180 MHz, 292, 307
VHF amplifier, 30 W at 175 MHz, 296
VHF amplifier, 75-ohm, 305
VHF amplifier, cascode, 200 MHz, 307, 309
VHF amplifier, RF, 200 MHz, 307, 308
VHF converter, 6 m, 301
VHF frequency doubler, 277
VHF front end, 305
VHF mixer, 244 MHz, 311
VHF oscillator, 47 MHz, 274
VHF power amplifier, 50 MHz, 243
VHF preamp, 2 m, 302
VHF tone-producing oscillator, 274
VHF-to-HF converter, 301
VHF-to-UHF tripler, 277, 278
VHF transmitter, 175 MHz, 280, 287
VHF TV hi-performance mixer, 257
VHF varactor tuner, 263
Vibrator, self-contained, 229
Vibrator, solid-state, 44

Vibrator, substitute, 36
Vibro-Keyer, 222
Viewfinder, camera, 252
Video, and deflection, 249
Video amplifier, 250, 252, 254
Video amplifier, cascode, 252
Video amplifier, FET and bipolar, 250
Video amplifier final, TV, 264
Video buffer, 258
Video IF amplifier, 258
Video IF amplifier, 44 MHz, 256
Video system power supply, 262
VLF power amplifier, 200 W, 292
VMP-1 audio alarm, 166
Voltage, current, continuity tester, 367
Voice-controlled relay, 232
Voltage amplifier, 20 dB audio, 322
Voltage checker, automotive, 368
Voltage-compensated dimmer, 100
Voltage-compensated phase control, 123
Voltage-compensating 3-phase control, 124
Voltage-compensating flash converter, 216
Voltage control, ac, 84, 89, 100
Voltage control, 1 kW ac, 128, 129
Voltage control, 400 Hz, 93
Voltage control, 800 W ac, 130
Voltage control, ac, low-cost, 99
Voltage control, hysteresis-free, 92, 129
Voltage control, by lamp brightness, 161
Voltage control, for motor, 91
Voltage control, with opto feedback, 90
Voltage control, RFI-proof ac, 125
Voltage control, for small motor, 111
Voltage-controlled amplifier, 253
Voltage-controlled oscillator, 230, 275
Voltage controller, ac and dc, 97
Voltage-differencing regulator, 10
Voltage divider, zener, 376
Voltage-feedback inverter, 53
Voltage loss alarm, 154
Voltage monitor, battery, 159
Voltage-reducing network, 376
Voltage-regulated inverter, 115 V, 45
Voltage regulator, 4
Voltage regulator, ac, 15
Voltage regulator, 12 V at 2 A, 10, 16
Voltage regulator, 12 V at 2 A dc, 11
Voltage regulator, 13 V at 6 A, 15
Voltage regulator, 22-30 V at 10 A, 12
Voltage regulator, 28 V at 0.5 A, 12
Voltage regulator, 90 V rms, 9
Voltage regulator, 100 V at 400 mA, 13
Voltage regulator, 1 kV at 0.1 A, 3
Voltage regulator, lamp, 16
Voltage safety alarm, 166
Voltage sensing alarm, 166

Voltage-sensing regulator protection, 11
Voltage source, 12 V at 1 A, 22
Voltage source, for car, 117 V ac, 42, 46
Voltage source, reference, 8
Voltage tester, automotive, 368
Voltmeter, dc, 372
Voltmeter, FET, 370
Voltmeter, FET dc, 373
Voltmeter, for RPM sensing, 156
VOM-to-VTVM converter, 372
VOX box, 232
VSWR bridge, 282
VTVM dc amplifier, 376
VTVM, FET, 370

Wailing electronic siren, 237, 238
Wall-mounting lamp dimmer, 126
Warning-light activator, 80
Warning-light flasher, 120 V ac, 182
Warning flasher, boat and car, 384
Water detector, 147
Water-level control, 148
Water-level control system, 149
Water-level sense and control, 150
Water-level sensing circuit, 149
Waveform, dimmer-circuit, 106
Waveform, Schmitt trigger, 72
Waveform generator, square, 44
Waveform phase control, 86
Waveform shaper, inverter, 55
Waveshaper for inverters, 55
Wetness sensor, 147
Wideband amplifier, 250
Wideband amplifier, multioctave, 298
Wideband audio amplifier, 4 W, 335
Wideband buffer, hi-Z and low-C, 224
Wideband crystal oscillator, 274
Wideband video amplifier, 250
Wideband VHF amplifier, 292
Wide-bandwidth temp regulator, 135
Wide-dynamic-range mike preamp, 223
Wide-frequency tone oscillator, 223
Wide-input-range regulator, 90 V ac, 4, 9
Wide-range audio preamp, 322
Wide-range crystal oscillator, 274
Wide-range dc converter, 39
Wide-range dimmer, 92, 129
Wide-range dimmer, 800 W, 130
Wide-range dimmer, 1 kW, 128, 129
Wide-range dip meter, 371
Wide-range inverter, 54
Wide-range inverter, 5-30 W, 57
Wide-range inverter, 80 W, 54
Wide-range kilowatt inverter, 49
Wide-range lamp dimmer, 124
Wide-range motor controller, 105

Wide-range motor speed control, 91
Wide-range timer, 203
Wide temperature-range dc converter, 34
Wide-tolerance ring counter, 193
Width control, sound-source, 365
Windows, SCR firing, 63
Wired-NOR, 62
Wireless mike, AM, 226, 228
Wireless mike, FM, 225
Wireline-controlled power switching, 178
Wireline-controlled switch, 174
Wireline sensing devices, 167
Wireline switching, multifunction, 175
Wire sensor for temp. control, 142
Whistle-on oscillator, 221, 224
Workman metronome, 236
Workman motor speed control, 109
Workman solar relays, 76
Wye-wired phase control, 112

Xenon flash trigger, 218

Zener-controlled regulator, 5
Zener-controlled regulator, 100 V, 13
Zener-controlled supply, 4-12 V, 20
Zener-controlled supply, 9 V, 30
Zener-controlled supply, 24 V, 18
Zener-controlled voltage reference, 8
Zener-protected switching converter, 34
Zener-regulated supply, 9 V, 24
Zener-regulated supply, 12 V, 22
Zener remote control circuit, 175
Zener-SBS "diac," 106
Zener-simulating LED supply, 26
Zener triggered relays, 175
Zener voltage divider, 376
Zero crossing, 62
Zero-crossing switch, 73
Zero-current ac switch, 71, 75
Zero-point switch, N.C., 154
Zero-point switch, 75, 135
Zero-voltage switching, see also entries under ZVS
Zero-voltage switching, 75, 85
ZVS, modulated SCR, 65
ZVS, normally closed, 154
ZVS, triac-controlled, 73, 75
ZVS circuits, 66
ZVS flame monitor, 151
ZVS optoelectronic circuit, 73
Zero-point switching, see entries under ZVS
ZVS resistive-load control, 135
ZVS temperature control, 140, 142
ZVS temperature controller, 132
ZVS triac thermostat, 138